Lecture Notes in Computer Science   1790

Edited by G. Goos, J. Hartmanis and J. van Leeuwen

T0223207

Springer
*Berlin
Heidelberg
New York
Barcelona
Hong Kong
London
Milan
Paris
Singapore
Tokyo*

Nancy Lynch   Bruce H. Krogh (Eds.)

Hybrid Systems: Computation and Control

Third International Workshop, HSCC 2000
Pittsburgh, PA, USA, March 23-25, 2000
Proceedings

Springer

Series Editors

Gerhard Goos, Karlsruhe University, Germany
Juris Hartmanis, Cornell University, NY, USA
Jan van Leeuwen, Utrecht University, The Netherlands

Volume Editors

Nancy Lynch
Massachusetts Institute of Technology
Laboratory for Computer Science
Cambridge, MA 02139, USA
E-mail: lynch@theory.lcs.mit.edu

Bruce H. Krogh
Carnegie Mellon University
Department of Electrical and Computer Engineering
Pittsburgh, PA 15235, USA
E-mail: krogh@ece.cmu.edu

Cataloging-in-Publication Data applied for

Die Deutsche Bibliothek - CIP-Einheitsaufnahme

Hybrid systems : computation and control ; third international
workshop ; proceedings / HSCC 2000, Pittsburgh, PA, USA, March, 23 -
25, 2000. Nancy Lynch ; Bruce H. Krogh (ed.). - Berlin ; Heidelberg ;
New York ; Barcelona ; Hong Kong ; London ; Milan ; Paris ; Singapore ;
Tokyo : Springer, 2000
    (Lecture notes in computer science ; Vol. 1790)
    ISBN 3-540-67259-1

CR Subject Classification (1991): C1.m, F.3, C.3, D.2.1, F.1.2, J.2, I.6.1

ISSN 0302-9743
ISBN 3-540-67259-1 Springer-Verlag Berlin Heidelberg New York

Springer-Verlag is a company in the BertelsmannSpringer publishing group.
© Springer-Verlag Berlin Heidelberg 2000
Printed in Germany

Typesetting: Camera-ready by author, data conversion by Firma Steingräber
Printed on acid-free paper   SPIN 10720042   06/3142   5 4 3 2 1 0

Preface

This volume contains the proceedings of the *Third International Workshop on Hybrid Systems: Computation and Control* (HSCC 2000), which was held on March 23-25, 2000, in Pittsburgh, Pennsylvania. The proceedings of the first two workshops in this series were published by Springer-Verlag, in the Lecture Notes in Computer Science series, as volumes 1386 and 1569.

The focus of the Hybrid Systems workshop series is on modeling, control, synthesis, design, and verification of *hybrid systems*. A hybrid system is a theoretical model for a computer controlled engineering system, with a dynamics that evolves both in a discrete state set and in a family of continuous state spaces. Hybrid systems research is motivated by, for example, control of electro-mechanical systems (robots), air traffic control, control of automated freeways, and chemical process control. The research area of hybrid systems overlaps both with computer science and with control theory. The workshop series is intended to foster the interaction between researchers from these fields in addressing problems in this new domain.

The scientific program of the workshop consisted of four invited talks and 32 contributed talks. The following researchers presented invited talks: K. Butts (Ford Research, USA), N. Leveson (MIT, USA), A. Sangiovanni-Vincentelli (U. California, Berkeley, USA), and B. Williams (MIT, USA). The contributed talks were based on the papers in these proceedings.

The program committee, chaired by the editors, selected the 32 contributed papers out of 71 submitted papers. The editors are grateful to the members of the program committee for their generous help in the reviewing and the selection process.

The editors are grateful to the speakers and all the other workshop participants, and to the sponsoring institutions whose support has made this event possible. Finally, they would like to thank George Woodzell for his system support, Drew Danielson for his help with local arrangements, and Joanne Talbot for all her hard work in assembling this proceedings volume.

March 2000   Nancy Lynch and Bruce Krogh

Organization

Steering Committee

Panos Antsaklis (University of Notre Dame)
Tom Henzinger (University of California, Berkeley)
Bruce Krogh (Carnegie Mellon University, Pittsburgh)
Nancy Lynch (Massachusetts Institute of Technology, Cambridge)
Oded Maler (Verimag, Gières)
Amir Pnueli (Weizmann Institute, Rehovot)
Alberto Sangiovanni-Vincentelli (University of California, Berkeley)
Shankar Sastry (University of California, Berkeley)
Jan van Schuppen (CWI, Amsterdam)
Frits Vaandrager (University of Nijmegen)

Program Committee

Bruce Krogh (co-chair) (Carnegie Mellon University)
Nancy Lynch (co-chair) (Massachusetts Institute of Technology)
Rajeev Alur (University of Pennsylvania)
Eugene Asarin (Institute for Information Transmission Problems, Moscow)
Marica Domenica Di Benedetto (University of Rome "La Sapienza")
Gautam Biswas (Stanford University)
Rene Boel (University of Ghent)
Michael Branicky (Case Western Reserve University)
Peter Caines (McGill)
Datta Godbole (Honeywell Technology Center)
Mark Greenstreet (University of British Columbia)
Stefan Kowalewski (Universität Dortmund, Chemietechnik)
Yassine Lakhnech (Institut für Informatik und Praktische Mathematik)
Michael Lemmon (University of Notre Dame)
Bengt Lennartson (Chalmers University of Technology)
Nancy Leveson (Massachusetts Institute of Technology)
Daniel Liberzon (Yale University)
John Lygeros (University of Cambridge, UK)
Oded Maler (Verimag, Gières)
Manfred Morari (Swiss Federal Institute of Technology)
Jöerge Raisch (Max-Planck Inst. für Dynamik Komplexer Techn. Sys., Germany)
Anders Rantzer (Lund Institute of Technology, Sweden)
Anders P. Ravn (DTU, Lyngby)
Alberto Sangiovanni-Vincentelli (Cadence European Laboratories)
Roberto Segala (University of Bologna)
Henny Sipma (Stanford University)
Eduardo Sontag (Rutgers University)
Claire Tomlin (Stanford University)

F.W. Vaandrager (University of Nijmegen)
H. Wong-Toi (Cadence, Berkeley)
Sergio Yovine (Verimag, Gières)
Feng Zhao (Xerox)

Additional Referees

Andrea Balluchi	Inseok Hwang	George Pappas
Luca Benvenuti	Gokhan Inalhan	Judi Romijn
Mireille Broucke	Bart Jacobs	Gerardo Schneider
Philippe Darondeau	Karl Johansson	Norihiko Shishido
Ansgar Fehnker	Anatoli Juditski	Joseph Sifakis
Elena De Santis	Salvatore La Torre	Geert Stremersch
Stefano Di Gennaro	Carl Livadas	Olaf Stursberg
Ronojoy Ghosh	Ian Mitchell	Rodney Teo
Radu Grosu	Pieter J. Mosterman	Rene Vidal
Ingo Hoffmann	Peter Niebert	
Thomas Hune	Meeko Oishi	

Sponsoring Institutions

Air Force Office of Scientific Research
IEEE Control Systems Society (Technical Co-sponsor)
Ford Motor Company
National Science Foundation
Dept. of Electrical and Computer Eng., CMU (Pittsburgh, PA, USA)
Dept. of Electrical Eng. and Computer Sci,, MIT (Cambridge, MA, USA)

Table of Contents

Invited Presentations

Selected Presentations

Hybrid Models for Automotive Powertrain Systems: Revisiting a Vision

Ken Butts

Powertrain Control Systems
Ford Research Laboratory
Dearborn, MI
kbutts1@ford.com

Abstract. Due to the persistent need to develop increasingly complex systems with improved quality and reduced development effort, automotive manufacturers are employing model-based development approaches wherever sensible. This is particularly true for powertrain control system development, as domain relevant computer-aided control system design tools have become commercially available. It is now possible to model and simulate the powertrain system dynamics in closed-loop with detailed behavioral models of the control algorithm. These control algorithm models capture nominal, initialization, diagnostic, and failure-mode-effects management modes of operation to the extent that simulation-based validation and verification procedures can be employed. These procedures help to ensure that the algorithm design and its associated software realization meet the system requirements with quality.

Simulation-based development (design, validation, and verification) methods only evaluate the system's behavior under the initial conditions, input scenarios, and parameter values as defined in the simulation test-suite. Thus, comprehensive validation and verification is expensive and time consuming. (Of course, exhaustive system validation and verification is impossible in a simulation-based development approach.) Importantly, given that powertrain models are being created to support mainline development, we now have an opportunity to go beyond simulation by employing systems analysis methodologies in the design process. The purpose of this talk is to describe hybrid systems analysis queries that, if answered in an efficient and intuitive way, would be a boon to the powertrain controller development community.

We begin by stating two analysis tool objectives that are derived from our experiences in using these models in a production development environment. First, wherever possible, we desire that our analysis methods be based on commercial tools. Second, we desire to analyze the models in the styles (one for physical plant models and another for control algorithm models) that they are currently built. We desire to analyze the models in the accepted styles because the model preparation requires significant engineering effort and it is unlikely that the organization could support the additional expertise, training, and effort required to specially prepare alternative analysis models.

Next we describe the modeling styles that are used in the production development process and provide an example based on an automatic

N. Lynch and B. Krogh (Eds.): HSCC 2000, LNCS 1790, pp. 1–2, 2000.

transmission control system. The physical plant model is comprised of a two-state engine model, a quasi-static torque converter model, the transmission dynamics for first gear, second gear and the one-to-two gearshift, and a simple longitudinal vehicle dynamics model. The discussion focuses on the hybrid nature of several components within the system. The associated control algorithm model is comprised of an abstract transmission shift scheduler and simplified shift control logic for the one-to-two gearshift and the one-to-two-to-one "change-of-mind" gearshift.

We list analysis queries that would enhance the powertrain controller development process if they were available. We also discuss the specific application of these queries to the transmission control example. These analyses include stability within a mode of operation, modal transition integrity, safety, and liveness.

We conclude with the remark that we have a wealth of new information available in the automotive powertrain controller development process: formal and detailed models of the system's behavior. We hope to be able to fully exploit these models through analysis.

Experiences in Designing and Using Formal Specification Languages for Embedded Control Software

Nancy G. Leveson

Aeronautics and Astronautics Department,
Massachusetts Institute of Technology
Cambridge, MA USA
leveson@mit.edu

Abstract. For the past ten years, I have been designing formal specification languages for specifying software requirements on complex systems. In order to understand what is needed in such languages, my students and I have been applying our ideas to real systems and using what we have learned to generate new hypotheses about what is needed to make such languages both useful and used. This research is part of a larger effort to assist in developing safety-critical embedded systems.

Some of the lessons we have learned:

1. Formal specifications can be practical in industry, but the notations need to be readable and reviewable by those who will be using them, not just by Ph.D. computer scientists. Most specification errors will be found by domain experts reading the specification, not by formal analysis tools (although tools can be useful, particularly in helping designers understand the specifications).
2. The problems involved in specifying large, complex systems are different than the problems involved in specifying the simple examples usually found in research papers. If we want our languages to be used, we need to start from real problems from the beginning and not simply eliminate all the parts of the problem we cannot handle.
3. Some common features of formal specification languages are very error-prone in use and should be eliminated from our languages.
4. Our languages must support building complex models. Support includes tools to assist in writing, visualizing, and validating such specifications.
5. Formal models and specifications are very expensive to produce. They will not be adopted by industry unless the payoff is worth it. To date, that has not been true. They will be used if we can solve problems with them that they cannot solve adequately in simpler or cheaper ways or that are important enough to them to be worth the investment.

Our current research goals include: integrating formal and informal specifications, adding "intent," supporting human problem solving (using what is known about this by cognitive psychologists), providing more assistance in building formal specifications, and devising analysis tools and algorithms to assist with important problems found in industrial projects. The talk will provide more details and examples.

N. Lynch and B. Krogh (Eds.): HSCC 2000, LNCS 1790, p. 3, 2000.

Model-Based Autonomous Systems
for Robotic Space Exploration

Brian Williams

Space Systems Laboratory, Massachusetts Institute of Technology
Cambridge, MA USA
williams@mit.edu

Abstract. A new generation of sensor rich, massively distributed systems is emerging that offers the potential for profound economic and environmental impact, including building energy systems, deep space probes and sensor webs that monitor the earth ecosystem. These robotic webs have the richness that comes from interacting with physical environments, together with the complexity of networked software systems. They must be efficient, capable and long lived, that is, able to survive decades of autonomous operation within unforgiving environments.

Model-based autonomy meets this challenge through two ideas. First, we note that programmers generate the desired function based on their commonsense knowledge of how the software and hardware modules behave. The idea of model-based programming is to exploit this modularity by having engineers program reactive systems by simply articulating and plugging together these commonsense models. The second challenge is the infeasibility of synthesizing a set of codes at compile time that envision all likely failure situations and responses. Our solution is to develop real time systems, called model-based executives that respond to novel situations on the order of hundreds of milliseconds, while performing extensive deduction, diagnosis and planning within their reactive control loop.

In this talk I will formulate a model-based executive as a deductive form of an optimal, model-based controller, in which models are specified through a combination of concurrent, probabilistic transition systems and propositional logic. This framework allows us to unify a diverse set of research results from model-based reasoning, planning, search, real-time propositional inference, and the theory of reactive languages. I will then discuss how reactivity is achieved using a high performance deductive kernel, called OPSAT that solves combinatorial optimization problems with constraints encoded in propositional logic. A first generation executive, called Livingstone, was demonstrated this year on NASA's first autonomous space probe, called Deep Space One, shortly before its asteroid encounter. Livingstone is also being demonstrated in a variety of space systems that include Mars rovers, Martian chemical plants, multi-spacecraft telescopes and the next generation shuttle. Finally, I will touch on future research that shifts from controlling the internals of single robotic systems to webs of robotic vehicles.

N. Lynch and B. Krogh (Eds.): HSCC 2000, LNCS 1790, p. 4, 2000.
© Springer-Verlag Berlin Heidelberg 2000

Models of Computation and Simulation of Hybrid Systems

Alberto Sangiovanni-Vincentelli

The Edgar L. and Harold H. Buttner Chair
of Electrical Engineering and Computer Science,
Department of EECS
University of California at Berkeley

Abstract. A design (at all levels of the abstraction hierarchy from functional specification to final implementation) is generally represented as a set of components, which can be considered as isolated monolithic blocks, which interact with each other and with an environment that is not part of the design. The model of computation defines the behavior and interaction of these blocks. Compactness of description, fidelity to design styles, ability to simulate, synthesize to an appropriate implementation and optimize its behavior are criteria to follow for the choice of an MOC to describe and manipulate a design. For example, some MOCs are suitable for describing complicated data transfer functions and completely unsuitable for complex control, while others are designed with complex control in mind.

We review the foundations of a theory of models of computation (MOC) (see Lee and Sangiovanni-Vincentelli, IEEE Trans. CAD, Dec. 1998). We will try to convey the basic notions and definitions to avoid ambiguity that often arises when MOCs are used in a non-rigorous fashion. We also believe that some degree of confusion has arisen in the hybrid system community due to an improper use of MOCs.

Hybrid systems in the general sense of the term could be considered as formalisms used to describe a complex system as combinations of MOCs where a single one is not powerful or expressive enough. When a hybrid system is simulated, the MOCs used to describe its behavior dictate the way the components of the system interact and execute. Since MOCs differ mostly for the way their components interact, the most difficult problem to solve when simulating them is to resolve the interfacing issue. We will review the issues and the ways used to cope with them. We will draw from the large bag of tricks developed over the years in the simulation community (especially for circuit simulation, e.g. SPICE, that exhibits some of the problems faced by the hybrid system community) to document difficulties and successes.

N. Lynch and B. Krogh (Eds.): HSCC 2000, LNCS 1790, p. 5, 2000.
© Springer-Verlag Berlin Heidelberg 2000

Modular Specification of Hybrid Systems in CHARON

Rajeev Alur, Radu Grosu, Yerang Hur, Vijay Kumar, and Insup Lee

Department of Computer and Information Science, University of Pennsylvania,
Philadelphia PA 19104-6389, USA,
{alur,grosu,yehur,kumar,lee}@cis.upenn.edu,
http://www.cis.upenn.edu/~alur,grosu,yehur,kumar,lee

Abstract We propose a language, called CHARON, for modular specification of interacting hybrid systems. For hierarchical description of the system architecture, CHARON supports building complex agents via the operations of instantiation, hiding, and parallel composition. For hierarchical description of the behavior of atomic components, CHARON supports building complex modes via the operations of instantiation, scoping, and encapsulation. Features such as weak preemption, history retention, and externally defined Java functions, facilitate the description of complex discrete behavior. Continuous behavior can be specified using differential as well as algebraic constraints, and invariants restricting the flow spaces, all of which can be declared at various levels of the hierarchy. The modular structure of the language is not merely syntactic, but can be exploited during analysis. We illustrate this aspect by presenting a scheme for modular simulation in which each mode can be compiled solely based on the locally declared information to execute its discrete and continuous updates, and furthermore, submodes can integrate at a finer time scale than the enclosing modes.

1 Introduction

A hybrid system typically consists of a collection of digital programs that interact with each other and with an analog environment. The design and implementation of hybrid systems remains a challenging task. We believe that availability of a specialized design language for hybrid systems will aid the developers significantly and lead to opportunities for greater design automation. Traditional tools for modeling and simulation of dynamical systems, such as MATLAB (see http://www.mathworks.com), provide little support for modular specifications. On the other hand, modern software design languages, such as STATECHARTS [10] and UML [6], provide no support for describing continuous behavior. In this paper, we introduce a language, called CHARON, for hierarchical specification of interacting hybrid systems. The design of our language was guided by two concerns. First, the language should support state-of-the-art modeling concepts such as encapsulation, reuse, preemption, and hierarchy. Second, it should be possible to give a modular formal semantics to the language which can be exploited during simulation, verification, and code generation.

N. Lynch and B. Krogh (Eds.): HSCC 2000, LNCS 1790, pp. 6–19, 2000.

In CHARON, a system is described as a collection of agents communicating via shared variables, and the behavior of each agent is specified by a hierarchical state machine. Key features of CHARON are summarized below.

Architectural hierarchy. The building block for describing the system architecture is an *agent* that communicates with its environment via shared variables. The language supports the operations of *composition* of agents to model concurrency, *hiding* of variables to restrict sharing of information, and *instantiation* of agents to support reuse.

Behavior hierarchy. The building block for describing flow of control inside an atomic agent is a *mode*. A mode is basically a hierarchical state machine, that is, a mode can have submodes and transitions connecting them. Variables can be declared locally inside any mode with standard scoping rules for visibility. Modes can be connected to each other only via well-defined entry and exit points. We allow *sharing* of modes so that the same mode definition can be instantiated in multiple contexts. Finally, to support *exceptions*, the language allows group transitions from default exit points that are applicable to all enclosing modes, and to support *history retention*, the language allows default entry transitions that restore the local state within a mode from the most recent exit.

Discrete updates. Discrete updates are specified by *guarded actions* labeling transitions connecting the modes. We assume *interleaving* semantics for concurrency (i.e., only one atomic agent is executed in a discrete round), *run-to-completion* semantics for individual agents (i.e., once an agent is chosen for discrete update, it keeps executing its transitions as long as there are enabled ones), and higher priorities for inner modes (i.e., group transitions from the default exit of a mode are examined only when there are no enabled transitions inside).

Continuous updates. Some of the variables in CHARON can be declared *analog*, and they flow continuously during continuous updates that model passage of time. The evolution of analog variables can be constrained in three ways: *differential* constraints (e.g. by equations such as $\dot{x} = f(x, u)$), *algebraic* constraints (e.g. by equations such as $y = g(x, u)$), and *invariants* (e.g. $|x - y| \leq \varepsilon$) which limit the allowed durations of flows. Such constraints can be declared at different levels of the mode hierarchy.

It should be noted that CHARON is a *modeling language*: it supports nondeterminism for both discrete and continuous updates, it is suitable for describing the system as well as the assumptions about the environment in which the system is supposed to operate, and for describing the same system at different levels of abstraction. The language constructs primarily facilitate the description of control flow, but it also supports calls to externally defined Java functions which can be used to write complex data manipulations.

After introducing the language in the next two sections, we proceed to illustrate how to exploit the modular structure during simulation. Since modes are hierarchical, multiple modes within an atomic agent can be active simultaneously, and a large number of transitions may be applicable in a given state.

In our modular scheme for discrete updates, each mode gets compiled into a function which gets control at one of its entry points along with an input global state, and returns the control at one of its exit points together with a modified global state. Such a modular scheme is possible since CHARON modes have explicit entry and exit points including the default ones, and inner transitions have higher priorities over the outer ones.

Introducing modularity in simulation of time rounds is more challenging. Since time is global, update of analog variables of all agents must be synchronized. Furthermore, within a single agent multiple modes are active, and the constraints on continuous update may be defined at any level of the hierarchy. This implies that simulating a flow requires solving constraints of all active modes of all agents simultaneously. In a modular scheme, we wish to compile each mode independently of the other.

Concurrency. To handle concurrency, we propose a scheme for distributed simulation in which each agent has its own local clock. The scheme ensures that the differences among local clocks are bounded.

Hierarchy. Each mode is responsible for integrating the variables whose update laws are defined locally, at a time scale of its own choice based on the local control laws and the invariants. A mode M is invoked from higher level with an input state, a bound δ on integration time, and an invariant constraint on the local variables of M. The integration within M assumes that the variables whose update laws are defined outside M stay unchanged. It can choose to integrate at time intervals shorter than δ, and can use integration routines of its submodes as black-boxes.

In summary, instead of solving the entire set of constraints simultaneously, the modular scheme computes the approximate solutions by layering the constraints as dictated by the modular specification.

Related work. Early formal models for hybrid systems include phase transition systems [13] and hybrid automata [1]. There has been a lot of research concerning analysis of hybrid automata leading to the model checker HyTech [5,11]. Models such as hybrid I/O automata [12] and hybrid modules [4] allow compositional treatment of concurrent hybrid behaviors. None of these models admit hierarchical specifications.

The notion of hierarchical state machines was introduced in STATECHARTS [10], and is present in many software design paradigms such as UML [6]. Our treatment of hierarchy is closest to hierarchical reactive modules [2] which shows how to define a modular semantics for hierarchical (discrete) modes.

The languages SHIFT [8] and HyCHARTS [9] allow hierarchical specifications of hybrid behavior, and STATEFLOW (see http://www.mathworks.com) allows hierarchical specifications of dynamic behavior. However, modular simulation has not been a concern in the design of these languages. Furthermore, CHARON supports new features such as preemption and reuse that are important from a programming perspective.

2   Language Overview

A hybrid system is described in CHARON by a set of *agents* communicating over a set of shared variables in an asynchronous way.

The agents may be grouped together in a hierarchical way into composed agents starting from the most primitive ones called atomic agents. Information flow inside a composed agent may be hidden to the outside world. The grouping of agents into composed agents gives the architecture of the hybrid system. A composed agent may also be understood as an architectural pattern that may be instantiated, i.e., reused in different contexts that match the pattern.

For example, at a lower level, a robot may be understood as the composition of a sensing agent, a controller agent, and an actuator agent. At a higher level, one may consider a team of cooperating robots, communicating with each other in order to achieve a common goal.

The behavior of an atomic agent is given by a set of *modes* that are linked together by a set of transitions. Each mode represents a particular behavior of the agent and has an associated dynamics given by a set of algebraic and differential constraints. The dynamics may be further constrained by a set of invariants. Modes may also be grouped together in a hierarchical way to form composed modes starting from the most primitive ones called leaf modes. Moreover, each mode may declare its own set of local variables that is hidden outside the mode, but is accessible to its submodes.

In other words, a mode is a sequential, communicating, hierarchical state machine with well defined dynamics, interfaces, and scoping rules for variables similar to structured programming languages. It may be also regarded as a behavioral pattern that may be instantiated.

For example, at a lower level, one may consider for a robot the modes walkForward, walkLeft, walkRight and walkBackward. At a higher level one may consider the modes avoidObstacle and trackWall.

Note that an atomic agent is nothing but a hierarchical mode. Its variables and behavior are completely determined by the mode. Moreover, a hierarchical agent is nothing but a set of hierarchical modes with local variables determined by the agent hierarchy. So why do we distinguish between modes and agents? The answer is that encapsulating modes inside agents prevents parallel composition inside modes, i.e., modes are entities composed in a purely hierarchical way.

Refer to [3] for more details and examples.

2.1   Variables

Discrete and analog variables. A hybrid agent has a finite set of typed variables denoted $A.V$. Some of these variables are updated in a discrete fashion and the others change in an analog fashion when time elapses. Accordingly, the set $A.V$ is partitioned in two sets, the set $A.dscV$ of *discrete variables* and the set $A.anaV$ of *analog variables*.

Differential and algebraic variables. In control theory it is common to compute the values of the analog variables $A.anaV$ by using algebraic and differential equations. For example, $\dot{x}=f(x,u)$ is a differential equation whereas $y=g(x,u)$

is an algebraic equation. Regarding f and g as functional blocks and x, y, u as wires, it is easy to see that the wire x is a feedback loop of f. As a consequence, the current value of the output x of f depends on the previous (infinitisimal) value of x. In contrast, the current value of the output y of g depends only on the current values of the inputs x and u. Hence, an algebraic equation is very similar to a combinational circuit whereas a differential equation is similar to a sequential circuit. In CHARON we generalize algebraic equations also to inequalities. We call the differential equations and algebraic equations generically as *constraints*. The variables defined by algebraic constraints are called *algebraic variables* and the variables defined by differential constraints are called *differential variables*. Hence, $A.anaV = A.diffV \cup A.algV$. We insist that $A.diffV \cap A.algV = \emptyset$. Note that hybrid automata do not make any distinction between these two kinds of variables.

Permitted read/write accesses. The variables $A.V$ of an agent A are classified according to their visibility and update permissions into three sets: the set $A.lclV$ of *local variables* that cannot be read or written by other agents, the set $A.wrtV$ of *write variables* that are written by A, and can be read by other agents, and the set $A.readV$ of *read variables* that are read by A, and may be written by other agents. The sets $A.readV$ and $A.wrtV$ need not be disjoint. Similarly, the set of local variables $A.lclV$ may be both read and written. The set of read and write variables $A.gblV = A.readV \cup A.wrtV$ is used for communication and it is called the set of *global variables*. The set $A.updV = A.wrtV \cup A.lclV$ of write and local variables is called the set of *updated variables*. Hence, our communication model is that of asynchronous communication over shared variables. This model is a very general and allows to define channels as a special case.

States and actions. Given a set V of typed variables, a *state* over V is a function mapping variables to their values. Given two sets V and W of variables, an *action* from V to W is a binary relation between the states over V and the states over W. In CHARON specifications, an action consists of an action *guard* over V and an action *body* from V to W. We say that an action is enabled (disabled) at a state s if its guard is true (false) at that state.

2.2 Hierarchical Modes

Hierarchy. A *mode* in CHARON has a very refined control structure, given by a hierarchical, hybrid state machine. It basically consists of a set of *submode references* connected by transitions such that at each moment of time only one of the submode references is active. A submode reference has associated again a mode and we require that the modes form an *acyclic* graph with respect to this association. By using modes and mode references several references may share the same mode. This is highly desirable because modes in a definition are never simultaneously active. A mode resembles an *or* state in STATECHARTS, but it has more powerful structuring mechanisms.

Variables. A mode has global as well as local variables. Global variables are used to share data with the environment of a mode , and are classified into the set *readV* of *read variables* and the set *wrtV* of *write variables*. The set

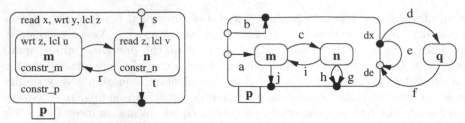

Fig. 1. Scoping rules and transition types

$gblV = readV \cup wrtV$ is called the set of *global variables*. The set of *local variables* $lclV$ of a mode is accessible only by its transitions and submodes. Thus, the scoping rules for variables are as in standard structured programming languages. For example, in Figure 1 left, the transitions of the mode p (like r, s, and t) may refer only to the variables x, y and z. These variables are global to the modes referred to by m and n. However, the variables local in the mode referred to by m may not be used in the mode referred to by n. For example, in Figure 1 left, the variable z may be accessed both in m and n but the variables u and v are private to m and n, respectively.

Dynamics. A mode has an associated set of *constraints*. These include differential equations, algebraic equations and *invariants* that are differential and algebraic equations or inequalities. The constraints define the flows of the mode, i.e., the way analog variables are updated while the agent is in this mode. The invariants define conditions that have to be satisfied by the variables in this mode, i.e., they define allowed durations. The scoping rules also apply for these constraints. For example, in Figure 1, constr_p may only refer to x, y, and z and constr_m may refer only to z and u. For each differential and algebraic variable updated by a mode we require that the variable is either updated by the mode itself or it is updated by all submodes of this mode. For example, in Figure 1, the local variable z is either updated by a constraint in the mode p or by constraints in both submodes m and n.

Interfaces. To obtain a modular language, we require the modes to have well defined *control points* classified into entry points (marked as white bullets) and exit points (marked as black bullets). The transitions connect the control points of a mode and its submode references to each other. For example, in Figure 1 right, a is an *entry* transition, g, h, and j are *exit* transitions, b is an *entry/exit* transition, and c and i are *internal* transitions. Between these transitions there is a subtle difference. Entry transitions initialize the local variables by reading only from the global variables. Exit transitions forget the values of the local variables by writing only to the global variables. It is only the internal transitions that may both read and write the local variables.

Preemption. To model preemption we use the special default exit point dx. A transition starting from the default exit point of a mode is called a *group transition*. It may be taken whenever the control is inside the mode and no internal transition is enabled. For example, in Figure 1 right, the group transition d is taken if it is enabled and all the transitions c, g, h, i, and j are disabled.

Hence, inner transitions have a higher priority than the group transitions, i.e., we use *weak preemption* (like the weak `kill` in Unix, versus the strong `kill -9`). This definition of priorities allows us to define in Section 4 a modular simulation.

History. To allow history retention, we use the special default entry point *de*. A transition entering the default entry point of a mode restores the values of all local variables along with the position of the control (a transition may enter a default entry of a mode only if the mode was left along its default exit). For example, both transitions e and f in Figure 1 right, enter the default entry point. The transition e is called a *self* group transition. A self transition (like e) or more generally a self loop like d, q, and f may be understood as an interrupt handling routine. While a self loop may be arbitrary complex, a self transition may do simple things like counting the number of occurrences of an event (e.g., clock events).

The *set of modes* in a CHARON specification is supposed to be globally accessible. Moreover, since a mode may refer to other modes we require that referencing forms an *acyclic graph*.

Leaf and top level modes. A *leaf mode* is a mode with no submodes and a default *identity* transition from its default entry point *de* to its default exit point *dx*. A *top-level* mode is a mode M with a single explicit entry point e and no exit points.

Mode operations. The mode definition can be viewed as an *encapsulation* operator over its submodes, and thus, modes are constructed from leaf-modes using encapsulation repeatedly in a non-recursive manner.

2.3   Hierarchical Agents

An atomic agent is basically a top level mode whose global variables are used for communication with other agents. As we already mentioned, atomic agents may be composed to form composed agents and communication inside composed agents may be hidden. Intuitively, composition of atomic agents is the union of their modes and hiding is a declaration of local variables. To make the operations over agents closed under composition and hiding, we define an agent as follows.

Definition 1. *(Agent) An agent P is a tuple consisting of*
   Modes. *A set of top-level modes M.*
   Local variables. *A set $lclV \subseteq \cup_{m \in M} m.gblV$ of local variables.*
   Global variables. *A set $gblV = (\cup_{m \in M} m.gblV) \setminus lclV$ of globals variables.*

Definition 2. *(Composition) If A and B are two agents, then the composition $A \| B$ is the agent with the set $lclV = A.lclV \cup B.lclV$ of local variables, the set $wrtV = A.wrtV \cup B.wrtV$ of write variables, the set $readV = A.readV \cup B.readV$ of read variables and the set $M = A.M \cup B.M$ of top level modes.*

Definition 3. *(Variable Renaming) Let A be an agent, $x \in A.gblV$ a global variable of the agent and $y \notin A.V$ a variable of the same type as x but not contained in A. Then the renaming $A[x := y]$ is the agent obtained by consistently renaming x by y in $A.V$ and in all modes $m \in A.M$.*

Definition 4. *(Variable Hiding) Let A be an agent, $x \in A.gblV$ a global variable of the agent. Then the variable hiding **hide** x **in** A is the agent obtained by replacing $A.gblV$ with $A.gblV \setminus \{x\}$ and $A.lclV$ with $A.lclV \cup \{x\}$.*

3 Global Semantics

One alternative in giving a semantics to a hierarchical system is to consider hierarchy as just a convenient syntactic abbreviation. This reduces the semantic definition to two considerably easier subproblems: a) show how to construct a flat system out of the hierarchical one and b) give a semantics to the flat system.

3.1 The Flattening Operation

Given a mode definition, the flattening operation recursively eliminates the submode references as follows: a) take for each reference m the associated definition, b) prefix all elements of the mode definition by m, c) continue recursively until all references point to a leaf mode definition. The set of elements obtained this way are taken as the elements of the flat mode.

As a consequence of flattening, all elements of the resulting mode are prefixed with a path $m_1:m_2:\ldots:m_k$ from the root mode reference m_1 down to the containing mode reference m_k of the original hierarchical mode. For example, a control point c has now the form $m_1:m_2:\ldots:m_k:c$. The set of local variables $flat(M).lclV$ of the flattened mode $flat(M)$ is the transitive closure of the local variables of M and the local variables of its submodes.

In the semantic definitions of the next section we model paths by *stacks*. Textually, we write stacks with the elements separated by colons and with the topmost element on the left. For example `s = a:b:s'` is the stack `s` with the top element `a`, the second element `b` and the rest of the stack `s'`. To show how stacks evolve in a pictorial way we use pattern matching. For example when we write `if ((as = a:b:as') & (bs = c:bs')) (as,bs) = (c:as', a:b:bs')` we mean that if the current value of the stack `as` has topmost elements `a` and `b` and the current value of the stack `bs` has the topmost element `c` then the next value of `as` has discarded `a` and `b` and pushed `c`, and the next value of `bs` has discarded `c` and pushed `a` and `b`.

3.2 Update Rounds

In an update round, the semantic function nondeterministically chooses one of the modes of the resulting flat agent and executes the discrete update on that mode. Using a pseudo-code like notation this can be described as shown below.

```
State updateRound (Agent a, State s){
    return forany (m in subModes(a)) discreteUpdate(m, s); }
```

The discrete update of a mode is a sequence of enabled implicit and explicit transitions starting at the default entry point of the mode and ending at the default exit point of the mode. The algorithm for generating this sequence is given below. In the first step it uses the global history variable `hs`, that is itself a stack, to execute a series of default entry transitions down to the last control point where the explicit execution got stuck, i.e., where all the explicit transitions were disabled. A default entry transition restores the saved submode and point by popping them from the history stack and pushing them on the control stack `ct`.

```
State discreteUpdate (Mode m, State s) {
   Stack ct = de:m:[]; State st = s;        //put de and m
   while (ct != dx:m:[]) {                   //while dx not reached
      while (ct = de:ct')                    //while de is the top point
         if (st.hs = pt:md:hs')
         (ct, st.hs) = (pt:md:ct', hs');   //default entry transition
         else ct = dx:ct';     //default identity transition for leaf mode
      while (enabledFanOut(ct, st) != {})
         (ct, st) = forany (t in enabledFanOut(ct, st)) t(ct, st);
      let (ct = pt:md:ct') in
         if (pt != de)
         (ct, st.hs) = (dx:ct', (pt=dx?de|pt):md:st.hs) }
   return st;}
```

If the history stack hs is empty and the top point on the control stack ct is
the default entry point de then a leaf mode has been reached and the identity
transition of the leaf mode is executed.

In the second step, the algorithm executes a sequence of explicit, enabled
transitions starting at the control point obtained in the previous step and ending
at the control point where all the explicit transitions are disabled. The enabled
fanout of a mode reference is the set of enabled transitions in the associated
mode definition, with source point pt and with source state st.

In the third step, the algorithm executes an implicit exit transition provided
that the last transition was not a self group transition (in this case, the top
point pt is equal to de). The default exit transition saves the relative value of
the control point from the previous step on the top of the history stack and passes
the control to the default exit of the parent mode. Note that, if the top point on
the control stack ct was the default exit point dx, then the exit transition saves
on the history stack hs the default entry point de. This assures that in the next
step, the deepest point is tried first.

Since the top of the control stack is dx and not de, the first step is skipped
when control is passed up to the parent mode. The second step in this case
amounts to executing a *group transition* if any enabled transition exists. If this
is not the case, the control is passed in the third step up again to the enclosing
parent mode and so on up to the top mode. If any of the group transitions
is enabled, then executing this transition (and possibly other), may return the
control to the default entry point de of the mode, and the algorithm proceeds
by skipping the third step and executing all the default entry transitions.

3.3   Time Rounds

In a time round, for a given state s_1, the semantic function executes for a time
interval d, and produces a new state $s_2 = s(d)$, where s is any flow that is a
solution of the active set of control constraints, not violating the current set of
invariants and such that $s(0) = s_1$. The semantic function is shown below, where
the type Constraints is assumed to contain a set of algebraic constraints, a set
of differential constraints and a set of invariants.

```
State timeRound (Agent a, State s) {
   Constraints c = agentConstraints(a, s);
   return forany ((f, d) in solution(c, s)) f(d); }
```

The set of active constraints for an agent is the union of the active constraints of each mode in the agent.

```
Constraints agentConstraints (Agent a, State s) {
   Constraints ac = {};
   forall (m in modes(a))
      ac = ac ∪ modeConstraints(m, s);
   return ac; }
```

For each mode, the set of active constraints is easily recovered form the history variable.

```
Constraints modeConstraints (Mode m, State s) {
   Constraints mc = getConstraints(m); Stack hs = s.hs
   while (hs = pt:md:hs') {
      mc = mc ∪ getConstraints(md);
      hs = hs'; }
   return mc; }
```

Hence, in a global semantics, the flows in all agents are synchronized with each other.

3.4   Global Execution

The semantic function for the execution of a hybrid agent nondeterministically chooses in each step either an update round or a time round, as shown by the following pseudo-code segment.

```
State macroStep (Agent a, State s) {
   [] return updateRound(a, s);
   [] return timeRound(a, s); }
```

4   Modular Simulation

The global semantics given in the previous section can be readily implemented in an algorithmic way to obtain a precise *simulation* for any hybrid system described in CHARON. However, such a simulation has a big disadvantage: it is *not modular*. In other words, one can not simulate the behavior of a mode in *isolation* independent of other modes or the mode hierarchy. The lack of modularity precludes efficient implementations. For example, all flows in the previous section are synchronized on the same clock.

In this section we present an alternative, *modular simulation* for hybrid agents. This simulation may have a very efficient implementation. However, its disadvantage is that it only approximates the conceptually ideal solution.

4.1   Update Rounds

In a modular simulation, the time and the update rounds of the mode of an atomic agent are constructed in a modular way from the time and the update rounds of its submodes. The state passed along the modes is *automatically co-erced* to the appropriate state for that mode, i.e., a mode can only access that part of the state that corresponds to its own variables. In programming languages terminology, the `discreteUpdate` and the `timeRound` functions are *polymorphic*.

In the modular version we do not have to work with path prefixed variables and points because the structure of a hierarchical mode is not destroyed (flat-tened). Moreover, in this case each mode has its own history variable, keeping a tuple: the last visited submode and its associated point. The modular version of the discrete update function is shown below. The initialization round of a mode is obtained by calling `discreteUpdate` at the initialization entry point.

```
Point×State discreteUpdate (Mode m, Point p, State s){
    Mode md = m; Point pt =p; State st = s;
    repeat {                           //loop
        if (md = m & pt = de)          //control is at default entry point
            (md, pt) = s.hs;           //execute default entry transition
        else                           //control is at regular entry
            (md, pt, st) = forany (t in enabledFanOut(md,pt, st))
            t(md, pt, st);             //execute transition
        if (md = m & pt in exitPts(m)) //control reached exit point
            return (pt, st);           //done
        else                           //control reached submode
            (pt, st) = discreteUpdate(md, pt, st);
    until (enabledFanOut(md, pt, st) = {}); }
    s.hs = (md, pt);                   //update history
    return (dx, st); }                 //done
```

4.2   Time Rounds

Taking the idea of modularity seriously, in a time round each agent should be able to integrate independently of the other agents, and the integration inside a submode should be done independently of its supermodes.

The independent integration of the subagents in a composite agent, or equiv-alently the integration of the top modes of the associated flattened agent, is the topic of the next section. In this section we are concerned with the hierarchical integration for a mode. The main goal is to allow the modes to integrate at different speeds without compromising too much the ideal solution.

Our main assumption is that the integration speed of the parent mode is of an order of magnitude *slower* than the integration speed of the submodes. In this case, we may assume that the values integrated in the parent mode, remain constant while the submodes perform their own integration. For example, in Figure 2, we assume that the integration speed for x is slower than the integration speed for y that is also slower than the integration speed for z. This idea is shown algorithmically below.

The time round function gets as input the mode, the state, the simplified invariants of its parent mode and the integration step of its parent mode. It first computes the current submodes and the set of invariants. Then it enters

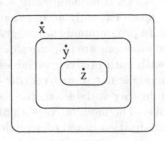

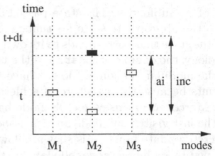

Fig. 2. Time round   **Fig. 3.** Global execution

the integration loop. In this loop it first simplifies the invariants according to the variables integrated in its supermode (their values are assumed to be fixed) and if the loop was traversed at least once, according to the variables declared in this mode or above but integrated in the submodes. Then it predicts its own integration step.

```
State×Time timeRound(Mode m, State s, Invariants i, Time t){
    State st; Mode md; Time d, dt;
    Invariants inv = getInv(m) ∪ i;    //get invariants
    (md, pt) = s.hs;                   //get active submode and point

    for (Time tm = 0; tm < t; tm = tm + dt) {        //while time left
        inv = simplify(s, inv);        //simplify invariants
        dt = predict(inv, s, getConstraints(m), tm);//predict dt
        (st, d) = timeRound(md, s, inv, dt);         //execute submode
        st = integrate(st, getConstraints(m), d);    //integrate
        if (d < dt | violated(inv, st, tm+d))
            return (st, tm + d); }        //violation return
    return (st, tm); }                   //normal return
```

Then it calls its current submode (known from the history variable) to execute a time round. It also constrains the integration time of the submode by passing its own simplified invariants. When the submode returns, the mode synchronizes its own differential variables with the differential variables owned by the submodes by performing the integration step. If the submode returned before the assigned integration time or the invariant of the mode was violated, the mode itself returns. Otherwise it returns normally. In this way, all variables are synchronized up to the top level.

4.3   Global Execution

In the modular simulation of the global execution we want to be able to integrate each subagent of a composite agent (or equivalently each mode of the corresponding flattened agent) at a possibly different speed and along intervals of different length. This however inevitably leads to an out of synchronization between the agents, because as long as an agent is integrating it cannot become aware of the changes produced by the other agents.

The main idea of our approach is to keep the out of synchronization interval between agents bounded, even if the agents proceed with different speeds. An intuitive analogy would be that of a rubber band that surrounds the agents and cannot be expanded more than a length, say dt.

For this purpose, each step in the global execution first picks up the modes with minimum and second minimum local time. For example, in Figure 3 we pick the modes M_2 and M_1. Then we compute the time round interval inc for the minimum mode such that its local time may exceed by at most dt the current local time of the second minimum mode. For example, in Figure 3, the increment is inc.

The time round may end before the time interval inc was finished if the invariants of M_2 get violated. Hence, the time round returns, as shown in Figure 3, with an actual time increment ai. In this case, the mode M_2 also executes an update round to synchronize the discrete variables with the analog ones. To be able to compute the minimum and the second minimum time values and their associated modes, we keep an array of current local times of modes. This idea is presented algorithmically below.

```
Time[]×State macroStep(Time[] mTms, Agent a, State s){
    Point p; Mode[] mds = modes(a);    //initialization
    int i = getMin(mTms);              //compute index for min.
    int j = get2ndMin(mTms);           //compute index for second min.
    m = mds[i];                        //select mode with min. time
    Time inc = mTms[j]-mTms[i]+dt;     //compute time interval

    (State s, Time ai) =
        timeRound(m, s, {}, inc);      //execute time round
    mTms[i] = mTms[i] + ai;            //update the actual time for m
    if (ai < inc) (p, s) =
        discreteUpdate(m, de, s);      //execute update round
    return (mTms, s); }                //make new state and time visible
```

5   Conclusion

In this paper, we have presented a language for specification of hybrid systems that supports concurrency and hierarchy in a modular fashion. We hope that CHARON is rich enough to support high-level modeling of embedded software, and is formal enough to support analysis. In this paper, we have proposed only a high-level outline for developing a modular simulator. We need to explore three orthogonal issues. First, finding a solution to a set of differential and algebraic constraints in presence of invariants requires careful detection of boundary crossings (see, for instance, [14]). Second, we handle concurrency by allowing agents to integrate separately based on their local clocks. When the guards and invariants of one agent depends on the values updated by the other agents, such a scheme may require detection and rollback. This is closely related to well understood problems concerning global states in distributed systems (see, for instance, [7]). Third, choosing different time scales for solving constraints at different levels of the mode hierarchy requires good heuristics to predict the step sizes. This can be done, in principle by determining the singular values of the linearized equations and scaling the equations appropriately. However, choosing a simple implicit integration scheme guarantees numerical stability and acceptable results, albeit with poor efficiencies [14].

Acknowledgments. We thank Joel Esposito and George Pappas for helpful discussions. Support from NSF grant CISE RI 9703220, NSF CAREER award CCR-9734115, DARPA/NASA grant NAG2-1214, DARPA grants ITO/MARS 130-1303-4-534328-xxxx-2000-0000, ATO/TMR DAAH04-96-1-0007, ARO grant MURI DAAH04-96-1-0007, ARO DAAG55-98-1-0393, and ARO DAAG55-98-1-0466 is gratefully acknowledged.

References

1. R. Alur, C. Courcoubetis, N. Halbwachs, T.A. Henzinger, P. Ho, X. Nicollin, A. Olivero, J. Sifakis, and S. Yovine. The algorithmic analysis of hybrid systems. *Theoretical Computer Science*, 138:3–34, 1995.
2. R. Alur and R. Grosu. Modular refinement of hierarchic reactive machines. In *Proceedings of the 27th Annual ACM Symposium on Principles of Programming Languages*, 2000. To appear.
3. R. Alur, R. Grosu, Y. Hur, V. Kumar, and I. Lee. CHARON: a language for modular specification of hybrid systems. Technical Report MS-CIS-2000-01, University of Pennsylvania, 2000.
4. R. Alur and T.A. Henzinger. Modularity for timed and hybrid systems. In *CONCUR '97: Eighth International Conference on Concurrency Theory*, LNCS 1243, pages 74–88. Springer-Verlag, 1997.
5. R. Alur, T.A. Henzinger, and P.-H. Ho. Automatic symbolic verification of embedded systems. *IEEE Transactions on Software Engineering*, 22(3):181–201, 1996.
6. G. Booch, I. Jacobson, and J. Rumbaugh. *Unified Modeling Language User Guide*. Addison Wesley, 1997.
7. D.P. Bertsekas and J. N. Tsitsiklis. *Parallel and Distributed Computation: Numerical Methods*. Athena Scientific, 1997.
8. A. Deshpande, A. Göllu, and L. Semenzato. The shift programming language and run-time systems for dynamic networks of hybrid automata. Technical Report UCB-ITS-PRR-97-7, University of California at Berkeley, 1997.
9. R. Grosu, T. Stauner, and M. Broy. A modular visual model for hybrid systems. In *Formal Techniques in Real Time and Fault Tolerant Systems (FTRTFT'98)*, LNCS 1486, pages 75–91. Springer-Verlag, 1998.
10. D. Harel. Statecharts: A visual formalism for complex systems. *Science of Computer Programming*, 8:231–274, 1987.
11. T.A. Henzinger, P. Ho, and H. Wong-Toi. HYTECH: the next generation. In *Proceedings of the 16th IEEE Real-Time Systems Symposium*, pages 56–65, 1995.
12. N. Lynch, R. Segala, F. Vaandrager, and H. Weinberg. Hybrid I/O automata. In *Hybrid Systems III: Verification and Control*, LNCS 1066, pages 496–510, 1996.
13. O. Maler, Z. Manna, and A. Pnueli. From timed to hybrid systems. In *Real-Time: Theory in Practice, REX Workshop*, LNCS 600, pages 447–484. Springer-Verlag, 1991.
14. W. Press, S. Teukolsky, W. Vetterling, and B. Flannery. *Numerical Recipes in FORTRAN*. Cambridge University Press, 1992.

Approximate Reachability Analysis of Piecewise-Linear Dynamical Systems[*]

Eugene Asarin[1], Olivier Bournez[2], Thao Dang[1], and Oded Maler[1]

[1] VERIMAG, Centre Equation, 2, av. de Vignate, 38610 Gières, France
{asarin,tdang,maler}@imag.fr
[2] LORIA, Campus Scientifique, BP 239, 54506 Vandoeuvre les Nancy, France
Olivier.Bournez@loria.fr

Abstract. In this paper we describe an experimental system called
$\boxed{\mathbf{d/dt}}$ for approximating reachable states for hybrid systems whose continuous dynamics is defined by linear differential equations. We use an
approximation algorithm whose accumulation of errors during the continuous evolution is much smaller than in previously-used methods. The
$\boxed{\mathbf{d/dt}}$ system can, so far, treat non-trivial continuous systems, hybrid
systems, convex differential inclusions and controller synthesis problems.

1 Introduction

The problem of calculating reachable states for continuous and hybrid systems has emerged as one of the major problems in hybrid systems research
[G96,GM98,DM98,KV97,V98,GM99,CK99,PSK99,HHMW99]. It constitutes a
prerequisite for exporting algorithmic verification methodology outside discrete
systems or hybrid systems with piecewise-trivial dynamics. For computer scientists it poses new challenges in treating continuous functions and their approximations and in applying computational geometry techniques to problems in
higher dimensional spaces. For control theorists and engineers the problem suggests a fresh way of looking at systems with under-specified inputs and increases
their awareness to some practical computational aspects of controller design.

In this paper we describe an experimental system called $\boxed{\mathbf{d/dt}}$ which can
approximate reachable states for hybrid systems whose continuous dynamics is
defined by linear differential equations. The performance is much better than
the more general method of "face-lifting" we have used in the past [DM98].

The rest of the paper is organized as follows. In section 2 we define the problem of calculating reachable states and suggest a general procedure which solves
it iteratively. The basic computation step of the procedure cannot be performed
exactly and in section 3 we describe an over-approximation scheme for linear systems, having the advantage of not propagating errors from one step to another.

[*] This work was partially supported by the European Community Esprit-LTR Project
26270 VHS (Verification of Hybrid systems), the French-Israeli collaboration project
970MAEFUT5 (Hybrid Models of Industrial Plants) and the Russian Foundation for
Basic Research under grant 97-01-00692.

N. Lynch and B. Krogh (Eds.): HSCC 2000, LNCS 1790, pp. 20–31, 2000.

Extensions of the algorithm to deal with hybrid systems, controller synthesis and continuous disturbances are described in section 4 along with several examples.

2   The Basic Problem

Let $T = \mathbb{R}_+$ be a *time domain*, let X be a bounded subset of $\mathbb{R}^n$ and consider a continuous dynamical system $\mathcal{A}$ over X defined by the equation $\dot{\mathbf{x}} = f(\mathbf{x})$. We use the notation $\mathbf{x} \xrightarrow{t} \mathbf{x}'$ to indicate that the solution α of the equation with $\mathbf{x}$ as an initial condition satisfies $\alpha[t] = \mathbf{x}'$. In words we say that $\mathbf{x}'$ is reachable from $\mathbf{x}$ in time t.

Definition 1 (Successors). *Let $\mathcal{A}$ be a dynamical system defined by $\dot{\mathbf{x}} = f(\mathbf{x})$. The successor operator $\delta : 2^X \to 2^X$ is defined for a subset F of X and an interval $I \subseteq T$ as:*

$$\delta_I(F) = \{\mathbf{x}' : \exists \mathbf{x} \in F \; \exists t \in I \; \mathbf{x} \xrightarrow{t} \mathbf{x}'\}$$

We use the notation δ_r for $\delta_{[r,r]}$ (states reachable after exactly r time), δ for $\delta_{[0,\infty)}$ (all states reachable after any non-negative amount of time) and $\delta_I(\mathbf{x})$ for $\delta_I(\{\mathbf{x}\})$. Note that δ has the semi-group property, i.e. $\delta_{I_2}(\delta_{I_1}(F)) = \delta_{I_1 \oplus I_2}(F)$ where $\oplus$ is the Minkowski sum, and in particular $\delta_{[0,r_2]}(\delta_{[0,r_1]}(F)) = \delta_{[0,r_1+r_2]}(F)$. In certain cases when the differential equation admits a closed-form solution, one may characterize $\delta(F)$ symbolically by a formula and then try to obtain a closed-form solution by quantifier elimination. However, this works in rather exceptional cases (see for example [CV95,PLY99]). Instead we propose a numerical algorithm which works by discretizing time into multiples of a fixed time step r. The abstract algorithm for calculating $\delta(F)$ is the following:

Algorithm 1 (Exact Calculation of $\delta(F)$)

$$
\boxed{
\begin{array}{l}
P^0 := F \\
\textbf{repeat} \\
\quad P^{k+1} := P^k \cup \delta_{[0,r]}(P^k) \\
\textbf{until } P^{k+1} = P^k
\end{array}
}
$$

In order for a function to be computable by a discrete device its domain and range need an effective representation as well as an effective and terminating procedure which takes the representation of any element of the domain and transforms it to a representation of its image by the function. For example, functions over the integers can be computed by applying well-known algorithms for addition and multiplication to unary, binary or decimal representations of numbers. The mathematical real numbers pose a special problem in this respect, a problem which we do not address here but assume to be solved for all practical purposes. Our main concern here is to compute functions over *subsets of X*. From this perspective Algorithm 1, when applied *exactly* suffers from the following two problems:

1. The exact calculation of $\delta_{[0,r]}$ is not more feasible than the calculation of the whole δ.
2. Even if $\delta_{[0,r]}$ was computable, the algorithm usually does not terminate after a finite number of steps.

To overcome these problems we resort to approximate calculation of $\delta_{[0,r]}$ and δ. In order to be effective, i.e. to do any computation at all, we can replace 2^X by a countable and effectively enumerable subset C whose union gives X, e.g. the set of all polyhedra with rational vertices. Elements of 2^X not in C are thus either under- or over-approximated (see Figure 1-(a)). The type of approximation which is used depends on the problem to be solved. If we want to characterize all the possible behaviors starting from a given initial set, an over-approximation is used. If we want to characterize the set of states from which a property can be satisfied, under-approximation is preferred.

An effective approximation of Algorithm 1 can thus be implemented by replacing all the operations (Boolean operations, equivalence testing and calculation of $\delta_{[0,r]}$) by their approximated versions.[1] If the approximate algorithm terminates, the result is an over-approximation of $\delta(F)$.

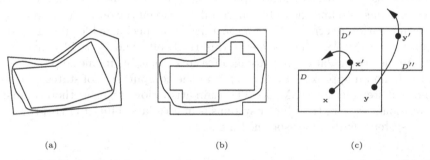

(a)   (b)   (c)

Fig. 1. (a) A set F and over- and under-approximated by polyhedra. (b) The same set approximated by orthogonal polyhedra. (c) Accumulation of errors in nave approximate computation.

The termination of the procedure, however, cannot be guaranteed since there are infinitely many polyhedral sets. Moreover, the implementation is very complicated because the sets P^k can be very complex non-convex polyhedra for which there is no useful canonical form and the test $P^{k+1} = P^k$ is very expensive. Hence we restrict further the class of sets to be what we call *griddy polyhedra*, i.e. 2^B where B is the set of all closed unit hypercubes with integer leftmost corners. Using this finite class of sets guarantees convergence of Algorithm 2 (provided we restrict our analysis to bounded domains) and allows

---

[1] Note that if the class C is closed under Boolean operations, only $\delta_{[0,r]}(F)$ needs to be approximated. This holds for arbitrary polyhedral sets but not for convex polyhedra or ellipsoids.

us to benefit from a relatively-efficient canonical representation for both convex and non-convex sets [BMP99], supported by an experimental software package. The price, however, for using orthogonal polyhedra is that the quality of the approximation they provide in terms of Haussdorf distance per vertex is poorer than that of arbitrary polyhedra (zero-order vs. first-order in the approximation jargon) but such a compromise seems unavoidable.

A nave approximate version of Algorithm 1 is guaranteed to converge to a superset of $\delta(F)$ after finitely many steps. However, the distance between the result and $\delta(F)$ might be too big for the result to be useful. The reason is that over-approximation errors accumulate dramatically as illustrated in Figure 1-(c) where we try to calculate successors of the set D. Since $\mathbf{x}'$ is reachable from $\mathbf{x}$ we must include the whole box D' in the set of successors. This box contains points such as $\mathbf{y}$ not really reachable from D, which bring in the next iteration new points, such as $\mathbf{y}'$, and we end up adding boxes such as D'' which are not reachable from D at all. This over-approximation error can propagate fast and the result might cover the whole space unless some hardly-formalizable hacking is used [DM98,GM99]. Similar phenomena are exhibited, for example, in abstract interpretation of programs over the integers [CC92] where over-approximation is called *widening*. This is why there is not much hope in finding finite quotients of continuous systems, except for special cases such as timed automata [AD94].

Here we need to find the right compromise between the desire to converge and the accumulation of errors. We propose a method, specialized for linear systems of the form $\dot{\mathbf{x}} = A\mathbf{x}$ which achieves this trade-off. The basic idea here is to separate the accumulation and storage of states reachable in one step (and those must contain approximation error) from the computation of states reachable in the next step (see also [GM99]). The main attraction of this method compared to traditional ways to treat linear systems is in its adaptability to hybrid systems and to systems with under-specified input.

3   The Approximate Method for Linear Systems

Let $conv(\{\mathbf{x}_1, \ldots, \mathbf{x}_m\})$ be the convex hull of a set of points, i.e. $\{\mathbf{x} : \mathbf{x} = \lambda_1\mathbf{x}_1 + \cdots, \lambda_m\mathbf{x}_m\}$ for non-negative λ_i whose sum is 1. For linear systems we have $\delta_t(\mathbf{x}) = e^{At}\mathbf{x}$ and the matrix exponential, as a linear operator, preserves convexity:

$$\delta_t(conv(\{\mathbf{x}_1, \ldots, \mathbf{x}_m\})) = conv(\{\delta_t(\mathbf{x}_1), \ldots, \delta_t(\mathbf{x}_m)\}).$$

This means that for a convex set $F = conv(\mathbf{V})$ where $\mathbf{V} = \{\mathbf{x}_1, \ldots, \mathbf{x}_m\}$, and for every t, the states reachable from F can be determined by the states reachable from $\mathbf{V}$ (see Figure 2-(a)). We exploit this property to approximate $\delta_{[0,r]}(conv(\mathbf{V}))$ based on the set of points $\mathbf{V} \cup \delta_r(\mathbf{V})$ where $\delta_r(\mathbf{V})$ is computed from $\mathbf{V}$ by a finite number of matrix exponentiations or numerical integration steps. Our approximation scheme consists of three steps:

1. Compute $G = conv(\mathbf{V}, \delta_r(\mathbf{V}))$ (see Figure 2-(b)). This set is an approximation of $\delta_{[0,r]}(conv(\mathbf{V}))$ but neither an over-approximation nor under-approximation. The convex-hull algorithm provides us with information concerning the orientation of the faces which is used in the next step.[2]

2. Push the faces of G outward to obtain a bloated convex polyhedron G' which is guaranteed to contain the required set (Figure 2-(c)). The amount of pushing is determined by the time step r and the matrix A (see the analysis in the appendix). Pushing inward will result in an under-approximation.

3. Over-approximate G' by a griddy polyhedron $\delta'_{[0,r]}(F)$ (Figure 2-(d)).

The approximate algorithm for calculating $\delta(F)$ for $F = conv(\mathbf{V})$ is defined below:

Algorithm 2 (Approximate Calculation of $\delta(F)$ for Linear Systems)

$$
\begin{array}{l}
P^0 := F;\ \mathbf{V}^0 := \mathbf{V};\ k{:=}0; \\
\textbf{repeat} \\
\quad k\quad := k+1; \\
\quad \mathbf{V}^k := \delta_r(\mathbf{V}^{k-1}); \\
\quad G^k := conv(\mathbf{V}^{k-1} \cup \mathbf{V}^k); \\
\quad G^k := bloat(G^k); \\
\quad G^k := griddy(G^k); \\
\quad P^k := P^{k-1} \cup G^k \\
\textbf{until}\ P^k = P^{k-1}
\end{array}
$$

There are two types of errors accumulated in the process of calculating P^k: from the actual set to its bloated convex hull and from there to the griddy polyhedron. However these errors do not propagate to the next step which computes P^{k+1} based on $\mathbf{V}^k \cup \mathbf{V}^{k+1}$ and not on P^k (Figure 2-(e)). Recall that our orthogonal polyhedra package [BMP99] maintains $\delta'_{[0,2r]}(F)$ as a *single* canonical object and *not* as a union of convex polyhedra or ellipsoids (Figure 2-(f)). The algorithm can be fine-tuned by changing the time step r and the size of the hypercubes.

Result 1 (Computation of Reachable States for Linear Systems) *There exists an implemented algorithm for over-approximating the reachable sets of systems defined by linear differential equations.*

The reason this result is not a theorem is due to the following facts:

1. There is always a trivial over-approximation of any subset F of X, namely X itself.

2. The smallest polyhedral or griddy set which contains $\delta(F)$ is as impossible to compute as $\delta(F)$.

---

[2] We use the convex-hull algorithm supplied with the LEDA library [MV99].

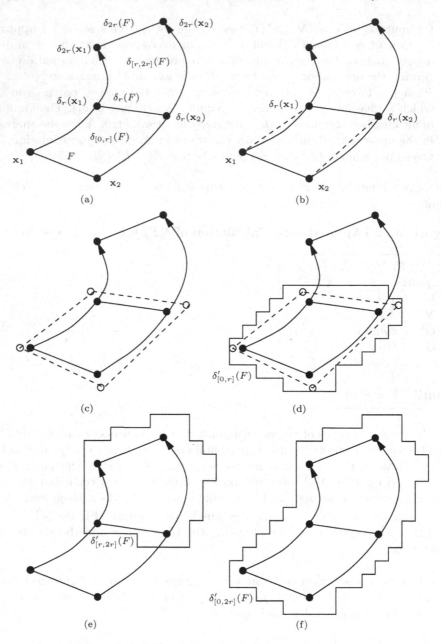

Fig. 2. (a) A set $F = conv(\{\mathbf{x}_1, \mathbf{x}_2\})$ and its exact successors for time intervals $[0, r]$ and $[r, 2r]$. (b) Approximating $\delta_{[0,r]}(F)$ by convex hull. (c) Bloating the convex polyhedron to obtain a polyhedral over-approximation. (d) Rectangulating the polyhedron into $\delta'_{[0,r]}(F)$. (e) Repeating the same procedure in the next time step to obtain $\delta'_{[r,2r]}(F)$. (f) The accumulated states $\delta'_{[0,2r]}(F) = \delta'_{[0,r]}(F) \cup \delta'_{[r,2r]}(F)$.

3. Like in many other numerical problems, the best upper-bounds which can be easily proved on the approximation error are much larger than what happens in practice.

So let us be content with the fact that the method gives reasonable approximation in rather short time. So far we were able to calculate rather easily the reachable states of non-trivial systems with up to 6 dimensions (in fact, the measure of complexity for such problems depends on the dimensionality, the coupling of the variables and the granularity of the discretization). Figure 3 shows the states reachable from

$$F = [0.025, 0.05] \times [0.1, 0.15] \times [0.05, 0.1]$$

by the 3-dimensional system defined by

$$A = \begin{pmatrix} -1.0 & -4.0 & 0.0 \\ 4.0 & -1.0 & 0.0 \\ 0.0 & 0.0 & 0.5 \end{pmatrix}$$

Fig. 3. Calculating reachable states for a 3-dimensional system.

4   Extensions and Applications

4.1   Piecewise-Linear Systems

For purely continuous linear systems there are classical methods, more efficient than ours, for solving certain problems such as stability or controller synthesis.

However the main advantage of our approach is manifested in the analysis and controller synthesis for linear hybrid automata which may switch between several "modes" and hence define piecewise-linear dynamical systems. We demonstrate the adaptation of our method to such systems informally using the hybrid automaton of Figure 4, which consists of two continuous variables, and two discrete states. In each discrete state the continuous variables evolve according to the corresponding linear dynamics and when some switching conditions (transition guards) are satisfied, the system moves from one state to another.

Starting from an initial set (q_0, F) the reachable states are calculated as follows: we apply our procedure to F with the A_0 dynamics and calculate forward $\delta^0(F)$. Then we calculate the intersection of the result with the guard to obtain a set F', move to state q_1 with F' as the set of initial states, calculate $\delta^1(F')$ and so on and so forth. This method is similar to the one used in tools such as KRONOS [DOTY96] for timed automata and HyTech [HHW97] for hybrid automata with constant derivatives [ACH+95].

The main technical difficulty in applying our vertex-based approximation technique to such systems is that not all trajectories departing from the vertices reach a transition guard simultaneously (some may not reach it at all). Hence we have to calculate $\delta(F)$ and intersect it with the guard to obtain the new initial set. Unfortunately, this set is already an over-approximation and, moreover, it might have many vertices and the reduction of their number might require further approximation. The bottom line is that we can avoid propagation of over-approximation errors during the continuous evolution but not while doing transitions.

An example run of $\boxed{\text{d/dt}}$ on the hybrid automaton of Figure 4 where

$$A_0 = \begin{pmatrix} -2.0 & -3.0 \\ 3.0 & -2.0 \end{pmatrix} \quad \text{and} \quad A_1 = \begin{pmatrix} 0.0 & -0.6 \\ 3.0 & 0.0 \end{pmatrix}$$

and the initial set is $F = \{q_0\} \times [0.3, 0.6] \times [-0.2, 0.2]$, appears in Figure 5. Initially the successors by A_0 (a "center" dynamics) are calculated until they all intersect the guard $x_1 \leq -0.15$ (a). Then dynamics A_1 is applied, shrinking the set until intersection with the guard $x_1 \geq -0.02$ (b). From this guard the dynamics A_0 induces a "ring" of states which stay in q_0 forever (c).

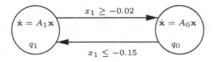

Fig. 4. A hybrid automaton.

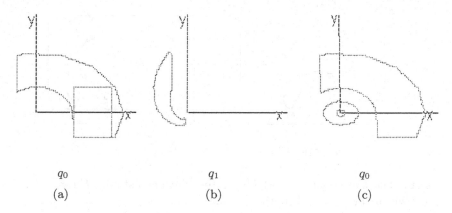

q_0

(a)

q_1

(b)

q_0

(c)

Fig. 5. The 3 stages in the calculation of $\delta(F)$ of the hybrid automaton of Figure 4.

4.2 Under-Approximation, Backward Reachability and Control

The δ operator is a basic ingredient in forward reachability analysis. Other verification and synthesis problems require different variants of this operator.

The reader might have guessed that calculating under-approximations is done by a slight variation of the algorithm, i.e. pushing the faces of the polyhedron *inside* and finding an orthogonal under-approximation. Backward reachability, that is, finding all the points from which a set F is reachable can be performed by computing δ for the reversed system $\dot{\mathbf{x}} = -A\mathbf{x}$.

For the purpose of controller synthesis for hybrid systems [ABDPM00] we need an under-approximation of the "F *Until* G" operator, which returns the points from which you can stay within the set F either forever or until you reach a set G (which is typically the guard of a transition to another state). A similar operator is needed for analyzing hybrid systems with invariants. Consider $F = [-0.1, 0.1] \times [-0.030.1]$, $G = [0.02, 0.06] \times [-0.05, -0.02]$ and a dynamics

$$A = \begin{pmatrix} -0.5 & 4.0 \\ -3.0 & -0.5 \end{pmatrix}$$

The two parts of F *Until* G, as calculated by $\boxed{\mathbf{d/dt}}$ appear in Figure 6.

4.3 Continuous Disturbances

Consider systems of the form $\dot{\mathbf{x}} = A\mathbf{x} + B\mathbf{u}$ where $\mathbf{u}$ ranges inside a convex set U. It has been suggested in [V98] to use the maximum principle from optimal control to find $\delta_r(F)$ of a convex set $F = conv(\mathbf{V})$ under all possible input signals. We have implemented this procedure and incorporated it into our system. We have tested it on a 4-dimensional example adapted from example 4.5.1 of [KV97], pp. 279-285, where ellipsoids are used instead of polyhedra. The system is defined

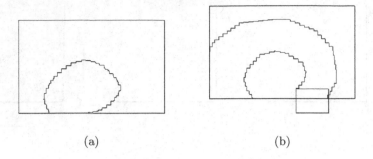

(a)   (b)

Fig. 6. The *Until* operator: (a) The states which can stay in F forever. (b) The states which can stay in F until reaching G.

by:

$$A = \begin{pmatrix} 0.0 & 1.0 & 0.0 & 0.0 \\ -8.0 & 0.0 & 0.0 & 0.0 \\ 0.0 & 0.0 & 0.0 & 1.0 \\ 0.0 & 0.0 & -4.0 & 0.0 \end{pmatrix} \qquad B = 1$$

$$F = [0.02.0] \times [-1.01.0] \times [0.0, 2.0] \times [-1.0, 1.0]$$

$$U = [-0.5, 0.5] \times [-0.005, 0.005] \times [-0.5, 0.5] \times [-0.005, 0.005]$$

In Figure 7 one can see the evolution of the projection on dimensions 3 and 4 over time, similar to the results in [KV97]. Further work on these technique might suggest effective methods for approximate strategies for differential games.

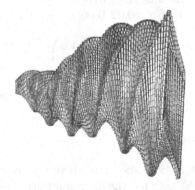

Fig. 7. The evolution of a 4-dimensional convex differential inclusion over time (projected on dimensions 3 and 4).

5   Discussion

In this work we have advanced the state-of-the-art in computer-aided reachability analysis for continuous and hybrid systems. We have implemented the tool $\boxed{\mathrm{d/dt}}$ and tested it over reproducible non-trivial examples. We are currently investigating various improvements and studying the trade-offs between accuracy and computational efficiency. We hope that such techniques and tools will be used in the future by control engineers.

References

ACH$^+$95.   R. Alur, C. Courcoubetis, N. Halbwachs, T.A. Henzinger, P.-H. Ho, X. Nicollin, A. Olivero, J. Sifakis and S. Yovine, The Algorithmic Analysis of Hybrid Systems, *Theoretical Computer Science* 138, 3–34, 1995.

AD94.   R. Alur and D.L. Dill, A Theory of Timed Automata, *Theoretical Computer Science* 126, 183–235, 1994.

ABDPM00.   E. Asarin, O. Bournez, T. Dang, A. Pnueli and O. Maler, Effective Synthesis of Switching Controllers for Linear Systems, submitted for publication, 2000.

BMP99.   O. Bournez, O. Maler and A. Pnueli, Orthogonal Polyhedra: Representation and Computation, in [VS99], 46-60.

CV95.   K. Cerans and J. Viksna, Deciding Reachability of Planar Multipolynomial systems, in R. Alur, T.A. Henzinger and E.D. Sontag (Eds.), *Hybrid Systems III*, 389-400, LNCS 1066, Springer, 1996.

CK99.   A. Chutinan and B.H. Krogh, Verification of Polyhedral Invariant Hybrid Automata Using Polygonal Flow Pipe Approximations, in [VS99] 76-90.

CC92.   P. Cousot and R. Cousot, Abstract Interpretation and Application to Logic Programs, *Journal of Logic Programming*, 103-179, 1992.

DM98.   T. Dang, O. Maler, Reachability Analysis via Face Lifting, in T.A. Henzinger and S. Sastry (Eds), *Hybrid Systems: Computation and Control*, LNCS 1386, 96-109, Springer, 1998.

DOTY96.   C. Daws, A. Olivero, S. Tripakis, and S. Yovine, The Tool KRONOS, in R. Alur, T.A. Henzinger and E. Sontag (Eds.), *Hybrid Systems III*, LNCS 1066, 208-219, Springer, 1996.

G96.   M.R. Greenstreet, Verifying Safety Properties of Differential Equations, in R. Alur and T.A. Henzinger (Eds.), *Proc. CAV'96*, LNCS 1102, 277-287, 1996.

GM98.   M.R. Greenstreet and I. Mitchell, Integrating Projections, in T.A. Henzinger and S. Sastry (Eds), *Hybrid Systems: Computation and Control*, LNCS 1386, 159-174, Springer, 1998.

GM99.   M.R. Greenstreet and I. Mitchell, Reachability Analysis Using Polygonal Projections, in [VS99] 76-90.

HHMW99.   T.A. Henzinger, B. Horowitz, R. Majumdar, and H. Wong-Toi, Beyond HyTech: Hybrid System Analysis Using Interval Numerical Methods, AAAI Spring Symposium on Hybrid Systems, Stanford University, 1999.

HHW97.   T.A. Henzinger, P.-H. Ho, and H. Wong-Toi, HyTech: A Model Checker for Hybrid Systems, *Software Tools for Technology Transfer* 1, 110-122, 1997.

KV97.   A. Kurzhanski ans I. Valyi, *Ellipsoidal Calculus for Estimation and Control*, Birkhauser, 1997.
MV99.   K. Mehlhorn and St. Nher, The LEDA Platform of Combinatorial and Geometric Computing, Cambridge University Press, 1999.
PLY99.   G. Pappas, G. Lafferriere and S. Yovine, A New Class of Decidable Hybrid Systems, in [VS99] 29-31.
PSK99.   J. Preussig, O. Stursberg and S. Kowalewski, Reachability Analysis of a Class of Switched Continuous Systems by Integrating Rectangular Approximation and Rectangular Analysis, in [VS99] 208-222.
VS99.   F. Vaandrager and J. van Schuppen (Eds.), *Hybrid Systems: Computation and Control*, LNCS 1569, Springer, 1999.
V98.   P. Varaiya, Reach Set Computation using Optimal Control, *Proc. KIT Workshop*, Verimag, Grenoble, 1998.

Appendix: Conservative Approximation

As we have already mentioned when describing the approximate method for linear systems, the set $G = conv(\mathbf{V}, \delta_r(\mathbf{V}))$ is not an over-approximation of $\delta_{[0,r]}(conv(\mathbf{V}))$ and should be replaced by its ϵ-neighborhood (or something bigger) in order to become such an over-approximation. Here we calculate the ϵ that should be used.

Consider an arbitrary point $p_0 \in conv(\mathbf{V})$ and a trajectory p_t starting from this point. We have $p_r = e^{rA}p_0$. This point belongs to $\delta_r(\mathbf{V})$ and hence to G. By convexity so does all the line segment $[p_0, p_r]$. Let us estimate now the distance between points of the true trajectory p_t for $t \in [0, r]$ and this line segment. In fact p_t may be approximated by linear interpolation between p_0 and p_r. The result of this interpolation is

$$\hat{p}_t = p_0 + \frac{t}{r}(p_r - p_0), \quad 0 \le t \le r$$

and by construction it belongs to the segment $[p_0, p_r]$. The error of this interpolation can be written as follows:

$$\epsilon(p_0, t) = ||\hat{p}_t - p_t|| = ||p_0 + \frac{t}{r}(e^{rA} - I)p_0 - e^{tA}p_0||.$$

Since

$$e^{tA} = I + At + \frac{1}{2}A^2t^2 + \sum_{i=3}^{\infty} \frac{1}{n!}A^i t^i$$

and $0 \le t \le r$ we find after obvious simplifications the bound of the error:

$$\epsilon(p_0, t) \le \epsilon = M\frac{1}{8}||A||^2 r^2 + M\sum_{i=3}^{\infty} \frac{1}{n!}||A||^i r^i,$$

where M is a constant bounding the norm $||p_0||$.

Hence, for every r, one can find a $\epsilon = O(r^2)$ such that all the points reachable from $conv(\mathbf{V})$ in time r are in ϵ-neighborhood of G. In order to over-approximate the set we just replace G by its ϵ-neighborhood.

Maximal Safe Set Computation for Idle Speed Control of an Automotive Engine

Andrea Balluchi[1], Luca Benvenuti[1,2], Maria D. Di Benedetto[2],
Guido M. Miconi[1], Ugo Pozzi[1], Tiziano Villa[1], Howard Wong-Toi[3], and
Alberto L. Sangiovanni–Vincentelli[1,4]

[1] PARADES, Via di S.Pantaleo, 66, 00186 Roma, Italy.
{alberto,balluchi,lucab,miconi,pozzi}@parades.rm.cnr.it
[2] Dip. di Ingegneria Elettrica, Università di L'Aquila, Poggio di Roio,
67040 L'Aquila, Italy, dibenede@giannutri.caspur.it
[3] Cadence Berkeley Labs, 2001 Addison St., Third Floor, Berkeley, CA 94704, USA.
howard@cadence.com
[4] Dept. of Electrical Engineering and Computer Sciences, University of California
at Berkeley, CA 94720, USA. alberto@eecs.berkeley.edu

Abstract. The specification for the idle control problem for automotive
engines is to maintain the crankshaft speed within a given range in the
presence of load changes. A new cycle-detailed hybrid model of the engine
that captures well the interactions between the discrete phenomena of
torque generation and spark ignition, and the continuous evolution of
the power-train and air dynamics, is proposed. The idle control problem
is formalized as a safety specification problem on the hybrid system. The
Tomlin-Lygeros-Sastry procedure [12] is applied to compute the maximal
controlled invariant set that satisfies the safety specification.

1 Introduction

The synthesis of a control strategy for an internal combustion engine in the idle
regime is one of the most challenging problems in engine control. The objective
is to maintain the engine speed as close as possible to the value that minimizes
fuel consumption, while preventing the engine from turning off when a sudden
load variation occurs. Load variations come from two sources: (1) from devices
powered by the engine, such as the air conditioning system and the steering
wheel servo-mechanism, or (2) from the driver changing the inertial load when
operating the clutch pedal. A survey on different engine models and control design methodologies for the idle control is given in [8]. Both time–domain (e.g. [5])
and crank–angle domain (e.g. [13]) average–value models have been proposed in
the literature. Several control design techniques have been applied to the idle
control problem, such as multivariable control [10], ℓ_1 control [5], H_∞ control [6],
μ-synthesis [7], sliding mode control [9] and LQ-based optimization [1].

In this paper, the idle control problem is specified as the one of keeping
the crankshaft speed within a specified range, robustly with respect to load
changes. The adoption of a hybrid formalism allows us to describe the cyclic

N. Lynch and B. Krogh (Eds.): HSCC 2000, LNCS 1790, pp. 32–44, 2000.

behavior of the engine, thus capturing the effect of each spark ignition on the generated torque, the interaction between the discrete torque generation and the continuous power-train and air dynamics, and the discrete changes in the power-train. The torque that is generated by each cylinder and applied to the engine crankshaft can be assumed to be a function of the spark ignition time, and of the air-fuel mixture mass loaded in the cylinder during the intake phase. Since the air-to-fuel ratio is assumed to be constant (at the stoichiometric value), then the mixture mass is controlled by the throttle plate position and is subject to the dynamics of the cylinder filling due to the intake manifold. Hence, the available controls for the idle problem are: the spark ignition time and the position of the throttle valve, which regulates the air inflow[1]. The problem of maintaining the crankshaft speed within a given range is formalized as a safety specification for the hybrid closed-loop system. A *safety specification* is a state-invariance property, specifying a set of good states within which the closed-loop system must remain. A systematic procedure for computing the maximal safe set has been recently proposed by Tomlin, Lygeros, and Sastry [12]. This set consists of all the hybrid states for which there exists a hybrid control strategy (the maximal controller) that maintains the state in the set of good states forever, in spite of any discrete and continuous disturbance. The procedure is not guaranteed to terminate in a finite number of steps.

By applying this procedure to the hybrid model of the engine, the maximal safe set for the idle control is determined. We also obtain as a by-product the entire set of possible controllers that satisfy the constraints. We are free to choose among them the ones that optimize some criteria of choice. Moreover, considering the amount of load torque as a parameter, we can determine the maximum value for which a non empty maximal safe set (and, hence, at least one controller that satisfies the constraints) exists[2]. For parameters corresponding to commercial cars, the procedure has terminated in a few steps (typically six).

To summarize our main contribution, the use of a hybrid framework, where discrete and continuous signals are modeled in a separate but integrated manner, is a definite advantage over other approaches since it allows us to solve exactly the control problem while other approaches, where the system is approximated by either continuous [11], or discrete sampled [13] representations, obtain approximate solutions. The paper is organized as follows: in Section 2, a description of the engine in the idle region of operation is offered and its hybrid features are exposed. In Section 3, a *hybrid automaton model of the engine for*

---

[1] The effect of a spark command on the torque generation is more visible than the one of a throttle plate command, since air inflow is subject to both the manifold dynamics and the delay due to the mix compression. Hence, sudden loads can be much better compensated with spark ignition than with air inflow, while air inflow can be used to control the engine in steady state. For simplicity, we do not consider the throttle valve actuation dynamics.

[2] Butts *et al.* [5] solve a sort of dual problem: given a bounded torque load accessible to measurement, synthesize a robust ℓ_1 controller for a discrete–time model of the engine that minimizes the excursion of the crankshaft speed for the system initially at rest.

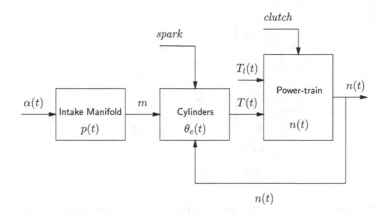

Fig. 1. The engine blocks and their communication topology.

the idle regime of operation is proposed for the *first* time. In Section 4, a general procedure for the calculation of the maximal controller is reviewed. In Section 5, the procedure is specialized to the idle control problem and some experimental results are described.

2   Description of the System

The overall system is composed of three main interacting blocks, namely the intake manifold, the cylinders and the power-train, as depicted in Figure 1. The manifold pressure p depends on the throttle valve angle α and determines the mass of air-fuel mixture m loaded by the cylinders. The torque T generated by the cylinders depends on both the mass m and the spark ignition time. Finally, the power-train dynamics and the crankshaft revolution speed n, controlled by the generated torque T, are subject to the sum of load torques T_l and the clutch position. In the sequel a detailed description of each block is reported.

Intake Manifold Dynamics. The mass of mix m entering a cylinder during the intake run is assumed to be proportional to the intake manifold pressure p at the end of the intake run. The pressure p is controlled by a throttle valve which changes the effective section of the intake manifold:

$$\dot{p}(t) = a_p p(t) + b_p \alpha(t) . \tag{1}$$

To prevent the choice of undesirable control laws that produce large excursions of the throttle valve, the throttle angle $\alpha(t)$ is constrained to belong to $[0, \alpha^{max}]$ with $\alpha^{max} = 20^0$.

Cylinders. In a 4–stroke combustion engine each cylinder cycles through the following four runs: intake (I), combustion (C), expansion (E) and exhaust (H).

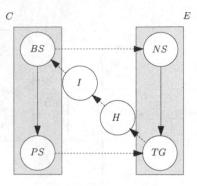

Fig. 2. Phases of a single cylinder: dashed lines denote transitions occurring when $\theta_c = 180$, solid lines denote transitions occuring when a spark ignition is given.

Ideally, spark ignition should occur exactly when the piston reaches the *top dead center* (TDC) configuration of the compression stroke. However, since combustion takes non-zero time to complete, it is convenient to produce a spark before the piston completes the compression stroke (*positive spark advance*), to achieve maximum fuel efficiency, i.e. the maximum torque generation. When a small value of torque is required, the spark can be ignited after the piston has completed the compression phase and is in the expansion stroke (*negative spark advance*).

Let θ_c denote the piston position, between two successive dead points, expressed in terms of the angle described by the crankshaft, obtained by the integration of the crankshaft velocity and by resetting θ_c to 0^0 when $\theta_c = 180^0$. The spark advance θ_s, defined as the angular distance from the TDC of the compression stroke, determines the crankshaft angular position at which the spark is given. It is positive for sparks given before the TDC ($\theta_s = 180^0 - \theta_c(t_s)$, where t_s is the ignition time), and negative otherwise ($\theta_s = -\theta_c(t_s)$). At the idle crankshaft speed, due to technological constraints, the feasible spark advances are $-15^0 \leq \theta_s \leq 20^0$. If the spark is ignited during the compression stroke, then C is split into two phases, namely BS (before spark) and PS (positive spark). Instead, if the spark is ignited during the expansion stroke, then E is split into two phases, namely NS (negative spark) and TG (torque generation). In the first case, the expansion stroke is represented by phase TG, while, in the second one, the compression stroke is represented by BS. The behavior of each piston is then characterized[3] by the six phases I, BS, PS, NS, TG and E, as shown in Figure 2.

When considering a 4-stroke internal combustion engine with four cylinders, the kinematics of the engine is such that, at any time, each cylinder is in a

---

[3] When the spark ignition is synchronous with the TDC, we assume the cylinder leads from BS to TG through the phase PS.

different stroke of the cycle. Since we assume that all the cylinders behave in the same way, then we can cluster all quadruples of cylinders' phases in only three engine phases, because it does not matter which cylinder is in a certain phase. Then, according to the ignition constraints $\theta_c \in [160^0, 180^0]$, in PS, and $\theta_c \in [0^0, 15^0]$, in NS, there are only three valid cylinder configurations and the discrete behavior of the system can be described by introducing the following three modes $S = (I, BS, TG, H)$, $S_+ = (I, PS, TG, H)$, $S_- = (I, BS, NS, H)$.

The transitions between S, S_+ and S_- are characterized as follows. In phase S, the cylinder in expansion is generating a torque (TG), and the cylinder in compression has not yet received the spark command (BS). If a spark ignition occurs before the end of the compression run, then the cylinder that is still in compression enters phase PS, which corresponds to the transition from S to S_+. Otherwise, at the TDC, the expansion phase starts (NS) and the transition from S to S_- takes place. In phase S_+, the spark command has been given for the cylinder in compression (PS), while the cylinder in expansion is generating a torque (TG). At the TDC, the cylinder which was in compression starts the expansion run entering phase TG, which corresponds to the transition from S_+ to S. In phase S_-, the cylinder in expansion is waiting for the spark command (NS), and the cylinder in compression has not received the spark command yet (BS). It is worth noting that no torque is generated in this case. When the spark ignition is given, the cylinder which is still in the expansion run changes from NS to TG, and the transition from S_- to S takes place.

The evolution of the torque, generated by each piston during the expansion phase, depends on the thermodynamics of the air-fuel mixture combustion. To simplify the model, we represent by T the average value of the torque generated over the expansion phase. Such value is proportional to the air m loaded in the cylinder during the intake phase and to the ignition efficiency (increasing) function $\eta(\theta_s) \leq 1$.

Since there is a delay from the time the air mass m is trapped in the cylinder and the time the torque is generated, the amount of loaded air mass must be stored for each cylinder. To this end, we introduce two variables, m_C and m_E denoting, respectively, the mass of air trapped at the end of the intake run in the cylinder starting the compression run, and the mass of air trapped at the end of the compression run in the cylinder starting the expansion run. Hence, the torque T produced by each piston is expressed as

$$T = \begin{cases} G\,\eta(\theta_s)\,m_C & \text{for positive spark advance,} \\ G\,\eta(\theta_s)\,m_E & \text{for negative spark advance.} \end{cases} \tag{2}$$

The torque $T(t)$ generated by the engine is obtained by applying a zero order hold block to each cylinder output T, and summing all the piecewise constant contributions of the pistons.

Power-train Dynamics. When the clutch is pushed, under the action of the torque $T(t)$ generated by the engine, the crankshaft speed evolution is determined by the following mechanical equations

$$\dot{n}(t) = a_n n(t) + b_n(T(t) - T_l(t)) \tag{3}$$
$$\dot{\theta}_c(t) = k_c n(t) \tag{4}$$

where $a_n = -B/J$ and $b_n = 1/J$, with J and B denoting the inertial momentum and the viscous friction coefficient of the segment of the power-train from the crankshaft to the clutch, respectively. If θ_c is in degrees and n is in revolutions per minute, then $k_c = 6$. When the clutch is released, the coefficients a_n and b_n are replaced by $a_n^L = -B/(J + J')$ and $b_n^L = 1/(J + J')$, where J' denotes the inertial momentum of the primary drive-line.

Finally, the torque load T_l is assumed to belong to $[0, T_l^{max}]$, where the value T_l^{max} is treated as a parameter for the control problem.

3   Hybrid Model of the Engine

We model the mixed discrete-continuous dynamics of the engine as a hybrid automaton. We consider a particular class of hybrid automata characterized by a set of discrete *locations* (also called *modes*) corresponding to the FSM states, a set of continuous variables and a set of piecewise-constant variables. The controller and the environment act on the system with two kinds of inputs: the continuous inputs affect the continuous dynamics; the discrete inputs determine the discrete mode transitions, the resetting of continuous variables and the setting of symbolic constants. This modeling formalism combines the features of [2] with elements of the hybrid dynamics of [12]. The formal definition and the behavior of this hybrid automaton is analogous to the one described in [4], with the separation between continuous variables and piecewise-constant variables explicitly introduced here. The hybrid model of the 4-stroke 4-cylinder internal combustion engine has six modes S_-, S, S_+, S_-^L, S^L, and S_+^L, derived from the three modes S_-, S, and S_+ of the four cylinders and the two discrete positions of the clutch, which can be either closed or open. Figure 3 shows the resulting hybrid automaton. Hence, we can formally write the engine hybrid automaton as a tuple $H = (\{Q, X, \Xi\}, \{\Sigma_c, U\}, \{M_c^{disc}, M_c^{cts}\}, \{\Sigma_e, D\}, \{M_e^{disc}, M_e^{cts}\}, \{f, \delta\})$, where:

- the **state space** is composed of the finite set of *modes* or *locations* Q, which consists of $S_-, S, S_+, S_-^L, S^L, S_+^L$, the space of *continuous variables* $X = \{(p, n, \theta_c) \mid (p, n, \theta_c) \in \mathbb{R}^3\}$, and the space of *piecewise-constant variables* $\Xi = \{(T, m_C, m_E, \theta_s) \mid (T, m_C, m_E, \theta_s) \in \mathbb{R}^4\}$. An element (q, x, ξ) in the space $Q \times X \times \Xi$ is called a *configuration*;

- the **control inputs** $\alpha(t)$ and σ_c can be described by means of the domain of continuous input values $U = [0, \alpha^{max}]$, the finite domain of *discrete control events* $\Sigma_c = \{spark\}$ with $\Sigma_c^\epsilon = \Sigma_c \cup \{\epsilon\}$ being the set of *discrete control moves* and the special ϵ move being the *silent* move, the *discrete controller feasible move function* $M_c^{disc} : Q \times X \times \Xi \rightarrow 2^{\Sigma_c^\epsilon} \setminus \{\}$ described as follows[4]:

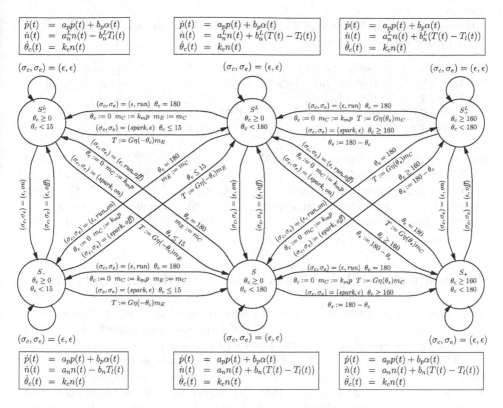

$$\dot{p}(t) = a_p p(t) + b_p \alpha(t)$$
$$\dot{n}(t) = a_n n(t) - b_n T_l(t)$$
$$\dot{\theta}_c(t) = k_c n(t)$$

$$\dot{p}(t) = a_p p(t) + b_p \alpha(t)$$
$$\dot{n}(t) = a_n n(t) + b_n(T(t) - T_l(t))$$
$$\dot{\theta}_c(t) = k_c n(t)$$

$$\dot{p}(t) = a_p p(t) + b_p \alpha(t)$$
$$\dot{n}(t) = a_n n(t) + b_n(T(t) - T_l(t))$$
$$\dot{\theta}_c(t) = k_c n(t)$$

Fig. 3. Hybrid model for the engine running at minimum.

$M_c^{disc}(S_+, \theta_c \leq 180^0) = \{\epsilon\}$ $M_c^{disc}(S_+^L, \theta_c \leq 180^0) = \{\epsilon\}$

$M_c^{disc}(S_-, \theta_c < 15^0) = \{\epsilon, spark\}$ $M_c^{disc}(S_-^L, \theta_c < 15^0) = \{\epsilon, spark\}$

$M_c^{disc}(S_-, \theta_c = 15^0) = \{spark\}$ $M_c^{disc}(S_-^L, \theta_c = 15^0) = \{spark\}$

$M_c^{disc}(S, \theta_c < 160^0) = \{\epsilon\}$ $M_c^{disc}(S^L, \theta_c < 160^0) = \{\epsilon\}$

$M_c^{disc}(S, 160^0 \leq \theta_c \leq 180^0) = \{\epsilon, spark\}$ $M_c^{disc}(S^L, 160^0 \leq \theta_c \leq 180^0) = \{\epsilon, spark\}$

and the *continuous controller feasible move function* $M_c^{cts} : Q \times X \times \varXi \to 2^U \setminus \{\}$
described as follows: $M_c^{cts}(q, x, \xi) = \{\alpha \mid \alpha \in [0, \alpha_{max} = 20^0]\}$, $\forall (q, x, \xi)$;

• the **disturbance inputs** $T_l(t)$ and σ_e can be described by means of the
domain of *continuous disturbance values* $D = [0, T_l^{max}]$, the finite set of *discrete
disturbance events* $\Sigma_e = \{on, off, run, run\_on, run\_off\}$ (where the events *on*
and *off* represent opening and closing the discrete position of the clutch, the
event *run* represents reaching the boundary $\theta_c = 180^0$ for the continuous state,
the events *run_on* and *run_off* represent the simultaneous occurrence of a clutch
operation and reaching the boundary, and $\Sigma_e^\epsilon = \Sigma_e \cup \{\epsilon\}$ is the set of *discrete*

---

[4] Notice that $M_c^{disc}(S_-, \theta_c = 15^0) = \{spark\}$ is a discrete control move required when
  the spark was not given yet and must be given now, since it is the last valid ignition
  time instant.

disturbance moves), the *discrete disturbance move function* $M_e^{disc} : Q \times X \times \Xi \rightarrow 2^{\Sigma_e^\epsilon} \setminus \{\}$ described as follows[5]:

$$M_e^{disc}(S_+, \theta_c < 180^0) = \{\epsilon, \mathit{off}\} \qquad M_e^{disc}(S_+^L, \theta_c < 180^0) = \{\epsilon, \mathit{on}\}$$
$$M_e^{disc}(S_+, \theta_c = 180^0) = \{\mathit{run}, \mathit{run\_off}\} \quad M_e^{disc}(S_+^L, \theta_c = 180^0) = \{\mathit{run}, \mathit{run\_on}\}$$
$$M_e^{disc}(S_-, \theta_c \leq 15^0) = \{\epsilon, \mathit{off}\} \qquad M_e^{disc}(S_-^L, \theta_c \leq 15^0) = \{\epsilon, \mathit{on}\}$$
$$M_e^{disc}(S, \theta_c < 180^0) = \{\epsilon, \mathit{off}\} \qquad M_e^{disc}(S^L, \theta_c < 180^0) = \{\epsilon, \mathit{on}\}$$
$$M_e^{disc}(S, \theta_c = 180^0) = \{\mathit{run}, \mathit{run\_off}\} \quad M_e^{disc}(S^L, \theta_c = 180^0) = \{\mathit{run}, \mathit{run\_on}\}$$

and the *continuous disturbance feasible move function*: $M_e^{cts} : Q \times X \times \Xi \rightarrow 2^D \setminus \{\}$ described as follows: $M_e^{cts}(q, x, \xi) = \{T_l \mid T_l \in [0, T_{lmax}]\}$, $\forall (q, x, \xi)$;

- the **transitions** are described by $f : Q \times X \times \Xi \times U \times D \rightarrow \mathbb{R}^n$ which models the time-invariant continuous dynamics, which depend on the mode[6] and the *transition function* $\delta : Q \times X \times \Xi \times \Sigma_c^\epsilon \times \Sigma_e^\epsilon \rightarrow 2^{Q \times X \times \Xi} \setminus \{\}$ modeling the discrete dynamics, as depicted in Fig. 3.

4 Synthesis of Hybrid Static State Feedback Controllers

The engine control problem at hand belongs to the class of *safety* problems. A *safety property* $\mathcal{P}$ asserts that nothing "bad" happens along trajectories and can be expressed by specifying a subset *Good* of the configuration space $(Q \times X \times \Xi)$. The co-set of *Good* is called *Bad*. The hybrid automaton H, with initial configurations $(Q \times X \times \Xi)_0 \subseteq \mathit{Good}$, is *safe* with respect to the safety property $\mathcal{P}$ if there exists a control strategy that guarantees all its trajectories that start in $(Q \times X \times \Xi)_0$ remain within *Good*. The maximal safe set, *Safe*, is the maximal subset $(Q \times X \times \Xi)_0$ of *Good* for which the hybrid automaton H is safe with respect to $\mathcal{P}$, i.e., the maximal robust-controlled invariant set of configurations contained in *Good*. The maximal controller is the class of all the hybrid static state-feedback control strategies that guarantee that all the trajectories starting in *Safe* remain within *Good*.

For the hybrid automaton described in Section 3, we define *Good* as the set of configurations for which the crankshaft speed is within the range $[770, 830]$, i.e., $\mathit{Good} = \{(q, x, \xi) \in Q \times X \times \Xi \mid 770 \leq n \leq 830\}$. The design of a controller requires the computation of the maximal safe set *Safe*.

Computing the maximal safe set [12]. This set is obtained by first overapproximating it with all the good configurations. Then all configurations are obtained from which the environment can drive the system into an unsafe configuration via either one discrete jump, or one continuous flow. These are the configurations from which the environment can push the system into *Bad* in one "step",

---

[5] Notice that $M_e^{disc}(S_+, \theta_c = 180^0) = \{\mathit{run}, \mathit{run\_off}\}$, $M_e^{disc}(S_+, \theta_c = 180^0) = \{\mathit{run}, \mathit{run\_on}\}$, $M_e^{disc}(S, \theta_c = 180^0) = \{\mathit{run}, \mathit{run\_off}\}$ and $M_e^{disc}(S^L, \theta_c = 180^0) = \{\mathit{run}, \mathit{run\_off}\}$ model discrete moves forced by the continuous state.

[6] We specify the continuous dynamics f by defining functions $f_q : X \times \Xi \times U \times D \rightarrow X$ for each $q \in Q$. The functions f_q, as specified in Figure 3, are taken from (1), (3) and (4).

and should be avoided by the controller. One iterates this computation, finding successively the configurations from which the environment can push the system into *Bad* in i steps. If the procedure terminates, we have determined the maximal safe set. The procedure is already described in full detail in [12,3]; here, we report only the definitions of the predecessor operators required to capture the previous notions.

Discrete uncontrollable predecessors operator $Pre_e : 2^{(Q \times X \times \Xi)} \to 2^{(Q \times X \times \Xi)}$:

$$Pre_e(K) = \{(q, x, \xi) \in Q \times X \times \Xi :$$
$$\forall \sigma_c \in M_c^{disc}(q, x, \xi) \, \exists \sigma_e \in M_e^{disc}(q, x, \xi) \text{ such that}$$
$$(\sigma_c, \sigma_e) \neq (\epsilon, \epsilon) \wedge \delta(q, x, \xi, \sigma_c, \sigma_e) \not\subseteq K\}.$$

Discrete controllable predecessors operator $Pre_c : 2^{(Q \times X \times \Xi)} \to 2^{(Q \times X \times \Xi)}$:

$$Pre_c(K) = \{(q, x, \xi) \in Q \times X \times \Xi :$$
$$\exists \sigma_c \in M_c^{disc}(q, x, \xi) \text{ such that } \forall \sigma_e \in M_e^{disc}(q, x, \xi)$$
$$(\sigma_c, \sigma_e) \neq (\epsilon, \epsilon) \wedge \delta(q, x, \xi, \sigma_c, \sigma_e) \subseteq K\}.$$

Continuous uncontrollable predecessor operator[7]
$$Unavoid\_Pre : 2^{(Q \times X \times \Xi)} \times 2^{(Q \times X \times \Xi)} \to 2^{(Q \times X \times \Xi)}:$$

$$Unavoid\_Pre(B, E) = \{(q, \hat{x}, \xi) \in Q \times X \times \Xi \mid$$
$$\forall u \in \mathcal{U} \, \exists \bar{t} > 0 \, \exists d \in \mathcal{D} \text{ such that, for the trajectory}$$
$$x(t) = \psi_q(u, d, \hat{x}, \xi)(t) \text{ we have: } (q, x(\bar{t}, \xi)) \in B \wedge$$
$$\forall \tau \in [0, \bar{t}) \, [u(\tau) \in M_c^{cts}(q, x(\tau), \xi) \wedge$$
$$d(\tau) \in M_e^{cts}(q, x(\tau), \xi) \wedge (q, x(\tau), \xi) \in Wait \, \cap \overline{E}] \}$$

Figure 4 shows the fixed-point computation to obtain the maximal safe set. The procedure successively prunes away configurations that are found to lead to a bad configuration upon one additional discrete step ($Pre_e(W^i)$), or a continuous step to a bad configuration ($Unavoid\_Pre(Pre_e(W^i) \cup \overline{W^i}, Pre_c(W^i))$. It is not guaranteed to stop in a finite number of steps.

A hybrid controller watches the entire state of the system at all times, and decides whether to (1) take discrete control actions that may cause an instantaneous change in the configuration, or to (2) let time pass under a continuous input u with the continuous variables evolving according to dynamics at the current mode. The formal definition of a safe hybrid controller and the rules which allow its extraction by the description of the maximal safe set are described in [4].

---

[7] We define $\mathcal{U} = \{u(.) \in PC^0 | u(t) \in U, \forall t \in \mathbb{R}\}$ and $\mathcal{D} = \{d(.) \in PC^0 | d(t) \in D, \forall t \in \mathbb{R}\}$, and we denote by *Wait* the set of configurations in which both players may choose not to play a discrete move, but instead wait for time to pass: $Wait = \{(q, x, \xi) \mid \epsilon \in M_c^{disc}(q, x, \xi) \text{ and } \epsilon \in M_e^{disc}(q, x, \xi)\}$. Trajectories at location q from initial state $(\hat{x}, \xi)$ following $u \in \mathcal{U}$ and $d \in \mathcal{D}$ are denoted $\psi_q(u, d, \hat{x}, \xi)$.

$$W^0 := Good$$
$$i := -1$$
repeat {
$$\quad i := i + 1$$
$$\quad W^{i+1} := W^i \setminus [Pre_e(W^i) \cup Unavoid\_Pre(Pre_e(W^i) \cup \overline{W^i}, Pre_c(W^i))]$$
} until $(W^{i+1} = W^i)$
$$Safe := W^i$$

Fig. 4. Computation of Maximal Safe Set [12].

5   Computation of the Maximal Safe Set for the Engine System

The parameters of the hybrid model M of the engine running at idle have been identified by measurements provided by the Powertrain Division of Magneti Marelli on a commercial engine.

In this section, for simplicity, we restrict the computation of the maximal safe set to the case in which only a positive spark advance is considered. In this case, the engine system is represented by the hybrid automaton M which consists of four discrete modes (S, S_+, S^L, S_+^L).

In M, there is a symmetry between the modes S, S_+ (subsystem M_2) and the mirror modes S^L, S_+^L (subsystem M_2^L). This allows the computation of the maximal safe set as follows: first, the maximal safe set is computed for the system M_2 representing the engine with the clutch open; then, using the previous results, the maximal safe set is derived for M_2^L representing the engine with the clutch closed; finally, the maximal safe set for the overall system M is obtained.

The procedure reported in Section 4 is quite complex when applied to M_2, due to the dimension of the continuous state space. However, since the set *Good* involves only the variable n, and variables (n, T, θ_c) are de-coupled from the remaining ones, we can apply a divide-and-conquer strategy: the procedure is applied first in the subspace (n, T, θ_c); then, using (2), the safe values for the variable m_C are obtained in terms of n, T, θ_c and θ_s. Finally, from the safe values of the air mass, the safe values for the manifold pressure p are obtained, so that the overall maximal safe set $Safe^{M_2}$ for the hybrid subsystem M_2 is derived. For the engine parameters of our model, this computation terminates in six iterations.

Due to symmetry, the maximal safe set for M_2^L is computed as the one for M_2, yielding a result differing only in the coefficients of the crankshaft speed: when the clutch pedal is released the crankshaft inertial load increases, and $a_n^L < a_n$; $b_n^L < b_n$. Comparing the maximal safe sets for M_2 and M_2^L, it holds:

$$Safe^{M_2}|_S \subset Safe^{M_2^L}|_{S^L},$$
$$Safe^{M_2}|_{S_+} \subset Safe^{M_2^L}|_{S_+^L}.$$

$$(5)$$

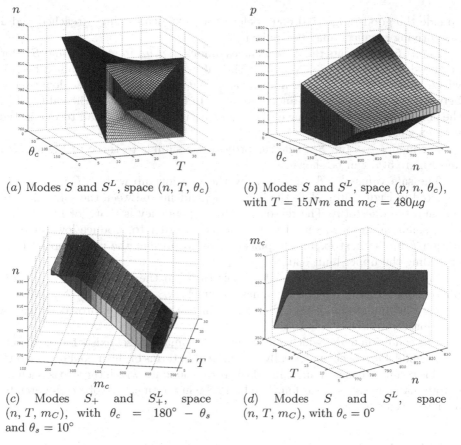

(a) Modes S and S^L, space (n, T, θ_c)

(b) Modes S and S^L, space (p, n, θ_c), with $T = 15Nm$ and $m_C = 480\mu g$

(c) Modes S_+ and S_+^L, space (n, T, m_C), with $\theta_c = 180° - \theta_s$ and $\theta_s = 10°$

(d) Modes S and S^L, space (n, T, m_C), with $\theta_c = 0°$

Fig. 5. Maximal Safe Set for the Engine System M

Finally, we consider the system M which contains the subsystems M_2 and M_2^L together with the uncontrollable transitions $S \leftrightarrows S^L$, $S_+ \leftrightarrows S_+^L$, $S \leftrightarrows S_+^L$ and $S^L \leftrightarrows S_+$. In summary, one can prove the following:

Proposition 1. *If $Safe^M$ is the maximal safe set for the engine system M,*

$$
\begin{aligned}
Safe^M|_S &= Safe^M|_{S^L} = Safe^{M_2}|_S, \\
Safe^M|_{S_+} &= Safe^M|_{S_+^L} = Safe^{M_2}|_{S_+}.
\end{aligned}
\tag{6}
$$

The maximal safe set for M is shown in Figure 5. Since the dimensions of the safe set are higher than three, it is difficult to visualize it. The three figures are projections of the safe set on different axes. The projection of the safe set on the subspace n, T, θ_c, shown in Figure 5(a) for modes S and S^L, does not depend on the value of the other state components p, m_C, θ_s. Note that the range of safe values for the speed n increases as θ_c increases, whenever a given torque T is considered. This corresponds to the fact that the greater the values of θ_c, the shorter is the interval of time on which the load torque may act to drive n

outside the good range [770, 830], before the next dead point is reached and a new driving torque T may be applied.

Figure 5(b) presents the dependencies among p, n, θ_c, in modes S and S^L, for given values of T and m_C. Note that the safe values for θ_s are obtained by (2), which holds in S and S^L. For a fixed value of θ_c, there is an inverse dependence between the safe values of p and those of n. In fact, the greater the values of n, the smaller the values of p have to be in order to produce a small torque in the next expansion phase.

For modes S_+ and S_+^L, safe set projections similar to those in Figure 5(a) and Figure 5(b) can be shown.

The safe values for θ_s are given by the relationship with n, T, m_C, and θ_c depicted in Figure 5(c). Note the inverse dependence between the safe values of m_C and the ones of n. The reason for this dependency is that, for high values of n, small values of m_C need to be applied in order to produce low values of torque T in the next expansion phase, so as to prevent the engine speed from exceeding the upper limit 830.

Figure 5(d) shows, for each safe couple (n, T) at the beginning of the engine phase $(\theta_c = 0)$, the interval of values to which m_C must belong to produce a torque that maintains the speed engine n in the good range, provided that an appropriate value of θ_s is chosen.

The condition under which a non-empty maximal safe set exists for the engine hybrid model has been analytically determined in terms of the model parameters. By this result, the maximum value of the torque load for which there exists a non-empty maximal safe set was found to be 12.8 Nm for an engine of a commercially available vehicle.

6   Conclusions

The problem of maintaining the crankshaft speed within a given range has been formalized for the first time as a safety specification for the closed-loop system modeled as a hybrid automaton, where continuous and discrete variables retain their distinctive nature. By applying a systematic procedure to the hybrid model of the engine, the maximal safe set for the idle control has been determined. We also obtained as a by-product the entire set of possible controllers that satisfy the constraints. This, in addition to the capability of modeling in a separate but integrated manner discrete and continuous signals, is a definite advantage over other approaches that approximate the system by relaxing it to continuous or discrete sampled representations. This result is the first of its kind in idle control, and allows us to determine tightly the maximum range of allowed torque disturbances, given the maximum interval of angular speed values. Further, from an application point of view, the relevance of this computation lies in the fact that it provides the upper bounds on the best performance achievable by an idle speed control strategy for a given engine.

The systematic procedure for the safe set computation cannot be guaranteed to converge in a finite number of steps in the general case. However, in the engine model available to us it converged in six steps. We are presently investigating the properties of the model and the range of parameters with respect to the size and the shape of the the the corresponding maximal safe set for the idle engine control problem.

Acknowledgments

This research is sponsored in part by CNR, the CNR MADESS Project, a grant from Magneti Marelli and from Cadence Design Systems. The support of and the interaction with Dr. Carlo Rossi and Dr. Gabriele Serra of Magneti Marelli, Power-train Division, is gratefully acknowledged.

References

1. M. Abate and V. Di Nunzio. Idle speed control using optimal regulation. Technical Report 905008, SAE, 1990.
2. E. Asarin, O. Maler, A. Pnueli, and J. Sifakis. Controller synthesis for timed automata. In *Proceedings of System Structure and Control*. IFAC, Elsevier, July 1998.
3. A. Balluchi, L. Benvenuti, H. Wong-Toi, T. Villa, and A. L. Sangiovanni-Vincentelli. A case study of hybrid controller synthesis of a heating system. In *Proc. 5th European Control Conference*, Karlsruhe, Germany, September 1999.
4. A. Balluchi, L. Benvenuti, H. Wong-Toi, T. Villa, and A. L. Sangiovanni-Vincentelli. Controller synthesis for hybrid systems with lower bounds on event separation. In *Proc. 37th IEEE Conference on Decision and Control*, December 1999.
5. K. R. Butts, N. Sivashankar, and J. Sun. Application of ℓ_1 optimal control to the engine idle speed control problem. *IEEE Trans. on Control Systems Technology*, 7(2):258–270, March 1999.
6. C. Carnevale and A. Moschetti. Idle speed control with H_∞ technique. Technical Report 930770, SAE, 1993.
7. D. Hrovat and B. Bodenheimer. Robust automotive idle speed control design based on μ-synthesis. In *Proc. IEEE American Control Conference*, pages 1778–1783, S. Francisco, CA, 1993.
8. D. Hrovat and J. Sun. Models and control methodologies for IC engine idle speed control design. *Control Engineering Practice*, 5(8), August 1997.
9. L. Kjergaard, S. Nielsen, T. Vesterholm, and E. Hendricks. Advanced nonlinear engine idle speed control systems. Technical Report 940974, SAE, 1994.
10. C. H. Onder and H. P. Geering. Model-based multivariable speed and air-to-fuel ratio control of a SI engine. Technical Report 930859, SAE, 1993.
11. D. Shim, J. Park, P. P. Khargonekar, and W. B. Ribbens. Reducing automotive engine speed fluctuation at idle. *IEEE Trans. on Control Systems Technology*, 4(4):404–410, July 1996.
12. C. Tomlin, J. Lygeros, and S. Sastry. Synthesizing controllers for nonlinear hybrid systems. In Thomas Henzinger and Shankar Sastry, editors, *First International Workshop, HSCC'98, Hybrid Systems: Computation and Control*, Lecture Notes in Computer Science 1386, pages 360–373, 1998.
13. S. Yurkovich and M. Simpson. Crank-angle domain modeling and control for idle speed. *SAE Journal of Engines*, 106(970027):34–41, 1997.

Optimization-Based Verification
and Stability Characterization
of Piecewise Affine and Hybrid Systems

Alberto Bemporad*, Fabio Danilo Torrisi, and Manfred Morari

Automatic Control Laboratory, Swiss Federal Institute of Technology
ETH Zentrum - ETL I24.2, CH 8092 Zürich, Switzerland
tel. +41-1-632 6679, fax +41-1-632 1211
bemporad,torrisi,morari@aut.ee.ethz.ch
http://control.ethz.ch/~hybrid

Abstract. In this paper, we formulate the problem of characterizing the stability of a piecewise affine (PWA) system as a verification problem. The basic idea is to take the whole $\mathbb{R}^n$ as the set of initial conditions, and check that all the trajectories go to the origin. More precisely, we test for semi-global stability by restricting the set of initial conditions to an (arbitrarily large) bounded set $\mathcal{X}(0)$, and label as "asymptotically stable in T steps" the trajectories that enter an invariant set around the origin within a finite time T, or as "unstable in T steps" the trajectories which enter a set $\mathcal{X}_{inst}$ of (very large) states. Subsets of $\mathcal{X}(0)$ leading to none of the two previous cases are labeled as "non-classifiable in T steps". The domain of asymptotical stability in T steps is a subset of the domain of attraction of an equilibrium point, and has the practical meaning of collecting the initial conditions from which the settling time to a specified set around the origin is smaller than T. In addition, it can be computed algorithmically in finite time. Such an algorithm requires the computation of reach sets, in a similar fashion as what has been proposed for verification of hybrid systems. In this paper we present a substantial extension of the verification algorithm presented in [6] for stability characterization of PWA systems, based on linear and mixed-integer linear programming. As a result, given a set of initial conditions we are able to determine its partition into subsets of trajectories which are asymptotically stable, or unstable, or non-classifiable in T steps.

1 Introduction

Hybrid models describe processes which evolve according to dynamics and logic rules. Hybrid systems have recently grown in interest not only for being theoretically challenging [10], but also for their impact on applications, for instance in the automotive industry [3].

---

* Corresponding author.

N. Lynch and B. Krogh (Eds.): HSCC 2000, LNCS 1790, pp. 45–58, 2000.
© Springer-Verlag Berlin Heidelberg 2000

An important class of hybrid systems are the so-called *Piecewise Affine* (PWA) systems. These are defined by partitioning the state-space into polyhedral regions, and associating with each region a different linear state-update equation. PWA systems can model a large number of physical processes, such as systems with static nonlinearities (for instance actuator saturation), and can approximate nonlinear dynamics with arbitrary accuracy via multiple linearizations at different operating points. The study of PWA systems is also motivated by the stability and performance analysis of high-performance controllers [20]. In particular, recently in [7] the authors show that a model predictive controller (MPC) for constrained linear systems can be explicitly expressed in closed-form as a continuous and piecewise affine state-feedback law. The resulting closed-loop system is therefore PWA, and criteria for proving stability and robust stability against disturbances and model uncertainties are of fundamental importance.

PWA systems are equivalent to interconnections of linear systems and finite automata, as pointed out by Sontag [26]. Based on different arguments, a similar result was proved constructively in [4], where the authors show that PWA systems are equivalent to the hybrid *mixed logical dynamical* (MLD) systems introduced in [5]. MLD systems are capable to model a broad class of systems arising in many applications: linear hybrid dynamical systems, hybrid automata, nonlinear dynamic systems where the nonlinearity can be approximated by a piecewise linear function, some classes of discrete event systems, linear systems with constraints, etc. Examples of real-world applications that can be naturally modeled within the MLD framework are reported in [5, 6]. The MLD framework allows specifying linear dynamics $x' = Ax + Bu$, any logic proposition, and the interaction between the two. The key idea of the approach consists of embedding the logic part in the state equations by transforming Boolean variables into 0-1 integers, and by expressing the relations as mixed-integer linear inequalities [5].

Despite the fact that PWA systems are just a simple extension of linear systems, they can exhibit very complex behaviors, as typical of nonlinear systems [24]. Blondel and Tsitsiklis [9] showed that even in the simple case of two component subsystems, verifying the stability of autonomous discrete-time PWA systems is either an $\mathcal{N}P$-hard problem (no polynomial-time algorithm), or undecidable. In view of these complexity results, no hope remains of finding criteria for stability of PWA systems as easy as for instance the Routh-Hurwitz rule for linear systems. Stability of each linear subsystem is not enough to guarantee stability of the overall system (and vice versa) [11, 28], as the switching rule between linear dynamics is fundamental for stability of the interconnection. Some criteria for stability of PWA systems were recently proposed, which are based on piecewise quadratic Lyapunov functions computed by solving linear matrix inequalities (LMI) [16], and multiple Lyapunov functions methods [11]. However, LMI based approaches have the drawback of being conservative, the more conservative the larger the number of regions in the polyhedral partition of the state space.

Complexity results were also shown in [4] for $\mathcal{N}P$-completeness of observability analysis, and undecidability of reachability in the context of *formal verifica-*

tion of hybrid automata is well known [1, 18]. The problem of formal verification can be simply stated as follows: For a given set of initial conditions and disturbances, certify that all possible trajectories never enter a set of unsafe states, or possibly provide a counterexample. In spite of this complexity, several tools for formal verification of hybrid systems have been proposed in the literature, mainly for linear hybrid automata [15, 19].

In this paper, we formulate the problem of characterizing the stability of a PWA system as a verification problem. The basic idea is to check for reachability from an (arbitrarily large) bounded set $\mathcal{X}(0)$ of initial conditions to (i) a set around the origin, and (ii) a set of very large (=unsafe) states. More precisely, we label as "asymptotically stable in T steps" the trajectories that enter an invariant set around the origin within a finite time T, or as "unstable in T steps" the trajectories which enter a (very large) set $\mathcal{X}_{\text{inst}}$. Subsets of $\mathcal{X}(0)$ leading to neither of the two previous cases are non-classified. Such a verification problem of "practical" stability is decidable. Many undecidable problems can be approximated by decidable ones which are equivalent from a practical point of view. The decidable algorithm shown in [4] for analysis of observability is another example of such a philosophy.

In order to solve the problem of verification of stability, we substantially extend the algorithm proposed in [6]. Safety tests and reach set computation are done via linear programming (LP), switching detection via mixed-integer linear programming (MILP), and approximation of the reach set by using tools from computational geometry. In particular, with respect to [6], we make the algorithm more efficient, and use an algorithm for arbitrarily precise inner and outer approximation of polyhedra [8].

The approach followed in this paper is related to the idea of *robust simulation* [17], which consist of simulating entire set evolutions rather than single trajectories for stability and performance analysis. In [17] the author tests for finite time stability by computing an outer approximation of the reach set via mathematical programming. However, an outer approximation is performed at each time step in order to bound the complexity of the reach set. It turns out that the approach provides only a sufficient condition to conclude about the stability of the initial set. On the contrary, in this paper an exact characterization of the initial set is obtained by first applying a verification algorithm to the system, and then by refining the results through linear programming. By removing all conservativeness, this allows partitioning the initial set into three subsets: (i) states belonging to the domain of asymptotic stability in T steps, (ii) states belonging to the domain of instability in T steps, and (iii) states which are non-classifiable in T steps.

2   Hybrid and Piecewise Affine Models

Several modeling frameworks were proposed in the literature. Two main categories were successfully adopted for analysis and synthesis purposes [10]: *hybrid control systems* [1, 2, 5, 21, 22], which consist of the interaction between continuous dynamical systems and discrete/logic automata, and *switched systems* [11, 16, 25], where the state-space is partitioned into regions, each one being associated to a different continuous dynamics.

Switched systems defined by a polyhedral partition of the state-space and linear dynamic equations are the so-called *piecewise affine* (PWA) systems

$$x(t+1) = A_i x(t) + B_i u(t) + f_i, \text{ for } x(t) \in C_i \triangleq \{x : H_i x \le K_i\} \quad (1)$$

where $x \in X \subseteq \mathbb{R}^n$, $u \in \mathbb{R}^m$, $\{C_i\}_{i=0}^{s-1}$ is a polyhedral partition of the sets of states X, and f_i is a constant vector. A *trajectory* is the collection of vectors $\{x(0), \dots, x(t), \dots\}$ satisfying the difference equation (1). Without additional hypotheses on continuity of the piecewise affine state-update mapping, definition (1) is not well posed in general, as the state-update function is twice (or more times) defined over common boundaries of sets C_i (the boundaries will be also referred to as *guardlines*). This is a technical issue which can be avoided as in [25].

In [4] the authors show that PWA systems are equivalent to the *mixed logic dynamical* (MLD) systems introduced in [5]. These are hybrid (control) systems defined by the interaction of logic, finite state machines, and linear discrete-time systems, defined by the equations

$$x(t+1) = \mathcal{A}x(t) + \mathcal{B}_1 u(t) + \mathcal{B}_2 \delta(t) + \mathcal{B}_3 z(t) \quad (2a)$$
$$\mathcal{E}_2 \delta(t) + \mathcal{E}_3 z(t) \le \mathcal{E}_1 u(t) + \mathcal{E}_4 x(t) + \mathcal{E}_5 \quad (2b)$$

where $x \in \mathbb{R}^{n_c} \times \{0,1\}^{n_\ell}$ is a vector of continuous and binary states, $u \in \mathbb{R}^{m_c} \times \{0,1\}^{m_\ell}$ are the inputs, and $\delta \in \{0,1\}^{r_\ell}$, $z \in \mathbb{R}^{r_c}$ represent auxiliary binary and continuous variables respectively, which are introduced when transforming logic relations into mixed-integer linear inequalities [23, 27], and $\mathcal{A}, \mathcal{B}_1, \mathcal{B}_2, \mathcal{B}_3, \mathcal{E}_1$, $\dots, \mathcal{E}_5$ are matrices of suitable dimensions. Throughout the paper, we will assume that both the PWA and the MLD forms are available. Their complementary role in the verification algorithm will be discussed later.

3   Stability Characterization Problem

As mentioned in the introduction, determining the stability of PWA systems can be a complex task. Nevertheless, we aim at estimating the domains of attraction of equilibrium points, and the set of initial conditions from which the state trajectory reaches magnitudes greater than an arbitrarily large value.

For simplicity of exposition, from now on we will assume that the system is piecewise linear ($f_i = 0$, for all $i = 0, \dots, s-1$), and autonomous ($B_i = 0$ for all $i = 0, \dots, s-1$)[1], and that the only equilibrium point (the origin) belongs to the

---

[1] Robust stability questions in the presence of disturbances $u(t) \in \mathcal{U}$, where $\mathcal{U}$ is a given bounded set, can be similarly formulated.

interior of one of the sets of the partition[2], which by convention will be referred to as $\mathcal{C}_0$. Denote by $\mathcal{D}_\infty(0) \subseteq \mathbb{R}^n$ the (unknown) domain of attraction of the origin (if the origin is unstable then $\mathcal{D}_\infty(0) = \{0\}$). Given an (arbitrarily large) bounded set $\mathcal{X}(0)$ of initial conditions, we want to characterize $\mathcal{D}_\infty(0) \cap \mathcal{X}(0)$.

A necessary condition for the origin to be asymptotically stable is that the matrix A_0 associated with the region $\mathcal{C}_0$ is strictly Hurwitz. Under this assumption, we can compute an invariant set in $\mathcal{C}_0$. In particular, we compute the *maximum output admissible set* (MOAS) $\mathcal{X}_\infty \subseteq \mathcal{C}_0$. $\mathcal{X}_\infty$ is the largest invariant set contained in $\mathcal{C}_0$, which by [14, Th.4.1] is a polyhedron with a finite number of facets, and is computed through a finite number of linear programs (LP's) [14][3].

In order to circumvent the undecidability of stability mentioned above, we define the following

Definition 1. *Consider the PWA system (1), and let the origin $0 \in \overset{\circ}{\mathcal{C}}_0 \triangleq \{x : H_0 x < K_0\}$, and A_0 be strictly Hurwitz. Let $\mathcal{X}_\infty$ be the maximum output admissible set (MOAS) in $\mathcal{C}_0$, which is an invariant for the linear system $x(t + 1) = A_0 x(t)$. Let T be a finite time horizon. Then, the set $\mathcal{X}(0) \subseteq \mathbb{R}^n$ of initial conditions is said to belong to the* domain of attraction in T steps $\mathcal{D}_T(0)$ *of the origin if $\forall x(0) \in \mathcal{X}(0)$ the corresponding final state $x(T) \in \mathcal{X}_\infty$.*

Note that $\mathcal{D}_T(0) \subseteq \mathcal{D}_{T+1}(0) \subseteq \mathcal{D}_\infty(0)$, and $\mathcal{D}_T(0) \to \mathcal{D}_\infty(0)$ as $T \to \infty$. The horizon T is a practical information about the speed of convergence of the PWA system to the origin.

Definition 2. *Consider the PWA system (1), and let $\mathcal{X}_{\text{inst}} \subseteq \mathbb{R}^n$ The set $\mathcal{X}(0) \subseteq \mathbb{R}^n$ of initial conditions is said to belong to the* domain of instability in T steps $\mathcal{I}_T(0)$ *if $\forall x(0) \in \mathcal{X}(0)$ there exists t, $0 \leq t \leq T$ such that $x(t) \in \mathcal{X}_{\text{inst}}$.*

In Definition (2), the set $\mathcal{X}_{\text{inst}}$ must be interpreted as a set of "very large" states. Although instability in T steps does not guarantee instability (for any finite T, a trajectory might reach $\mathcal{X}_{\text{inst}}$ and converge back to the origin), it has the practical meaning of labeling as "unstable" the trajectories whose magnitude is unacceptable, for instance because the PWA system is no longer valid as a model of the real system. Instability in T steps represents a condition of *loss of safety* for the PWA system.

As $\mathcal{D}_T(0)$ and $\mathcal{I}_T(0)$ can have a nonempty intersection, we introduce the following

---

[2] The hypothesis of having equilibria only in the interiors of sets $\mathcal{C}_i$, although restrictive, is certainly satisfied when (1) is the result of the linearization of a nonlinear system around different equilibria, and is needed later for easily computing nonempty invariant sets. Moreover, the approach of this paper can be straightforwardly extended to handle multiple equilibria of the PWA system which are not on the border of the polyhedral partition. These can be easily detected by standard linear analysis, and a maximum output admissible sets can be computed for each equilibrium.

[3] If the effect of perturbations $u(t) \in \mathcal{U} \subseteq \mathbb{R}^m$, where $\mathcal{U}$ is a given bounded set of disturbances and $B_0 \neq 0$, has to be taken into account $\mathcal{X}_\infty$ is the largest invariant set under disturbance excitation, and can be computed as proposed in [13].

Definition 3. *Consider the PWA system (1). The set $\mathcal{X}(0) \subseteq \mathbb{R}^n$ of initial conditions is said to belong to the* domain of safe stability in T steps $\mathcal{S}_T(0)$ *if $\mathcal{S}_T(0) \subseteq \mathcal{D}_T(0)$ and $\mathcal{S}_T(0) \cap \mathcal{I}_T(0) = \emptyset$.*

Definition 3 describes trajectories which asymptotically converge to the origin without crossing the set $\mathcal{X}_{\text{inst}}$.

Given a set of initial conditions $\mathcal{X}(0)$, we aim at finding subsets of $\mathcal{X}(0)$ which are safely asymptotically stable ($\mathcal{X}(0) \cap \mathcal{S}_T(0)$), and subsets which lead to practical instability in T steps ($\mathcal{X}(0) \cap \mathcal{I}_T(0)$). Subsets of $\mathcal{X}(0)$ leading to none of the two previous cases are labeled as *non-classifiable in T steps* As we will use linear optimization tools, we assume that $\mathcal{X}(0)$ and $\mathbb{R}^n \backslash \mathcal{X}_{\text{inst}}$ are convex polyhedral sets. Typically, non-classifiable subsets shrink and eventually disappear for increasing T.

3.1   Switching Sequences

The evolution of the PWA system (1) for $u(t) = 0$, $f_i = 0$, $\forall i = 0, \ldots, s-1$, is given by

$$x(t) = A_{i(t-1)} A_{i(t-2)} \cdots A_{i(0)} x(0) \qquad (3)$$

where in (3) $i(k) \in \{0, \ldots, s-1\}$ is the index such that $H_{i(k)} x(k) \leq K_{i(k)}$, $k = 0, \ldots, t-1$, is satisfied. The previous questions of practical stability can be answered once all the switching sequences $I(t) \triangleq \{i(0), \ldots, i(t-1)\}$ leading to $\mathcal{X}_\infty$ or $\mathcal{X}_{\text{inst}}$ from $\mathcal{X}(0)$ are known. In fact, for safe stability in T steps it is enough to check that the reach set at time T, $\mathcal{X}(T, \mathcal{X}(0)) \triangleq A_{i(T-1)} A_{i(T-2)} \cdots A_{i(0)} \mathcal{X}(0)$, satisfies the set inclusion $\mathcal{X}(T, \mathcal{X}(0)) \subseteq \mathcal{X}_\infty$ for all admissible switching sequences $I(T)$. However, the number of all possible switching sequences $I(T)$ is combinatorial with respect to T and s, and any enumeration method would be impractical. In the next section we show that a verification algorithm can be used to avoid such an enumeration.

4   Verification

In order to determine admissible switching sequences $I(t)$, we need to exploit the special structure of PWA systems (1). This allows an easy computation of the reach set, as long as the evolution remains within a single region $\mathcal{C}_i$. Whenever the reach set crosses a guardline and enters a new region $\mathcal{C}_j$, a new reach set computation based on the j-th linear dynamics is computed, as shown in Fig. 1(a).

Let $\mathcal{X}(0)$ be a convex polyhedral set, and partition it into subregions $\mathcal{X}_i(0) \triangleq \mathcal{X}(0) \cap \mathcal{C}_i$, $i = 0, \ldots, s-1$. For all nonempty sets $\mathcal{X}_i(0)$, computing the evolution $\mathcal{X}(T, \mathcal{X}_i(0))$ requires: (i) the reach set $\mathcal{X}(t, \mathcal{X}_i(0)) \cap \mathcal{C}_i$, i.e. the set of evolutions at time t in $\mathcal{C}_i$ from $\mathcal{X}_i(0)$; (ii) crossing detection of the guardlines $\mathcal{P}_h \triangleq \mathcal{X}(t, \mathcal{X}_i(0)) \cap \mathcal{C}_h \neq \emptyset$, $\forall h = 0, \ldots, i-1, i+1, \ldots, s-1$; ($iii$) elimination of redundant constraints and approximation of the polyhedral representation of

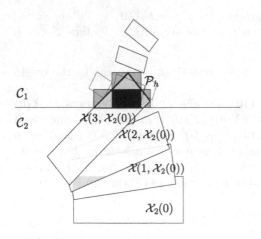

$\mathcal{C}_1$

$\mathcal{C}_2$

$\mathcal{X}(3, \mathcal{X}_2(0))$

$\mathcal{X}(2, \mathcal{X}_2(0))$

$\mathcal{X}(1, \mathcal{X}_2(0))$

$\mathcal{X}_2(0)$

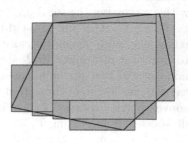

(a) Reach set evolution, guardline crossing, outer approximation of a new intersection

(b) Outer rectangular approximation of a polytope

Fig. 1. Reachability Analisys

the new regions $\mathcal{P}_h$ (approximation is desirable, as the number of facets of $\mathcal{P}_h$ can grow linearly with time); (iv) detection of emptiness of $\mathcal{X}(t, \mathcal{P}_h)$ (emptiness happens when all the evolutions have crossed the guardlines), detection of safe stability $\mathcal{X}(t, \mathcal{P}_h) \subseteq \mathcal{X}_\infty$, detection of practical instability $\mathcal{X}(t, \mathcal{P}_h) \subseteq \mathcal{X}_{\text{inst}}$ (these three will be referred to as *fathoming* conditions).

4.1 Reach Set Computation

Let the set of initial conditions be defined by the polyhedral representation $\mathcal{X}(0) \triangleq \{x : S_0 x \leq T_0\}$. The subset S of $\mathcal{X}(0)$ whose evolution lies in $\mathcal{C}_i$ for t steps is given by

$$S = \left\{ x \in \mathbb{R}^n : \begin{array}{l} S_0 x \leq T_0 \\ H_i A_i^k x \leq K_i, \ k = 0, \ldots, t \end{array} \right\} \tag{4}$$

As S is a polyhedral set, the reach set $\mathcal{X}(t, \mathcal{X}_i(0)) \bigcap \mathcal{C}_i = A_i^t S$ is a polyhedral set as well. In the presence of input disturbances and nonzero offsets f_i, $S = \{x \in \mathbb{R}^n : S_0 x \leq T_0, \ H_i(A_i^k x + \sum_{j=0}^{k-1} A_i^j [B_i u(k-1-j) + f_i]) \leq K_i, \ k = 0, \ldots, t\}$, which is a polyhedron in the augmented space of tuples $(x, u(0), \ldots, u(t-1))$. A compact representation of the set $\mathcal{X}(t, \mathcal{X}_i(0)) \bigcap \mathcal{C}_i$ (as inequalities over the final state $x(t)$) can be computed by a geometric projection procedure, for which efficient tools exist, e.g. [12].

4.2 Guardline Crossing Detection

Switching detection amounts to finding all possible new regions $\mathcal{C}_h$'s entered by the reach set at the next time step, i.e. nonempty sets $\mathcal{P}_h \triangleq \mathcal{X}(t, \mathcal{X}_i(0)) \bigcap \mathcal{C}_h$,

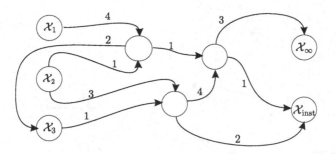

Fig. 2. Graph of evolution G

$h \neq i$. Rather then enumerating and checking nonemptiness for all $h = 0, \ldots, i - 1, i + 1, \ldots, s - 1$, we can exploit the equivalence between PWA systems and MLD models (2), and solve the switching detection problem via mixed-integer linear programming. More in detail, in the MLD form the condition $x(t) \in C_h$ is associated to the condition $\delta(t) = \delta_h \in \{0, 1\}^{r_\ell}$, for instance $x(t) \in C_5 \Leftrightarrow \delta(t) = [1\ 0\ 1]'$. Switching detection amounts to finding all feasible vectors $\delta(t) \in \{0, 1\}^{r_\ell}$ which are compatible with the constraints in (2) plus the constraint $x(t-1) \in \mathcal{X}(t-1, \mathcal{X}_i(0)) \cap C_i$. Such a problem is a mixed-integer linear feasibility test (MILFT), and can be efficiently solved through standard recursive branch and bound procedures. Thus, in the average case the MLD form (through the branch and bound algorithm) requires only a very small number of feasibility tests, while the PWA form would require for enumerating and solving a feasibility test for all the possible regions.

4.3 Approximation of Intersection

The computation of the reach set proceeds in each region C_h from each new intersection $\mathcal{P}_h$. A new reach set computation is started from $\mathcal{P}_h$, unless $\mathcal{P}_h$ is contained in some larger subset of C_h which has already been explored. As in principle the number of facets of $\mathcal{P}_h$ grows linearly with time, we need to approximate $\mathcal{P}_h$ so that its complexity is bounded (and therefore reach set computation from $\mathcal{P}_h$ has a limited complexity with respect to the initial region), and checking for set inclusion is a simple task. Hyper-rectangular approximations are the best candidates, as set inclusion between hyper-rectangles reduces to a simple comparison of the coordinates of the vertices. On the other hand, a crude rectangular outer approximation of $\mathcal{P}_h$ can lead to explore large regions which are not reachable from the initial set $\mathcal{X}(0)$, as they are just introduced by the approximation itself. In [8] the authors propose an iterative method for inner and outer approximation which is based on linear programming, and approximates with arbitrary precision polytopes by a collection of hyper-rectangles, as depicted in Fig. 1(b).

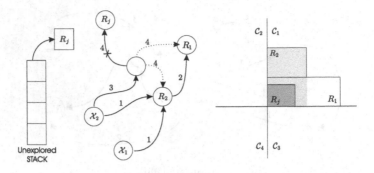

Fig. 3. Adding and removing nodes to the graph G

4.4 Fathoming

In Sect. 4.1 we showed how to compute the evolution of the reach set $\mathcal{X}(t, \mathcal{P}_h)$ inside a region $\mathcal{C}_i$. The computation is stopped once one of the following happens:

1. The set $\mathcal{X}(t, \mathcal{P}_h) \bigcap \mathcal{C}_i$ is empty. This means that the whole evolution has left region $\mathcal{C}_i$.
2. $\mathcal{X}(t, \mathcal{P}_h) \subseteq \mathcal{X}_\infty$, i.e. all possible evolutions from $\mathcal{P}_h$ are safely stable.
3. $\mathcal{X}(t, \mathcal{P}_h) \subseteq \mathcal{X}_{\text{inst}}$, i.e. all possible evolutions from $\mathcal{P}_h$ have violated the condition for safe stability.
4. The time $t > T$.

These conditions can be checked through linear programming.

4.5 Graph of Evolution

The result of the exploration algorithm detailed in the previous sections can be conveniently represented on a graph G (Fig. 2). The nodes of G represent sets from which a reach set evolution is computed, and an oriented arc of G connects two nodes if a transition exists between the two correspoding sets. Each arc has an associated weight which represents the time-steps needed for the transition. The graph has initially no arc, and nonempty initial sets $\mathcal{X}_i(0)$ and $\mathcal{X}_\infty$, $\mathcal{X}_{\text{inst}}$ as nodes. As long as a new intersection $\mathcal{X}(t, \mathcal{X}_i(0)) \bigcap \mathcal{C}_h$ is detected, it is approximated by a collection of hyper-rectangles, as described in Sect. 4.3. Each hyper-rectangle becomes a new node in G, and is connected by a weighted arc from $\mathcal{X}_i(0)$. In addition, each hyper-rectangle is pushed on a stack of sets to be explored.

Before starting a new reach set computation from a set R_j extracted from the stack, we check for inclusion of R_j in other nodes of G. If this happens, say $R_j \subseteq R_1$ and $R_j \subseteq R_2$ as in Fig. 3, the node associated with R_j is removed from G, and all arcs pointing to R_j are directed to both R_1 and R_2 (dotted arrows). Finally, whenever the reach set hits $\mathcal{X}_\infty$ (or $\mathcal{X}_{\text{inst}}$), an arc is drawn from $\mathcal{P}_h$ to $\mathcal{X}_\infty$ (or $\mathcal{X}_{\text{inst}}$).

After the verification algorithm terminates, the oriented paths on G from initial nodes $\mathcal{X}_i(0)$ to terminal nodes $\mathcal{X}_\infty$ and $\mathcal{X}_{\text{inst}}$ determine a superset of feasible switching sequences $I(t) = \{i(0), \ldots, i(t-1)\}$. In fact, because of the outer approximation of new intersections $\mathcal{P}_h$, not all switching sequences are feasible. Nevertheless, feasibility can be simply tested via linear programming. Once all feasible switching sequences $I(t)$ have been identified, the partition of the initial set into safely stable and unstable regions is determined by the sets $A_{i(t-1)} A_{i(t-2)} \cdots A_{i(0)} \mathcal{X}(0)$, $t \leq T$.

Algorithm 1.

1   initialize GRAPH with nonempty initial nodes $\mathcal{X}_i(0)$, $i = 0, \ldots, n_0$,
        and disjoint final nodes $\mathcal{F}_j$, $j = 1, \ldots, n_f$;

2   **push** in STACK $\mathcal{X}_i(0)$, $i = 0, \ldots, n_0$;

3   **while** STACK nonempty **do**

4   **pop** region R_j from STACK, and **let** i such that $R_j \subseteq \mathcal{C}_i$;

5   **if** no region in GRAPH includes R_j **then**

6            $t \leftarrow t^* \triangleq$ minimum arrival time from initial nodes to R_j;

7   **for** $j = 1, \ldots, n_f$ **do**

8   **if** $\mathcal{X}(t, R_j) \subseteq \mathcal{F}_j$ **then go to** 20;

9   **if** $\mathcal{X}(t, R_j) \cap \mathcal{F}_j \neq \emptyset$ **then**

10   connect R_j to $\mathcal{F}_j$ with weight $t - t^*$;

11            $t \leftarrow t + 1$;

12            $\mathcal{X}(t, R_j) = A_i \mathcal{X}(t-1, R_j) + B_i \mathcal{U} + \{f_i\}$;

13   **for all** $h \neq i$ such that $\mathcal{P}_h \triangleq \mathcal{C}_h \cap \mathcal{X}(t, R_j) \neq \emptyset$ **do**

14   insert $\mathcal{P}_h$ in GRAPH and connect R_j to $\mathcal{P}_h$ with weight $t - t^*$;

15   **push** $\mathcal{P}_h$ on STACK;

16            $\mathcal{X}(t, R_j) \leftarrow \mathcal{X}(t, R_j) \cap \mathcal{C}_j$;

17   **if** $\mathcal{X}(t, R_j) \neq \emptyset$ and $t < T$ **then go to** 9;

18   **else**

19   redirect all arcs to R_j to all regions R_h in GRAPH, $R_h \supseteq R_j$;

20   **end** .

4.6   Verification Algorithm

The techniques proposed in the previous sections for verification of PWA systems are summarized in Algorithm 1. In step 1, $\mathcal{F}_1 = \mathcal{X}_\infty$ and $\mathcal{F}_2 = \mathcal{X}_{\text{inst}}$. Step 6 is computed by standard techniques for shortest path computation, while step 13 by branch and bound. In step 14, the collection of hyper-rectangles computed by outer approximating $\mathcal{P}_h$ are put on the stack, rather than $\mathcal{P}_h$.

Note that Algorithm 1 can be generalized to verification purposes, by interpreting $\mathcal{F}_1$ as a set of target states, and $\mathcal{F}_2$ as a set of unsafe states. Moreover,

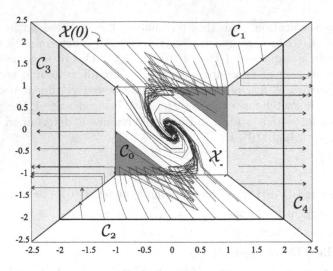

Fig. 4. PWA system (5), initial region $\mathcal{X}(0)$, MOAS $\mathcal{X}_\infty$, and trajectories of the system

linear programs can be performed during reach set computation in order to determine the range of given state components. The algorithm can be extended to include disturbances $u(t) \in \mathcal{U}$, where $\mathcal{U}$ is a given bounded polyhedral set, at the price of more complicate computations (see footnote 3).

We finally remark that the termination of Algorithm 1 after a finite time is guaranteed because no exploration is performed for $t > T$ (step *17*).

5   An Example

Consider the PWA system

$$
x(t+1) = \begin{cases}
\begin{bmatrix} 0 & -.5 \\ 1 & 1 \end{bmatrix} x(t) & \text{if } \begin{bmatrix} 1 & 0 \\ 0 & 1 \\ -1 & 0 \\ 0 & -1 \end{bmatrix} x(t) \le \begin{bmatrix} 1 \\ 1 \\ 1 \\ 1 \end{bmatrix} & (\mathcal{C}_0) \\[18pt]
\begin{bmatrix} .9 & .1 \\ 0 & .8 \end{bmatrix} x(t) & \text{if } \begin{bmatrix} 0 & -1 \\ 1 & -1 \\ -1 & -1 \end{bmatrix} x(t) \le \begin{bmatrix} -1 \\ 0 \\ 0 \end{bmatrix} & (\mathcal{C}_1) \\[18pt]
\begin{bmatrix} .9 & .1 \\ 0 & .8 \end{bmatrix} x(t) & \text{if } \begin{bmatrix} 0 & 1 \\ -1 & 1 \\ 1 & 1 \end{bmatrix} x(t) \le \begin{bmatrix} -1 \\ 0 \\ 0 \end{bmatrix} & (\mathcal{C}_2) \\[18pt]
\begin{bmatrix} 2 & 0 \\ 0 & 1 \end{bmatrix} x(t) & \text{if } \begin{bmatrix} 1 & 0 \\ 1 & -1 \\ -1 & 0 \end{bmatrix} x(t) \le \begin{bmatrix} -1 \\ 0 \\ 0 \end{bmatrix} & (\mathcal{C}_3) \\[18pt]
\begin{bmatrix} 2 & 0 \\ 0 & 1 \end{bmatrix} x(t) & \text{if } \begin{bmatrix} 1 & 0 \\ -1 & 1 \\ -1 & -1 \end{bmatrix} x(t) \le \begin{bmatrix} -1 \\ 0 \\ 0 \end{bmatrix} & (\mathcal{C}_4)
\end{cases}
\tag{5}
$$

and let $\mathcal{X}(0) = \{x \in \mathbb{R}^2 : \|x\|_\infty \le 2\}$, $\mathcal{X}_{\text{inst}} = \{x \in \mathbb{R}^2 : \|x\|_\infty \ge 10\}$. The origin is asymptotically stable, as A_0 has eigenvalues $\frac{1}{2} \pm j\frac{1}{2}$. The corresponding maximum output admissible set in $\mathcal{C}_0$

$$
\mathcal{X}_\infty = \left\{ x \in \mathbb{R}^2 : \begin{bmatrix} 1 & 0 \\ -1 & 0 \\ 0 & 1 \\ 0 & -1 \\ 1 & 1 \\ -1 & -1 \end{bmatrix} x \le \begin{bmatrix} 1 \\ 1 \\ 1 \\ 1 \\ 1 \\ 1 \end{bmatrix} \right\}
\tag{6}
$$

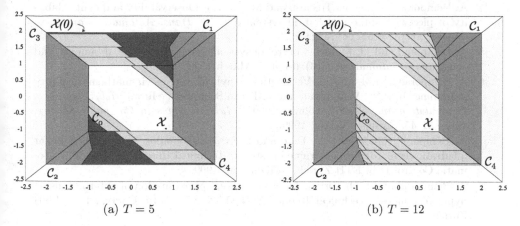

(a) $T = 5$ (b) $T = 12$

Fig. 5. Stability characterization of system (5)

was computed by the algorithm in [14]. A simulation of the system from different initial conditions is depicted in Fig. 4, which shows that the trajectories either converge to the origin or diverge to infinity. We characterize the set of initial conditions by running Algorithm 1. The results are shown in Fig. 5. With the time horizon $T = 5$, not all the set of initial conditions is classified for stability (the darkest subsets are non-classifiable in 5 steps). By augmenting the time horizon, the region of states which are non-classifiable in T steps shrinks, and disappears for $T = 12$. Algorithm 1 is implemented in Matlab 5.3 on a Pentium II 400, and requires 57 s to produce the plot in Fig. 5(b) ($T = 12$).

Acknowledgments

The authors thank the partners of the Esprit Project 26270 and Giancarlo Ferrari Trecate for interesting discussions. This research has been supported by the Swiss National Science Foundation.

References

[1] R. Alur, C. Courcoubetis, T.A. Henzinger, and P.-H. Ho. Hybrid automata: an algorithmic approach to the specification and verification of hybrid systems. In A.P. Ravn R.L. Grossman, A. Nerode and H. Rischel, editors, *Hybrid Systems*, volume 736 of *Lecture Notes in Computer Science*, pages 209–229. Springer Verlag, 1993.

[2] A. Asarin, O. Maler, and A. Pnueli. On the analysis of dynamical systems having piecewise-constant derivatives. *Theoretical Computer Science*, 138:35–65, 1995.

[3] A. Balluchi, M. Di Benedetto, C. Pinello, C. Rossi, and A. Sangiovanni-Vincentelli. Hybrid control for automotive engine management: the cut-off case. In T.A. Henzinger and S. Sastry, editors, *Hybrid Systems: Computation and Control*, volume 1386 of *Lecture Notes in Computer Science*, pages 13–32. Springer Verlag, 1998.

[4] A. Bemporad, G. Ferrari-Trecate, and M. Morari. Observability and controllability of piecewise affine and hybrid systems. *IEEE Trans. Automatic Control*, to appear. http://control.ethz.ch/.

[5] A. Bemporad and M. Morari. Control of systems integrating logic, dynamics, and constraints. *Automatica*, 35(3):407–427, March 1999.

[6] A. Bemporad and M. Morari. Verification of hybrid systems via mathematical programming. In F.W. Vaandrager and J.H. van Schuppen, editors, *Hybrid Systems: Computation and Control*, volume 1569 of *Lecture Notes in Computer Science*, pages 31–45. Springer Verlag, 1999.

[7] A. Bemporad, M. Morari, V. Dua, and E. N. Pistikopoulos. The explicit linear quadratic regulator for constrained systems. Technical Report AUT99-16, Automatic Control Lab, ETH Zürich, Switzerland, 1999.

[8] A. Bemporad and F.D. Torrisi. Inner and outer approximation of polytopes using hyper-rectangles. Technical Report AUT00-02, Automatic Control Lab, ETH Zurich, 2000.

[9] V.D. Blondel and J.N. Tsitsiklis. Complexity of stability and controllability of elementary hybrid systems. *Automatica*, 35:479–489, March 1999.

[10] M. S. Branicky. *Studies in hybrid systems: modeling, analysis, and control*. PhD thesis, LIDS-TH 2304, Massachusetts Institute of Technology, Cambridge, MA, 1995.

[11] M. S. Branicky. Multiple Lyapunov functions and other analysis tools for switched and hybrid systems. *IEEE Trans. Automatic Control*, 43(4):475–482, April 1998.

[12] K. Fukuda. *cdd/cdd+ Reference Manual*. Institute for operations Research ETH-Zentrum, ETH-Zentrum, CH-8092 Zurich, Switzerland, 0.61 (cdd) 0.75 (cdd+) edition, December 1997.

[13] E.G. Gilbert and I. Kolmanovsky. Maximal output admissible sets for discrete-time systems with disturbance inputs. In *Proc. American Contr. Conf.*, pages 2000–2005, 1995.

[14] E.G. Gilbert and K. Tin Tan. Linear systems with state and control constraints: the theory and applications of maximal output admissible sets. *IEEE Trans. Automatic Control*, 36(9):1008–1020, 1991.

[15] T.A. Henzinger, P.-H. Ho, and H. Wong-Toi. HYTECH: a model checker for hybrid systems. *Software Tools for Technology Transfer*, 1:110–122, 1997.

[16] M. Johannson and A. Rantzer. Computation of piece-wise quadratic Lyapunov functions for hybrid systems. *IEEE Trans. Automatic Control*, 43(4):555–559, 1998.

[17] M. Kantner. Robust stability of piecewise linear discrete time systems. In *Proc. American Contr. Conf.*, pages 1241–1245, Evanston, IL, USA, 1997.

[18] Y. Kesten, A. Pnueli, J. Sifakis, and S. Yovine. Integration graphs: a class of decidable hybrid systems. In R.L. Grossman, A. Nerode, A.P. Ravn, and H. Rischel, editors, *Hybrid Systems*, volume 736 of *Lecture Notes in Computer Science*, pages 179–208. Springer Verlag, 1993.

[19] S. Kowalewski, O. Stursberg, M. Fritz, H. Graf, I. Hoffmann, J. Preußig, M. Remelhe, S. Simon, and H. Treseler. A case study in tool-aided analysis of discretely controlled continuos systems: the two tanks problem. In *Hybrid Systems V*, volume 1567 of *Lecture Notes in Computer Science*, pages 163–185. Springer-Verlag, 1999.

[20] D. Liberzon and A.S. Morse. Basic problems in stability and design of switched systems. *IEEE Control Systems Magazine*, 19(5):59–70, October 1999.

[21] J. Lygeros, D.N. Godbole, and S. Sastry. A game theoretic approach to hybrid system design. In R. Alur and T. Henzinger, editors, *Hybrid Systems III*, volume 1066 of *Lecture Notes in Computer Science*, pages 1–12. Springer Verlag, 1996.

[22] J. Lygeros, C. Tomlin, and S. Sastry. Controllers for reachability specifications for hybrid systems. *Automatica*, 35(3):349–370, 1999.

[23] R. Raman and I. E. Grossmann. Relation between milp modeling and logical inference for chemical process synthesis. *Computers Chem. Engng.*, 15(2):73–84, 1991.

[24] E. Sontag. From linear to nonlinear: Some complexity comparisons. In *Proc. 34th IEEE Conf. on Decision and Control*, pages 2916–2920, 1995.

[25] E. D. Sontag. Nonlinear regulation: The piecewise linear approach. *IEEE Trans. Automatic Control*, 26(2):346–358, April 1981.

[26] E.D. Sontag. Interconnected automata and linear systems: A theoretical framework in discrete-time. In R. Alur, T.A. Henzinger, and E.D. Sontag, editors, *Hybrid Systems III - Verification and Control*, number 1066 in Lecture Notes in Computer Science, pages 436 448. Springer-Verlag, 1996.

[27] M.L. Tyler and M. Morari. Propositional logic in control and monitoring problems. *Automatica*, 35(4):565–582, 1999.

[28] V. I. Utkin. Variable structure systems with sliding modes. *IEEE Trans. Automatic Control*, 22(2):212–222, April 1977.

Invariant Sets and Control Synthesis for Switching Systems with Safety Specifications *

Luca Berardi[1], Elena De Santis[2], and Maria Domenica Di Benedetto[2,3]

[1] University of Rome "La Sapienza", Dipartimento di Informatica e Sistemistica
berardi@eecs.berkeley.edu
[2] University of L'Aquila, Dipartimento di Ingegneria Elettrica
{desantis,dibenede}@ing.univaq.it
[3] EECS Department, University of California, Berkeley

Abstract. A structural procedure is proposed for solving the problem of maximal safe-set determination based on maximal controlled invariant sets. However, the procedure is not guaranteed to converge in a finite number of steps. The procedure is made computationally appealing first by linearizing and discretizing the dynamical systems and, second, by using an inner approximation of these sets that, together with the classical outer approximation, yields tight bounds for an error due to the truncation of the procedure after a finite number of steps. The theory is applied to idle-speed regulation in engine control.

1   Introduction

Hybrid systems have been the subject of intensive study in the past few years. In particular, emphasis has been placed on solving problems with safety specifications, which are described by giving a set of good states within which the controlled hybrid system should evolve. The set of all initial states guaranteeing that the evolution of the system remains in the good set is the maximal controlled invariant set contained in the set of good hybrid states. This set is called *maximal safe set* and the set of all control strategies which make this set invariant is the *maximal controller*. A systematic procedure for solving problems with safety specifications has been proposed in [21], [18]. The procedure is not, however, guaranteed to converge in a finite number of steps and is computationally complex.

In [4] and [5] we analyzed this problem for a restricted class of hybrid systems, called switching systems, with the goal of obtaining computationally efficient procedures. *Switching systems* are characterized by a finite state machine (FSM) and a set of dynamical systems, each corresponding to a state of the FSM. The transitions between two different states of the FSM are determined by external uncontrollable events which act as discrete disturbances. The motivation to study

---

* Research supported in part by DARPA under grant F33615-98-C-3614 administered through the Air Force Research Laboratory, in part by M.U.R.S.T. and in part by Magneti-Marelli

N. Lynch and B. Krogh (Eds.): HSCC 2000, LNCS 1790, pp. 59–72, 2000.

this class of systems came from the application of hybrid systems techniques to the automotive engine control problem [2]. An algorithm for the determination of the safe set was proposed that presents an important computational advantage over the general procedure of [21], [18] obtained by *exploiting the structure of the FSM*. The problem was decomposed into a number of different sub-problems that consist of finding a robust controlled invariant set for a given dynamical system. The theory was applied to idle-speed regulation for automotive engine control. To do so, we gave a procedure that follows the essential ideas formalized in this paper: we linearized and sampled the nonlinear dynamical system describing the engine behavior and we remarked that these ideas have general applicability.

In this paper, we show how our procedure can be generalized to the case of a general hybrid system. Since no general procedure is known for the determination of maximal controlled invariant sets for nonlinear dynamical systems, we propose to linearize and use a discrete-time representation of the nonlinear dynamical systems as an important step towards a computationally efficient approach. In fact, for discrete-time linear systems and polyhedral constraining sets, several results for the computation of maximal controlled invariant sets have been reported in the literature (see e.g. [6], [11], [12], [15], [17]). We then propose numerical methods for the computation of controlled invariant sets for discrete-time linear systems and polyhedral constraining sets based on the results of [9]. Even in this simpler case, the procedure for the computation of the maximal controlled invariant set may not converge in a finite number of steps. Hence, we propose a procedure for approximating the maximal controlled invariant set and we show how to obtain an accurate bound of the error by combining inner and outer approximations. We then proceed to show how to choose a discretization in order to obtain a precise relation between the invariant sets associated with general continuous-time dynamical systems and those of the corresponding discrete-time systems. Finally, we solve the idle control problem for automotive engines using our approach.

2   Switching Systems

Switching systems can be considered indexed collections of dynamical systems, each determining the evolution of the system, except during those instants of time in which there is a "jump" between two different dynamical systems. This jump is uniquely determined by external events, which act as discrete disturbances. Switching systems can be defined following the general model of hybrid automata given in [18].

Definition 1. *(Switching Systems) A switching system is a tuple:*
$\mathcal{H} = (Q, X, U, Y, S_C, \mathcal{S}, E, R)$, *(or, respectively,* $\mathcal{H} = (Q, X, U, Y, S_D, \mathcal{S}, E, R)$ *)*
where:

- Q *is a finite collection of discrete state variables taking values in the set of discrete states* $\mathbf{Q} = \{q_1, q_2, ..., q_N\}$;

- X *is a finite collection of continuous state variables taking values in the continuous state space* $\mathbf{X} = \mathbb{R}^n$;
- U *is a finite collection of input variables. We assume* $U = U_D \cup U_C \cup U_d$ *where* U_D *contains discrete and* $U_C \cup U_d$ *contains continuous variables. Variables in* U_D *take values in the set* $\mathbf{U}_D$ *and they are regarded as discrete disturbance variables.* $\mathbf{U}_D$ *contains a special element which we denote by* ϵ. *Variables in* U_C *(resp.* U_d*) take values in the set* $\mathbf{U}_C = \mathbb{R}^m$ *(resp.* $\mathbf{U}_d \subset \mathbb{R}^r$ *) and they are regarded as control (resp. disturbance) variables. Moreover we denote by* $\mathcal{U}_C$ *the class of (continuous) control functions and by* $\mathcal{U}_d$ *the class of continuous disturbance functions.*
- Y *is a finite collection of continuous output variables, taking values in the set* $\mathbf{Y} = \mathbb{R}^p$.
- $\mathcal{S}_C$ *is the class of continuous time dynamical systems defined by the equations:*

$$\begin{aligned} \dot{x}(t) &= f_i(x(t), u(t), \delta(t)) \\ y(t) &= h_i(x(t), u(t)) \end{aligned} \quad i \in J \tag{1}$$

where $t \in \mathbb{R}$, $J = 1, ..., N \subset \mathbb{N}$, *and* f_i *is such that,* $\forall u(\cdot) \in \mathcal{U}_C$, $\forall \delta(\cdot) \in \mathcal{U}_d$, *the solution* $x(t)$ *of each differential equation, for* $i \in J$, *exists and is unique. (or, respectively,* $\mathcal{S}_D$ *is the class of discrete time dynamical systems defined by the equations:*

$$\begin{aligned} x(t+1) &= f_i(x(t), u(t), \delta(t)) \\ y(t) &= h_i(x(t), u(t)) \end{aligned} \quad i \in J \tag{2}$$

where $t \in \mathbb{Z}$, $J = 1, ..., N \subset \mathbb{N}$, *and* f_i *is such that,* $\forall u(\cdot) \in \mathcal{U}_C$, $\forall \delta(\cdot) \in \mathcal{U}_d$, *the solution* $x(t)$ *of each difference equation, for* $i \in J$, *exists and is unique).*
- $\mathcal{S} : \mathbf{Q} \to \mathcal{S}_C$ *(or, respectively,* $\mathcal{S} : \mathbf{Q} \to \mathcal{S}_D$*) is a mapping associating to each discrete state of the switching system a continuous time (or, respectively, a discrete time) dynamical system.*
- $E \subset \mathbf{Q} \times \mathbf{U}_D \times \mathbf{Q}$ *is a collection of discrete transitions; E is such that: 1)* $(q, \epsilon, q) \in E$, $\forall q \in \mathbf{Q}$; *2) if* $(q, \epsilon, q') \in E$ *then* $q = q'$; *3) if* $(q, \sigma, q') \in E$, $\sigma \neq \epsilon$, *then* $q \neq q'$.
- $R : E \times \mathbf{X} \to \mathbf{X}$ *assigns to each* $(q, \sigma, q') \in E$ *a reset function.*

The triple (Q, U_D, E) can be viewed as an FSM having state set Q, inputs U_D (external events) and transitions defined by E. In our case, which event in the set Σ determines a switching doesn't play a direct role. We call the pair $(q, \mathcal{S}(q))$ consisting of the discrete state q and the associated dynamical system $\mathcal{S}(q)$ a switching system **configuration**.

We now define the evolution in time of a switching system. First, following [18], we introduce the concept of hybrid time basis for the temporal evolution of switching systems. Let us denote by T the set $\mathbb{R}^+$, or respectively the set $\mathbb{Z}^+$, depending on whether we are considering switching systems with $\mathcal{S} : \mathbf{Q} \to \mathcal{S}_C$ or: $\mathcal{S} : \mathbf{Q} \to \mathcal{S}_D$.

Definition 2. *(Hybrid Time Basis) [18] A hybrid time basis* τ *is a finite or infinite sequence of sets* I_i, $i \in \mathbb{N}$ *satisfying the following conditions:*

– I_i is of the form $I_i = \{t \in T : t_i \le t \le t_i'\}$ unless τ is a finite sequence and I_L is the last set of the sequence, in which case it can be of the form $I_L = \{t \in T : t \ge t_L\}$;
– For all i, $t_i \le t_i'$ and for $i > 0$, $t_i = t_{i-1}'$.

We denote by $\mathcal{T}$ the set of all hybrid time bases. We define now an execution of a switching system, which describes its evolution in time.

Definition 3. *(Switching System Execution) Let a function $\varphi : \mathbf{Q} \to T$ be given. An execution χ of a switching system $\mathcal{H}$ is a collection $\chi = (\tau, q, x, \sigma, u, \delta, y)$ with $\tau \in \mathcal{T}$, $q : \tau \to \mathbf{Q}$, $x : \tau \to \mathbf{X}$, $\sigma : \tau \to \mathbf{U}_D$, $u : \tau \to \mathbf{U_C}$ with $u(\cdot) \in \mathcal{U}_C$ $\delta : \tau \to \mathbf{U_d}$ with $\delta(\cdot) \in \mathcal{U}_d$, and $y : \tau \to \mathbf{Y}$ satisfying:*

– *Minimum permanence time in each configuration: $\tau = \{I_i\}$ is such that: $t_i' - t_i \ge \varphi(q(t_i))$.*
– *Continuous evolution: For all i with $t_i < t_i'$:*
  - *x, u, δ, y are continuous and q is constant over I_i .*
  - *$\forall t \in I_i$, $x(t)$ is the (unique) solution of $\mathcal{S}(q(t_i))$ with initial condition $x(t_i)$, given some control input $u(\cdot)$ and some continuous disturbance $\delta(\cdot)$, and $y(t)$ is the corresponding output.*
– *Discrete evolution: For all i, $e_i = (q(t_i'), \sigma(t_i'), q(t_{i+1})) \in E$ and $x(t_{i+1}) = R(e_i, x(t_i'))$.*

Problem 1. Consider the switching system of Definition 1. For $i = 1, ..., N$, let $\Omega(q_i) \subset \mathbb{R}^p$ be a given set associated with state $q_i \in \mathbf{Q}$. Let q_0 be an element of $\mathbf{Q}$. Find the set $X_0 \subset \mathbb{R}^n$ of all possible continuous initial states such that $\exists \hat{u}(\cdot) \in \mathcal{U}_C$ such that for any execution with $q(t_0) = q_0$, $x(t_0) \in X_0$, $u(\cdot) = \hat{u}(\cdot)$, the following constraints are satisfied, $\forall \delta(\cdot) \in \mathcal{U}_d$:

$$y(t) \in \Omega(q(t)), \quad \forall t \ge t_0 \tag{3}$$

In what follows, we set for simplicity, and w.l.o.g., $\Omega(q_i) = \Omega$, $\forall i = 1, ..., N$. Moreover, for the sake of notational simplicity, we set $\Delta_i = \varphi(q_i)$, $q_i \in \mathbf{Q}$.

Remark 1. It is possible to use the formulation of Problem 1 to deal with approximate output tracking problems. Given a reference trajectory $y_R(t)$ for the output, we require that the output of our switching system *at any instant of time*, differs from the given reference for at most a prescribed quantity ε. If the reference trajectory is the output of an exosystem:

$$\begin{cases} \dot{w} = s(w) \\ y_R(t) = q(w), \quad w(0) = w_0 \end{cases} \tag{4}$$

(where, for simplicity, only the case of continuous time systems has been considered), incorporate (4) into the model of the switching system and define a new output function as $y_E(t) = y(t) - y_R(t)$. Moreover, set $\Omega = \{y_E : -\varepsilon \mathbf{1} \le y_E \le \varepsilon \mathbf{1}\}$. Then, the approximate tracking problem is formulated in the form of Problem 1.

3   Problem Solution

To state the main results, the following definitions are needed.

Definition 4. *A set $\Sigma \subset \mathbb{R}^n$ is robustly controlled invariant with respect to configurations $\{(q, \mathcal{S}(q)), \; q \in \mathbf{Q}' \subset \mathbf{Q}\}$ and constraints (3) if: $\forall q \in \mathbf{Q}'$, $\forall x \in \Sigma$, $\exists u(\cdot) \in \mathcal{U}_C$ such that the solution $x(t)$ of $\mathcal{S}(q)$ with $x(t_0) = x$ is such that $x(t) \in \Sigma$ and $y(t) \in \Omega$, $\forall t > t_0$, $\forall \delta(\cdot) \in \mathcal{U}_d$.*

Definition 5.   *Given some set $\Lambda \subset \mathbb{R}^n$, define*

$$\Omega_{ix}^{\Delta}(\Lambda) = \left\{ \begin{array}{c} x \in \mathbb{R}^n : \exists u(\cdot) \in \mathcal{U}_C \text{ such that } y(t) \in \Omega, \\ \text{for all } t_0 \leq t \leq t_0 + \Delta \text{ and} \\ x(t_0 + \Delta) \in \Lambda, \; \forall \delta(\cdot) \in \mathcal{U}_d \end{array} \right\}$$

where $y(t)$ is the output of $\mathcal{S}(q_i)$ with $x(t_0) = x$.

Let $\mathcal{I}_i(\Lambda)$ be the maximal robust controlled invariant set with respect to configuration $(q_i, \mathcal{S}(q_i))$ and constraint (3) contained in the set Λ.

Definition 6. *Given some set $\Lambda \subset \mathbb{R}^n$, define*

$$R_{ij}^{-1}(\Lambda) = \left\{ \begin{array}{c} x \in \mathbb{R}^n : \exists \sigma \in \mathbf{U}_D \text{ such that} \\ R((q_i, \sigma, q_j), x) \in \Lambda \end{array} \right\}.$$

If $R((q_i, \sigma, q_j), x) = x$ for all $(q_i, \sigma, q_j) \in E$, we say that R is the identity reset function.

A connected FSM can be decomposed into its strongly connected components (maximal sets of mutually reachable states) $F_1, F_2, \cdots, F_M$ and there is a partial ordering among the strongly connected components. The strongly connected components of F determine a Directed Acyclic Graph (DAG), T, where the nodes correspond to $F_1, F_2, \cdots, F_M$. Without loss of generality, we assume that the DAG is *rooted*, e.g., there is only one node that has no incoming arc.

Definition 7. *A node $q_i \in \mathbf{Q}$ is closed if the set that solves Problem 1 with $q(t_0) = q_i$, denoted Σ_i, has been found. Otherwise, $q_i \in \mathbf{Q}$ is open.*

Algorithm 1: Structural algorithm for the determination of the safe set

MAIN:

Init: Set $\Sigma_i = \emptyset, i = 1, ..., N$. Let A be the set of nodes belonging to strongly connected components containing only one node. Let M be the set of strongly connected components containing two or more nodes.

Repeat

Do Find an open node $v = q_i \in A$ such that either it has no successors or all its successors are closed. Solve the node v applying procedure *Star(v)* and mark it as closed.

While *no such node could be found.*

Do Find a strongly connected component $F \in M$ such that each node belonging to F either has no successors not belonging to F or all its successors not belonging to F are closed. Let $I_F = \{q_1, q_2, ..., q_{N_F}\} \subset \mathbf{Q}$ be the set of nodes belonging to F. If R restricted to F is the identity reset function and $\Delta_i = 0$ for $i = 1, ..., N_F$, solve the strongly connected component F by applying procedure $Strongly\_Connected\_Simp(F)$ and mark all its nodes as closed. Otherwise, solve the strongly connected component F by applying procedure $Strongly\_Connected(F)$ and mark all its nodes as closed.

While *no such strongly connected component could be found.*

Until *the root node has been marked as closed.*

SUB $Star(\ v \ )$ If $v = q_i$ has no successors, set $\Sigma_i = \mathcal{I}_i (\mathbb{R}^n)$. Otherwise, let $I_{succ} = \{q_{i_1}, ..., q_{i_{mi}}\} \subset \mathbf{Q}$ be the set of nodes that are successors of the node v. Let $\Sigma_i = \Omega_{ix}^{\Delta_i} \left(\mathcal{I}_i \left(\bigcap_{j=1}^{i_{mi}} R_{i,j}^{-1}(\Sigma_j) \right) \right)$. If $\Sigma_i = \emptyset$ then **EXIT**.

SUB $Strongly\_Connected(\ F \ )$

For all $q_i \in I_F$, let $I_{Fsucc} = \{q_{i_1}, ..., q_{i_{mi}}\} \subset \mathbf{Q}$ be the set of nodes that are successors of the node q_i. In general, the set I_{Fsucc} contains nodes not belonging to F. If $q_r \in I_{Fsucc}$ and $q_r \notin I_F$ then, in what follows, we let $\Sigma_r^k = \Sigma_r, \forall k$, the set Σ_r being well defined since q_r is closed.

Init: $\Sigma_i^0 = \mathcal{I}_i (\mathbb{R}^n)$, $i = 1, ..., N_F$, $k = 0$ **Repeat**
    For $i = 1, ..., N_F$
$$\Sigma_i^{k+1} = \Omega_{ix}^{\Delta_i} \left(\mathcal{I}_i \left(R_{i,i_1}^{-1} \left(\Sigma_{i_1}^k \right) \cap ... \cap R_{i,i_{m1}}^{-1} \left(\Sigma_{i_{m1}}^k \right) \right) \right)$$
    End For
    $k = k + 1$
Until a set $\{\Sigma_i\}_{i \in I_F}$ of fixed points has been found.
If such a set cannot be found, then **EXIT**.

End SUB

SUB $Strongly\_Connected\_Simp(\ F \ )$

For all $q_i \in I_F$, let $\{q_{i_1}, ..., q_{i_{mi}}\} \subset \mathbf{Q}$ be the set of nodes that are successors of the node q_i. In general, the set $\{q_{i_1}, ..., q_{i_{mi}}\}$ may contain nodes not belonging to F. Find the maximal robust controlled invariant set Σ^* with respect to configurations $(q_1, \mathcal{S}(q_1)), \cdots, (q_{N_F}, \mathcal{S}(q_{N_F}))$ and constraint (3) contained in the set $\Sigma_{i_1} \cap \Sigma_{i_2} \cap \cdots \cap \Sigma_{i_{mi}}$. If such a set cannot be found, then **EXIT**. Let $\Sigma_i = \Sigma^*$, $i = 1, ..., N_F$ and return.

End SUB

Proposition 1. *Consider a switching system with E described by a general connected FSM. Let $\Sigma_i, i = 1, \cdots, N$ be the sets found by Algorithm 1. If $q(t_0) = q_i$, for some $i = 1, \cdots, N$, then $X_0 = \Sigma_i$.*

The structural approach proposed with Algorithm 1 decomposes the original problem into a number of different sub-problems, each consisting of finding a maximal (robustly) controlled invariant set in a given constraining set for a continuous state dynamical system or for a finite set of dynamical systems. There are essentially two levels of computation involved: a higher level corresponding to the steps of Algorithm 1, and a lower level, called by the higher level, corresponding to the computation of the invariant sets. The structure of the FSM has been exploited in order to achieve maximal computational efficiency when solving the continuous sub-problems. Assuming that the lower level converges appropriately, Algorithm 1 is guaranteed to converge to the exact solution, if it exists. Moreover, in the case of acyclic FSM or if $\Delta_i = 0$ for $i = 1, ..., N_F$, it converges to the exact solution in a finite number of steps. In order to better understand the improvements in computational efficiency, we compare our procedure with the one described in literature (see e.g. [21]), on a switching system described by:

$$Q_1 \rightarrow Q_2 \rightarrow ... \rightarrow Q_N$$

Applying our procedure to this example, we have:

$\Sigma_N = \mathcal{I}_N(\mathbb{R}^n)$
For $i \leftarrow N - 1$ to 1
    $\Sigma_i = \mathcal{I}_i(\Sigma_{i+1})$
End For

The solution needs N iterations, and the computation of N controlled invariant sets. Applying the procedure described in [21] to the same system, we have:

$$W^{i+1} = W^i - Reach(Pre_d(W^i), \emptyset)$$

where W^i is the safe-set approximation found at iteration i, in the mixed discrete-continuous state space, and: $W^0 = \{(Q_i, \Omega_{ix})\}_{i=1...N}$.
Let: $w_k^i = W^i\big|_{q=Q_k}$. The preceding computation is equivalent to:

For $i \leftarrow 1$ to N
    For $k \leftarrow 1$ to $N - i + 1$
        $w_k^i = \mathcal{I}_k(w_k^{i-1} \cap ... \cap w_N^{i-1})$
    End For
End For

In this case, the procedure of [21] needs the computation of $N(N-1)/2$ maximal controlled invariant sets. The two procedures require the same computation only in the case of a switching system described at the top level by a strongly connected FSM.

Algorithm 1 can be extended to the more general class of hybrid systems, where "invariance" conditions and controllable transitions are present in addition to uncontrollable switchings. We give two examples to illustrate how the procedure works. The extension to more complex FSM topologies can be handled similarly and can be found in [3]. Assume to have a system described by the following discrete structure:

$$Q_1 \rightarrow Q_2$$

where the transition from Q_1 to Q_2 is forced when the system ceases to satisfy an "invariance" condition of the form: $x \in \Gamma_1$.

- The safe set for configuration Q_2 is given by: $\Sigma_2 = \mathcal{I}_2(\mathbb{R}^n)$, i.e. the maximal controlled invariant set, with respect to configuration Q_2 satisfying the output constraints.
- The safe-set for configuration Q_1 is given by:

$$\Sigma_1 = \mathcal{I}_1(\Gamma_1) \bigcup \mathcal{Z}_1 \left(\Gamma_1, \overline{\Gamma}_1 \cap R_{12}^{-1}(\Sigma_2) \right)$$

where:

$$\mathcal{Z}_1(\mathcal{A}, \mathcal{B}) = \left\{ \begin{array}{c} x : \exists t_F, u(\bullet) \in \mathcal{U} \text{ such that: if for some} \\ t_0 < t_F \quad x(t_0) = x, \quad x(t) \in \mathcal{A} \cap \Omega_{1x} \\ \text{for } t_0 \leq t < t_F \text{ and } x(t_F) \in \mathcal{B} \end{array} \right\} \cup (\mathcal{B} \cap \Omega_{1x})$$

and $x(t)$ is the state evolution for configuration Q_1.

The interpretation for the formula above is as follows: for each x in the safe set for Q_1, there should be a control law such that the corresponding trajectory never goes outside the invariant set Γ_1, or, if it does, the reset function maps the actual state x in a "safe state" for configuration Q_2. Moreover, the constraints on the output must always be satisfied.

The case of controllable transitions follows the same logic. Suppose that the transition from Q_1 to Q_2 is completely controllable. Then:

- the safe-set for configuration Q_2 is given by: $\Sigma_2 = \mathcal{I}_2(\mathbb{R}^n)$;
- the safe-set for configuration Q_1 is given by:

$$\Sigma_1 = \mathcal{I}_1(\mathbb{R}^n) \bigcup \mathcal{Z}_1 \left(\mathbb{R}^n, R_{12}^{-1}(\Sigma_2) \right)$$

In this case the trajectory can or cannot be entirely in the controlled invariant set $\mathcal{I}_1(\mathbb{R}^n)$, but if it goes outside it must end into a set which is the reverse image of the safe-set for Q_2, and when it happens the discrete controller forces a switch from Q_1 to Q_2. Also in this case constraints on the output must constantly be satisfied.

4   Construction of Invariant Sets and Convergence Properties

In general, the computation of a controlled invariant set is an open problem. In fact, while conditions such that a given set enjoys the controlled invariance property have been extensively studied in the context of viability theory (see e.g. [1]), there are no implementable results applicable to general nonlinear systems. In the case of continuous-time linear systems, Dorea and Hennet in [13] characterize controlled invariance for general convex polyhedral sets with computable conditions. In addition, they show that no iterative formulas exist to exactly compute some maximal controlled invariant set for general continuous-time systems. Fortunately, methods for the computation of maximal controlled invariant sets for discrete-time linear systems and polyhedral constraining sets are well known in the literature (see e.g. [6], [11], [12], [15], [17]). In all of these papers, recursive algorithms are given that converge to the exact required set, if it exists. Hence linearizing and discretizing general dynamical systems is certainly a feasible path towards a computationally efficient approach for maximal controlled invariant set computation.

Given the i-th discrete-time linear system:

$$x\left(t+1\right) = A_i\, x\left(t\right) + B_i\, u\left(t\right) + F_i\delta\left(t\right)$$

this set can be computed by means of the following backward procedure (see e.g. [6]):

Algorithm 2: Maximal Controlled Invariant Set

$$\mathcal{I}^0 \leftarrow \Lambda$$
$$k \leftarrow 0$$

Repeat
$$\mathcal{I}^{k+1} \leftarrow \Omega_{ix}^1(\mathcal{I}^k)$$
Until $\mathcal{I}^{k+1} = \mathcal{I}^k$
$$\mathcal{I}_i(\Lambda) \leftarrow \mathcal{I}^k$$

where Λ is the initial constraint set in the state-space, and:

$$\Omega_{ix}^1(\Lambda) = \{x \in \mathbb{R}^n : \exists u : A_i x + B_i u + F_i\delta \in \Lambda \text{ and } y \in \Omega, \forall \delta \in \mathbf{U}_d\}$$

In general, this recursive algorithm converges to the solution asymptotically. If Λ, Ω, and $\mathbf{U}_d$ are polyhedral, so are the sets $\mathcal{I}^k$, but not necessarily the limit of the sequence for k which tends to infinity. In general, the maximal controlled invariant algorithm is not guaranteed to terminate in a finite number of steps. To the best of our knowledge, the only result that gives a sufficient condition for controller synthesis decidability can be found in [20].

At each step, the computation of $\mathcal{I}^k$ in Algorithm 2 involves a projection procedure and the elimination of redundant constraints. The approaches presented in the literature essentially differ for the algorithm used to project a polyhedron on a given subspace. The classical Fourier Motzkin elimination method and its modified versions (see [19] and [17]) can be used to perform this task. In our example, we adopted the algorithm developed in [10], which has the advantage of identifying and removing redundant inequalities at every step. To the best of our knowledge, a systematic comparison among the projection algorithms has not been done so far.

In this section we make the following assumptions:

- In each configuration of our switching system, the dynamical system is a discrete-time linear system, i.e. we have: $\mathcal{S} : \mathbf{Q} \rightarrow S_D$, where

$$\mathcal{S}\left(q_i\right) = \begin{cases} x\left(t+1\right) = A_i\, x\left(t\right) + B_i\, u\left(t\right) + F_i\delta\left(t\right) \\ y\left(t\right) = C_i\left(x\left(t\right) + D_i u\left(t\right)\right) \end{cases} \quad i \in J$$

– Constraints are given as linear inequalities and affect both the continuous state and the control input of the system:

$$\begin{cases} W_i x \le M_i & \text{(a)} \\ D_i u \le d_i & \text{(b)} \end{cases} \tag{5}$$

Since the output y represents the controlled output, we can choose the matrices $\{C_i\}_{i=1\ldots n}$ and $\{D_i\}_{i=1\ldots n}$ so as to specify both constraints on the state and control in terms of constraints on the output:

$$C_i = \begin{bmatrix} I & 0 \\ 0 & 0 \end{bmatrix} \text{ and: } D_i = \begin{bmatrix} 0 & 0 \\ 0 & I \end{bmatrix}$$

Then (5) is equivalent to:

$$y \in \Omega \text{ where: } \Omega = \left\{ y : \begin{bmatrix} W_i \\ D_i \end{bmatrix} y \le \begin{bmatrix} M_i \\ d_i \end{bmatrix} \right\}$$

The continuous disturbance $\delta(.)$ takes values in a bounded polyhedral set (polytope), $\mathbf{U}_d$, that is a set described by linear inequalities:

$$G\delta \le H$$

The reset function $R((q_i, \sigma, q_j), x)$ is an affine function:

$$R((q_i, \sigma, q_j), x) = R_{ij} x + p_{ij}$$

In order to apply the algorithm described in Section 3, we need to give a numerical implementation of the following operators:

$$\text{a) } R_{ij}^{-1}(\bullet); \text{ b) } \Omega_{ix}^{\Delta}(\bullet); \text{ c) } \mathcal{I}_i(\bullet);$$

The computation of the set $R_{ij}^{-1}(\Lambda)$ is straightforward if Λ is a polyhedral set, i.e., $\Lambda = \{x : Vx \le N\}$. In fact we have:

$$R_{ij}^{-1}(\Lambda) = \{x : V(R_{ij}x + p_{ij}) \le N\} = \{x : V R_{ij} x \le N - V p_{ij}\}$$

The implementation of both operators $\Omega_{ix}^{\Delta}(\bullet)$ and $\mathcal{I}_i(\bullet)$ is based on the implementation of the operator:

$$\Omega_{ix}^{1}(\bullet) = \Omega_{ix}^{\Delta}(\bullet) \text{ for } \Delta = 1.$$

Then,

$$\Omega_{ix}^{1}(\Lambda) = \{x \in \mathbb{R}^n : \exists u : A_i x + B_i u + F_i \delta \in \Lambda \text{ and } y \in \Omega, \forall \delta \in \mathbf{U}_d\}$$

The procedure used to compute the set $\Omega_{ix}^{1}(\Lambda)$ can be found in [3]. At each step, redundant inequalities are eliminated by an appropriate algorithm. The numerical implementation of $\Omega_{ix}^{\Delta}(\bullet)$ is the composition of the operator $\Omega_{ix}^{1}(\bullet)$ repeated Δ times. If the set Λ is polyhedral, the set $\Omega_{ix}^{\Delta}(\Lambda)$ is also polyhedral and, given the implementation of Ω_{ix}^1, it is described by a set of non-redundant linear inequalities.

The computation of the maximal controlled invariant set contained in a polyhedron Λ is similar, since we repeatedly apply $\Omega_{ix}^1(\bullet)$ halting when the set found at iteration k is the same as the one found at iteration $k-1$.

Since, in general, the maximal controlled invariant algorithm is not guaranteed to terminate in a finite number of steps and, in addition, it is not possible to give an a priori bound on the number of inequalities characterizing it, a basic question is whether it is possible to find a good approximation. If the algorithm terminates, then the set found after the last iteration is the maximal controlled invariant set. Otherwise, the algorithm, as it progresses, computes an increasingly better approximation of this set. Further, since the numerical implementation of the operator $\Omega_{ix}^1(\bullet)$ eliminates redundant inequalities, the algorithm gives also a representation of the set $\mathcal{I}_i(\bullet)$ that is minimal in terms of the number of linear inequalities.

Algorithm 2 builds, recursively, outer approximations of the maximal controlled invariant set. Hence at each step, although we can go as close as we want to the exact solution, we obtain sets that are not invariant. Our idea is to construct also inner approximations to the maximal controlled invariant set by building recursively sets that, at each step, are controlled invariant. Then, we can approximate the maximal controlled invariant set with the inner approximation and combine both outer and inner approximations to quantify the error associated with the inner approximation. The evaluation of this error is of paramount importance in our approach, where the error is not confined in the computation of just one maximal controlled invariant set, but it propagates backward, as the higher level algorithm proceeds, and one could easily find an empty set at some step of the recursion, even if the given problem has a solution.

Suppose that no disturbance is active on the system ($\mathbf{U}_d = \{0\}$). For recursively constructing an inner approximation, if the set Λ is convex, bounded, containing the origin in its nonempty interior, the algorithm used is identical to Algorithm 2, except for the initialization of $\mathcal{I}^0$, which has to be set equal to a starting controlled invariant set and not to the set of state-constraints:

- if the system is controllable, set $\mathcal{I}^0 = \{0\}$;
- if the system is asymptotically stabilizable, set $\mathcal{I}^0 = X$, where X is a controlled invariant set with non-empty interior contained in the set of constraints Λ .

For finding the controlled invariant set X, an ellipsoidal $\bar{\lambda}$-contractive set contained in the set of constraints can be obtained easily. Then, by applying the procedure described in [6] for a value λ' of the parameter λ greater than $\bar{\lambda}$, a polyhedral λ-contractive set, $\lambda' < \lambda < 1$, containing the maximal λ'-contractive set is obtained in a finite number of steps. This set is obviously controlled invariant and has nonempty interior, because the maximal λ'-contractive set contains the ellipsoidal $\bar{\lambda}$-contractive set. Alternatively, an invariant controlled set contained in the constraint set can be obtained by using the constructive result of [14].

We can show that [3] the algorithm terminates if a fixed point is found and converges to the maximal controlled invariant set. This result, in the case of a controllable system, generalizes the one in [16], where the invertibility of A_i is required.

The maximal controller is the set of all control laws $\hat{u}(\cdot)$ that make all constraints satisfied, for any allowed switching and disturbance, starting from the set X_0, found by Algorithm 1. When the discrete location is q_i, if we apply a control law in the maximal controller, the state, starting from Σ_i, reaches a maximal controlled invariant set in Δ_i steps of time. It remains in this set until a switching occurs (since the chosen control makes this set invariant). A control law which makes a given set invariant depends on the set itself. If the set is a polyhedron containing the origin and if the dynamical system is a linear discrete-time system, it can be a piecewise linear state feedback control law that can be determined by using a technique introduced in [15] for the polytopic case and generalized to the polyhedral case in [13]. The polyhedron is partitioned in a certain number of subsets, and for each subset a different linear state feedback law is applied. The number of these subsets may be arbitrarily large, and hence there is no bound on the complexity of the control law, which has to be computed on-line. To reduce the on-line computational effort, the result of [8] can be used. Blanchini shows that if a controlled invariant polytope is approximated by a suitable smooth domain, a simpler control law exists.

If the dynamical systems corresponding to each location of the FSM are continuous and not discrete-time, we have to choose the discrete-time system that corresponds to the given continuous-time one so that a precise relation can be established between the invariant sets computed as previously illustrated and those of the continuous-time system. One way of obtaining such a relation is to consider the Euler Approximating System (EAS) (see [8] and [7]). As the parameter τ characterizing the Euler Approximation tends to zero, better and better approximations are obtained for the maximal controlled invariant set of the continuous-time system. Moreover, these sets are invariant for the continuous-time system. The same result holds if the sampled-data system is used, but in that case the approximations are not invariant [8].

5   Application to Engine Idle-Speed Regulation

The idle-speed control problem deals with the task of maintaining, while in the idle mode, the engine speed into a given range, rejecting torque disturbances due to accessory loads (such as the air-conditioning system and the steering wheel servo-mechanism), preventing the engine from switching off. The power-train model used for idle-speed control is:

$$\begin{cases} \dot{p} = a_m n p + b_m \alpha \\ \dot{n} = \frac{a_n}{J_{eq}} n + b_n \left(T - T_{LOAD} \right) \end{cases} \tag{6}$$

where n is the engine speed expressed in RPM (Revolutions Per Minute); p is the manifold pressure expressed in mbar; J_{eq} is the momentum of inertia for the transmission chain (kg m^2); a_m, b_m, a_n, b_n, are constants; $T = k_1\eta\,(AV)\,p$ is the torque produced by the engine given the spark advance angle AV, the efficiency function $\eta\,(AV)$ and the constant k_1; T_{LOAD} is the torque disturbance from accessory loads. The two control inputs are the throttle opening angle α and the spark advance angle AV. Clutch insertion or release has the effect of modifying the parameter J_{eq}, the momentum of inertia, causing a sudden unpredictable change in the power-train parameters.

We assume that no information is available about minimum permanence times in each configuration, so we let $\Delta_i = 0$, $i = 1, 2$. Moreover, the reset function is the identity.

The problem is to find under which conditions it is possible to maintain the engine speed into the desired range $800 \pm 30\ RPM$, satisfying the constraints on control inputs: $0° \leq \alpha \leq 20°$, $0° \leq AV \leq 20°$.

Defining the output of the system as $y = \begin{bmatrix} n\ p\ \alpha\ AV \end{bmatrix}'$, we can express the constraints on the state and the inputs as output constraints. By applying the method described in Section 4, we determined the safe set and we found that the maximum value allowed for the continuous disturbance T_{LOAD} is 12 Nm. The maximal controller is described by linear inequalities, where the bound vector depends on the state. A controller may be chosen among all possible ones by introducing an optimality criterion. Simulations carried on the switching nonlinear model show the effectiveness of the proposed approach (see [3] for more details on this application).

6   Conclusions

We proposed a structural procedure for the determination of the maximal safe set for hybrid systems. While demonstrably more efficient than the elegant procedure of [21], the procedure still suffers from computational complexity stemming from the computation of maximal controlled invariant sets of general dynamical systems. The procedure is made computationally more appealing by linearizing the nonlinear dynamics and using a discrete-time equivalent model, since procedures for the computation of maximal controlled invariant sets for discrete-time linear systems are well-known in the literature. Even for this case, the procedure for the determination of the maximal controlled invariant set may not converge in a finite number of steps. We propose an inner approximation algorithm that together with the classical outer approximation yields tight bounds for an error due to the truncation of the procedure after a finite number of steps. The theory has been applied to idle-speed regulation in engine control to demonstrate its power.

References

1. Aubin, J.P. Viability theory, Birkhauser, Boston, (1991).
2. Balluchi, A., Di Benedetto, M.D., Pinello, C., Rossi, C., Sangiovanni-Vincentelli, A. "Hybrid Control in Automotive Applications: the Cut-off Control", *Automatica*, Special Issue on Hybrid Systems, vol. 35, March (1999).

3. Berardi, L., De Santis, E., Di Benedetto, M.D. "Invariant sets and control synthesis for switching systems with safety specifications", Department of Electrical Engineering, University of L'Aquila, Research Report no. 99-35, October 1999.
4. Berardi, L., De Santis, E., Di Benedetto, M.D. "Control of switching systems under state and input constraints", *European Control Conference 1999*, August 31 - Sept. 3 (1999).
5. Berardi, L., De Santis, E., Di Benedetto, M.D. "Control of switching systems under state and input constraints", *38th IEEE Conference on Decision and Control*, Phoenix, AZ, Dec. 7-10, (1999).
6. Blanchini, F. "Ultimate Boundedness Control for Uncertain Discrete-Time Systems via Set-Induced Lyapunov Functions", *IEEE Trans. on Automatic Control*, **AC-39**, pp. 428-433, (1994).
7. Blanchini, F. "Nonquadratic Lyapunov functions for robust control", Automatica 31, pp. 451-461, (1995).
8. Blanchini, F., Miani, S. "Constrained stabilization via smooth Lyapunov functions", *Systems and Control Letters*, 35, pp. 155-163, (1998).
9. d'Alessandro, P., De Santis, E. "General Closed loop optimal solutions for linear dynamic systems with linear constraints", *J. of Mathematical Systems, Estimation and Control*, vol. 6, no. 2, (1996).
10. d'Alessandro, P., A conical approach to linear programming, scalar and vector optimization problems, Gordon and Breach Science Publishers, 1997.
11. De Santis, E. "On maximal invariant sets for discrete time linear systems with disturbances". *Proc. 3rd IEEE Med. Symposium*, Cyprus, 1995.
12. Dorea, C. E. T., Hennet, J. C. "Computation of Maximal Admissible Sets of Constrained Linear Systems", *Proc. of 4th IEEE Med. Symposium*, Krete, pp. 286-291, (1996).
13. Dorea, C. E. T., Hennet, J. C. "(A,B)-Invariance conditions of polyhedral domains for continuous-time systems", *European J. of Control*, vol.5, pp. 70-81, (1999).
14. Farina, L., Benvenuti, L. "Invariant polytopes of linear systems", IMA, vol.15, pp.233-240, (1998).
15. Gutman, P.O., Cwikel, M. "Admissible Sets and Feedback Control for Discrete-Time Linear Dynamical Systems with Bounded Controls and States", *IEEE Transactions on Automatic Control*, **AC-31**, No. 4, pp. 373-376, (1986).
16. Gutman, P.O., Cwikel M. "An Algorithm to Find Maximal State Constraint Sets for Discrete Time linear Dynamical Systems with Bounded Controls and States", *IEEE Transactions on Automatic Control*, **AC-32**, No. 3, pp. 251-254, (1987).
17. Keerthi, S.S., Gilbert E.G. "Computation of Minimum-Time Feedback Control Laws for Discrete-Time Systems with State-Control Constraints", *IEEE Trans. on Automatic Control*, **AC-32**, pp. 432-435, (1987).
18. Lygeros, J., Tomlin, C., Sastry, S. "Controllers for Reachability Specifications for Hybrid Systems", *Automatica*, Special Issue on Hybrid Systems, vol. 35, (1999).
19. Murty, K.G. Linear Programming. New York: J. Wiley, (1983).
20. Shakernia, O., Pappas, J.P., Sastry, S. "Decidable Controller Synthesis for a Class of Linear Systems", This Conference.
21. Tomlin, C., Lygeros, J., Sastry, S. "Synthesizing controllers for nonlinear hybrid systems", First International Workshop, HSCC'98, Hybrid Systems: Computation and Control, *Lecture Notes in Computer Science*, vol.1386, pp. 360-373, (1998).

Verification of Hybrid Systems
with Linear Differential Inclusions
Using Ellipsoidal Approximations *

Oleg Botchkarev** and Stavros Tripakis

The University of California at Berkeley
195M Cory Hall Berkeley CA 94720
Phone: (510) 642-5649, Fax: (510) 642-6330
{olegb,stavros}@eecs.berkeley.edu

Abstract. A general verification algorithm is described. It is then shown how ellipsoidal methods developed by A. B. Kurzhanski and P. Varaiya can be adapted to the algorithm. New numerical algorithms that compute approximations of unions of ellipsoids and intersections of ellipsoids and polyhedra were developed. The presented techniques were implemented in the verification tool called VeriSHIFT and some practical results are discussed.

Keywords: hybrid systems, verification, reachability analysis, ellipsoidal approximations.

1   Introduction

A number of application domains, such as car manufacturing, robotics, chemical process control, or avionics, involve *controllers*, consisting of: (a) a set of *sensors* and *actuators*, representing the interface between the controller and its environment; (b) a *control logic* (implemented as one or more circuits or as one or more pieces of software running concurrently), which represents the way the controller should act on the environment.

A promising model for describing such systems is *hybrid automata* [8]. Hybrid automata are finite-state machines equipped with continuous variables. Each discrete state of an automaton has a system of differential equations that govern its continuous variables. Most correctness criteria for such systems can be stated as a *safety* property: the system must never reach an "unsafe" (or a "bad") state.

Ensuring correctness of the model is often not a trivial task. Simulation of the system is not adequate, since it can only help examine a limited number of trajectories. Analytical methods are often not applicable, considering the complex interaction of continuous and discrete dynamics. An alternative is *reachability analysis*. It consists of computing the set of all reachable states of the system and

---

* Research supported by National Science Foundation Grant ECS 9725148 and ONR Contract 11
** Corresponding author.

N. Lynch and B. Krogh (Eds.): HSCC 2000, LNCS 1790, pp. 73–88, 2000.
© Springer-Verlag Berlin Heidelberg 2000

then checking that no "bad" state belongs to the reachable set. The reachability problem has been shown to be undecidable, even for models of hybrid automata with simple dynamics (e.g., $\dot{x} \in [a, b]$). Moreover, the so-called *state explosion problem* (the machine representation of the set of reachable states is too large) often limits the applicability of the method, even for decidable sub-classes of the model.

Approximations have been used as a remedy to both the undecidability and the state-explosion problems. Computing an over- or under-approximation (i.e., external or internal approximation) of the exact set of reachable states can be, first, decidable, and second, less expensive, in terms of time and memory. The price to pay is accuracy: what does it mean for a "bad" state to be reachable in the approximative analysis?

This paper presents a new reachability technique for systems of hybrid automata with linear dynamics, expressed as *differential inclusions*: $\dot{x} \in Ax + U$. The basic model of hybrid automaton and its semantics are presented in section 2.

The algorithm performs reachability analysis for bounded time. Reachability for bounded time means that the set of states reachable in Δ time units is computed, where Δ is a parameter supplied by the user. The skeleton of the algorithm, correctness, and trade-offs between accuracy and efficiency are discussed in section 3. A generalization of the algorithm is presented in [1].

The algorithm is based on the ability to approximate: (a) the reachable set of a linear differential inclusion (time propagation); (b) intersections of convex sets; (c) unions of convex sets; (d) linear transformations and geometric sums of convex sets.

Among methods of reachability analysis are those based on ellipsoidal techniques. The presented work is an attempt to use some of the methods described in [5,6].

New methods for computing over-approximations of unions of ellipsoids and intersections of ellipsoids and polyhedrons have also been devised. They are presented in section 4.

These reachability techniques have been implemented in a prototype tool called *VeriSHIFT* (section 5). The tool accepts systems of hybrid automata, communicating by input/output variables and synchronous message passing. Dynamic creation and reconfiguration of automata is also supported.

2   The Model

In this section we present the model of a single hybrid automaton. We consider the extension to systems of communicating hybrid automata in [1].

Preliminaries Let R be the set of real numbers, $\mathsf{R}^{m \times n}$ the set of $m \times n$ real matrices, $\mathcal{C}^n$ the set of convex closed subsets of R^n, and $\mathcal{C}_b{}^n$ the set of convex compact subsets of R^n. $\mathsf{B}_\epsilon^n(x)$, the *ball of dimension n with center x and radius ϵ*, is defined to be the convex set $\{y \in \mathsf{R}^n \mid |x - y| \le \epsilon\}$.

Given a set $P \in C^n$ and a matrix $A \in \mathsf{R}^{m \times n}$, AP is the *linear transformation* of P, that is, a set from C^m defined as $AP = \{Ax \mid x \in P\}$.

Given two sets $P_1, P_2 \in C^n$, let $P_1 + P_2$ denote the *geometric (Minkowski) sum* of P_1, P_2, defined as:

$$P_1 + P_2 = \{x \mid \exists x_1 \in P_1, x_2 \in P_2, x = x_1 + x_2\}.$$

A *flow* in R^n is defined as a triple $F : (A, I, U)$, where $A \in \mathsf{R}^{n,n}$ and $I \in C$, $U \in C_b{}^n$. F defines the following system of differential equations:

$$\dot{x}(t) = Ax(t) + u(t)$$
$$x(t) \in I, \ u(t) \in U.$$

Given points $x_0, x_1 \in \mathsf{R}^n$, we say that x_1 is *F-reachable from x_0 at time t*, denoted $x_0 \overset{t}{\leadsto}_F x_1$, if there exist functions of time $x(\cdot)$ and $u(\cdot)$ such that $x(0) = x_0$, $x(t) = x_1$, and for all $\tau \in [0, t]$, $x(\tau) \in I$, $u(\tau) \in U$ and $\dot{x}(\tau) = Ax(\tau) + u(\tau)$.

Given a set $X_0 \in C_b{}^n$, the *reachable set of flow F from X_0 at time t*, denoted $\mathcal{X}_F(X_0, t)$, is the set of all points that are F-reachable from points of X_0 at time t.

A Hybrid Automaton. We define a hybrid automaton $\mathcal{A}$ with *linear differential inclusions* to be a tuple (Q, X, F, T, G, R), where:

- Q is a finite set of *discrete states* (or *locations*, or *modes*).
- X is a set of n *continuous variables* taking values in R.
- $F : Q \rightarrow \mathsf{R}^{n \times n} \times C^n \times C_b{}^n$ associates with each discrete state q a flow (A, I, U). I is called the *invariant* of q and will be denoted as $I(q)$.
- $T \subseteq Q \times Q$ is a set of *discrete transitions*.
- $G : T \rightarrow C^n$ associates with each discrete transition a *guard*.
- $R : T \rightarrow \mathsf{R}^{n \times n} \times C_b{}^n$ associates with each transition a pair (B, P). This pair defines the *reset* of the continuous variables[1]: $x := Bx + P$.

Given a discrete state q, let out(q) be its set of *out-going transitions*, $\{(q, q') \in T\}$.

We now turn to the semantics of a hybrid automaton like $\mathcal{A}$. A *state* of $\mathcal{A}$ is a pair $(q, x) \in Q \times \mathsf{R}^n$ such that $x \in I(q)$.

Given a state (q, x) and a *delay* $\delta \in \mathsf{R}$, we say that there is a *time transition* from (q, x) to a state (q, y), denoted $(q, x) \overset{\delta}{\leadsto} (q, y)$, if $x \overset{\delta}{\leadsto}_{F(q)} y$.

Given a state (q, x) and a discrete transition $a = (q, q') \in T$, such that $R(a) = (B, P)$, we say that there is a *discrete jump* from (q, x) to a state (q', y), denoted $(q, x) \overset{a}{\longrightarrow} (q', y)$, if $x \in G(a)$, $y \in I(q')$ and $y \in Bx + P$.

---

[1] If P contains a single point, $P = \{y\}$ the reset is deterministic, that is, each x is mapped to a unique $x' = Bx + y$. If P is not a singleton, then the reset is non-deterministic.

Reachability. Given a set of initial states S_0, we say that a state s is *reachable* from S_0 if there exists $s_0 \in S_0$ and a sequence

$$s_0 \overset{\delta_1}{\rightsquigarrow} s_1' \overset{a_1}{\longrightarrow} s_1 \overset{\delta_2}{\rightsquigarrow} s_2' \overset{a_2}{\longrightarrow} \cdots \overset{\delta_k}{\rightsquigarrow} s_k' \tag{1}$$

such that $s_k' = s$. The sequence (1) and the corresponding trajectory of continuous variables $x(\cdot)$ is called an *execution* of the hybrid automaton, and k is called the *length* of the execution. We say that s is reachable from S_0 *in time Δ* if

$$\delta_1 + \delta_2 + \cdots + \delta_k \le \Delta.$$

Given a discrete state q, we say that q is reachable from S_0 (in time Δ) if there exists a state (q, x) which is reachable from S_0 (in time Δ).

3   Reachability Using Convex Approximations

Given an automaton A and a set S_0 of initial states, we want to verify whether a discrete state q_{bad} is *not* reachable from S_0 in time Δ.

In this section we describe the skeleton of the reachability algorithm. The algorithm is based on the ability to: (a) effectively represent convex compact sets $X \in \mathcal{C}_b{}^n$; (b) compute an over-approximation $\mathcal{X}_F^+(X_0, t) \supseteq \mathcal{X}_F(X_0, t)$ of the reachable set of a linear flow F from a convex compact set X_0 at time t; (c) check whether the intersection of two convex sets is non-empty; (d) compute over-approximations of intersections, unions and geometric sums[2] of convex sets; (e) compute linear transformations of convex sets. Section 4 deals with points (a) – (d) in detail. In this section, we assume that an effective representation of convex sets and the above operations are available. First we present the basic structure of the reachability algorithm. Then we discuss alternatives and their impact on accuracy and efficiency.

3.1   The Basic Algorithm

The algorithm maintains a table $\mathcal{T}$ of tuples of the form: (q, X, τ), where $q \in Q$, $X \in \mathcal{C}_b{}^n$ and $\tau \in [0, \Delta]$. (q, X, τ) is supposed to represent a set of *unexplored* states $(q, x), x \in X$. A state $s = (q, x)$ is unexplored in the sense that q_{bad} might be reachable from s in time $\Delta - \tau$. An invariant of the algorithm is that if a state (q, y) is reachable in time Δ then at some point the table will contain a tuple (q, X, τ), where $\tau \le \Delta$ and, either $y \in X$, or there exist $x \in X$ and $t \in \mathbb{R}$ such that $x \overset{t}{\rightsquigarrow}_{F(q)} y$.

$\mathcal{T}$ is initialized to S_0 (the set of initial states), such that for all $(q, X, \tau) \in \mathcal{T}$, $\tau = 0$. The algorithm essentially repeats three steps. First, it chooses an unexplored tuple (q, X, τ). Second, it propagates X in time, until time reaches

---

[2] The geometric sum $P_1 + P_2$ can be computed exactly if at least one of P_1, P_2 is a singleton. Consequently, the reset of a set X, $BX + P$, can be computed exactly if the reset is deterministic (i.e., P is a singleton).

Δ; meanwhile, it computes the intersection of the reachable tube from X with each of the out-going guards of q. Third, for each intersection V with a guard, the algorithm computes the reset of V with respect to the corresponding discrete transition, and adds a new (unexplored) tuple to the table.

The second step involves computing the reachable set of $F(q)$ from X at time t, for $t \in [0, \Delta - \tau]$. Since it is not possible to compute this set for infinitely many time values, we have to discretize time, that is, we compute $\mathcal{X}_F^+(X_0, t)$ for $t = k\delta$, where $k = 0, ..., \lceil \frac{\Delta - \tau}{\delta} \rceil$. The *time step* δ is a parameter of the algorithm, given by the user. In order not to "miss" a guard during the propagation of X in discrete time steps, we "enlarge" the reachable set at each time step by a ball of radius ϵ (see step 2, below). ϵ can be effectively computed as a function of $F(q)$ and δ, so that correctness of the over-approximation is ensured:

Lemma 1. *The following estimate is true for all $t \in [0, \delta]$:*

$$\mathcal{X}_F(X_0, t) \subseteq X_0 + B_\epsilon(0) \tag{2}$$

where $\mathcal{X}_F(X_0, t)$ denotes the reachable set of differential inclusion

$$F : \dot{x} \in Ax + U, \ U \in \mathcal{C}_b{}^n,$$

$B_\epsilon(0) \in \mathcal{C}_b{}^n$ *is a ball of radius ϵ with the center in 0, and*

$$\epsilon = (e^{N_A \delta} - 1)D + e^{N_A \delta} N_U \delta,$$

$$N_A = \|A\| = \max_{\|x\|=1} \|Ax\|,$$

$$D = \max_{x \in X_0} \|x\|, \ N_U = \max_{u \in U} \|u\|.$$

Proof To ensure that inclusion (2) holds it is enough to take ϵ equal to the Hausdorff semidistance between X_0 and $\mathcal{X}_F(X_0, t)$:

$$\epsilon = h_+(\mathcal{X}_F(X_0, t), X_0) = \max_{y \in \mathcal{X}_F(X_0, t)} \min_{x \in X_0} \|x - y\|.$$

Then it is not difficult to see that $h_+(\mathcal{X}_F(X_0, t), X_0 + B_\epsilon(0)) = 0$ which implies $\mathcal{X}_F(X_0, t) \subseteq X_0 + B_\epsilon(0)$.

Let us estimate this Hausdorff distance:

$$
\begin{aligned}
h_+(\mathcal{X}_F(X_0, t), X_0) &= \max_{y \in \mathcal{X}_F(X_0, t)} \min_{x \in X_0} \|y - x\| \\
&= \max_{y \in X_0} \max_{u(\cdot) \in U} \min_{x \in X_0} \|x_F(y, t, u(\cdot)) - x\| \\
&\leq \max_{y \in X_0} \max_{u(\cdot) \in U} \|x_F(y, t, u(\cdot)) - y\| \\
&= \max_{y \in X_0} \max_{u(\cdot) \in U} \left\| e^{At}y - y + \int_0^t e^{A(t-s)} u(s) ds \right\| \\
&\leq \max_{y \in X_0} \|(e^{At} - I)y\| + \max_{u(\cdot) \in U} \left\| \int_0^t e^{A(t-s)} u(s) ds \right\|
\end{aligned}
$$

$$\leq \max_{y \in X_0} \|(e^{At} - I)y\| + \int_0^t \max_{u \in U} \|e^{A(t-s)}u\|ds$$
$$\leq (e^{N_A \Delta} - 1)D + e^{N_A \Delta}N_U \Delta.$$

∎

Now we are ready to detail the steps of the algorithm:

1. If the table T is empty, stop and announce that q_{bad} is unreachable in time Δ. Otherwise, if there exists a tuple $(q_{bad}, \_, \_) \in T$, stop and announce that q_{bad} is *possibly* reachable in time Δ. Otherwise, choose a tuple $(q, X, \tau) \in T$ with minimal τ, remove it from the table and proceed to step 2.
2. Let (q, X, τ) be the tuple chosen in step 1, $F = F(q)$ and $out(q) = \{a_1, ..., a_l\}$. Also, for $i = 1, ..., l$, let $a_i = (q, q_i)$, $R(a_i) = (B_i, P_i)$ and $G_i = G(a_i)$.
   (a) Compute $\mathcal{X}_F^+(X_0, k\delta) + B_\epsilon^n(0)$, for $k = 0, ..., m$, where m is the minimum between $\lceil \frac{\Delta - \tau}{\delta} \rceil$ and the smallest $k \leq \lceil \frac{\Delta - \tau}{\delta} \rceil$ such that $(\mathcal{X}_F^+(X_0, k\delta) + B_\epsilon^n(0)) \cap I(q) = \emptyset$.
   (b) For each $i = 1, ..., l$, let k_i be the first time the reachable set intersects the guard G_i, that is, $k_i = \min\{k \mid (0 \leq k \leq m) \wedge ((\mathcal{X}_F^+(X_0, k\delta) + B_\epsilon^n(0)) \cap G_i \neq \emptyset)\}$. If the reachable set never intersects G_i, we set $k_i = m + 1$. Let $\tau_i = k_i\delta$. If $B_i \neq 0$ (i.e., the reset is not constant), then we compute:

$$V_i \supseteq \bigcup_{j=k_i}^{m} \left((\mathcal{X}_F^+(X_0, k\delta) + B_\epsilon^n(0)) \cap G_i\right). \tag{3}$$

   That is, V_i is (an over-approximation of) the union of intersections of the reachable set and the guard at times $t \geq \tau_i$. If $B_i = 0$ the computation of V_i is unnecessary.
3. For each $\tau_i \leq \Delta$, computed at step 2, we add a tuple $(q_i, X_i', \tau + \tau_i)$ to the table T, where,

$$\begin{aligned} X_i' &= P_i, \text{ if } B_i = 0, \\ X_i' &\supseteq B_i V_i + P_i, \text{ otherwise.} \end{aligned} \tag{4}$$

   Go back to step 1.

We should point out that in step 2(b), we do not need to compute (an over-approximation of) the intersection of the reachable set and the guard at each time step to check whether it is non-empty. Instead, we can have a procedure that *checks*, given two convex sets, whether their intersection is non-empty. If it is, then we can compute it.

Correctness and Termination We now state the main properties of the algorithm.

Lemma 2. *Let $X_0 \in \mathcal{C}^n$, $G \in \mathcal{C}^n$, $F : \dot{x}(t) \in Ax + U$ - some linear flow and $\delta > 0$. If ϵ is choosen accordingly to lemma (1), then*

$$\bigcup_{\tau=0}^{\Delta}(\mathcal{X}_F^+(X_0, \tau) \cap G) \subset \bigcup_{i=0}^{k}((\mathcal{X}_F^+(X_0, i\delta) + B_\epsilon(0)) \cap G), \tag{5}$$

where

$$k \in \mathbf{Z}, \; (k-1)\delta < \Delta \leq k\delta.$$

The **proof** is a consequence of lemma (1).

Theorem 1. *If the algorithm terminates doing reachability analysis of a hybrid automaton A for time horizon Δ and reports that the state q_{bad} is unreachable (step 1), and a state (q^*, x^*) is reachable at time $t \leq \Delta$ as result of a sequence of time and discrete transitions ending with a discrete transition, then at some step of execution the table $\mathcal{T}$ contained a tuple (q^*, X, τ) such that $x^* \in X$ and $\tau \leq t$.*

Proof Let us suppose that

$$s_0 \overset{\delta_1}{\underset{F(q_0)}{\rightsquigarrow}} s_1' \overset{a_1}{\longrightarrow} s_1 \overset{\delta_2}{\underset{F(q_1)}{\rightsquigarrow}} \cdots s_{N-1}' \overset{a_{N-1}}{\longrightarrow} s_{N-1} \overset{\delta_N}{\underset{F(q_{N-1})}{\rightsquigarrow}} s_N' \overset{a_N}{\longrightarrow} s_N, \qquad (6)$$

$$s_{N-1} = (q', x'), \; s_N' = (q', x''), \; s_N = (q^*, x^*), \; \delta_1 + \delta_2 + \cdots + \delta_N = t,$$

is an execution of length N that leads to the state (q^*, x^*) at time t. Let G_i and $R_i = (B_i, P_i)$ denote the guard set and the reset relation that correspond to transition $q' \longrightarrow q^*$ in sequence (6).

The theorem is obviously true for the states that can be reached by executions of length 0.

Suppose that the theorem is true for states that can be reached in time Δ by executions of length $N-1$. Let us prove that it is true for executions of length N as well. If the theorem is true for executions of length $N-1$, then at some step of execution the table $\mathcal{T}$ must contain a tuple $r' = (q', X', \tau')$ such that $x' \in X'$ and $\tau' \leq \delta_1 + \delta_2 + \cdots + \delta_{N-1} = t'$. Since the tuple r' appeared in the table, step 2 of the algorithm was applied to it. Using lemma (2) we can conclude that the set V_i constucted by (3) contains the point x''. Hence, the set X_i' resulting from the reset relation (4) contains the point x^*. That proves the theorem.   ∎

Theorem 2. *If the algorithm terminates and reports that the state q_{bad} is unreachable in time Δ (step 1), then the state q_{bad} is unreachable in time Δ.*

The **proof** is a direct consequence of theorem (1).

Termination of the algorithm is not guaranteed for systems that may present so-called *zeno* behavior: an infinite number of discrete jumps in a finite amount of time. The following theorem states that termination is guaranteed when the time "consumed" by each loop of discrete transitions of the automaton is bounded from below by a positive number (in our case, at least by δ).

Theorem 3. *If, for any loop $a_1 = (q_1, q_2), a_2 = (q_2, q_3), ..., a_k = (q_k, q_1) \in \mathcal{T}$, there exists $i \in [1, k]$ such that the following conditions are satisfied:*

1. $R(a_i) = (0, P)$, that is, the reset of a_i is constant.
2. $\big(P + B_\epsilon^n(0)\big) \cap G(a_{i+1}) = \emptyset$ (by convention, a_{k+1} is taken to be a_1).

then the algorithm terminates.

The **proof** is quite obvious and is based in the fact that any cycle in the transition graph takes at least one intergration step δ.

3.2 Possible Modifications

Alternative choices could be made at some points in the algorithm. We discuss these possibilities below and comment on their impact on the accuracy and the efficiency of the algorithm.

1. If the table contains two tuples (q, X_1, τ_1) and (q, X_2, τ_2) with the same discrete state, then we replace these tuples by a single tuple $(q, X_1 \cup X_2, \min\{\tau_1, \tau_2\})$. This decreases the size of the table and results in fewer tuples to be explored. On the other hand, since we can only compute an over-approximation of the union $X_1 \cup X_2$, the accuracy of the algorithm might be compromised. Correctness is not affected.
2. At step 1, instead of removing the chosen tuple (q, X, τ) from the table $\mathcal{T}$ we mark the tuple *explored*. Only unexplored tuples are chosen in step 1. Moreover, before adding to $\mathcal{T}$ a new (unexplored) tuple (q, X', τ'), $\mathcal{T}$ is searched for a tuple (q, X'', τ'') such that $X' \subseteq X''$ and $\tau' \geq \tau''$. If such a tuple exists then (q, X', τ') is not added. The status of (q, X'', τ'') (explored or not) is not changed. The correctness of the algorithm is not affected. The size of $\mathcal{T}$ could increase since explored tuples are not removed. On the other hand, new tuples are not added to the table when not necessary, which results in fewer tuples to explore and shorter running time.

4 Ellipsoidal Approximations

The reachability algorithm described in the previous section can work with any representation of convex compact sets, as long as the operations used by the algorithm can be performed effectively on the chosen representation.

The verification tool described here uses ellipsoidal techniques for approximation and reachability analysis. One of the advantages of ellipsoidal methods is that an ellipsoid in $\mathcal{C}_b^n$ can be described as a pair $(x, P) \in R^n \times R^{n \times n}$, that is, using only $O(n^2)$ space. Time complexity of ellipsoidal operations is also polynomial.[3] The numerical methods that have been used are directly taken from or based on the results described in publications [5,6].

In these works it is shown that ellipsoidal over-approximations of reachable sets can be expressed through ordinary differential equations with coefficients given in explicit analytical form. Other results include parametric representation of ellipsoidal over-approximations of geometric sums and intersections of

---

[3] As a comparison, the worst-case complexity of polyhedral operations is exponential.

ellipsoids. The reader is referred to the above-mentioned publications for the details of the methods.

Here, we present new techniques that we have developed for operations on ellipsoids and polyhedra and for unions of ellipsoids.

Definition of Ellipsoids Let $\langle l, x \rangle$, $l, x \in \mathsf{R}^n$, denote the inner product of l and x. An *ellipsoid* $\mathcal{E}(p, P)$, $p \in \mathsf{R}^n$, $P \in \mathsf{R}^{n \times n}$, $P = P' \geq 0$ [4] is a convex compact set described by the *support function*[5]

$$\rho(l \mid \mathcal{E}(p, P)) = \langle l, p \rangle + \langle l, Pl \rangle^{1/2}.$$

If the matrix P is non-degenerate, the ellipsoid $\mathcal{E}(p, P)$ can alternatively be defined as a level set of a quadratic function:

$$\mathcal{E}(p, P) = \{x \mid \langle x - p, P^{-1}(x - p) \rangle \leq 1\}.$$

Approximation of Unions of Ellipsoids The reachability algorithm of section 3 uses union of convex sets in step 2(b), equation (3). Union is needed also if tuples (q, X_1, τ_1), $(q, X_2, \tau_2) \in \mathcal{T}$ are replaced by a single tuple $(q, X_1 \cup X_2, \min\{\tau_1, \tau_2\})$, as described in one of the alternative heuristics. Here we describe the algorithm for over-approximating the union of two ellipsoids by an ellipsoid. Such an algorithm should be efficient, since it is likely to be the bottle neck of the basic verification algorithm. It should also exploit the fact that the reachable set changes only slightly in one time-propagation step.

Let us suppose that we want to approximate $\mathcal{E}(p, P) \cup \mathcal{E}(r, R)$, where P and R are non-degenerate matrices.

The algorithm builds an increasing sequence of ellipsoids:

$$\mathcal{E}(p, P) = \mathcal{E}(p_0, P_0), \mathcal{E}(p_1, P_1), ..., \mathcal{E}(p_k, P_k), \tag{7}$$

until $\mathcal{E}(p_k, P_k) \supseteq \mathcal{E}(r, R)$.

$\mathcal{E}(p_{i+1}, P_{i+1})$ is obtained from $\mathcal{E}(p_i, P_i)$ as follows. Given P_i and R, we compute matrices L_i, V_i, D_i and S_i. L_i is a lower triangular matrix that is the result of Cholesky decomposition of the matrix P_i^{-1}: $L_i L_i' = P_i^{-1}$. V_i is a matrix of eigenvectors and D_i is a diagonal matrix of eigenvalues of matrix $C_i = L_i^{-1} R_i^{-1} L_i'^{-1}$ such that $C_i = V_i D_i V_i'$. We denote

$$S_i = L_i V_i \tag{8}$$

and $y_i = S_i'(r - p_i)$.

Then we find a vector x_i^* that is the solution to the non-convex optimization problem

$$J_i(x) = \langle x - y_i, D_i(x - y_i) \rangle \rightarrow \max$$

---

[4] P' is the transpose of matrix P. Similarly, x' is the transpose of vector x.
[5]   The support function $\rho(l \mid X)$ of $X \in \mathcal{C}_b{}^n$ is defined as $\rho(l \mid X) = \max_{x \in X} \langle l, x \rangle$. Inversely, a support function uniquely defines a convex compact set.

with the constraint $\|x\| \leq 1$. In [12] it is shown how the problem can be solved and it is proved that the result is the global maximum.

We will denote $l_i^* = \frac{S_i^{-1'} x_i^*}{\|S_i^{-1'} x_i^*\|}$.

If $J_i(x_i^*) \geq 1$, it means that $\mathcal{E}(p_i, P_i) \supseteq \mathcal{E}(r, R)$ and the algorithm terminates. If $J_i(x_i^*) < 1$, we compute:

$$p_{i+1} = S_i'^{-1} \left(y_i + \frac{d_i}{2} x_i^* \right) + p_i, \tag{9}$$

and

$$P_{i+1} = S_i'^{-1} \left((1 + \alpha_i^{-1}) D_i^{-1} + (1 + \alpha_i) \frac{d_i^2}{4} x_i^* x_i^{*'} \right) S_i^{-1}, \tag{10}$$

where

$$d_i = 1 - \sqrt{\langle x_i^*, D_i^{-1} x_i^* \rangle} - \langle x_i^*, y_i \rangle + \epsilon, \tag{11}$$

$$\alpha_i = \frac{2\sqrt{\langle x_i^*, D_i^{-1} x_i^* \rangle}}{d_i \langle x_i^*, x_i^* \rangle^2}, \tag{12}$$

and ϵ is any non-negative number.

Lemma 3. *(See [5]) Let $\mathcal{E}_1 = \mathcal{E}(q_1, Q_1)$ and $\mathcal{E}_2 = \mathcal{E}(q_2, Q_2)$.*

1. *The ellipsoid $\mathcal{E} = \mathcal{E}(q_1 + q_2, Q(\beta))$, where $\beta > 0$ and*

$$Q(\beta) = (1 + \beta^{-1})Q_1 + (1 + \beta)Q_2,$$

   is properly defined and is an external approximation of the geometrical sum $\mathcal{E}_1 + \mathcal{E}_2$, i.e.

$$\mathcal{E}_1 + \mathcal{E}_2 \subseteq \mathcal{E}(q_1 + q_2, Q(\beta))$$

   for any $\beta > 0$.
2. *With vector $l \in \mathbb{R}^n, \|l\| = 1$, given, the equality*

$$p = \sqrt{\frac{\langle Q_1 l, l \rangle}{\langle Q_2 l, l \rangle}}$$

   defines a scalar parameter p, such that

$$\rho(l|\, \mathcal{E}(q_1 + q_2, Q(\beta))) = \rho(l|\, \mathcal{E}(q_1, Q_1) + \mathcal{E}(q_2, Q_2)).$$

   (The approximation $\mathcal{E}(q_1 + q_2, Q(\beta))$ touches the exact sum in direction l.)

Lemma 4. *Given any $\epsilon > 0$, for the ellipsoid $\mathcal{E}(p_{i+1}, P_{i+1})$ in sequence (7) the following holds:*

1.

$$\mathcal{E}(p_i, P_i) \subset \mathcal{E}(p_{i+1}, P_{i+1}); \tag{13}$$

2.

$$\rho(l_i^* | \mathcal{E}(p_{i+1}, P_{i+1})) = \rho(l_i^* | \mathcal{E}(r, R)) + \epsilon. \qquad (14)$$

Proof The linear transformation

$$x' = S_i'(x - p_i),$$

transforms the ellipsoid $\mathcal{E}(p_i, P_i)$ into the unit ball $\mathcal{E}(0, I_n) = k\{x| \langle x, x \rangle \le 1\}$ [6], and transforms the ellipsoid $\mathcal{E}(r, R)$ into the ellipsoid $\mathcal{E}(y_i, D_i)$. The next ellipsoid in the sequence $\mathcal{E}(p_{i+1}, P_{i+1})$ as specified by (9) and (10) is the result of the reverse transformation applied to the ellipsoid

$$\mathcal{E}' = \mathcal{E}\left(y_i + \frac{d_i}{2}x_i^*, (1 + \alpha_i^{-1})D_i^{-1} + (1 + \alpha_i)\frac{d_i^2}{4}x_i^* x_i^{*\prime}\right)$$

which, according to lemma (3), is an external approximation of the sum

$$\mathcal{E}(y_i, D_i) + \mathcal{E}\left(\frac{d_i}{2}x_i^*, \frac{d_i^2}{4}x_i^* x_i^{*\prime}\right)$$

It is not difficult to see that $0 \in \mathcal{E}\left(\frac{d_i}{2}x_i^*, \frac{d_i^2}{4}x_i^* x_i^{*\prime}\right)$, which ensures that $\mathcal{E}' \supseteq \mathcal{E}(y_i, D_i)$, and $x_i^* \in \mathcal{E}\left(\frac{d_i}{2}x_i^*, \frac{d_i^2}{4}x_i^* x_i^{*\prime}\right)$. Also, the choice of parameters α_i and d_i ensures that

$$\rho(x_i^* | \mathcal{E}(0, I_n)) + \epsilon = \rho\left(x_i^* \left| \mathcal{E}(y_i, D_i) + \mathcal{E}\left(\frac{d_i}{2}x_i^*, \frac{d_i^2}{4}x_i^* x_i^{*\prime}\right)\right.\right) = \rho(x_i^* | \mathcal{E}').$$

The later implies (14). ∎

Theorem 4. *For any $\epsilon > 0$ the algorithm always terminates in a finite number of steps.*

Proof Because the support functions $\rho(l| \mathcal{E}(p_{i+1}, P_{i+1}))$ and $\rho(l| \mathcal{E}(r, R))$ are continuous, for any $\epsilon > 0$ there is $\delta > 0$ such that for any $l : \|l - l_i^*\| \le \delta$ the following inequality holds

$$\rho(l| \mathcal{E}(p_{i+1}, P_{i+1})) > \rho(l| \mathcal{E}(r, R)),$$

which means that no l_j can belong to the set $\{l| \|l - l_i^*\| \le \delta\}$. If at some step l_j belongs to this set, $J_j(x_j^*) > 1$ and the algorithm terminates.

There can be only a finite number of $l_i : \|l_i\| = 1$ such that for any i, j: $\|l_i - l_j\| > \delta$. Therefore, the algorithm always terminates in a finite number of steps. ∎

In practice, if the ellipsoid $\mathcal{E}(p, P)$ has to be extended just "slightly" in order to contain $\mathcal{E}(r, R)$, which is the case when approximating the union of reachable sets at successive time steps, it is likely that the union algorithm terminates after a single step, i.e., $\mathcal{E}(p_1, P_1) \supseteq \mathcal{E}(r, R)$.

----

[6] Here $I_n \in \mathbb{R}^{n \times n}$ denotes the identity matrix

Intersections of Ellipsoids and Polyhedra Guards of discrete transitions are usually given in terms of conjunctions of linear inequalities, which define a polyhedron. Here we discuss a method for approximating intersections of ellipsoids and polyhedra.

In order to approximate the intersection of an ellipsoid $\mathcal{E}$ and a polyhedron H, we first compute ellipsoidal over-approximations of the intersection of $\mathcal{E}$ with each of the facets of H. Then, we compute the intersection of the resulting ellipsoids. Since each facet of H is a half-space, we now show how to approximate the intersection of an ellipsoid and a half-space.

Theorem 5. *Suppose that the ellipsoid*

$$\mathcal{E}(q, Q) = \{x|\ \langle x - q,\ Q^{-1}(x - q)\rangle \leq 1\},$$

where $Q = Q^T > 0$, and the half-space

$$S = \{x|\ \langle b,\ x\rangle \geq \alpha\}$$

have a non-empty intersection.

1. *Then for any $p \in [0, \frac{\alpha'+1}{2})$ the ellipsoid*

$$\mathcal{E}(q_+(p),\ Q_+(p)) = \{x|\ \langle x - q_+(p),\ Q_+^{-1}(p)(x - q_+(p))\rangle \leq 1\}$$

is an external approximation of the intersection $\mathcal{E}(q, Q) \cap S$, where

$$q_+(p) = q + pP^{-1}e_1,\ Q_+^{-1}(p) = PCP^T,$$

and

$$P = VD^{1/2}B,\ e_1 = (1, 0, ..., 0)^T,$$

$$C = \begin{pmatrix} \beta_1 & 0 & \cdots & 0 \\ 0 & \beta & \cdots & 0 \\ \vdots & \vdots & \ddots & \vdots \\ 0 & 0 & \cdots & \beta \end{pmatrix},$$

$$\beta_1 = \frac{1}{(p-1)^2},\ \beta = \frac{\alpha'+1-2p}{(\alpha'+1)(p-1)^2},\ \alpha' = \frac{\alpha - \langle b, q\rangle}{\sqrt{\langle b, b\rangle}}, \tag{15}$$

$$D = \begin{pmatrix} \lambda_1 & 0 & \cdots & 0 \\ 0 & \lambda_2 & \cdots & 0 \\ \vdots & \vdots & \ddots & \vdots \\ 0 & 0 & \cdots & \lambda_n \end{pmatrix},\ V = (v_1\ v_2\ ...\ v_n),$$

where $\lambda_1, ..., \lambda_n$ are eigenvalues of matrix Q^{-1} and $v_1, ..., v_n$ are corresponding eigenvectors that are linearly independent and $\|v_i\| = 1$, $i = \overline{1, n}$,

$$B = (b'\ b_2\ b_3\ ...\ b_n),\ b' = \frac{1}{\sqrt{\langle b, b\rangle}}D^{-1/2}V^Tb,$$

$\{b_2, b_3, ..., b_n\}$ is an orthonormal basis of subspace $\{x|\ \langle x, b'\rangle = 0\}$.

2. *The ellipsoid* $\mathcal{E}(q_+(p), Q_+(p))$ *touches* $\mathcal{E}(q, Q)$ *at point*

$$x^* = B^T D^{1/2} V^T(e_1 - q). \tag{16}$$

3.

$$\bigcap_{p \in [0, \frac{\alpha'+1}{2})} \mathcal{E}(q_+(p), Q_+(p)) = \mathcal{E}(q, Q) \cap S. \tag{17}$$

Proof The fact that matrix Q (and Q^{-1} as well) is self-adjoint and positive definite gives us the following equalities:

$$V^T V = V V^T = I, \; B^T B = B B^T = I.$$

Also note that

$$Q^{-1} = V D V^T.$$

Let us apply a linear transformation:

$$x = V D^{-1/2} B x' + q, \tag{18}$$

then the following holds:

$$\langle x - q, Q^{-1}(x - q) \rangle = \\ \langle V D^{-1/2} B x', Q^{-1} V D^{-1/2} B x' \rangle = \\ \langle x', B^T D^{-1/2} V^T Q^{-1} V D^{-1/2} B x' \rangle = \langle x', x' \rangle.$$

Thus, transformation (18) converts the ellipsoid $\mathcal{E}(q, Q)$ to the unit ball

$$B_0 = \{x | \langle x, x \rangle = 1\}.$$

At the same time transformation (18) converts the half-space S to the half-space

$$S' = \{x' | \langle x', B^T D^{-1/2} V^T b \rangle \le \alpha - \langle b, q \rangle\} = \{x' | \langle x', \frac{1}{\sqrt{\langle b, b \rangle}} B^T D^{-1/2} V^T b \rangle \le \alpha'\}.$$

Due to the selection of the matrix B:

$$\frac{1}{\sqrt{\langle b, b \rangle}} B^T D^{-1/2} V^T b = e_1,$$

thus

$$S' = \{x' | \langle x', e_1 \rangle \ge \alpha'\}.$$

Now we take an ellipsoid that is defined by the matrix

$$C^{-1}(p) = \begin{pmatrix} \beta_1 & 0 & \cdots & 0 \\ 0 & \beta & \cdots & 0 \\ \vdots & \vdots & \ddots & \vdots \\ 0 & 0 & \cdots & \beta \end{pmatrix},$$

where β_1 and β are determined by (15) and center

$$c(p) = (p, 0, 0, ..., 0).$$

If $p \in [0, \frac{\alpha'+1}{2})$ the ellipsoid is defined and it is not difficult to see that

$$\mathcal{E}(c(p), C(p)) \supset B_0 \cap S'.$$

Moreover,

$$\mathcal{E}(c(p), C(p)) \cap S' = B_0 \cap S'$$

and the ellipsoid $\mathcal{E}(c(p), C(p))$ touches B_0 at point $e_1 = (1, 0, 0, ..., 0)$.
It is not difficult to see also that

$$\mathcal{E}(c(p), C(p)) \subset \{(x_1, 0, 0, ..., 0) | x_1 \geq ?p - 1\}.$$

If $p \to \frac{\alpha'+1}{2}$ then $2p - 1 \to \alpha'$. Also notice that if $p = 0$, $\mathcal{E}(c(p), C(p)) = B_0$. Therefore

$$\bigcap_{p \in [0, \frac{\alpha'+1}{2})} \mathcal{E}(c(p), C(p)) = B_0 \cap S'.$$

If we apply the reverse transformation

$$x' = B^T D^{1/2} V^T (x - q),$$

to the ellipsoid $\mathcal{E}(c(p), C(p))$, we will get the ellipsoid $\mathcal{E}(q_+(p), Q_+(p))$ as defined above. Properties (16) and (17) will be satisfied and point e_1 at which $\mathcal{E}(c(p), C(p))$ touches B_0 will be converted to point x^* as defined by (16) and ellipsoids will touch each other at the point x^*. ∎

Intersection Check The problem of checking whether two non-degenerate ellipsoids $E(p_1, P_1)$ and $E(p_2, P_2)$ intersect is equivalent to a convex quadratic optimization problem:

$$J(x) = \langle x - p_1, P_1^{-1}(x - p_1) \rangle \to \min$$

with constraint

$$\langle x - p_2, P_2^{-1}(x - p_2) \rangle \leq 1.$$

If x^* is the solution to the problem and $J(x^*) > 1$, then the ellipsoids do not intersect. Otherwise, they do intersect.

In order to check whether an ellipsoid and a polyhedron intersect, it is possible to check whether the ellipsoid intersects with all half-spaces that form the faces of the polyhedron. If it does not intersect at least with one of them, the ellipsoid does not intersect the polyhedron. Of course, this method is quite coarse but it is simple and effective. If the intersection is actually empty and the above check did not find that out, that still can be discovered during computation of the intersection.

5   Implementaion of the Tool

The techniques are implemented in the verification tool called *VeriSHIFT* [7]. The tool is a C++ library that consists of all necessary numerical algorithms: ellipsoidal and polyhedral [8] representation of convex sets and operations on them, reachability algorithms, verification algorithms described in this paper, etc. The user of the tool writes C++ code in order to describe a model: for each class of hybrid automaton the user writes a definition of a C++ class derived from the special class *HybridObject* provided by the library. The model can be defined in terms of high level notions such as discrete states, transitions, input/output continuous variables, events, bound convex sets, as described in [1]. Each of above notions is defined as a class in the library. Actions taken upon discrete transitions can be described as C++ functions.

The library provides the notion of discrete configuration also implemented as a class. A discrete configuration contains a set of objects with their discrete states, dataflow and configuration connections and is accompanied by a bound convex set containing possible valuations of the continuous variables.

Together with discrete states, continuous variables and events classes describing hybrid automata can contain variables of any possible C++ type including other hybrid automata classes. There is only one requirement: classes have to able to create their copies in other discrete configurations. Objects within the same discrete configuraion can use any mechanism provided by C++ for communicating with each other as long as that does not interfere with the mechanism provided by the *VeriSHIFT* library. In a function called upon a discrete transition of an object the object can modify its private data, can create/destroy other objects or can call methods of other objects.

Execution of a typical program using *VeriSHIFT* starts with creating an initial discrete configuration: creating an empty configuration, creating new objects within the configuration, setting up connections between objects. Once the initial configuration is set up, verification is started by calling a special library function.

In order to verify the properties of the model, the user can observe what discrete states the objects enter or can assign special actions to transitions they are interested in. Also it is possible to examine reachable sets of continuous variables at any phase of the execution.

Acknowledgements

We would like to thank Alexander B. Kurzhanski and Pravin Varaiya for making this work possible.

---

[7]   The source code of the tool along with the documentation and examples can be found at http://robotics.EECS.Berkeley.EDU/~olegb/VeriSHIFT/

[8]   The Polyhedral Library 2.0 by Doran Wilde and Herve Le Verge has been used.

References

1. Botchkarev O., Ellipsoidal Techniques for Verification of Hybrid Systems, 2000. Available at: http://robotics.eecs.berkeley.edu/~olegb/VeriSHIFT/.
2. Dang T. and Maler O., Reachability Analysis via Face Lifting. In T.A. Henzinger and S. Sastry (Eds), Hybrid Systems: Computation and Control , 96-109, LNCS 1386, Springer, 1998.
3. Henzinger T, Ho P. and Wong-Toi H., HyTech: A Model Checker for Hybrid Systems. In *Software Tools for Technology Transfer*, 1, 1997.
4. Kurzhanski A. B. and Filippova T. F., On the Theory of Trajectory Tubes: a Mathematical Formalism for Uncertain Dynamics, Viability and Control, in: *Advances in Nonlinear Dynamics and Control*, ser. PSCT 17, pp.122 - 188, Birkhäuser, Boston, 1993.
5. Kurzhanski A. B. and Vályi I. *Ellipsoidal Calculus for Estimation and Control*, Birkhäuser, Boston, ser.SCFA, 1996.
6. Kurzhanski A. B. and Varaiya P., Ellipsoidal Techniques for Reachability Analysis, 2000. *Proceedings of this conference*.
7. Puri A., Borkar V. and Varaiya P., ϵ-Approximations of Differential Inclusions, in: R.Alur, T.A.Henzinger, and E.D.Sonntag eds., *Hybrid Systems*, pp. 109 – 123, LNCS 1201, Springer, 1996.
8. Puri A. and Varaiya P., Decidability of Hybrid Systems with Rectangular Differential Inclusions, in D.Dill ed., *Proc. CAV'94*, LNCS 1066, Springer, 1966.
9. Rockafellar, R. T., *Convex Analysis*, Princeton University Press, 1970.
10. Varaiya P., Reach Set Computation Using Optimal Control, in *Proc.of KIT Workshop on Verification of Hybrid Systems*, Verimag, Grenoble, 1998.
11. *VeriSHIFT*   Home   Page.   http://robotics.EECS.Berkeley.EDU/~olegb/VeriSHIFT/.
12. Ye Y., On Affine Scaling Algorithms for Non-Convex Quadratic Programming. In Mathematical Programming, 56(1992), pp. 285 – 300.

Theory of Optimal Control Using Bisimulations

Mireille Broucke[1], Maria Domenica Di Benedetto[1,2], Stefano Di Gennaro[2], and
Alberto Sangiovanni-Vincentelli[1]

[1] Dept. of Electrical Engineering and Computer Sciences
University of California at Berkeley, CA 94720, USA
Tel: +11 510 642-1792; Fax: +11 510 643-5052;
[2] Dip. di Ingegneria Elettrica, Università di L'Aquila,
Poggio di Roio, 67040 L'Aquila, Italy
{mire,marika,alberto}@eecs.berkeley.edu, digennar@dis.uniroma1.it

Abstract. We consider the synthesis of optimal controls for continuous
feedback systems by recasting the problem to a hybrid optimal control problem: to synthesize optimal enabling conditions for switching between locations in which the control is constant. An algorithmic solution
is obtained by translating the hybrid automaton to a finite automaton
using a bisimulation and formulating a dynamic programming problem
with extra conditions to ensure non-Zenoness of trajectories. We show
that the discrete value function converges to the viscosity solution of the
Hamilton-Jacobi-Bellman equation as a discretization parameter tends
to zero.

1 Introduction

The goal of this paper is the development of a computationally appealing tech-
nique for synthesizing optimal controls for continuous feedback systems $\dot{x} = f(x, u)$, by reducing substantially the complexity of the problem. This goal is
achieved by virtue of recasting the problem to a hybrid optimal control problem.
The hybrid problem is obtained by approximating the control set $U \subset \mathbb{R}^m$ by a
finite set $\Sigma \subset U$ and defining vector fields for the locations of the hybrid system
of the form $f(x, \sigma)$, $\sigma \in \Sigma$; that is, the control is constant in each location. The
hybrid control problem is, then, to synthesize an optimal switching rule between
locations, or equivalently, optimal enabling conditions, such that a target set
$\Omega_f \subset \Omega$ is reached while a hybrid cost function is minimized, for each initial
condition in a specified set $\Omega \subset \mathbb{R}^n$.

Casting the problem into the domain of hybrid control is not appealing per
se, on the contrary! Algorithmic approaches for solving the controller synthesis
problem for specific classes of hybrid systems have appeared [8,12] but no general, efficient algorithm is yet available. Hence, to be able to solve the (nonlinear)
hybrid optimal control problem, we must exploit some additional property. We
have a feasible and quite appealing approach if we can translate the problem
to an equivalent discrete problem, which abstracts completely the continuous
behavior. This translation is possible if we can construct a finite *bisimulation*

N. Lynch and B. Krogh (Eds.): HSCC 2000, LNCS 1790, pp. 89–102, 2000.
© Springer-Verlag Berlin Heidelberg 2000

defined on the hybrid state set. The bisimulation can be constructed using the geometric approach reported in [4], based on the following key assumption: $n-1$ *local (on Ω) first integrals can be expressed analytically for each vector field* $f(x, \sigma)$, $\sigma \in \Sigma$. This assumption is imposed in the transient phase of a feedback system's response, when the vector field is non-vanishing and local first integrals always exist, though analytical expressions for them may not be readily computable.

If the assumption is met, then we can transform the hybrid system to a finite automaton. The control problem posed on the finite automaton is to synthesize a discrete supervisor, providing a switching rule between automaton locations, that minimizes a discrete cost function approximating the original cost function, for each initial discrete state. We provide a dynamic programming solution to this problem, with extra constraints to ensure non-Zenoness of the closed-loop trajectories. By imposing non-Zeno conditions on the synthesis we obtain piecewise constant controls. The discrete value function depends on the discretizations of U and of Ω using the bisimulation. We quantify these discretizations by parameters δ and δ_Q, respectively. The main theoretical contribution is to show that as $\delta, \delta_Q \to 0$, the discrete value function converges to the unique viscosity solution of the Hamilton-Jacobi-Bellman (HJB) Equation.

There is a similarity between our approach to optimal control and *regular synthesis*, introduced in [2], in the sense that both restrict the class of controls to a set that has some desired property and both use a finite partition to define switching behavior. Our work provides a constructive approach to obtain the cell decomposition by using a finite bisimulation, which further allows us to formulate the synthesis problem on its quotient system - a finite automaton. The idea of using a time abstract model formed by partitioning the continuous state space has been pursued in a number of papers recently. Lemmon, Antsaklis, Stiver and coworkers [10] use a partition of the state space to convert a hybrid model to a discrete event system (DES). This enables them to apply controller synthesis for DES's to synthesize a supervisor. While our approach is related to this methodology, it differs in that we have explicit conditions for obtaining the partition. In [9] hybrid systems consisting of a linear time-invariant system and a discrete controller that has access to a quantized version of the linear system's output is considered. This approach suffers from spurious solutions that must be trimmed from the automaton behavior. Hybrid optimal control problems have been studied in papers by Witsenhausen [11] and Branicky, Borkar, Mitter [3]. These studies concentrate on problems of well-posedness, necessary conditions, and existence of optimal solutions but do not provide algorithmic solutions.

2   Optimal Control Problem

Notation. $1(\cdot)$ is the indicator function. $cl(A)$ denotes the closure of set A. $\| \cdot \|$ denotes the Euclidean norm. Let $C^1(\mathbb{R}^n)$ and $\mathcal{X}(\mathbb{R}^n)$ denote the sets of continuously differentiable real-valued functions and smooth vector fields on $\mathbb{R}^n$, respectively. $\phi_t(x_0, \mu)$ denotes the trajectory of $\dot{x} = f(x, \mu)$ starting from x_0 and using control $\mu(\cdot)$.

Let U be a compact subset of $\mathbb{R}^m$, Ω an open, bounded, connected subset of $\mathbb{R}^n$, and Ω_f a compact subset of Ω. Define $\mathcal{U}_m$ to be the set of meansurable functions mapping $[0, T]$ to U. We define the minimum hitting time $T : \mathbb{R}^n \times \mathcal{U}_m \to \mathbb{R}^+$ by

$$T(x, \mu) := \begin{cases} \infty & \text{if } \{t \mid \phi_t(x, \mu) \in \Omega_f \} = \emptyset \\ \min\{t \mid \phi_t(x, \mu) \in \Omega_f\} & \text{otherwise.} \end{cases} \tag{1}$$

A control $\mu \in \mathcal{U}_m$ specified on $[0, T]$ is *admissible* for $x \in \Omega$ if $\phi_t(x, \mu) \in \Omega$ for all $t \in [0, T]$. The set of admissible controls for x is denoted $\mathcal{U}_x$. Let

$$\mathcal{R} := \{ x \in \mathbb{R}^n \mid \exists \mu \in \mathcal{U}_x. \ T(x, \mu) < \infty \}.$$

We consider the following optimal control problem. Given $y \in \Omega$,

$$\text{minimize} \qquad J(y, \mu) = \int_0^{T(y,\mu)} L(x(t), \mu(t))dt + h(x(T(y, \mu))) \tag{2}$$

$$\text{subject to} \qquad \dot{x} = f(x, \mu), \qquad a.e. \ t \in [0, T(y, \mu)] \tag{3}$$

$$x(0) = y \tag{4}$$

among all admissible controls $\mu \in \mathcal{U}_y$. $J : \mathbb{R}^n \times \mathcal{U}_m \to \mathbb{R}$ is the *cost-to-go* function, $h : \mathbb{R}^n \to \mathbb{R}$ is the *terminal cost*, and $L : \mathbb{R}^n \times \mathbb{R}^m \to \mathbb{R}$ is the *instantaneous cost*. At $T(y, \mu)$ the terminal cost $h(x(T(y, \mu)))$ is incurred and the dynamics are stopped. The control objective is to reach Ω_f from $y \in \Omega$ with minimum cost.

Assumption 2.1.

(1) $f : \mathbb{R}^n \times \mathbb{R}^m \to \mathbb{R}^n$ satisfies $\|f(x', u') - f(x, u)\| \leq L_f[\|x' - x\| + \|u' - u\|]$ for some $L_f > 0$. Let M_f be the upper bound of $\|f(x, u)\|$ on $\Omega \times U$.

(2) $L : \mathbb{R}^n \times \mathbb{R}^m \to \mathbb{R}$ satisfies $|L(x', u') - L(x, u)| \leq L_L[\|x' - x\| + \|u' - u\|]$ and $1 \leq |L(x, u)| \leq M_L$, $x \in \Omega$, $u \in U$, for some $L_L, M_L > 0$.

(3) $h : \mathbb{R}^n \to \mathbb{R}$ satisfies $|h(x') - h(x)| \leq L_h \|x' - x\|$ for some $L_h > 0$, and $h(x) \geq 0$ for all $x \in \Omega$. Let M_h be the upper bound of $|h(x)|$ on Ω.

The *value function* or optimal cost-to-go function $V : \mathbb{R}^n \to \mathbb{R}$ is given by

$$V(y) = \inf_{\mu \in \mathcal{U}_y} J(y, \mu)$$

for $y \in \Omega \setminus \Omega_f$, and by $V(y) = h(y)$ for $y \in \Omega_f$. A control μ is called ϵ-*optimal* for x if $J(x, \mu) \leq V(x) + \epsilon$. It is well-known [7] that V satisfies the *Hamilton-Jacobi-Bellman* (HJB) equation

$$-\inf_{u \in U} \left\{ L(x, u) + \frac{\partial V}{\partial x} f(x, u) \right\} = 0 \tag{5}$$

at each point of $\mathcal{R}$ at which it is differentiable. The HJB equation is an infinitesimal version of the equivalent *Dynamic Programming Principle* (DPP) which says that

$$V(x) = \inf_{\mu \in \mathcal{U}_x} \left\{ \int_0^t L(\phi_s(x, \mu), \mu(s))ds + V(\phi_t(x, \mu)) \right\}, x \in \Omega \setminus \Omega_f$$

$$V(x) = h(x) \qquad\qquad\qquad\qquad\qquad x \in \Omega_f.$$

The subject of assiduous effort has been that the HJB equation may not have a C^1 solution. This gap in the theory was closed by the inception of the concept of viscosity solution [6], which can be shown to provide the unique solution of (5) without any differentiability assumption. In particular, a bounded uniformly continuous function V is called a *viscosity solution* of HJB provided, for each $\psi \in C^1(\mathbb{R}^n)$, the following hold:

(i) if $V - \psi$ attains a local maximum at $x_0 \in \mathbb{R}^n$, then

$$- \inf_{u \in U} \left\{ L(x_0, u) + \frac{\partial \psi}{\partial x}(x_0) f(x_0, u) \right\} \leq 0,$$

(ii) if $V - \psi$ attains a local minimum at $x_1 \in \mathbb{R}^n$, then

$$- \inf_{u \in U} \left\{ L(x_1, u) + \frac{\partial \psi}{\partial x}(x_1) f(x_1, u) \right\} \geq 0.$$

Assumption 2.2. For every $\epsilon > 0$ and $x \in \mathcal{R}$, there exists $N_\epsilon > 0$ and an admissible piecewise constant ϵ-optimal control μ having at most N_ϵ discontinuities and such that $\phi_t(x, \mu)$ is transverse to $\partial \Omega_f$.

The transversality assumption implies that the viscosity solution is continuous at the boundary of the target set, a result needed in proving uniform continuity of V. The finite switching assumption holds under mild assumptions such as Lipschitz continuity of the vector field and cost functions, and is based on approximating measurable functions by piecewise constant functions.

3 Hybrid System

The approach we propose for solving the continuous optimal control problem first requires a mapping to a hybrid system and, second, employs a bisimulation of the hybrid system to formulate a dynamic programming problem on the quotient system. In this section we define the hybrid optimal control problem. First, we discretize U by defining a finite set $\Sigma_\delta \subset U$ which has a mesh size

$$\delta := \sup_{u \in U} \min_{\sigma \in \Sigma_\delta} \|u - \sigma\|.$$

We define the hybrid automaton $H := (\Sigma \times \mathbb{R}^n, \Sigma, D, E_h, G, R)$ with the following components.

State set $\Sigma \times \mathbb{R}^n$ consists of the finite set $\Sigma = \Sigma_\delta \cup \{\sigma_f\}$ of control locations and n continuous variables $x \in \mathbb{R}^n$. σ_f is a terminal location when the continuous dynamics are stopped (in the same sense that the dynamics are "stopped" in the continuous optimal control problem).

Events $\Sigma = \Sigma_\delta \cup \{\sigma_f\}$ is a finite set of control event labels.

Vector fields $D : \Sigma \to \mathcal{X}(\mathbb{R}^n)$ is a function assigning an autonomous vector field to each location. We use the notation $D(\sigma) = f_\sigma$.

Control switches $E_h \subset \Sigma \times \Sigma$ is a set of control switches. $e = (\sigma, \sigma')$ is a directed edge between a source location σ and a target location σ'. If $E_h(\sigma)$ denotes the set of edges that can be enabled at $\sigma \in \Sigma$, then $E_h(\sigma) := \{(\sigma, \sigma') \mid \sigma' \in \Sigma \setminus \sigma\}$ for $\sigma \in \Sigma_\delta$ and $E_h(\sigma_f) = \emptyset$. Thus, from a source location not equal to σ_f, there is an edge to every other location (but not itself), while location σ_f has no outgoing edges.

Enabling conditions $G : E_h \to \{g_e\}_{e \in E_h}$ is a function assigning to each edge an enabling (or guard) condition $g \subset \mathbb{R}^n$. We use the notation $G(e) = g_e$.

Reset conditions $R : E_h \to \{r_e\}_{e \in E_h}$ is a function assigning to each edge a reset condition, $r_e : \mathbb{R}^n \to 2^{\mathbb{R}^n}$, where we use the notation $R(e) = r_e$.

Semantics. A state is a pair (σ, x), $\sigma \in \Sigma$ and $x \in \mathbb{R}^n$. In location $\sigma \in \Sigma_\delta$ the continuous state evolves according to the vector field $f(x, \sigma)$. In location σ_f, the vector field is $\dot{x} = f(x, \mu_f)$ where μ_f is the (not necessarily constant) control of the terminal location. Trajectories of H evolve in *steps* of two types. A σ-*step* is a binary relation $\overset{\sigma}{\to} \subset (\Sigma \times \mathbb{R}^n) \times (\Sigma \times \mathbb{R}^n)$, and we write $(\sigma, x) \overset{\sigma'}{\to} (\sigma', x')$ iff (1) $e = (\sigma, \sigma') \in E_h$, (2) $x \in g_e$, and (3) $x' = r_e(x)$. The transition $(\sigma, x) \overset{\sigma'}{\to} (\sigma', x')$ is taken at the first time in location σ when the control event label is σ' and $x \in g_e$ for $e = (\sigma, \sigma')$. A t-*step* is a binary relation $\overset{t}{\to} \subset (\Sigma \times \mathbb{R}^n) \times (\Sigma \times \mathbb{R}^n)$, and we write $(\sigma, x) \overset{t}{\to} (\sigma', x')$ iff (1) $\sigma = \sigma'$, (2) at $t = 0, x' = x$, and (3) for $t \geq 0$, $x' = \phi_t(x, \sigma)$, where $\phi_t(x) = f(\phi_t(x, \sigma), \sigma)$. A *hybrid control* is a finite or infinite sequence of labels $\omega = \omega_1 \omega_2 \ldots$, with $\omega_i \in \Sigma \cup \mathbb{R}^+$. $\omega_i \in \mathbb{R}^+$ is the duration of the t-step at step i. The set of hybrid controls is denoted $\mathcal{S}$. A *hybrid trajectory* π over $\omega \in \mathcal{S}$ is a finite or infinite sequence $\pi : (\sigma_0, x_0) \overset{\omega_1}{\to} (\sigma_1, x_1) \overset{\omega_2}{\to} (\sigma_2, x_2) \overset{\omega_3}{\to} \ldots$ where $(\sigma_i, x_i) \in \Sigma \times \mathbb{R}^n$. Trajectory π is *accepted* by H iff $\forall i$, $(\sigma_i, x_i) \overset{\omega_{i+1}}{\to} (\sigma_{i+1}, x_{i+1})$ is either a t-step or σ-step of H. Let π be the trajectory (not necessarily accepted by H) starting at $(\sigma, x) \in \Sigma \times \Omega$ and defined over $\omega \in \mathcal{S}$. We say ω is *admissible* for (σ, x) on interval $[0, T]$ if (1) π remains in $\Sigma \times \Omega$ for $t \in [0, T]$, and (2) corresponding to ω is a piecewise constant control $\mu_\omega(t)$ (with a finite number of discontinuities in finite time). Let $\mathcal{S}_{(\sigma, x)}$ be the set of admissible controls for (σ, x).

3.1   Hybrid Optimal Synthesis

We want to synthesize enabling conditions so that for each $y \in \mathcal{R}$, the cost-to-go from y well-approximates the viscosity solution at y of HJB. This requires posing a hybrid optimal synthesis problem. We define a *hybrid cost-to-go function* $J_H : \Sigma \times \mathbb{R}^n \times \mathcal{S} \to \mathbb{R}$ as follows. For $\omega \in \mathcal{S}_{(\sigma, x)}$,

$$J_H((\sigma, x), \omega) = J(x, \mu_\omega).$$

The *hybrid value function* $V_H : \Sigma \times \mathbb{R}^n \to \mathbb{R}$ is

$$V_H((\sigma, x)) = \inf_{\omega \in \mathcal{S}_{(\sigma, x)}} J_H((\sigma, x), \omega).$$

Hybrid optimal synthesis problem:
Given H and $0 < \epsilon^1 < \epsilon^2$, synthesize g_e, $e \in E_h$, subject to:

1. $g_e = \Omega_f$ if $e = (\sigma, \sigma_f)$, $\sigma \in \Sigma_\delta$.
2. For each $e \in E_h$, $g_e \subseteq \Omega$.
3. For all $\omega \in S$ and $(\sigma, x) \in \Sigma \times \Omega$ such that $V_H((\sigma, x)) < \infty$, $\pi_{(\sigma,x)}$ is accepted by H if ω is admissible and ϵ^1-optimal for (σ, x).
4. For all $\omega \in S$ and $(\sigma, x) \in \Sigma \times \Omega$, $\pi_{(\sigma,x)}$ is not accepted by H if either ω is not admissible for (σ, x), ω is not ϵ^2-optimal for (σ, x), or $V_H((\sigma, x)) = \infty$.

4   Construction of Bisimulation

We propose to solve the hybrid optimal control problem using the bisimulation of H. In this section we define bisimulation and the quotient system that is obtained from it.

Let λ represent a t-step corresponding to some $t \in \mathbb{R}^+$. A *bisimulation* of H is an equivalence relation $\simeq \subset (\Sigma_\delta \times \mathbb{R}^n) \times (\Sigma_\delta \times \mathbb{R}^n)$ such that for all states $p_1, p_2 \in \Sigma_\delta \times \mathbb{R}^n$, if $p_1 \simeq p_2$ and $\sigma \in \Sigma_\delta \cup \{\lambda\}$, then if $p_1 \xrightarrow{\sigma} p_1'$, there exists p_2' such that $p_2 \xrightarrow{\sigma} p_2'$ and $p_1' \simeq p_2'$. If $\simeq$ is finite, the quotient system is a finite automaton.

Since the dynamics are restricted to the set Ω, the set of interesting equivalence classes of $\simeq$, denoted Q, are those that intersect $\Sigma_\delta \times cl(\Omega)$. For each $q \in Q$ we define a distinguished point $(\sigma, \xi) \in q$. We associate q with its distinguished point by the notation $q = [(\sigma, \xi)]$. It is now possible to define the enabling and reset conditions of H in terms of Q. In particular, the enabling conditions of H are synthesized as subsets of Q while the reset conditions are defined as follows. For $e = (\sigma, \sigma')$

$$r_e(x) = \{ y \mid \exists \xi. [(\sigma, x)] = [(\sigma, \xi)] \wedge [(\sigma', \xi)] = [(\sigma', y)] \}. \tag{6}$$

That is, $r_e(x)$ is the projection to $\mathbb{R}^n$ of the set of equivalence classes $[(\sigma', y)]$ such that the projection to $\mathbb{R}^n$ of $[(\sigma', y)]$ and $[(\sigma, x)]$ have nonempty intersection. This definition in effect gives an over-approximation of the identity map in terms of the equivalence classes of $\simeq$ and will introduce non-determinacy in the finite automaton. Notice also that (6) encodes information about the bisimulation in H. This sequence of steps is not typical; it is characteristic of our synthesis procedure. We define a mesh size on Q by $\delta_Q = \max_{q \in Q} \sup_{(\sigma,x),(\sigma,y) \in q} \{\|x - y\|\}$. Finally, for each $q = [(\sigma, \xi)] \in Q$ we associate the duration τ_q, the maximum time to traverse q using constant control σ. That is, $\tau_q = \sup_{(\sigma,x),(\sigma,y) \in q} \{ t \mid y = \phi_t(x, \sigma) \}$.

Geometric construction. We give a brief review of the method developed in [4] for obtaining bisimulations. We require the following (related) assumptions on the vector fields on $cl(\Omega)$.

Assumption 4.1.

(1) $n - 1$ first integrals can be defined analytically on Ω for each $f(x, \sigma)$, $\sigma \in \Sigma_\delta$.
(2) There exists $m_f > 0$ such that $\|f(x, u)\| \geq m_f$ for all $x \in cl(\Omega)$, $u \in U$.

A bisimulation of $\Sigma_\delta \times \mathbb{R}^n$ is constructed using a set of simple, co-dimension one tangential foliations with associated submersions $\gamma_i^\sigma(x) = y_i^\sigma$, $i = 1, \ldots, n-1$ and a simple co-dimension one transversal foliation with submersion $\gamma_n^\sigma = y_n^\sigma$, such that $(y_1^\sigma, \ldots, y_n^\sigma)$ form a set of euclidean coordinates for each $\sigma \in \Sigma_\delta$. We *discretize* the foliations by selecting a finite set of leaves. Fix $k \in \mathbb{Z}^+$ and let $\Delta = \frac{1}{2^k}$. Define

$$C_k = \{0, \pm\Delta, \pm 2\Delta, \ldots, \pm 1\}. \tag{7}$$

Each $y_i^\sigma = c$ for $c \in C_k$, $i = 1, \ldots, n$ defines a hyperplane denoted $\tilde{W}_{i,c}^\sigma$, and a submanifold $W_{i,c}^\sigma = (\gamma^\sigma)^{-1}(\tilde{W}_{i,c}^\sigma)$. The collection of submanifolds for $\sigma \in \Sigma_\delta$ is

$$\mathcal{W}_k^\sigma = \{ W_{i,c}^\sigma \mid c \in C_k, i \in \{1, \ldots, n\} \}. \tag{8}$$

$\Omega \setminus \mathcal{W}_k^\sigma$ is the union of $2^{n(k+1)}$ disjoint open sets $\mathcal{V}_k^\sigma = \{V_j^\sigma\}$. We define the equivalence relation $\simeq$ on $\Sigma_\delta \times \mathbb{R}^n$ as follows: $(\sigma, x) \simeq (\sigma', x')$ iff (1) $\sigma = \sigma'$ and (2) $x \in W$ iff $x' \in W$, and $x \in V$ iff $x' \in V$, for all $W \in \mathcal{W}_k^\sigma$ and $V \in \mathcal{V}_k^\sigma$.

5   Discrete Problem

In this section we transform the hybrid optimal control problem to a dynamic programming problem on a non-deterministic finite automaton, for which an algorithmic solution may be found. Consider the class of non-deterministic automata with cost structure represented by the tuple $A = (Q, \Sigma_\delta, E, obs, Q_f, \hat{L}, \hat{h})$. Q is the state set, as above, and Σ_δ is the set of control labels as before. $obs : E \to \Sigma_\delta$ is a map that assigns a control label to each edge and is given by $obs(e) = \sigma'$, where $e = (q, q')$, $q = [(\sigma, \xi)]$ and $q' = [(\sigma', \xi')]$. Q_f is the target set given by the over-approximation of Ω_f, $Q_f = \{q \in Q \mid \exists x \in \Omega_f . (\sigma, x) \in q \}$.

$E \subseteq Q \times Q$ is the transition relation encoding t-steps and σ-steps of H. A will be used to synthesize g_e of H, so E includes all possible edges between locations. The synthesis procedure on A will involve trimming undesirable edges. Thus, $(q, q') \in E$, where $q, q' \in Q$, $q = [(\sigma, \xi)]$ and $q' = [(\sigma', \xi')]$ if either (a) $\sigma = \sigma'$, there exists $x \in \Omega$ such that $(\sigma, x) \in q$, and there exists $\tau > 0$ such that $\forall t \in [0, \tau]$, $(\sigma, \phi_t(x, \sigma)) \in q$ and $(\sigma, \phi_{\tau+\epsilon}(x, \sigma)) \in q'$ for arbitrarily small $\epsilon > 0$, or (b) $\sigma = \sigma'$, there exists $x \in \Omega$ such that $(\sigma, x) \in q$, and there exists $\tau > 0$ such that $\forall t \in [0, \tau)$, $(\sigma, \phi_t(x, \sigma)) \in q$ and $(\sigma, \phi_\tau(x, \sigma)) \in q'$, or (c) $\sigma \neq \sigma'$ and there exists $x \in \Omega$ such that $(\sigma, x) \in q$ and $(\sigma', x) \in q'$. Cases (a) and (b) say that from a point in q, q' is the first state (different from q) reached after following the flow of $f(x, \sigma)$ for some time. Case (c) says that an edge exists between q and q' if their projections to $\mathbb{R}^n$ have non-empty intersection.

Let $e = (q, q')$ with $q = [(\sigma, \xi)]$ and $q' = [(\sigma', \xi')]$. $\hat{L} : E \to \mathbb{R}$ is the *discrete instantaneous cost* given by

$$\hat{L}(e) := \begin{cases} \tau_q L(\xi, \sigma) & \text{if } \sigma = \sigma' \\ 0 & \text{if } \sigma \neq \sigma'. \end{cases} \tag{9}$$

This definition reflects that no cost is incurred for control switches. $\hat{h} : Q \to \mathbb{R}$ is the *discrete terminal cost* given by

$$\hat{h}(q) := h(\xi).$$

The domain of $\hat{h}$ can be extended to Ω, with a slight abuse of notation, by $\hat{h}(x) := \hat{h}(q)$ where $q = \arg\min_{q'}\{\|x - \xi'\| \mid q' = [(\sigma',\xi')]\}$.

5.1   Semantics

A transition or *step* of A from $q = [(\sigma,\xi)] \in Q$ to $q' = [(\sigma',\xi')] \in Q$ with observation $\sigma' \in \Sigma_\delta$ is denoted $q \xrightarrow{\sigma'} q'$. If $\sigma \neq \sigma'$ the transition is referred to as a *control switch*; otherwise, it is referred to as a *time step*. If $E(q)$ is the set of edges that can be enabled from $q \in Q$, then for $\sigma \in \Sigma_\delta$,

$$E_\sigma(q) = \{e \in E(q) \mid obs(e) = \sigma\}.$$

If $|E_\sigma(q)| > 1$, then we say that $e \in E_\sigma(q)$ is *unobservable* in the sense that when control event σ is issued, it is unknown which edge among $E_\sigma(q)$ is taken. If $\sigma = \sigma'$, then $|E_\sigma(q)| = 1$, by the uniqueness of solutions of ODE's and by the definition of bisimulation.

A *control policy* $c : Q \to \Sigma_\delta$ is a map assigning a control event to each state; $c(q) = \sigma$ is the control event issued when the state is at q. A *trajectory* π of A over c is a sequence $\pi = q_0 \xrightarrow{\sigma_1} q_1 \xrightarrow{\sigma_2} q_2 \xrightarrow{\sigma_3} \ldots, q_i \in Q$. A trajectory is *non-Zeno* if between any two non-zero duration time steps there are a finite number of control switches and zero duration time steps. Let $\Pi_c(q)$ be the set of trajectories starting at q and applying control policy c, and let $\tilde{\Pi}_c(q)$ be the set of trajectories starting at q, applying control policy c, and eventually reaching Q_f. If for every $q \in Q$, $\pi \in \Pi_c(q)$ is non-Zeno then we say c is an *admissible control policy*. The set of all admissible control policies for A is denoted $\mathcal{C}$.

A control policy c is said to have a *loop* if A has a trajectory $q_0 \xrightarrow{c(q_0)} q_1 \xrightarrow{c(q_1)} \ldots \xrightarrow{c(q_{m-1})} q_m = q_0$, $q_i \in Q$. A control policy has a *Zeno loop* if it has a loop made up of control switches and/or zero duration time steps only. One can show that a control policy is admissible iff it has no Zeno loops.

5.2   Dynamic Programming

In this section we formulate the dynamic programming problem on A. This involves defining a cost-to-go function and a value function that minimizes it over control policies suitable for non-deterministic automata.

Suppose $\pi = q_0 \xrightarrow{\sigma_1} q_1 \to \ldots \to q_{N-1} \xrightarrow{\sigma_N} q_N \in \Pi$, where $q_i = [(\sigma_i,\xi_i)]$ and π takes the sequence of edges $e_1 e_2 \ldots e_N$. We define a *discrete cost-to-go* $\hat{J} : Q \times \mathcal{C} \to \mathbb{R}$ by

$$\hat{J}(q,c) = \begin{cases} \max_{\pi \in \tilde{\Pi}_c(q)}\left\{\sum_{j=1}^{N_\pi} \hat{L}(e_j) + \hat{h}(q_{N_\pi})\right\} & \text{if } \Pi_c(q) = \tilde{\Pi}_c(q) \\ \infty & \text{otherwise} \end{cases}$$

where $N_\pi = \min\{j \geq 0 \mid q_j \in Q_f\}$. We take the maximum over $\tilde{\Pi}_c(q)$ because of the non-determinacy of A: it is uncertain which among the (multiple) trajectories allowed by c will be taken so we must assume the worst-case situation. The *discrete value function* $\hat{V} : Q \to \mathbb{R}$ is

$$\hat{V}(q) = \min_{c \in C} \hat{J}(q, c)$$

for $q \in Q \setminus Q_f$ and $\hat{V}(q) = \hat{h}(q)$ for $q \in Q_f$. We show in Proposition 1 that $\hat{V}$ satisfies a DPP that takes into account the non-determinacy of A and ensures that optimal control policies are admissible. This DPP describes the accumulation of cost over one step to be the worst case cost among edges that have the same label. Let $\mathcal{A}_q$ be the set of control assignments $c(q) \in \Sigma_\delta$ at q such that c is admissible.

Proposition 1. $\hat{V}$ *satisfies*

$$\hat{V}(q) = \min_{c(q) \in \mathcal{A}_q} \left\{ \max_{e=(q,q') \in E_{c(q)}(q)} \{ \hat{L}(e) + \hat{V}(q') \} \right\}, \quad q \in Q \setminus Q_f \qquad (10)$$

$$\hat{V}(q) = \hat{h}(q), \qquad\qquad\qquad\qquad\qquad\qquad q \in Q_f. \qquad (11)$$

5.3 Synthesis of g_e

The synthesis of enabling conditions or *controller synthesis* is typically a post-processing step of a backward reachability analysis (see, for example, [12]). This situation prevails here as well: equations (10)-(11) describe a backward analysis to construct an optimal policy $c \in \mathcal{C}$. Once c is known the enabling conditions of H are extracted as follows.

Consider each $e = (\sigma, \sigma') \in E$ of H with $\sigma \neq \sigma'$. There are two cases. If $\sigma' \neq \sigma_f$ then $g_e = \{ x \mid (\sigma, x) \in q, q \in Q \wedge c(q) = \sigma' \}$. That is, if the control policy designates switching from $q \in Q$ with label σ to $q' \in Q$ with label σ', then the corresponding enabling condition in H includes the projection to $\mathbb{R}^n$ of q. The second case when $\sigma' = \sigma_f$ is for edges going to the terminal location of H. Then $g_e = \{ x \mid (\sigma, x) \in q, q \in Q_f \}$.

6 Main Result

We will prove that $\hat{V}$ converges to V, the viscosity solution of the HJB equation, as $\delta_Q, \delta \to 0$. The proof will be carried out in three steps. In the first step we consider restricting the set of controls to piecewise constant functions, whose constant intervals are a function of the state. In the second step we introduce the discrete approximations of L and h. In the last step we introduce the discrete states Q and consider the non-determinacy of A.

In the sequel we make use of a filtration of control sets $\Sigma_k \equiv \Sigma_{\delta_k}$ corresponding to a sequence $\delta_k \to 0$ as $k \to \infty$, in such a manner that $\Sigma_k \subset \Sigma_{k+1}$. Considering (8), we define a filtration of families of submanifolds such that $\mathcal{W}_k^\sigma \subset \mathcal{W}_{k+1}^\sigma$, for each $\sigma \in \Sigma_k$.

Step 1: piecewise constant controls.

In the first step we define a class of piecewise constant functions that depend on the state and show that the value function which minimizes the cost-to-go over this class converges to the viscosity solution of HJB as $\delta_k \to 0$. The techniques of this step are based on those in Bardi and Capuzzo-Dolcetta [1] and are related to those in [5].

We consider the optimal control problem (2)-(4) when the set of admissible controls is $\mathcal{U}_k^1$, piecewise constant functions consisting of finite sequences of control labels $\sigma \in \Sigma_k$ and each σ is applied for a time $\tau(\sigma, x)$. Let $(\sigma, x) \in q$ for some $q \in Q$ and define $\tau(\sigma, x)$ to be the minimum of the time it takes the trajectory starting at x and using control $\sigma \in \Sigma_k$ to reach (ta) $\partial \Omega_f$, and (tb) some x' such that $(\sigma, x') \notin q$. If a trajectory is at x_i at the start of the $(i+1)$th step, then the control σ_{i+1} is applied for time $\tau_{i+1} := \tau(\sigma_{i+1}, x_i)$ and $x_{i+1} = \phi_{\tau_{i+1}}(x_i, \sigma_{i+1})$.

Let

$$\mathcal{R}_k^1 := \{\, x \in \mathbb{R}^n \mid \exists \mu \in \mathcal{U}_k^1 \,.\, T(x, \mu) < \infty \,\}.$$

We define the cost-to-go function $J_k^1 : \Omega \times \mathcal{U}_k^1 \to \mathbb{R}$ as follows. For $x \in \Omega$ and $\mu = \sigma_1 \sigma_2 \ldots \in \mathcal{U}_k^1$, if $T(x, \mu) < \infty$ then

$$J_k^1(x, \mu) = \sum_{j=1}^N \int_0^{\tau(\sigma_j, x_{j-1})} L(\phi_s(x_{j-1}, \sigma_j), \sigma_j) ds + h(x_N)$$

where $N = \min\{j \geq 0 \mid x_j \in \partial \Omega_f\}$. $J_k^1(x, \mu) = \infty$, otherwise. We define the value function $V_k^1 : \mathbb{R}^n \to \mathbb{R}$ as follows. For $x \in \Omega \setminus \Omega_f$,

$$V_k^1(x) = \inf_{\mu \in \mathcal{U}_k^1} J_k^1(x, \mu) \qquad (12)$$

and for $x \in \Omega_f$, $V_k^1(x) = h(x)$.

$\{V_k^1\}$ forms a family of equibounded, locally equicontinuous functions. It can then be shown that, along some subsequence k_n, $V_{k_n}^1$ converges to a continuous function V_*. Moreover, the following holds:

Proposition 2. V_* is the unique viscosity solution of HJB.

Step 2: approximate cost functions.

In this step we keep the semantics on piecewise constant controls of Step 1 but replace cost functions L and h by approximations L^2 and $\hat{h}$. We define the cost-to-go function $J_k^2 : \Omega \times \mathcal{U}_k^1 \to \mathbb{R}$ as follows. First, we define an approximate instantaneous cost $L^2 : \Omega \times \Sigma_k \to \mathbb{R}$ given by

$$L^2(x, \sigma) := \hat{L}(q) \qquad (13)$$

where $(\sigma, x) \in q$. For $x \in \Omega$ and $\mu = \sigma_1 \sigma_2 \ldots \in \mathcal{U}_k^1$, if $T(x, \mu) < \infty$ then

$$J_k^2(x, \mu) = \sum_{j=1}^N L^2(x_{j-1}, \sigma_j) + \hat{h}(x_N)$$

where $N = \min\{j \geq 0 \mid x_j \in \partial\Omega_f\}$. We define a value function $V_k^2 : \mathbb{R}^n \to \mathbb{R}$ as follows. For $x \in \Omega \setminus \Omega_f$,

$$V_k^2(x) = \inf_{\mu \in \mathcal{U}_k^1} J_k^2(x, \mu) \tag{14}$$

and for $x \in \Omega_f$, $V_k^2(x) = \hat{h}(x)$. For $x \in \Omega$ such that $V_k^2(x) < \infty$, V_k^2 satisfies the DPP $V_k^2(x) = \min_{\sigma \in \Sigma_k}\{L^2(x, \sigma) + V_k^2(\phi_{\tau(\sigma,x)}(x, \sigma))\}$.
Remark 6.1.

For each $x \in \cup_k \mathcal{R}_k^1$ and $\epsilon > 0$ there exists $m \in \mathbb{Z}^+$ and $\mu \in \mathcal{U}_m^1$ such that μ is an ϵ-optimal control for x w.r.t. V^1 satisfying Assumptions 2.2. This follows from Assumptions 2.2, $V_k^1(x) \geq V(x)$, and the fact that we can well-approximate an ϵ-optimal control for V by a control in $\mathcal{U}_m^1$, for large enough m.

Proposition 3. *Let $k_0 \in \mathbb{Z}^+$, $x \in \mathcal{R}_{k_0}^1$, and $\mu \in \mathcal{U}_{k_0}^1$ be an ϵ-optimal control for x. Then $|J_k^1(x, \mu) - J_k^2(x, \mu)| \to 0$ as $k \to \infty$.*

Proof. First, we require two facts which are stated without proof, for brevity.
Fact 1. *If $\delta_k < \frac{m_f}{L_f}$, then for all $q \in Q$,*

$$\tau_q \leq \frac{\delta_k}{m_f - L_f \delta_k}. \tag{15}$$

For the next fact, we require a definition. let C_k be as in (7) and γ_n^σ the transversal foliation of $\dot{x} = f(x, \sigma)$. For $\sigma \in \Sigma_k$, define the region in $\mathbb{R}^n$

$$M_c^\sigma := \{ x \in (\gamma_n^\sigma)^{-1}(c) \mid c \in C_k \}.$$

Fact 2. *Let $x, x' \in M_c^\sigma$ for some $c \in C_k$ and $\sigma \in \Sigma_k$. Let τ, τ' be times such that $\phi_\tau(x, \sigma), \phi_{\tau'}(x', \sigma) \in M_{c+\Delta}^\sigma$. Then $|\tau - \tau'| \leq c_\gamma \tau \delta_k$ for some $c_\gamma > 0$.*
Now we have

$$|J_k^1(x, \mu) - J_k^2(x, \mu)| \leq \left| \sum_{j=1}^{N} \left[\int_0^{\tau(\sigma_j, x_{j-1})} L(\phi_s(x_{j-1}, \sigma_j), \sigma_j) ds \right] + h(x_N) \right.$$
$$\left. - \sum_{j=1}^{N} [\tau_{q_{j-1}} L(\xi_{j-1}, \sigma_j)] - \hat{h}(x_N) \right|$$

where $(x_{j-1}, \sigma_j) \in q_{j-1}$ and $q_{j-1} = [(\xi_{j-1}, \sigma_j)]$. There exists ξ_N such that $\hat{h}(x_N) = h(\xi_N)$ and $\|x_N - \xi_N\| \leq \delta_k$. Also, using the Mean Value Theorem, there exists $\tilde{t}$ with $\tilde{x} = \phi_{\tilde{t}}(x_{j-1}, \sigma_j)$ and $\|\tilde{x} - \xi_{j-1}\| \leq \delta_k$ such that

$$|J_k^1(x, \mu) - J_k^2(x, \mu)| \leq \sum_{j=1}^{N} |\tau(\sigma_j, x_{j-1}) L(\tilde{x}, \sigma_j) - \tau_{q_{j-1}} L(\xi_{j-1}, \sigma_j)|$$
$$+ |h(x_N) - \hat{h}(x_N)|$$
$$\leq \sum_{j=1}^{N} \tau_{q_{j-1}} L_L \delta_k + \sum_{j=1}^{N} [\tau_{q_{j-1}} - \tau(\sigma_j, x_{j-1})] L(\tilde{x}, \sigma_j) + L_h \delta_k.$$

Using Fact 1 the first term on the r.h.s. decreases linearly as δ_k. Call the second term on the r.h.s. "B". Splitting B into sums over control switches and time steps, we have

$$B \leq M_L \sum_{j=2}^{N} [\tau_{q_{j-1}} - \tau(\sigma_j, x_{j-1})] \mathbf{1}(\sigma_j = \sigma_{j-1})$$

$$+ M_L \sum_{j=1}^{N} [\tau_{q_{j-1}} - \tau(\sigma_j, x_{j-1})] \mathbf{1}(\sigma_j \neq \sigma_{j-1})$$

$$\leq M_L \sum_{j=2}^{N} c_{j-1} \tau_{q_{j-1}} \delta_k + M_L \sum_{j=1}^{N} \tau_{q_{j-1}} \mathbf{1}(\sigma_j \neq \sigma_{j-1})$$

for some $c_{j-1} \in \mathbb{R}$. In the second line we used Fact 2 and the fact that $\tau_{q_{j-1}} \geq \tau(\upsilon_j, x_{j-1})$. Using Fact 1 the first term on the r.h.s. decreases linearly as δ_k. The second term on the r.h.s. goes to zero since μ has a fixed number of control switches for all $k \geq k_0$. $\qquad\square$

Step 3: discrete states and non-determinacy.

We define $\hat{V}_k(x) := \min_{\sigma \in \Sigma_k} \{ \hat{V}_k(q) \mid (\sigma, x) \in q \}$. Also let $\hat{\mathcal{R}}_k = \{x \in \Omega \mid \hat{V}_k(x) < \infty\}$ and $\hat{\mathcal{R}} = \cup_k \hat{\mathcal{R}}_k$.

Remark 6.2.

(a) By Remark 6.1 and $V_k^1(x) \leq V_k^2(x)$, for each $x \in \cup_k \mathcal{R}_k^1$ and $\epsilon > 0$ there exists $m_\epsilon \in \mathbb{Z}^+$ and $\mu \in \mathcal{U}_{m_\epsilon}^1$ such that μ is an ϵ-optimal control for x w.r.t. V_k^2 satisfying Assumptions 2.2.

(b) $\hat{\mathcal{R}} \subset \cup_k \mathcal{R}_k^1$, but the converse is not true, in general.

(c) If μ is an ϵ-optimal control for x w.r.t. V_k^2, then we can assume $\phi_t(x, \mu)$ does not self-intersect, for if it did we can find $\tilde{\mu}$, also ϵ-optimal, which eliminates loops in $\phi_t(x, \mu)$.

(d) $\|x - y\| \to 0$ as $k \to \infty$ for all $y \in r_e(x)$ and all edges e of H_k, the hybrid automaton defined using Σ_k and C_k given in (7).

Proposition 4. *For all $x \in \hat{\mathcal{R}}$, $|\hat{V}_k(x) - V_k^2(x)| \to 0$ as $k \to \infty$.*

Proof. Fix $\epsilon > 0$ and $x \in \hat{\mathcal{R}}$. By Remark 6.2(a) there exists $m_\epsilon > 0$ and an ϵ-optimal control $\mu \in \mathcal{U}_{m_\epsilon}^1$ for x. Let us denote μ as an open loop control $\mu = ((\sigma_1, \tau_1), \ldots, (\sigma_N, \tau_N))$, where τ_i is the time σ_i is applied. If c is a policy derived using δ_k and C_k, for $k \geq m_\epsilon$, then $0 \leq \hat{V}_k(q) - V_k^2(x) \leq \hat{J}_k(q, c) - J^2(x, \mu) + \epsilon$, where $q = [(\sigma_1, x)]$. If we can show there exists $\bar{k} \geq m_\epsilon$ such that for $k > \bar{k}$, there exists a policy $\bar{c}$ such that $\hat{J}_k(q, \bar{c}) - J_k^2(x, \mu) < \epsilon$ and using the fact that $|\hat{V}_k(q) - \hat{V}_k(x)| \to 0$ as $k \to \infty$, then the result follows.

Consider the set Ψ_k of (discontinuous) trajectories $\phi_t(x, \tilde{\mu})$ where $\tilde{\mu} \in \mathcal{U}_k^1$ is denoted $((\sigma_1, \tilde{\tau}_1, \ldots, (\sigma_N, \tilde{\tau}_N))$. Also $x_j^- = \phi_{\tilde{\tau}_j}(x_{j-1}, \sigma_j)$ and $x_j \in r_e(x_j^-)$, where $e = (\sigma_j, \sigma_{j+1})$ is an edge of H_k, defined in Remark 6.2(d). We can find $\bar{k}_1 \geq m_\epsilon$ such that, by Remark 6.2(d) and the transversality of $\phi_t(x, \mu)$ with the submanifolds where it switches controls and with Ω_f, there exists $\tilde{\mu} \in \mathcal{U}_k^1$ such that $\phi_t(x, \tilde{\mu}) \in \Psi_k$ switches controls on the same (transversal) submanifolds and reaches Ω_f. Let $W_k^2(\phi) = \sum_{j=1}^N L^2(x_{j-1}, \sigma_j) + \hat{h}(x_N)$. We observe that for $\phi, \phi' \in \Psi_k$ and $\mu \in \mathcal{U}_{\overline{k}_1}^1$, $|W_k^2(\phi) - W_k^2(\phi')| \to 0$ as $k \to \infty$, using Lipschitz continuity of L and h, Remark 6.2(d), and the fact that μ is fixed for all $k > \overline{k}_1$. Notice that $J^2(x, \mu) = W_k^2(\phi^2)$ for some $\phi^2 \in \Psi_k$. We can define the control policy $\overline{c}$ such that automaton A accepts the time abstract trajectory starting at q corresponding to each trajectory of Ψ_k and with all other control assignments of $\overline{c}$ as time steps. $\overline{c}$ is admissible because otherwise some $\phi' \in \Psi_k$ would have a Zeno loop. Since ϕ' approaches $\phi_t(x, \mu)$ as $k \to \infty$, this would imply $\phi_t(x, \mu)$ has a loop, contradicting Remark 6.2(c). Now we observe that $\hat{J}(q, \overline{c}) = \max_{\phi \in \Psi_k} W_k^2(\phi) := W_k^2(\overline{\phi})$. Thus, $\hat{J}_k(q, \overline{c}) - J_k^2(x, \mu) \leq |W_k^2(\overline{\phi}) - W_k^2(\phi^2)| \to 0$ as $k \to \infty$.   □

Theorem 1. *For all $x \in \hat{\mathcal{R}}$, $\hat{V}_k(x) \to V(x)$ as $k \to \infty$.*

7 Conclusion

In this paper we have developed a methodology for the synthesis of optimal controls based on hybrid systems and bisimulations. The idea is to translate an optimal control problem to a switching problem on a hybrid system whose locations describe the dynamics when the control is constant. When the vector fields for each location of the hybrid automaton have local first integrals which can be expressed analytically we are able to define a finite bisimulation using the approach of [4]. From the finite bisimulation we obtain a (time abstract) finite automaton upon which a dynamic programming problem can be formulated that can be solved efficiently.

We are presently working on three topics that will enhance considerably the significance of our work:

- The dynamic programming problem is equivalent to a shortest path problem on a non-deterministic graph. We are in the process of carrying through the implementation issues to obtain an algorithmic solution.
- Throughout the paper we have assumed that, once the bisimulation is expressed using first integrals, the corresponding finite automaton can be constructed directly. In fact, this task is not so straightforward. We are working on the automatic generation of finite automata that give time abstract behavior of vector fields.
- If it is not possible to obtain a finite bisimulation, one may still be able to construct a finite automaton that approximates the continuous and discrete behavior of the hybrid system. But this automaton will have non-deterministic behavior that results in spurious solutions, not corresponding to the true dynamics of the hybrid system. We are working on a procedure to eliminate these spurious solutions.

References

1. M. Bardi and I. Capuzzo-Dolcetta. *Optimal control and viscosity solutions of Hamilton-Jacobi-Bellman equations.* Birkhäuser, Boston, 1997.
2. V.G. Boltyanskii. Sufficient conditions for optimality and the justification of the dynamic programming method. *SIAM Journal of Control,* 4, pp. 326-361, 1966.
3. M. Branicky, V. Borkar, S. Mitter. A unified framework for hybrid control: model and optimal control theory. *IEEE Trans. AC,* vol. 43, no. 1, pp. 31-45, January, 1998.
4. M. Broucke. A geometric approach to bisimulation and verification of hybrid systems. In *Hybrid Systems: Computation and Control,* F. Vaandrager and J. van Schuppen, eds., LNCS 1569, p. 61-75, Springer-Verlag, 1999.
5. I. Capuzzo Dolcetta and L.C. Evans. Optimal switching for ordinary differential equations. *SIAM J. Control and Optimization,* vol. 22, no. 1, pp. 143-161, January 1984.
6. M Crandall, P. Lions. Viscosity solutions of Hamilton-Jacobi equations. *Trans. Amer. Math. Soc.,* vol. 277, no. 1, pp. 1-42, 1983.
7. W.H. Fleming, R.W. Rishel. *Deterministic and stochastic optimal control.* Springer-Verlag, New York, 1975.
8. O. Maler, A Pnueli, J. Sifakis. On the synthesis of discrete controllers for timed systems. In *Proc. STACS '95,* E.W. Mayr and C. Puech, eds. LNCS 900, Springer-Verlag, p. 229-242, 1995.
9. J. Raisch. Controllability and observability of simple hybrid control systems-FDLTI plants with symbolic measurements and quantized control inputs. *International Conference on Control '94,* IEE, vol. 1, pp. 595-600, 1994.
10. J. Stiver, P. Antsaklis, M. Lemmon. A logical DES approach to the design of hybrid control systems. *Mathemtical and computer modelling.* vol. 23, no. 11-1, pp. 55-76, June, 1996.
11. H.S. Witsenhausen. A class of hybrid-state continuous-time dynamic systems. *IEEE Trans. AC,* vol. 11, no. 2, pp. 161 - 167, April, 1966.
12. H. Wong-Toi. The synthesis of controllers for linear hybrid automata. In *Proc. 36th IEEE Conference on Decision and Control,* pp. 4607-4612, 1997.

Behavior Based Robotics
Using Hybrid Automata

Magnus Egerstedt*

Division of Optimization and Systems Theory
Royal Institute of Technology
SE - 100 44 Stockholm, Sweden
magnuse@math.kth.se

Abstract. In this article, we show how a behavior based control system for autonomous robots can be modeled as a hybrid automaton, where each node corresponds to a distinct robot behavior. This type of construction gives rise to chattering executions, but we show how regularized automata suggest a solution to this problem. We also discuss some design and implementation issues.

1 Introduction

For mobile, autonomous robots the ability to function in, and interact with a dynamic, changing environment is of key importance. A successful way of structuring the control system in order to deal with this problem is within a *behavior based* control architecture [3]. The main idea is to identify different controllers, responses to sensory inputs, with desired robot behaviors. A behavior could, for instance, be obstacle avoidance in which sonar information about a close obstacle should result in a movement away from that obstacle. This way of structuring the control system into separate behaviors, dedicated to performing certain tasks such as avoid obstacles or traverse doors, has turned out to be a successful design. It has the major advantage that it makes the system modular, which both simplifies the design process as well as offers a possibility to add new behaviors to the system without causing any major increase in complexity.

The suggested outputs from the different, concurrently active behaviors are fused together according to some action coordination rule, and this makes it easy to stress such questions as safety explicitly, since, for example, an avoidance behavior can just be given higher priority than a reach target behavior.

However, within this framework, a number of design issues still need to be addressed. Those range from questions concerning the design of the individual behaviors to action coordination issues [5]. For instance, given a reactive obstacle avoidance behavior, modeled as a repulsive field surrounding the obstacle, how should an approach target behavior be designed so that it takes advantage of

---

* This work was sponsored in part by the Swedish Foundation for Strategic Research through its Centre for Autonomous Systems at KTH.

N. Lynch and B. Krogh (Eds.): HSCC 2000, LNCS 1790, pp. 103–116, 2000.

the fact that it is going to run in parallel with an obstacle avoidance behavior? Furthermore, how should these behaviors be combined?

What will be investigated in this article is how a behavior based system can be modeled as a *hybrid automaton* with each of the discrete nodes corresponding to a distinct behavior. If the system where to be described by such an automaton it would hopefully help us understand and explain some of the so called *emergent* phenomena that complex robotics systems can give rise to. We will furthermore see that questions concerning safety and optimality can be addressed nicely within this framework.

The outline of the article is as follows: First, in Section 2, we discuss some of the properties of a behavior based robotics system, and we show how this can be modeled as a hybrid automaton. Some regularization techniques are then exploited in order to get rid of potential chattering in the automaton. In the next section, some control design issues are discussed, and we describe a heuristic method for constructing behaviors that are safe at the same time as they are close to optimal with respect to a given performance evaluation functional. We conclude, in Section 4, with a brief discussion about a proposed, systematic strategy for implementing the hybrid automata.

2   Behavior Based Robotics

As already mentioned, for autonomous robots operating in a partially unknown, dynamic environment a successful way of structuring the controllers is within a behavior based framework [3],[10]. Different robot behaviors are identified, e.g. obstacle avoidance or reach target, and their functionality is defined by a tight mapping from sensory data to a desired action. Typically, in a so called *reactive* behavior based system, no representation of the world is contained in this mapping, while a *deliberative* system exploits planning or world models in the control loops.

The desired output actions are then normally fused together by an arbitration mechanism, as seen in Figure 1, where a wide-spread solution to the action fusion problem, used for example in the schema theoretical paradigm, is to represent the goals, targets and obstacles by weighted attractive or repulsive potential fields, resulting in weighted, desired orientation vectors. The action coordination is then simply done using vector summation. This way of letting behaviors be active simultaneously is desirable in many situations. For instance, while approaching a target an obstacle avoidance behavior has to be active for safety reasons while the performance is improved if the robot tries to approach the goal at the same time as it is avoiding obstacles. This calls for a fused, coordinated control scheme [2],[3].

2.1   Obstacle Negotiation

The specific problem that will be investigated in this article is how to move a robot between two points. This *point-to-point motion* should be done so that

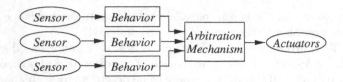

Fig. 1. Block diagram of the behavior based control architecture.

the detection of an obstacle results in a repulsive potential field, acting on the platform when the robot is closer to the obstacle than a desired safety distance, d_{OA}, where the subscript stands for obstacle avoidance. This behavior is an example of a so called *reactive* obstacle avoidance behavior. The word reactive, a commonly used one in the robotics community, is used here since the behavior can be thought of as a reflex. When the robot moves too close to an obstacle, it is forced to change the motion in order to avoid hitting the obstacle. This is a reasonable safety strategy since the robot may be moving around in a highly unstructured world, where the occurrence of unpredicted, or unmodeled obstacles is very likely.

We now assume that we have direct access to the robot's longitudinal velocity, v, at the same time as the heading of the robot, ϕ, can be controlled directly as

$$\dot{\phi} = \omega.$$

Furthermore, if the sonars on the robot, with center of gravity at (x, y) and heading ϕ, detect a point-obstacle at (x_{ob}, y_{ob}) that is closer to the robot than d_{OA}, the reactive control response will be given by a vector field acting on the robot as

$$\omega = C_{OA} W_{OA}(d)(\tilde{\phi} - \phi), \tag{1}$$

$$d = \sqrt{(x - x_{ob})^2 + (y - y_{ob})^2},$$

where   and
$$W_{OA}(d) = \begin{cases} \frac{1}{d^2} & \text{if } d < d_{OA} \\ 0 & \text{if } d \geq d_{OA}, \end{cases}$$

$$\tilde{\phi} = \pi + \text{atan2}(y_{ob} - y, x_{ob} - x), \tag{2}$$

as seen in Figure 2. Here, C_{OA} is just a constant weight, and d_{OA} is the fixed distance from the obstacle where the behavior becomes active.

Since a real, extended obstacle cannot be considered to be a point, in the actual implementation of the avoidance behavior, the desired heading needs to be calculated as the orientation of the sum of the weighted vectors that each individual sonar reading contributes with. For a Nomad 200, that is going to be our experimental platform, this corresponds to taking the sum over 16 elements since the Nomad is equipped with 16 ultrasonic sensors.

A standard kinematic model of the mobile robot [1] gives that

$$\dot{x} = v \cos \phi$$
$$\dot{y} = v \sin \phi, \tag{3}$$

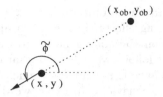

Fig. 2. The general idea behind a repulsive obstacle avoidance behavior.

and if we set $z = (x, y, \phi)$, we let $\dot{z} = f_{OA}(z)$ denote the full state, closed-loop obstacle avoidance behavior where we initially let v be constant.

2.2 Hybrid Automata

When adding a goal attraction behavior, defined in the same way as the obstacle avoidance behavior except that we now have an attractive instead of a repulsive field, we get two different possible hybrid automata for describing the situation. This depends on whether the two behaviors are active simultaneously or not, as seen in Figure 3. If one chooses to work with fused, concurrently active behaviors, then different controllers affect the system simultaneously, resulting in a smooth overall performance [11]. But in that case, however, the system does not correspond to an automaton where each node represents a single behavior. This would make the automata approach meaningless since we would then just "hide" all of the difficulties that the complex control system gives rise to in the individual nodes of the automaton.

On the other hand, the other possible solution to the coordination problem, corresponding to hard switches between the different behaviors, has the major disadvantage that it both affects the performance in a negative way, not allowing for the smooth performance that fused behaviors produce, and that it increases the risk of introducing chattering into the system. Therefore our idea is to impose hard switches on the behavior based system in such a way that we can model

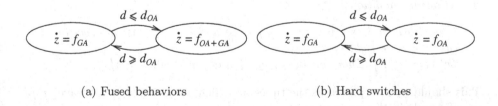

(a) Fused behaviors   (b) Hard switches

Fig. 3. The two possible goal attraction and obstacle avoidance automata. Here, d_{OA} is the fixed distance from the obstacle where the obstacle avoidance behavior becomes active.

each behavior as a node in an automaton, at the same time as we want to avoid the negative, chattering effects that such an approach could potentially give rise to. This will be done by adding nodes to the automaton as a way of regularizing it, and in what follows we will show that even though we introduce hard switches, the performance is not affected much when using a regularized automaton instead of fused behaviors. In other words, what we want to do is to remove some of the so called *Zeno*[1] properties of the system. What this corresponds to is a hybrid system that exhibits an infinite number of discrete transitions in finite time.

Even though the main focus in this article is not going to be on hybrid automata theory, we need to include some initial definitions. This is necessary in order to be able to state what we mean by a Zeno hybrid automaton as well as to capture the hybrid aspects of a behavior based robotic system.

The following brief definitions are based on [6],[8],[14].

Definition 1 (Hybrid Automaton). *A hybrid automaton is considered to be a collection (Q, X, I, f, E) where Q and X are sets of discrete and continuous variables respectively. I is a set of initial states, while f describes the continuous and E the discrete evolution of the states.*

A discrete state combined with the continuous dynamics connected to that state will be referred to as a *node* in the automaton. The general idea behind this construction can be seen in Figure 4.

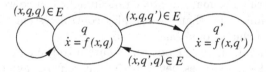

Fig. 4. The basic structure of a hybrid automaton.

Definition 2 (Hybrid Time Trajectory). *A hybrid time trajectory τ is a finite or infinite sequence of intervals of the real line, $\tau = \{I_i\}$, $i \in \mathbb{N}$, satisfying the following conditions:*

- *I_i is closed, unless τ is a finite sequence and I_i is the last interval in which case it can be right open.*
- *Let $I_i = [\tau_i, \tau_i']$. Then for all i, $\tau_i \leq \tau_i'$ and for $i > 0$, $\tau_i = \tau_{i-1}'$.*

This should be interpreted as the times at which we arrive (τ_i) and leave (τ_i') a specific node in the automaton.

---

[1] The name Zeno refers to the philosopher Zeno of Elea (500–400 B.C.), whose major work consisted of a number of famous paradoxes. They were designed to explain his view that the ideas of motion and evolving time lead to contradictions. An example is Zeno's Second Paradox of Motion, in which Achilles is racing against a tortoise.

Note that hybrid time trajectories can extend to infinity if τ is an infinite sequence or if it is a finite sequence ending with an interval of the form $[\tau_N, \infty)$.

Definition 3 (Execution). *An execution χ of a hybrid automaton H is a collection $\chi = (\tau, q, x)$, satisfying*

- *Initial Condition: $(q(\tau_0), x(\tau_0)) \in I$.*
- *Discrete Evolution: $(x(\tau'_{i-1}), q(\tau'_{i-1}), q(\tau_i)) \in E$, for all i.*
- *Continuous Evolution: for all i with $\tau_i < \tau'_i$, x and q are continuous over $[\tau_i, \tau'_i]$ and for all $t \in [\tau_i, \tau'_i)$, we have $\frac{d}{dt} x(t) = f(q(t), x(t))$.*

Furthermore, an execution $\chi = (\tau, q, x)$ is called infinite, if τ is an infinite sequence, or $\sum_i (\tau'_i - \tau_i) = \infty$. We use $\mathcal{H}_{(q_0, x_0)}$ to denote the set of all infinite executions of H with initial condition $(q_0, x_0) \in I$. An execution is *admissible* if $\sum_i (\tau'_i - \tau_i) = \infty$, and it is *Zeno* if it is infinite but not admissible. For a Zeno execution $\chi = (\tau, q, x)$ we define the Zeno time as $\tau_\infty = \sum_i (\tau'_i - \tau_i) < \infty$. What this means is that the hybrid system makes an infinite number of discrete transitions in finite time, $[\tau_0, \tau_\infty]$, and we finally state the following definition.

Definition 4 (Zeno Hybrid Automaton). *A hybrid automaton H is called Zeno, if there exists $(q_0, x_0) \in I$ such that $\mathcal{H}_{(q_0, x_0)}$ contains a Zeno execution.*

2.3   Regularization

It is clear that a Zeno hybrid automaton has the undesirable property that it blocks time. For the type of automata that we will encounter here, the infinite number of discrete transitions, made in finite time, is caused by the fact that the underlying system that the automaton tries to model is a switched system that exhibits sliding in the sense of Filippov [7]. They thus form a special class of Zeno hybrid automata since they, in theory, make an infinite number of transitions in zero time.[2] The underlying, switched systems have continuous flows that point toward the switching surface, resulting in a new, induced flow on that surface. In these cases, the automaton can be regularized by the introduction of a new node with the continuous flow given by the Filippov solution [8],[15]. The general idea behind this construction can be seen in Figure 5.

If we now assume that C_{OA} in (1) is large enough so that the heading of the robot can be considered to be more or less instantaneously driven to its desired configuration, the hybrid automaton in Figure 3(b) can admit Zeno executions. This obvious fact is best illustrated by Figure 6, where the extra node that needs to be added in order to regularize the automaton can easily be identified as well. The extra node is just a node containing the sliding dynamics that is defined on the boundary between the two behaviors.

When an obstacle is closer to the robot than d_{OA}, the obstacle avoidance behavior becomes active. Since the repulsive potential field from that behavior

---

[2] The other class of Zeno automata has a slightly more complex dynamics. Here the automaton changes nodes faster and faster, with the jump times converging to the Zeno time, τ_∞ [8].

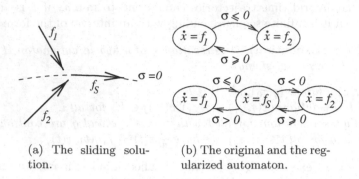

(a) The sliding solution.

(b) The original and the regularized automaton.

Fig. 5. Regularization of a Filippov type Zeno hybrid automaton.

is orthogonal to the surface on which the behavior becomes active, the sliding solution is just

$$f_S = \alpha f_{OA} + (1 - \alpha) f_{GA},$$

where GA stands for goal attraction, and $\alpha \in [0, 1]$ is chosen so that $f_S \perp f_{OA}$. Adding this type of information about the different behaviors makes it possible to generate the extra node in the automaton automatically. It furthermore suggests that our method would scale when more that two behaviors affect the motion of the robot, as long as an automatic procedure for designing the sliding solutions could be identified for the new behaviors as well.[3]

Fig. 6. Goal attraction together with obstacle avoidance results in a Filippov type Zeno automaton. The grey region around the obstacle corresponds to the region where obstacle avoidance is active. The arrows correspond to the different vector fields that are acting on the robot.

The assumption about instantaneous heading control is obviously a simplification but it still gives a model that is rich enough to capture the, from our point of view, relevant phenomena. In fact, in real life we have a possibility of chattering that here reveals itself as a Zeno execution.

---

[3] This typically depends on whether we have access to a geometric description of the switching surface or not.

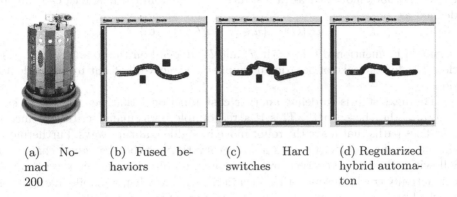

(a)   No-mad 200   (b) Fused be-haviors   (c)   Hard switches   (d) Regularized hybrid automaton

Fig. 7. Simulation of b) fused behaviors, c) hard switches, and d) a regularized automaton on the Nomad simulator, the `Nserver`.

The regularized point-to-point automaton was implemented and tested on a Nomad 200 mobile robot. In Figure 7, the results from running the system on the `Nserver`, the Nomad simulation package, can be seen. In (7b) fused behaviors are displayed, resulting in a smooth movement around the obstacle, while the chattering solution in (7c) corresponds to hard switches. The reason why we do not have sliding in this case is due to the part of the dynamics of the robot that was ignored in the analysis. It is still clear that from a performance perspective, (7c) is an unsuccessful design. In (7d) the result from using a regularized automaton can be seen, and even though we only have one behavior active at a time, the performance is satisfactory.

3   Controller Design

Given the reactive obstacle avoidance behavior from the previous section, the main question that we want to address here is: *How do we construct an appropriate approach target behavior?* Obviously, we can do better than to just use an attractive potential field, and it will turn out that our automata approach allows us to explicitly deal with safety and optimality.

What we want to do is to produce a robot behavior that satisfies the safety specifications at the same time as the solution is close to optimal with respect to a given performance evaluation functional, and a first formulation, inspired by [19], of what we want to accomplish could be the following. If we let our admissible controls be $u \in \mathcal{U}$, and define a safety functional

$$\mathcal{J}_s(u) = \min_{t \geq 0} \left\{ (x(t) - x_{ob})^2 + (y(t) - y_{ob})^2 \right\}, \tag{4}$$

where the dependence on the control, u, is given implicitly by the controlled system dynamics from the previous section. The set of controls, $\mathcal{U}_s(C)$, that make the robot move at least a distance C away from the obstacle, can thus be defined as

$$\mathcal{U}_s(C) = \{u \in \mathcal{U} \ : \ \mathcal{J}_s(u) \geq C\}. \tag{5}$$

It should be mentioned that both $\mathcal{J}_s$ and $\mathcal{U}_s$ depend on the robot's initial position, but for the sake of notational simplicity we leave that out from the definitions.

The next step is to define another cost functional that penalizes high curvature of the chosen path. This is a reasonable performance criterion since it penalizes paths that make the robot move in sudden, abrupt ways. Furthermore, this smoothness objective gives a trajectory that a robot has good chances of following when it is governed by physical limits on what signals the actuators can actually track. In some other situations, such as when a mobile manipulator is asked to carry a cup of coffee, the smoothness of the curve is absolutely crucial and is obviously of key importance to a successful, "non-spilling" execution of the task.

The idea now is to choose the control candidates for minimizing this new performance functional from the set of safe controls, $\mathcal{U}_s(C_s)$, where C_s is our preferred safety margin.

Unfortunately it turns out that this is a very hard problem to solve numerically (not to mention analytically) [19], which implies that in this formulation, it is not suitable for situations where on-line computations are necessary. However, the underlying approach could suggest a way for producing a solution to the obstacle negotiation problem that is both safe, computationally feasible, and makes the system behave in a satisfactory way with respect to keeping the curvature of the produced path small.

The main idea is that instead of focusing on the hard optimal control problem, we should concentrate on just producing optimal (or close to optimal) geometric trajectories that lead around the obstacle. This way we do not have to deal with the actual kinematics of the robot in the optimization formulation. Instead we add the kinematics when we track the produced path. This means that we cannot be sure that we actually find the optimal controller, but rather that we find one that is reasonably close to the optimal one as long as we have a good enough trajectory tracker.

The desired overall behavior that these heuristics give rise to (under the assumption of perfect tracking), together with the corresponding automaton, is depicted in Figure 8.

3.1   Path Planning

One first observation is that for a path produced by a scalar function $y_d = f(x_d)$, the curvature is given by

$$\kappa(x_d) = \frac{f''(x_d)}{(1 + f'(x_d)^2)^{3/2}}, \tag{6}$$

where the subscript d stands for the desired robot position.

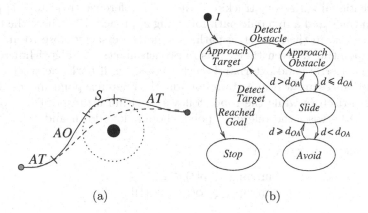

(a)   (b)

Fig. 8. In the left figure, an optimal path is planned and followed by an Approach Target behavior until an obstacle is detected. Then an Approach Obstacle behavior follows another path to the region where the regularized sliding behavior becomes active. When the target can be reached by an optimal path, not intersecting the safety region around the obstacle (called Detect Target in the right figure), Approach Target becomes active again. In the right figure, the corresponding automaton is depicted.

Thus, if we minimize $f''(x_d)^2$ instead of $\kappa(x_d)^2$ we make $\kappa(x_d)^2$ small automatically, which is a desired feature, as seen in the previous paragraph.

Since we, by following this proposed route, minimize the L^2-norm of the second derivative, the resulting curve will be a *cubic spline*. This is a fortunate fact since it means that we will not be forced to relay on extensive world information or to do any heavy computations on-line which tends to be the case when more sophisticated planning algorithms are used [12],[13],[18]. It is thus an almost trivial task to generate the splines that connect the robot and the target in the approach target behavior, and the robot and the obstacle in the approach obstacle behavior, as seen in Figure 8.

3.2   Tracking

We now have an on-line method for producing low curvature paths around detected obstacles, and hence our next task is to find a good tracking algorithm so that the robot follows the proposed path robustly.

We let the general reference path, parameterized by s, be given by

$$\begin{aligned} x_d &= p(s) \\ y_d &= q(s) \end{aligned} \quad 0 \le s \le s_f, \tag{7}$$

where the idea is to let the motion of the reference point be governed by a differential equation containing error feedback. It can be viewed as a combination of the conventional trajectory tracking, where the reference trajectory is parameterized in time, and a dynamic path following approach [17], where the criterion is to stay close to the geometric path, but not necessarily close to an *a priori* specified point at a given time. This approach makes our algorithm robust to measurement errors and external disturbances since, if both the tracking errors and disturbances are within certain bounds, the reference point moves along the reference trajectory while the robot follows it within a prespecified look-ahead distance. Otherwise, the reference point should slow down and "wait" for the robot.

Our control objectives are

$$\limsup_{t \to \infty} \rho(t) \le \epsilon_\rho$$
$$\limsup_{t \to \infty} |\phi(t) - \phi_d(t)| \le \epsilon_\phi, \tag{8}$$

where ϵ_ρ and ϵ_ϕ are positive numbers that can be made arbitrarily small, $\rho(t) = \sqrt{(x_d - x)^2 + (y_d - y)^2}$, where (x,y) is the actual position of the robot, and ϕ and ϕ_d are actual and desired robot orientations.

From (7) we directly get that $\dot{x}_d = p'(s)\dot{s}$, $\dot{y}_d = q'(s)\dot{s}$, which implies that if the robot would track the path perfectly, we would have

$$\dot{s} = \frac{p'(s)}{p'^2(s) + q'^2(s)}\dot{x} + \frac{q'(s)}{p'^2(s) + q'^2(s)}\dot{y}, \tag{9}$$

since this corresponds to $\dot{x} = \dot{x}_d$ and $\dot{y} = \dot{y}_d$. On the other hand, (9) does not contain any position error feedback, which is important for the robustness. Therefore we propose our dynamics for the reference point as follows:

$$\dot{s} = \frac{ce^{-\alpha\rho}v_0}{\sqrt{p'^2(s) + q'^2(s)}}, \tag{10}$$

where v_0 is the desired speed at which one wants the vehicle to track the path, and α and c are appropriate, positive numbers.

We now let our control algorithm be as follows:

$$v = \gamma\rho\cos(e_\phi)$$
$$\omega = ke_\phi + \dot{\phi}_d, \quad k > 0, \tag{11}$$

where both γ and k are positive, $e_\phi = \phi_d - \phi$, and $\phi_d = \arctan2(y_d - y, x_d - x)$.

In [4],[5] it was shown that for the platform model (3), governed by the control (11), the steady state tracking error, ρ, can be made as small as one wants while ϕ tends to ϕ_d exponentially. Furthermore, in steady state we have that $v \approx v_0$. Thus we, by using the control law (11), meet the control objectives defined in (8).

We thus have a way of both producing and tracking paths, and we now combine these two together into the path following behavior that moves the robot safely around the obstacles at the same time as its executed trajectories are not too far from optimal with respect to curvature. As seen in Figure 9, where real experimental data are displayed, the method seems to work well.

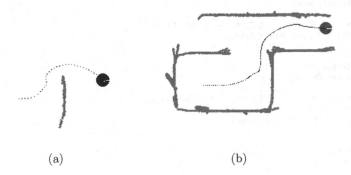

(a)   (b)

Fig. 9. The results from implementing our ideas on the Nomad 200 can be seen. The reason why the sonar readings seem rather inaccurate is due to the fact that the robot has some drift in the odometry at the same time as the sonar resolution is rather coarse.

4   Implementation

There is a need to be able to define the hybrid automata in a structured and systematic way, making it easy to reuse and reorder nodes in different configurations. Therefore, at the Centre for Autonomous Systems (CAS) [2] at KTH, a programming environment for doing mobile manipulation[4] within the hybrid automata framework has been developed [16]. It is called the *MMCA*, the Mobile Manipulation Control Architecture, and the core of the MMCA is an engine that executes hybrid automata, where, as mentioned, the nodes corresponds to different behaviors. The architecture is designed to be open and allows the user to experiment with the contents of the behaviors freely, e.g. internal representations and algorithms, as long as the behaviors contain:

(*i*) A function returning the desired state (in our case joint angles and platform pose)

(*ii*) Conditions for when to make the discrete transitions

A program written in the MMCA language begins with a specification of the initial node. Then all the nodes in the automaton are listed, where each node is specified by name, type, parameters and transitions. The transitions refer to the other nodes, or to itself, and the type of a node determines its functionality, such as what type of controller it is using, and it also defines which parameters or initial values that can be passed to the node.

A sample file that defines the task of opening a door might look like

```
INTERFACE = Puma560_XR4000;
INITNODE = Approach;
BEGIN
  NAME = Approach;
```

---

[4] From our point of view, this simply means that the mobile, behavior based platform has been augmented by the addition of a robotic arm, mounted on top [9].

```
    TYPE = Visual_Servo;

    Object = Door_Handle;     % Servo on a door handle
    TRANSITION[End_Position] = Grasp;
END
BEGIN
    NAME = Grasp;
    TYPE = Grasp_Object;

    Object = Door_Handle;     % Grasp a door handle
    TRANSITION[Got_Grip]  = Pull;
    TRANSITION[Lost_Grip] = Approach;
END
BEGIN
    NAME = Pull;
    TYPE = Follow_Arc;

    Radius = 0.8;             % Estimate of the arc radius
    Angle  = 90;              % Open door 90 deg.
    TRANSITION[Ready] = End;  % Terminates the control cycle
END
```

5   Conclusions

In this article, it is shown that a behavior based control system can be modeled as a hybrid automaton, where each node corresponds to a distinct robot behavior. In order to achieve this, we have to impose hard switches on the transitions between the different behaviors, resulting in a potentially chattering overall behavior. We furthermore show how regularization techniques can be used to solve this problem by adding extra nodes to the automaton. Those extra nodes correspond to the sliding dynamics on the boundary between the different behaviors. The performance aspect of this approach is verified experimentally on a Nomad 200 mobile platform.

We also propose a heuristic method for designing reach target behaviors in such a way that questions concerning safety and optimality can be addressed explicitly. Our proposed method is based on a combination of path planning and trajectory tracking techniques, placing it in the deliberative part of the behavior based control architecture spectrum. Furthermore, we show that this approach works well in practice on our experimental platform.

We conclude the article with a brief presentation of a programming environment, the MMCA, for defining hybrid automata in a systematic and structured way.

Acknowledgments

The author would like to thank Karl Henrik Johansson, John Lygeros, and Shankar Sastry for valuable comments about the regularization aspects of hybrid automata. He would also like to thank Xiaoming Hu and Henrik Christensen for their ideas on autonomous robotics, and Lars Petersson for helping developing the MMCA programming environment.

References

1. J. Ackermann. *Robust Control*. Springer-Verlag, London, 1993.
2. M. Andersson, A. Orebäck, M. Lindström, and H.I. Christensen. *Intelligent Sensor Based Robotics*. Ch. ISR: An Intelligent Service Robot, Lecture Notes in Artificial Intelligence, Heidelberg: Springer Verlag, 1999.
3. R.C. Arkin. *Behavior-Based Robotics*. The MIT Press, Cambridge, Massachusetts, 1998.
4. M. Egerstedt, X. Hu, and A. Stotsky. Control of a Car-Like Robot Using a Virtual Vehicle Approach. Proceedings of the *37th IEEE Conference on Decision and Control*, pp. 1502–1507, Tampa, Florida, USA, Dec. 1998.
5. M. Egerstedt, X. Hu, and A. Stotsky. A Hybrid Control Approach to Action Coordination for Mobile Robots. Proceedings of *IFAC'99:14th World Congress*, Beijing, China, Jul., 1999.
6. M. Egerstedt, K. Johansson, J. Lygeros, and S. Sastry. Behavior Based Robotics Using Regularized Hybrid Automata. Proceedings of *CDC'99*, Phoenix, Arizona, Dec, 1999.
7. A.F. Filippov. *Differential Equations with Discontinuous Righthand Sides*. Kluwer Academic Publishers, 1988.
8. K. Johansson, M. Egerstedt, J. Lygeros, and S. Sastry. Regularization of Zeno Hybrid Automata. *Systems and Control Letters*, 1999. Accepted for publication in 1999 Special Issue on Hybrid Systems.
9. O. Khatib, K.Yokoi, K.Chang, D.Ruspini, R.Holmberg, A.Casal and A.Baader: Force Strategies for Cooperative Tasks in Multiple Mobile Manipulation Systems. *International Symposium of Robotics Research*, Munich, October 1995.
10. D. Kortenkamp, R.P. Bonasso, and R. Murphy, Eds. *Artificial Intelligence and Mobile Robots*. The MIT Press, Cambridge, Massachusetts, 1998.
11. J. Košecká: *A Framework for Modeling and Verifying Visually Guided Agents: Design, Analysis and Experiments*. Dissertation, Grasp Lab, March 1996.
12. B. Krogh and C. Thorpe. Integrated Path Planning and Dynamic Steering Control for Autonomous Vehicles, Proceedings of the *1986 IEEE International Conference on Robotics and Automation*, San Francisco, CA, pp. 1664-1669, 1986.
13. J.C. Latombe. *Robot Motion Planning*, Kluwer Academic Publishers, 1991.
14. J. Lygeros, C. Tomlin, and S. Sastry. Controllers for Reachability Specifications for Hybrid Systems. *Automatica*, Vol. 35, No. 3, March 1999.
15. J. Malmborg. *Analysis and Design of Hybrid Control Systems*. PhD thesis, Department of Automatic Control, Lund Institute of Technology, Lund, Sweden, May 1998.
16. L. Petersson, M. Egerstedt, and H.I. Christensen. A Hybrid Control Architecture for Mobile Manipulation. Proceedings of the *IEEE/RSJ International Conference on Intelligent Robots and Systems*, Kyongju, Korea, Oct. 1999.
17. N. Sarkar, X. Yun, and V. Kumar. Dynamic Path Following: A New Control Algorithm for Mobile Robots. *Proceedings of the 32nd Conference on Decision and Control*, San Antonio, Texas, Dec. 1993.
18. A. Stenz. Optimal and Efficient Path Planning for Partially Known Environments, Proceedings of the *1994 IEEE International Conference on Robotics and Automation*, 1994.
19. C. Tomlin, G. Papas, J. Košecká, J. Lygeros, and S.S. Sastry. Advanced Air Traffic Automation: A Case Study in Distributed Decentralized Control, *Control Problems in Robotics*, Lecture Notes in Control and Information Sciences 230, Springer-Verlag, London, 1998.

Hybrid Controllers for Hierarchically Decomposed Systems*

Kagan Gokbayrak and Christos G. Cassandras

Department of Manufacturing Engineering, Boston University
Boston, MA 02215
{kgokbayr, cgc}@bu.edu

Abstract. We consider hybrid systems consisting of a lower-level component with *time-driven* dynamics interacting with a higher-level component with *event-driven* dynamics. These typically arise in manufacturing environments where the lower-level component represents physical processes and the higher-level component represents events related to these physical processes. We formulate an optimization problem which aims at jointly optimizing the performance of both hierarchical components and present a hybrid controller for accomplishing this task. A numerical example is given to illustrate the operation of the hybrid controller.

1 Introduction

The term "hybrid" is used to characterize systems that combine *time-driven* and *event-driven* dynamics. The former are represented by differential (or difference) equations, while the latter may be described through various frameworks used for Discrete Event Systems (DES), such as timed automata, max-plus equations, or Petri nets (see [5]). Broadly speaking, two categories of modeling frameworks have been proposed to study hybrid systems: Those that extend event-driven models to include time-driven dynamics; and those that extend the traditional time-driven models to include event-driven dynamics; for an overview, see [1][2][3][11].

The hybrid system modeling framework we will consider in this paper is largely motivated by the structure of many manufacturing systems. In these systems, discrete entities (referred to as *jobs*) move through a network of workcenters which process the jobs so as to change their physical characteristics according to certain specifications. Associated with each job is a *temporal* state and a *physical* state. The temporal state of a job evolves according to event-driven dynamics and includes information such as the waiting time or departure time of the job at the various workcenters. The physical state evolves according to time-driven dynamics modeled through differential (or difference) equations which, depending on the particular problem being studied, describe changes in

---

* This work is supported in part by NSF under grants EEC-9527422 and ACI-9873339, AFOSR under grant F49620-98-1-0387, AFRL under contract F30603-99-C-0057 and EPRI/DOD under contract WO8333-03.

N. Lynch and B. Krogh (Eds.): HSCC 2000, LNCS 1790, pp. 117–129, 2000.

such quantities as the temperature, size, weight, chemical composition, or some other measure of the "quality" of the job. The interaction of time-driven with event-driven dynamics leads to a natural trade-off between temporal requirements on job completion times and physical requirements on the quality of the completed jobs. For example, while the physical state of a job can be made arbitrarily close to a desired "quality target", this usually comes at the expense of long processing times resulting in excessive inventory costs or violation of constraints on job completion deadlines. Our objective, therefore, is to formulate and solve optimal control problems associated with such trade-offs.

In this paper, we formulate and analyze a large class of optimal control problems for hybrid systems viewed as consisting of two hierarchical components. The lower-level component represents physical processes characterized by time-driven dynamics, and the higher-level component represents events related to these physical processes. In the manufacturing context, jobs undergo various physical processes taking place at workcenters which are supervised through events such as starting and stopping the processes at appropriate times. Unlike earlier work in [9], which assumes a constant control input for each job and focuses on the optimization of the higher level component, we design a *hybrid controller* which has the task of communicating with both components and jointly solving coupled optimization problems, one for each component, hence outperforming the previous methods. To accomplish this objective, we will utilize techniques from classical optimal control theory (see [4] [12]) for the lower-level, along with recently developed optimization techniques (see [9]) for the higher-level viewed as a DES. A key difficulty we face for the latter is the presence non-differentiabilities in the event-driven state dynamics which limit the use of classical gradient-based techniques. Recently, however, it has been shown that approximating the event-driven dynamics using surrogate functions this difficulty can be overcome (see [6]).

2   Problem Formulation

The general hybrid system model we consider is illustrated in Fig. 1. A system is initially at some *physical* state ζ_1 at time x_0 and subsequently evolves according to the *time-driven* dynamics

$$\dot{z}_1 = g_1(z_1, u_1, t), \quad z_1(x_0) = \zeta_1 \tag{1}$$

where u_1 is a control. In general, we write $u_i(t)$ to allow for explicit dependence on time, but omit it here for notational simplicity. At time x_1, a switch (event) takes place causing the physical state to become $z_2(x_1) = \zeta_2$. In general, we allow for $z_2(x_1) \neq z_1(x_1)$, and the physical state subsequently evolves according to new time-driven dynamics with this initial condition. The time of this switch, which we refer to as the *temporal* state of the system, depends on *event-driven* dynamics of the form

$$x_1 = f_0(x_0, z_1, u_1, t) \tag{2}$$

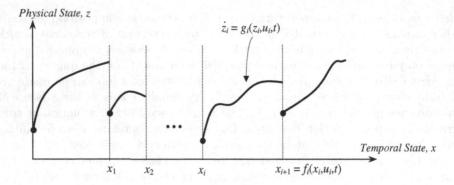

Fig. 1. Hybrid System Framework

In general, after the ith switch, the time-driven dynamics characterizing the physical state z_i are given by

$$\dot{z}_i = g_i(z_i, u_i, t), \quad z_i(x_i) = \zeta_i \tag{3}$$

and the event-driven dynamics characterizing the switching times (temporal states) x_i are given by

$$x_i = f_i(x_{i-1}, z_i, u_i, t) \tag{4}$$

Note that the choice of control following the ith switch affects both the physical state z_i and the next temporal state x_{i+1}. Thus, the switches at times $x_1, x_2, \ldots$ are generally *not* exogenous events that dictate changes in the state dynamics, but rather temporal states intrically connected to the control of the system. We emphasize this fact since it is one of the crucial elements of a "hybrid" system. In some applications, the event-driven dynamics (4) may be viewed as exogenous switching times, substantially simplifying the analysis; this is not the case in the problems we tackle in what follows.

In the context of manufacturing systems, the switches in Fig. 1 correspond to jobs that we index by $i = 1, \ldots, N$. We shall limit ourselves to a single-stage process modeled as a single-server queueing system. The objective is to process N total jobs. The server processes one job at a time on a first-come first-served non-preemptive basis (i.e., once a job begins service, the server cannot be interrupted, and will continue to work on it until the operation is completed). Jobs arriving when the server is busy wait in a queue whose capacity is larger than N.

As job i is being processed, its *physical* state, denoted by $z_i \in \mathbb{R}$ (chosen scalar for simplicity), will be assumed to evolve according to LTI time-driven dynamics

$$\dot{z}_i = g_i(z_i, u_i, t) = az_i + bu_i, \quad z_i(\tau_i) = \zeta_i \tag{5}$$

where τ_i is the time processing begins and ζ_i is the initial state at that time. The control variable u_i (assumed here to be scalar for simplicity) is used to attain a final desired physical state corresponding to a target "quality level". On the other hand, the *temporal* state of the ith job is denoted by x_i and represents the time when the job completes processing and departs from the system. Letting α_i be the arrival time of the ith job, the event-driven dynamics describing the evolution of the temporal state are given by the following "max-plus" recursive equation:

$$x_i = f(x_{i-1}, u_i, t) = \max(x_{i-1}, \alpha_i) + s_i(u_i) \tag{6}$$

where we set $x_0 = -\infty$ in which case $x_1 = \alpha_1 + s_1(u_1)$ and the first job begins service as soon as it arrives. It is assumed that the job arrival sequence $\{\alpha_1, \ldots, \alpha_N\}$ and the initial conditions ζ_i for $i = 1, .., N$ are given. The case where the order of jobs is not given is an alternative problem which we do not address in this paper. The recursive relationship (6) is known in queueing theory as the Lindley equation [5] and is the specific form of the event-driven dynamics (4) applicable to this particular hybrid system.

This system is *hybrid* in the sense that it combines the time-driven dynamics (5) with the event-driven dynamics (6), the two being coupled through the choice of the control sequence $\{u_1, \ldots, u_N\}$ where $u_i(t)$ is defined over an interval $[\max(x_{i-1}, \alpha_i), x_i)$ which depends on the choice of $u_i(t)$. The deterministic optimal control problem we consider has the general form

$$\min_{u_1, \ldots, u_N} J = \sum_{i=1}^{N} L_i(x_i, u_i) \tag{7}$$

subject to (5) and (6), where $L_i(x_i, u_i)$ is a cost function associated with job i.

We will concentrate on a family of problems for which the cost functions $L_i(x_i, u_i)$ are separable in the sense that

$$L_i(x_i, u_i) = \phi_i(u_i, s_i) + \psi_i(x_i) \tag{8}$$

The term $\psi_i(x_i)$ is the cost related to the ith job departing at time x_i. This cost may be associated with inventory level or tardiness of the job with respect to a required "due date." For example, $\psi_i(x_i) = (x_i - x_{id})^2$ defines a cost where departing after the due date x_{id} incurs a tardiness cost and completing the job before due date incurs an inventory (backlog) cost. The term $\phi_i(u_i, s_i)$ includes the cost due to applying control u_i for s_i units of time required to bring the physical state of the job as close as possible to a targeted "quality level" represented by a desired final state. Unlike earlier work (e.g. [9], [8], [13]), we do not constrain the final physical state. Instead, the deviation of the departing job's physical state from the desired "quality level" incurs a cost which is included in $\phi_i(u_i, s_i)$. Thus, the optimization problem of interest is

$$\min_{u_1, \ldots, u_N} \sum_{i=1}^{N} [\phi_i(u_i, s_i) + \psi_i(x_i)] \tag{9}$$

subject to (5) and (6).

In earlier work (e.g. [9], [8], [13]) the final state was fixed, therefore for given u_i, the processing time s_i was uniquely determined. This simplified the analysis because the cost on control could be written in terms of u_i only. However, in this paper, since the final physical state is not constrained, the service time s_i is not uniquely determined by u_i. In what follows, we consider cost functions of the form

$$\phi_i(u_i, s_i) = \frac{1}{2}h(z_{fi} - z_{di})^2 + \int_0^{s_i} \frac{1}{2}ru_i^2(t)dt \qquad (10)$$

where a quadratic cost is imposed on the deviation of the final state from the desired value and on the control applied over a processing interval $[0, s_i)$. An additional quadratic cost on the physical state $z_i(t)$ for $t \in [0, s_i)$ may also be included; in manufacturing applications, however, it is typical that only the final state z_{fi} is of interest.

In this setting, (9) is not easy to solve. Our approach is to uniquely determine u_i given s_i by decomposing the hybrid system as explained next.

Let us decompose the hybrid system hierarchically into two levels: At the lower level reside the time-driven dynamics based on which we need to control the physical state of each job to attain a target "quality level." At the higher level reside the event-driven dynamics based on which service times are controlled over all N jobs. This decomposition is convenient because the optimization at the lower level can be done one job at a time (or in parallel over all N jobs), whereas at the higher level the optimization involves the coordination over all N jobs simultaneously.

Lower-level problem. At the lower level, we consider a quadratic cost

$$\theta_i(s_i) = \frac{1}{2}h(z_{fi} - z_{di})^2 + \int_0^{s_i} \frac{1}{2}ru_i^2(t)dt \qquad (11)$$

which we view as a function of s_i, the time horizon available, i.e., the service time to be allocated to the ith job. Note that z_{di} is a desired final physical state, z_{fi} is the actual final state, and h, r are weights associated with the terminal cost and control cost respectively. We choose the notation $\theta_i(s_i)$ to differentiate this cost function from $\phi_i(u_i, s_i)$ in (10), since we will now seek to optimize over a given s_i. In particular, we face an optimization problem for each $i = 1, \ldots, N$:

$$\min_{u_i} \theta_i(s_i)$$

$$\text{s.t.} \quad \dot{z}_i = az_i + bu_i, \quad z_i(0) = \zeta_i \qquad (12)$$

This problem can be solved as a function of s_i, so that the optimal control is parameterized by s_i. Once the solution is obtained, we can evaluate the cost as a function of s_i to get

$$\theta_i^*(s_i) = \min_{u_i} \frac{1}{2}h(z_{fi}^* - z_{di})^2 + \int_0^{s_i} \frac{1}{2}ru_i^{*2}(t)dt \qquad (13)$$

Higher-level problem. If the higher level is provided with the information $u_i^*(s_i)$, then $\phi_i(u_i, s_i)$ in (9) becomes $\phi_i(u_i^*(s_i), s_i)$, a function of the service time

s_i. Let us denote this function by $\tilde{\phi}_i(s_i)$ and we are then faced with the following problem:

$$\min_{s_1,\ldots,s_N} \sum_{i=1}^{N} \left[\tilde{\phi}_i(s_i) + \psi_i(x_i)\right]$$

$$s.t. \quad x_i = \max(x_{i-1}, a_i) + s_i \tag{14}$$

When this problem is solved, then the optimal values $s_1^*, \ldots, s_N^*$ can be communicated to the lower level. Therefore, the optimal control values become $u_i^*(s_i^*)$ for all $i = 1, \ldots, N$. Note that $\tilde{\phi}_i(s_i) = \theta_i^*(s_i)$ for all $i = 1, \ldots, N$.

3   Hybrid Controller

The hybrid controller we propose for coordinating the two problems (12) and (14) outputs to the lower level the optimal controls $u_i^*(s_i^*)$ for all $i = 1, \ldots, N$, and to the higher level the optimal service times s_i^* for all $i = 1, \ldots, N$. The operation of the controller is overviewed next in terms of its four basic steps. Note that, for simplicity, we assume that the desired final physical state for each job is z_d and the initial physical state is z_0.

Step 1: System Identification. The values of a and b in the physical state equation $\dot{z} = az + bu$, the cost associated with the physical process ϕ, the desired final physical state z_d, and the initial state z_0 are input to the controller from the lower level. Similarly, the arrival sequence $\{\alpha_i\}$, $i = 1, \ldots, N$, and the cost associated with the temporal states ψ are also input to the controller.

Step 2: Lower level controller evaluates $\theta_i^*(s_i)$ and $u_i^*(s_i, z_d)$ for all $i = 1, \ldots, N$. The problem (12) is solved and the values of $\theta_i^*(s_i)$ and $u_i^*(s_i, z_d)$ depend on the specific constraints imposed on the controls. If, for example, the controller can output arbitrary values at any time, the optimal control for this process can be obtained as the (transient) solution of a standard LQ problem (details are omitted) to give

$$u_i^*(t) = 2abe^{-at}h\frac{z_d - z_0 e^{as_i}}{2rae^{-as_i} + e^{as_i}b^2h - b^2he^{-as_i}} \tag{15}$$

and the optimal final state is

$$z_{fi}^* = \frac{2z_0 ra + e^{as_i}b^2hz_d - b^2he^{-as_i}z_d}{2rae^{-as_i} + e^{as_i}b^2h - b^2he^{-as_i}} \tag{16}$$

Therefore,

$$\theta_i^*(s_i) = \min_{u(t)}\left[\frac{1}{2}h(z_{fi}^* - z_{di})^2 + \int_0^{s_i}\frac{1}{2}ru_i^{*2}(t)dt\right]$$

$$= h\frac{ra(z_d - z_0 e^{as_i})^2}{2ra + e^{2as_i}b^2h - b^2h}$$

By way of comparison, if the control is constrained to be constant (which is sometimes the case in manufacturing applications), its optimal value can be obtained as

$$u_i^* = hba(e^{as_i} - 1)\frac{z_d - z_0 e^{as_i}}{rs_i a^2 + hb^2(e^{as_i} - 1)^2} \qquad (17)$$

which yields

$$z_{fi}^* = z_0 e^{as_i} + \frac{bu}{a}(e^{as_i} - 1)$$

$$= \frac{z_0 e^{as_i} \frac{rs_i a^2}{hb^2(e^{as_i}-1)^2} + z_d}{\frac{rs_i a^2}{hb^2(e^{as_i}-1)^2} + 1}$$

and

$$\theta_i^*(s_i) = \frac{1}{2}h\frac{(z_d - z_0 e^{as_i})^2}{\left(\frac{hb^2(e^{as_i}-1)^2}{rs_i a^2} + 1\right)} \qquad (18)$$

Other types of controllers, such as P, PI, and PID, are also applicable. They will have corresponding $u_i^*(s_i, z_d)$ and $\theta_i^*(s_i)$ values depending on the solution of (12) using feasible controller outputs.

Step 3: Higher level controller evaluates s_i^* for all $i = 1, \ldots, N$. Once the cost of service $\tilde{\phi}_i(s_i)$, which is equal to $\theta_i^*(s_i)$, is known for some i, one can solve (14) and get the optimal service times s_i^*. The solution to this problem is the topic of ongoing research with some results reported in [5],[8],[7]. Although the problem appears similar to classical discrete-time optimal control problems commonly found in the literature (e.g., [4]), there are two issues to address. First, the index $i = 1, \ldots, N$ does not count time steps, but rather asynchronously departing jobs. Second, the presence of the "max" function in the state equation (6) prevents us from using standard gradient-based techniques, since it introduces a non-differentiability at the point where $a_i = x_{i-1}$. Regarding the first issue, although the absence of a synchronizing clock presents a difficulty encountered in all DES, note that the mathematical treatment of the recursive equation (6) is in fact no different than that of any other similar recursion where the index represents synchronized time steps as in classical discrete-time optimal control problems. Therefore, this issue is not really problematic. Regarding the second issue, recent work in [8],[7],[6] has led to the development of efficient algorithms that make use of non-smooth optimization techniques and exploit the structure of the problem. Alternatively, as described in [6], it is possible to approximate the corner of the max function with a differentiable surrogate, leading to very efficient numerical solutions with little loss of accuracy. In particular, we "smooth" the max function by fitting it with a Bezier function at the neighborhood of the corner (see Fig. 2) and solve the resulting 'Two Point Boundary Value Problem' (TPBVP).

124   K. Gokbayrak and Ch.G. Cassandras

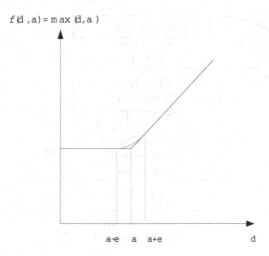

Fig. 2. Bezier approximation of a max function.

A Bezier function is constructed using $n+1$ "control points" represented by vectors $\mathbf{v}_1, \ldots, \mathbf{v}_n$ and is parametrically given by

$$\mathbf{v}(t) = \sum_{i=0}^{n} \mathbf{v}_i B_{i,n}(t)$$

where

$$B_{i,n}(t) = \frac{n!}{i!(n-i)!} t^i (1-t)^{n-i}$$

The control points define a "characteristic polygon" and the Bezier function has the property that it is contained within the convex hull of this characteristic polygon. In our case, there are three obvious control points to use: the point (a, a) where the max function is not differentiable and two points $(a - \epsilon, a)$ and $(a + \epsilon, a + \epsilon)$ which define a neighborhood of a on the d-axis in Fig. 2. The Bezier function in the neighborhood can therefore be formulated as

$$\mathbf{v}(t) = (a + \epsilon, a + \epsilon)t^2 + 2(a, a)t(1 - t) + (a - \epsilon, a)(1 - t)^2$$
$$= (a + \epsilon(2t - 1), a + \epsilon t^2)$$

where $t \in [0, 1]$. The derivative of the Bezier function is

$$\frac{d(a + \epsilon t^2)/dt}{d(a + \epsilon(2t - 1))/dt} = t$$

i.e., it starts at $\mathbf{v}(0) = (a - \epsilon, a)$ with derivative 0 and ends at $\mathbf{v}(1) = (a + \epsilon, a + \epsilon)$ with derivative 1, coinciding with the derivatives of the max function at these control points. Note that the derivative of the Bezier approximation of the max function stays between 0 and 1 inside the characteristic polygon.

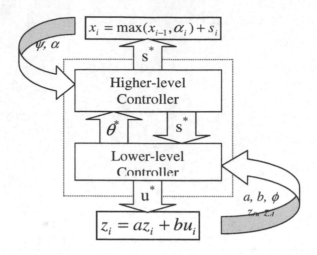

Fig. 3. Hybrid Controller Operation

The value $\epsilon > 0$ determines how tightly the surrogate function fits and can be adjusted during the execution of the TPBVP solver to achieve any desired accuracy. Selecting a very small ϵ at the beginning of the algorithm may result in chattering; therefore, it is more desirable to gradually decrease ϵ as the algorithm approaches the optimal. In the limit, the solution obtained using this approach converges to the true optimal.

Step 4: Optimal controls are output to lower level. Once the optimal service times s_i^*, $i = 1, \ldots, N$, are determined, they are provided to the lower level controller that obtained $u_i^*(s_i, z_d)$ at Step 2. The final values of the optimal controls are $u_i^*(s_i^*, z_d)$, $i = 1, \ldots, N$, and these are issued to the lower level for controlling the physical processes.

During normal system operation, the controller supplies the optimal control sequence $\{u_i^*(s_i^*, z_d)\}$ to the physical (lower) level and the optimal service time sequence $\{s_i^*\}$ to the higher level (see Fig. 3). Based on $\{s_i^*\}$ and the arrival time sequence $\{\alpha_i\}$, the higher level can signal the lower level when to start and when to stop the ith process.

4   Numerical Example

In order the illustrate the operation of the hybrid controller, we consider a single-stage hybrid manufacturing system which incurs the cost

$$J(\mathbf{u}, \mathbf{x}) = \min \sum_{i=1}^{N} \phi_i(u_i, s_i) + \psi_i(x_i)$$

$$\text{s.t. } x_i = \max(x_{i-1}, \alpha_i) + s_i$$

while processing N jobs.

In this example, we will assume that, for $i = 1, ..., N$, the cost $\psi_i(x_i)$ associated with the temporal state x_i is

$$\psi_i(x_i) = \beta(x_i - \alpha_i)^2$$

which penalizes the system time of the ith job. The cost of processing the ith job, $\phi_i(u_i, s_i)$, is in the form of (10). If, in addition, the physical state of the ith job evolves according to

$$\dot{z}_i = u_i , \qquad z_i(x_i) = q$$

where we assume the final state to be fixed, then the lower-level cost $\theta_i(s_i)$ in (11) becomes

$$\theta_i(s_i) = \int_0^{s_i} \frac{1}{2} r u_i^2$$

For simplicity, we assume fixed initial state $\zeta_i = 0$ for $i = 1, ..., N$. The arrival sequence $\{\alpha_i\}$ is also given.

Step 1: System Identification The values $a(= 0), b(= 1)$, the cost functions $\theta(s_i)$ and $\psi(x_i)(= \beta(x_i - \alpha_i)^2)$ and the sequence $\{\alpha_i\}$ are passed to the hybrid controller.

Step 2: Lower level controller evaluates $\theta_i^*(s_i)$ and $u_i^*(s_i, z_d)$ for all $i = 1, \ldots, N$. The Hamiltonian for the lower-level component is defined as

$$H(t) = \frac{1}{2} r u_i^2(t) + p(t) u_i(t)$$

where $p(t)$ is the co-state, hence the necessary conditions for optimality are

$$\dot{z}_i^*(t) = \frac{\partial H}{\partial p} = u_i^*(t)$$

$$\dot{p}^*(t) = -\frac{\partial H}{\partial z_i} = 0$$

$$0 = \frac{\partial H}{\partial u_i} = r u_i^*(t) + p^*(t)$$

Therefore,

$$u_i^*(t) = -\frac{p}{r} \triangleq u_i^* \text{ (constant)}$$

Integrating the state equation $\dot{z}_i = u_i$ gives

$$u_i^* = \frac{q}{s_i}$$

The optimal control, therefore, will incur a cost

$$\tilde{\phi}(s_i) = \theta^*(s_i) = \int_0^{s_i} \frac{1}{2} r u_i^2 dt = \frac{1}{2} r u_i \frac{q}{s_i} s_i = \gamma u_i$$

where $\gamma = \frac{1}{2} r q$.

Step 3: Higher level controller evaluates s_i^* **for all** $i = 1, \ldots, N$. The higher level optimization problem becomes

$$J(\mathbf{u}, \mathbf{x}) = \min \sum_{i=1}^{N} \gamma u_i + \beta(x_i - \alpha_i)^2$$

subject to

$$x_i = \max(x_{i-1}, \alpha_i) + \frac{q}{u_i}$$

Let us form the augmented cost

$$\bar{J}(\mathbf{u}, \mathbf{x}, \lambda) = \min \sum_{i=1}^{N} \gamma u_i + \beta(x_i - \alpha_i)^2 + \lambda_i(\max(x_{i-1}, \alpha_i) + \frac{q}{u_i} - x_i)$$

The optimality equations are

$$\frac{\partial \bar{J}}{\partial u_i} = 0, \quad \frac{\partial \bar{J}}{\partial x_i} = 0, \text{ for } i = 1, ..., N$$

which yield

$$\frac{\partial \bar{J}}{\partial u_i} = \gamma - \lambda_i \frac{q}{u_i^2} = 0 \Rightarrow u_i^2 = \frac{q\lambda_i}{\gamma}$$

$$\lambda_i = \lambda_{i+1} \frac{\max(x_i, \alpha_{i+1})}{\partial x_i} + 2\beta(x_i - \alpha_i) \text{ for } i < N$$

$$\lambda_N = 2\beta(x_N - \alpha_N)$$

Using the Bezier approximation approach described in the previous section, this TPBVP can be solved effectively to evaluate the optimal service time sequence $\{s_i^*\}$.

Step 4: Optimal controls are output to lower level. The optimal control input $u_i^* = \frac{q}{s_i}$ is fed to the system while processing the ith job (during $[\max(x_{i-1}, \alpha_i), x_i)$ interval) which departs at time x_i.

Example 1. Consider the one-stage system where $N = 10$ jobs all arrive at time $t = 0$. If $r = 6$, $q = 10$, $\beta = 1$ then the optimal controls and service times are as follows.

Job	Service Time	Optimal Control	Departure Time
1	1.35	7.43	1.35
2	1.36	7.37	2.70
3	1.38	7.25	4.08
4	1.42	7.06	5.50
5	1.47	6.79	6.97
6	1.55	6.44	8.52
7	1.67	5.98	10.20
8	1.86	5.38	12.06
9	2.18	4.58	14.24
10	2.95	3.39	17.19

This control results in a cost of $J^* = 2774.83$.

Note: The time requirement of our algorithm on a standard PC is of the order of seconds. Readers are referred to the web site

$$http://vita.bu.edu/cgc/newhybrid/onestage.html$$

to reproduce this example or try different examples by interactively varying the arrival sequence and other problem parameters.

5   Conclusions

In this paper, we considered hybrid systems modeled as a two-level hierarchy and hybrid controllers that were designed to jointly optimize the performance of both levels. The lower-level optimization problem, i.e., the determination of the optimal control $u_i^*(t)$ for the physical process of each job i when the service time s_i is known, employs classical control techniques. The higher-level optimization problem, i.e., the determination of the optimal service time sequence $\{s_i\}$ when the cost of each service time $\tilde{\phi}_i(s_i)$ is known, employs recently developed optimization techniques for DES (see [9]). The result of one optimization problem is the input to the other, therefore these optimization problems are highly coupled. The key to the decoupling process is the following: Since the lower-level controller knows the form of the optimal control solution for the deterministic process, it passes the cost information $\tilde{\phi}_i(s_i)$ to the higher-level controller. The higher-level controller can then determine the optimal service sequence $\{s_i\}$ which is passed to the lower-level controller for determination of optimal controls $u_i^*(t)$ to the physical processes.

This decomposition method relies highly on the deterministic structure of the physical processes. In the case where the arrival sequence is not known, one can start with the mean value information and resolve the optimization problem as the arrivals are observed. The speed of the solution algorithm in such a method is a key issue and is the subject of ongoing research. Another interesting case where the physical processes are stochastic was considered in [10] and is also a topic of ongoing research.

The idea of decomposition is not limited to the specific class of problems presented in this paper. The event driven dynamics at the higher-level and the time driven dynamics at the lower-level can be arbitrary.

References

1. A. Alur, T. A. Henzinger, and E. D. Sontag, editors. *Hybrid Systems*. Springer-Verlag, 1996.
2. P. Antsaklis, W. Kohn, M. Lemmon, A. Nerode, and S. Sastry, editors. *Hybrid Systems*. Springer-Verlag, 1998.
3. M. S. Branicky, V. S. Borkar, and S. K. Mitter. A unified framework for hybrid control: Model and optimal control theory. *IEEE Tr. on Automatic Control*, 43 (1):31-45, 1998.

4. A. E. Bryson and Y. C. Ho. *Applied Optimal Control.* Hemisphere Publishing Co., 1975.
5. C. G. Cassandras. *Discrete Event Systems: Modeling and Performance Analysis.* Irwin Publ., 1993.
6. C. G. Cassandras, Q. Liu, K. Gokbayrak, and D. L. Pepyne. Optimal control of a two-stage hybrid manufacturing system model. In *Proceedings of 38th IEEE Conf. On Decision and Control,* pages 450-455, Dec. 1999.
7. C. G. Cassandras, D. L. Pepyne, and Y. Wardi. Generalized gradient algorithms for hybrid system modele of manufacturing systems. In *Proc. Of 37th IEEE Conf. On Decision and Control,* pages 2627-2632, December 1998.
8. C. G. Cassandras, D. L. Pepyne, and Y. Wardi. Optimal control of systems with time-driven and event-driven dynamics. In *Proc. Of 37th IEEE Conf. On Decision and Control,* pages 7-12, December 1998.
9. C. G. Cassandras, D. L. Pepyne, and Y. Wardi. Optimal control of a class of hybrid systems. *submitted for publication,* 1999.
10. K. Gokbayrak and C. G. Cassandras. Stochastic optimal control of a hybrid manufacturing system model. In *Proceedings of 38th IEEE Conf. On Decision and Control,* pages 919-924, Dec. 1999.
11. R. L. Grossman, A. Nerode, A. P. Ravn, and H. Rischel, editors. *Hybrid Systems - Vol. 736 of Lecture Notes in Computer Science.* Springer-Verlag, 1993.
12. D. E. Kirk. *Optimal Control Theory.* Prentice-Hall, 1970.
13. D. L. Pepyne and C. G. Cassandras. Modeling, analysis, and optimal control of a class of hybrid systems. *Journal of Discrete Event Dynamic Systems: Theory and Applications,* 8(2):175-201, 1998.

Beyond HyTech: Hybrid Systems Analysis Using Interval Numerical Methods*

Thomas A. Henzinger[1], Benjamin Horowitz[1], Rupak Majumdar[1], and
Howard Wong-Toi[2]

[1] Department of Electrical Engineering and Computer Sciences
University of California at Berkeley
{tah,bhorowit,rupak}@eecs.berkeley.edu
[2] Cadence Berkeley Laboratories, Berkeley, CA
howard@cadence.com

Abstract. Since hybrid embedded systems are pervasive and often safety-critical, guarantees about their correct performance are desirable. The hybrid systems model checker HyTech provides such guarantees and has successfully verified some systems. However, HyTech severely restricts the continuous dynamics of the system being analyzed and, therefore, often forces the use of prohibitively expensive discrete and polyhedral abstractions. We have designed a new algorithm, which is capable of directly verifying hybrid systems with general continuous dynamics, such as linear and nonlinear differential equations. The new algorithm conservatively overapproximates the reachable states of a hybrid automaton by using interval numerical methods. Interval numerical methods return sets of points that enclose the true result of numerical computation and, thus, avoid distortions due to the accumulation of round-off errors. We have implemented the new algorithm in a successor tool to HyTech called HyperTech. We consider three examples: a thermostat with delay, a two-tank water system, and an air-traffic collision avoidance protocol. HyperTech enables the direct, fully automatic analysis of these systems, which is also more accurate than the use of polyhedral abstractions.

1 Introduction

In a hybrid system, digital controllers interact with a continuous environment. Because of the increasing ubiquity of embedded real-time systems, hybrid systems directly control many of the devices in our daily lives. Moreover, hybrid systems are often components of safety- or mission-critical systems. For these reasons, it is necessary to have rigorous guarantees about the correct performance of hybrid systems.

---

* This research was supported in part by the DARPA (NASA) grant NAG2-1214, the DARPA (Wright-Patterson AFB) grant F33615-C-98-3614, the ARO MURI grant DAAH-04-96-1-0341, and the NSF CAREER award CCR-9501708.

N. Lynch and B. Krogh (Eds.): HSCC 2000, LNCS 1790, pp. 130–144, 2000.

Hybrid automata [1] provide a modeling paradigm for hybrid systems. In a hybrid automaton, the discrete state and dynamics are modeled by the vertices (called *locations*) and edges of a graph, respectively, and the continuous state and dynamics are modeled by points in $\mathbb{R}^n$ and differential equations, respectively. Symbolic model checking on a hybrid automaton provides correctness guarantees. HYTECH [10] is a model checker for hybrid systems that has been successful in analyzing many hybrid systems of practical interest [2,5,13,14,15,16,23,26,27].

Despite its successes, HYTECH has several shortcomings. It restricts the dynamical model of the automaton being analyzed to that of linear hybrid automata. In linear hybrid automata, the continuous dynamics are governed by polyhedral differential inclusions, and all trajectories are composed of lines with piecewise constant slopes. These limitations force the verifier to approximate the complex dynamics of a hybrid system in a less expressive dynamical model. This approximation may take the form of *rate translation* [11], in which the first derivative of every continuous variable is bounded above and below by constants. *Location splitting* may be used to make the approximation arbitrarily accurate: each location can be split into many new locations; in these new locations, the dynamics may be bounded more precisely. However, location splitting leads to state explosion, as accuracy in the model comes at the price of a large number of new locations. Thus, the restrictive input language often forces the use of prohibitively large approximate models.

A second deficiency of HYTECH is that arithmetic overflows frequently occur in the course of HYTECH's computation. To explain this problem, we briefly describe the basic algorithm underlying HYTECH. A state s of a hybrid automaton has two types of successors: *flow* successors, which are the states reachable from s by letting time progress; and *jump* successors, which are the states reachable from s if the automaton undergoes a change of location. Call a set of states a *polyhedral region* if its continuous part is a polyhedron. For linear hybrid automata, the flow and jump successors of a polyhedral region form again polyhedral regions. The computation engine of HYTECH computes the set of states that can be reached from an initial polyhedral region by any number of flows and jumps. The iterated computation of flow and jump successors continues until either a target state is reached or no new states are generated.[1] The polyhedral manipulations for computing flow and jump successors use exact computation over rationals stored as integer pairs. However, these repeated computations quickly generate rationals with very large representations, leading to arithmetic overflows.

We have implemented the program HYPERTECH, which addresses both inadequacies of HYTECH. First, HYPERTECH supports the analysis of hybrid automata with much more general dynamics. In particular, HYPERTECH can analyze automata whose continuous dynamics are given by differential equations of the form $dx_i/dt = f(x_1, \ldots, x_n)$, where f is a composition of polynomials, exponentials, and trigonometric functions. This class of hybrid systems includes all multi-modal linear systems, i.e., systems whose continuous dynamics are given

---

[1] In general, the computation may fail to terminate, because the reachability problem for linear hybrid automata is undecidable [1].

by matrix differential equations of the form $dx/dt = Ax + Bu$ (where u represents the control input or disturbance input). HYPERTECH's more permissive input language enables a direct modeling of the continuous dynamics of hybrid systems. Since the need for introducing abstractions (in the form of rate translation and location splitting) is removed, input automaton models can be much more compact than with HYTECH.

Second, HYPERTECH uses interval numerical methods [20,22] to compute an overapproximation of the set of reachable states of a hybrid automaton. In interval methods, the computed solution to a numerical problem, e.g., an initial value problem, is guaranteed to enclose the true solution. This is in contrast to conventional numerical methods, in which the accumulation of round-off errors may cause a computed solution to deviate from the real solution. The analysis engine of HYPERTECH, like that of HYTECH, starts with an initial region and iteratively adds flow and jump successors. However, HYPERTECH uses an interval ordinary differential equation (ODE) solver, instead of polyhedral manipulations, to compute an overapproximation of the flow successors of a set of states. It is the use of *interval* numerical methods which guarantees that the reachable states of a hybrid automaton H are contained in the set of states computed by HYPERTECH when run on H. All regions resulting from interval methods are *rectangular*, i.e., a product of intervals. Since geometrically manipulating rectangles is simpler than manipulating arbitrary polyhedra, the internal representations of HYPERTECH's rectangles never grow very large. In this way, HYPERTECH avoids the numeric overflow errors of HYTECH.

In essence, while the restrictive dynamics of HYTECH force an approximation in the model (*static approximation*), the permissive dynamics of HYPERTECH allow approximation to occur only during the computation of reachable states (*dynamic approximation*). Despite the fact that the dynamic approximation using rectangular regions seems rough, we demonstrate it to be superior to static approximation using polyhedral differential inclusions, on three examples: a thermostat with delay, a two-tank water system, and an air-traffic collision avoidance protocol.

In traditional numerical integration, the accumulation of round-off errors may cause the computed solution to a numerical problem to differ widely from the real solution. In the context of hybrid systems analysis, the loss of precision caused by the accumulation of round-off errors in the numerical integration process may lead the analyzer, whether human or computer, to overlook potentially hazardous events. In contrast to hybrid systems simulators (see [21] for a survey) and reach-set computation tools [3,4,6,8,26] which use traditional (not interval-based) numerical methods, HYPERTECH is guaranteed not to miss any events. Thus, if an unsafe (target) state of a hybrid automaton is reachable, HYPERTECH is guaranteed to note its reachability.

The rest of the paper is organized as follows. In Section 2, we describe the syntax and semantics of the hybrid automaton model, and define the reachability problem. In Section 3, we describe in detail the algorithm implemented in

HyperTech. In Section 4, we describe the results of running HyperTech on three examples, and compare the results with HyTech.

2   Hybrid Automata

To model hybrid systems, we use hybrid automata [1]. Let $\mathbb{R}^n$ be the n-dimensional Euclidean space. A *rectangle of dimension* n is a subset of $\mathbb{R}^n$ that is the Cartesian product of (possibly unbounded) intervals, all of whose finite endpoints are rational. For a positive integer n, let $\mathcal{R}^n$ denote the set of all n-dimensional rectangles. An *axis-parallel hyperplane* $h \subseteq \mathbb{R}^n$ is a set of points $\{\mathbf{x} \mid \mathbf{x}_i = a\}$ for some $i \in \{1, \dots, n\}$ and some rational number a. Let a *dynamical equation* be an expression A generated by the grammar

$$A := (\dot{x}_1 = B_1 \wedge \dot{x}_2 = B_2 \wedge \cdots \wedge \dot{x}_i = B_i), \ B := x_j \mid [a, b] \mid B_1 \ \text{op} \ B_2 \mid \mathsf{f}(B)$$

where i and j are positive integers, a and b are any rational numbers such that $a \leq b$, op is one of the arithmetic operations $+$ (addition), $-$ (subtraction), $\cdot$ (multiplication), $/$ (division), or $\hat{\ }$ (exponentiation), and f is one of the functions sin, cos, tan, or exp. We shall use conventional mathematical notation for dynamical equations whenever possible, and if $a = b$ we shall often omit the square braces. For example, $\dot{x}_1 = \frac{1}{2}\sqrt{x_1 - x_2^2} \wedge \dot{x}_2 = \sqrt{x_1^2 + x_2^2}$ is a dynamical equation. For a positive integer n, let $\mathcal{E}^n$ denote the set of all dynamical equations in which for each subexpression of the form $\dot{x}_i$ or x_j, both $i \leq n$ and $j \leq n$. The above example of a dynamical equation is a member of $\mathcal{E}^2$.

2.1   Syntax

A *hybrid automaton* H consists of the following components:

- A finite set $X = \{x_1, \dots, x_n\}$ of real-valued variables. A valuation of these variables represents a continuous state of a hybrid system.
- A finite directed multigraph (V, E). The vertices in V (called *control locations*) represent the discrete state of a hybrid system. The edges in E (*control switches*) represent transitions between discrete states.
- Three functions $inv : V \to \mathcal{R}^n$, $init : V \to \mathcal{R}^n$, and $flow : V \to \mathcal{E}^n$. Each invariant $inv(v)$ represents a condition that must be satisfied if the automaton is to remain in location v. Each initial condition $init(v) \subseteq inv(v)$ represents the continuous states in which the hybrid automaton may begin executing, when control starts at location v. Each flow condition $flow(v)$ constrains the continuous dynamics of the hybrid system at location v.
- Two functions $pre : E \to \mathcal{R}^n$ and $post : E \to \mathcal{R}^n$. For each edge $e = (v, v')$ in E, we require that $pre(e) \subseteq inv(v)$ and that $post(e) \subseteq inv(v')$. Intuitively, $pre(e)$ represents the condition on the continuous state that must hold if control is to pass from v to v', and $post(e)$ constrains the possible values of the variables after the transfer of control from v to v'.

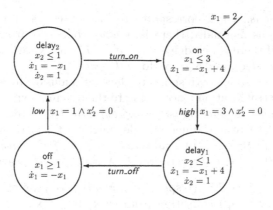

Fig. 1. Thermostat with delay

– A function $update : E \rightarrow 2^{\{1,\dots,n\}}$ that assigns to each edge $e = (v, v') \in E$ a subset $update(e) \subseteq \{1, \dots, n\}$. After traversing e, if the index i is in $update(e)$, then the variable x_i gets nondeterministically reset so as to lie in the i-th projection of $post(e)$, whereas if $i \notin update(e)$, then x_i remains unchanged.
– A finite set Σ of events, and a function $event$ that assigns to each edge $e \in E$ an event.

As an example, consider the hybrid automaton of Figure 1, which models a thermostat system with delays: after the thermometer detects that the temperature is too low or too high, there may be a delay of up to one second before the appropriate control action (turn the heater on or off, respectively) is taken. The variable x_1 measures the temperature. Initially, $x_1 = 2$ and the heater is on. The temperature rises according to the differential equation $\dot{x}_1 = -x_1 + 4$. Eventually, the temperature reaches three degrees; after a delay of one second in location delay_1, the thermostat sends a *turn_off* signal to the heater. The variable x_2 measures the delay. The temperature then falls according to the equation $\dot{x}_1 = -x_1$ until $x_1 = 1$. One second after the temperature reaches one degree, the thermostat sends a *turn_on* signal to the heater, and the run of the automaton continues.

2.2   Semantics

We now give a formal definition of the semantics of a hybrid automaton. A *state* of a hybrid automaton is a pair $(v, \mathbf{x})$, with location $v \in V$, continuous state $\mathbf{x} \in \mathbb{R}^n$, and $\mathbf{x}$ satisfying $inv(v)$. The *state space* of a hybrid automaton is the set of its states. If $\mathbf{u} \in \mathbb{R}^n$ is a vector, we denote by $X := \mathbf{u}$ the interpretation for the variables in X in which $x_i = \mathbf{u}_i$ for $i = 1, \dots, n$. A hybrid automaton has two types of transitions:

- *Jump transitions,* which correspond to instantaneous transitions between control locations. Formally, there is a *jump transition* from state $(v, \mathbf{x})$ to state $(v', \mathbf{x}')$ if there is an edge $e = (v, v') \in E$ with $\mathbf{x}$ satisfying $pre(e)$, and $\mathbf{x}'$ satisfying $post(e)$, and $\mathbf{x}'_i = \mathbf{x}_i$ for $i \notin update(e)$.
- *Flow transitions,* which correspond to the continuous evolution of the system at a single control location v according to the dynamics specified by $flow(v)$. Formally, there is a *flow transition of duration* $t \geq 0$ from state $(v, \mathbf{x})$ to state $(v, \mathbf{x}')$ if there is a differentiable function $f : [0, t] \to \mathbb{R}^n$ such that: (1) $f(0) = \mathbf{x}$, $f(t) = \mathbf{x}'$; (2) for all reals $t' \in [0, t]$, $f(t') \in inv(v)$; and (3) for all reals $t' \in [0, t]$, the interpretation $X, \dot{X} := f(t'), \dot{f}(t')$ satisfies $flow(v)$.

We say that $(v', \mathbf{x}')$ is a *flow* (respectively *jump*) *successor* of $(v, \mathbf{x})$ if there is a flow (respectively jump) transition from $(v, \mathbf{x})$ to $(v', \mathbf{x}')$. A *run* of a hybrid automaton is an infinite sequence of states $(v_0, \mathbf{x}^0), (v_1, \mathbf{x}^1), \ldots$ such that $\mathbf{x}^0 \in init(v_0)$, and for all $i \geq 0$, $(v_{i+1}, \mathbf{x}^{i+1})$ is a jump or flow successor of $(v_i, \mathbf{x}^i)$.

2.3   Reachability Problem

The fundamental verification problem for hybrid automata is *safety verification*: given a partition of the state space into "safe" states and "unsafe" states, verify that each execution of the hybrid automaton does not reach the unsafe states. Dually, one may look at the *reachability* question: does any run of the hybrid automaton ever reach an unsafe state? Formally, given a hybrid automaton H and a subset S of its state space, the reachability problem asks if there is a run $(v_0, \mathbf{x}^0), (v_1, \mathbf{x}^1), \ldots$ of H such that $(v_i, \mathbf{x}^i) \in S$ for some i. If there is such a run, we say that the set S is *reachable*. Clearly, a solution to the reachability problem gives a solution to the safety verification problem as well. The reachability problem is undecidable even for simple subclasses of hybrid automata [12]. However, semidecision procedures —for example, the algorithm of HyTech— often terminate on specific problems of practical interest.

3   The HyperTech Algorithm

3.1   Interval Numerical Methods

In numerical computations, such as the numerical solution of ODEs, rounding errors may distort the accuracy of a sequence of calculations. Thus, ordinary numerical methods cannot provide fully rigorous guarantees about the safety of dynamical systems. Interval numerical methods [20] address this problem by computing sets of points that contain the true solutions to a numerical problem. In particular, interval ODE solvers find guaranteed bounds for the solutions to initial value problems.

In interval methods, the fundamental object of computation is not a floating point number, but rather an interval. An *interval* $[\underline{x}, \overline{x}]$ is a nonempty set of real numbers $\{x \in \mathbb{R} \mid \underline{x} \leq x \leq \overline{x}\}$, where $\underline{x} \leq \overline{x}$ are both real numbers. One can extend to intervals the usual arithmetic operations over reals: if op is an arithmetic operation, then $[\underline{x}, \overline{x}] \text{ op } [\underline{y}, \overline{y}] = \{x \text{ op } y \mid x \in [\underline{x}, \overline{x}], y \in [\underline{y}, \overline{y}]\}$. The operations $+, -, \cdot$, and $/$ on intervals may be seen to satisfy the following identities:

$$[\underline{x}, \overline{x}] + [\underline{y}, \overline{y}] = [\underline{x} + \underline{y}, \overline{x} + \overline{y}]$$
$$[\underline{x}, \overline{x}] - [\underline{y}, \overline{y}] = [\underline{x} - \overline{y}, \overline{x} - \underline{y}]$$
$$[\underline{x}, \overline{x}] \cdot [\underline{y}, \overline{y}] = [\min(\underline{x} \cdot \underline{y}, \underline{x} \cdot \overline{y}, \overline{x} \cdot \underline{y}, \overline{x} \cdot \overline{y}), \max(\underline{x} \cdot \underline{y}, \underline{x} \cdot \overline{y}, \overline{x} \cdot \underline{y}, \overline{x} \cdot \overline{y})]$$
$$1 / [\underline{x}, \overline{x}] = [1/\overline{x}, 1/\underline{x}] \quad \text{if } 0 \notin [\underline{x}, \overline{x}]$$

A computer implementation of these operations sets the processor's rounding mode to round down when computing the lower bound of the result, and round up when computing the upper bound. This guarantees that the computed result always encloses the result that would have been obtained using exact arithmetic calculation. In a similar fashion, one can implement interval versions of standard functions (e.g., $\sin x$, e^x, etc.) so that the computed result contains the exact result. Several interval arithmetic packages exist, either as libraries [18] or as extensions to regular programming languages [17].

Interval methods to solve initial value problems use as primitives the interval operations $+$, $-$, $\cdot$, and $/$ defined above, plus interval implementations of standard functions such as sine and cosine. From an initial condition (a rectangle r_0 at time 0), these methods usually compute a rough enclosure $r_{\Delta t}$ of the solution at time Δt, where Δt is an input parameter to the program. This rough enclosure, which is a rectangle, is usually narrowed by a pruning procedure that reduces the accumulation of numerical errors, and mitigates the wrapping effect. (The *wrapping effect* is the error resulting from enclosing a nonrectangular region by a rectangle.) This iteration —computing $x_{i\Delta t}$ using $x_{(i-1)\Delta t}$ by finding, and then pruning, a rough enclosure at time $i\Delta t$— continues for a number of steps which is specified by another input parameter.

Several implementations of interval ODE solvers are publicly available, for example [19,24]. These typically use Picard iteration to prove the existence and uniqueness of a solution, and to find a rough enclosure. This enclosure is then pruned both by using a mean value method and by bounding the error term in a truncated Taylor expansion. To reduce the wrapping effect, local coordinate transforms may be applied. For a variety of examples, these implementations find fairly tight solution enclosures. In our implementation, we have used the ADIODES library [24]. Our choice of this library is independent of the other parts of HYPERTECH; thus, any other interval ODE solver, e.g., AWA [19], may be used in place of ADIODES.

3.2   Overapproximating Reachable States

For a complex hybrid automaton H, precise analytic or closed-form descriptions of the reachable states of H may not exist or may be extremely difficult to find. In such cases, one must seek feasibly computable approximations of the reachable states. An *overapproximation* of the reachable states of H is a superset T of the reachable states of H. For analysis of the safety of a hybrid automaton, such an approximation may be useful, since if no unsafe state is in T, then no unsafe state is reachable. However, since there may be states in T which are not reachable states of H, the presence of an unsafe state in T does not necessarily imply that an unsafe state of H is reachable. In such cases, one could try to refine the automaton under consideration.

Alternatively, one could compute overapproximations of the states which are backward reachable from the intersection of T and the unsafe states. If no initial state is contained in this backward approximation, then no unsafe state is reachable. This process may be iterated to find closer approximations of the reachable unsafe states [7,9]. It is an interesting question in its own right to determine whether an error run produced by an overapproximative algorithm is an actual error run.

3.3   Overapproximation Using Interval Numerical Methods

For a hybrid automaton with discrete state set V and n real-valued variables, let a *region* be a set of states of the form $\{v\} \times U$, where U is a rectangle in $\mathbb{R}^n$. For any control location $v \in V$, let $\mathcal{H}_v$ be the set $\bigcup \{\{v\} \times pre(v, v') \mid (v, v') \in E\}$. For a rectangle U and a location v, let $U_{\Delta t,v}$ be the points $\mathbf{x}' \in \mathbb{R}^n$ such that there is a flow transition from $(v, \mathbf{x}) \in U$ to $(v, \mathbf{x}')$ of duration Δt. The procedure of HyperTech, which is presented in Figure 2, works as follows. It maintains two sets of regions: *Reached*, the explored set of regions, and *Frontier*, the set of regions that still need to be explored. As long as *Frontier* $\neq \emptyset$, one member $\{v\} \times U$ of *Frontier* is selected and removed from *Frontier*. The rectangle U is propagated according to the dynamics of v. An overapproximation of the set of reachable states $(v, \mathbf{x})$ is added to *Reached*, and an overapproximation of the set of reachable states $(v', \mathbf{x})$ (with $(v, v') \in E$) is added to *Frontier*.

The subroutine *Propagate*$_1$ first computes Y, a rectangular overapproximation of $U_{\Delta t,v}$. HyperTech uses an interval ODE solver to compute this overapproximation. The size of Δt must be determined by the user. Let S be the set of points reachable for some $\Delta \tau \leq \Delta t$, i.e., $S = \bigcup_{0 \leq \Delta \tau \leq \Delta t} U_{\Delta \tau,v}$. In addition to Y, the interval numerical method generates a rectangle T that contains the set S. In the procedure, *New-Reached* gets set to an overapproximation of the set of states in $\{v\} \times S$. Moreover, *New-Frontier* gets set to an overapproximation of the jump successors of states in $\{v\} \times S$. Notice that for large values of Δt, this bound on S may be quite coarse, and may not suffice to prove the safety property of interest. In that case, we have to reduce Δt and run the procedure again. Thus, whereas computations will be faster for larger values of Δt, more accurate analysis may require smaller values. (This speed/accuracy tradeoff is illustrated in Figure 6.) We wish to emphasize that our procedure is sound regardless of which $\Delta t > 0$ is chosen.

Theorem 1. *Let H be a hybrid automaton, and let $(v, \mathbf{x})$ be a reachable state of H. If the procedure Reachable-States (using subroutine Propagate$_1$) terminates on H, then $(v, \mathbf{x}) \in \bigcup Reached$.*

Reachable-States(H : hybrid automaton)
  Initialization: *Frontier* := $\{\{v\} \times init(v) \mid v \in V\}$;
                Reached := *Frontier*;
  while *Frontier* $\neq \emptyset$ do
        pick $(\{v\} \times U) :\in$ *Frontier*;
        Frontier := *Frontier* $\setminus (\{v\} \times U)$;
        (*New-Reached*, *New-Frontier*) := *Propagate* (v, U);
        Reached := *Reached* $\cup$ *New-Reached*;
        Frontier := *Frontier* $\cup$ *New-Frontier*;
  endwhile;

Propagate$_1$(v : location, U : rectangle)
  $Y :=$ a rectangular overapproximation of $U_{\Delta t,v}$;
  $T :=$ a rectangle which contains $inv(v) \sqcap \left(\bigcup_{0 \leq \Delta\tau \leq \Delta t} U_{\Delta\tau,v} \right)$;
  New-Reached := $\{\{v\} \times T\}$;
  New-Frontier :=
        (*Unexplored-Jump-Successors*(v,T)) $\cup (\{v\} \times (inv(v) \cap Y))$;
  return (*New-Reached*, *New-Frontier*);

Unexplored-Jump-Successors(v : location, T : rectangle)
  return $\{\{v'\} \times Z \mid (v, v') \in E, Z = Update(T \cap pre(v, v')),$
              $Z \neq \emptyset, (\{v'\} \times Z) \not\subseteq \bigcup Reached\}$;

Fig. 2. HYPERTECH's procedure for reach-set computation

While *Propagate*$_1$ performs only one time step computation, under additional assumptions it is possible to group together multiple time step computations. The resulting procedure, called *Propagate*$_2$, is shown in Figure 3. In order for the subroutine *Propagate*$_2$ to function correctly, the hybrid automaton H must satisfy the following conditions: (1) for each edge $(v, v') \in E$, the rectangle $pre(v, v')$ is a boundary of the invariant $inv(v)$; and (2) for each control location v and each point $\mathbf{x} \in inv(v)$, there exists a unique edge $e = (v, v')$ such that, under the dynamics $flow(v)$, the point $\mathbf{x}$ moves strictly monotonically towards the hyperplane $pre(v, v')$, and $\mathbf{x}$ eventually crosses $pre(v, v')$. For a large class of examples, including the hybrid automata in this paper, these two conditions hold.

Note that the above conditions imply that transitions are *urgent* —they must be taken as soon as they are enabled. Thus, *Propagate*$_2$ needs only to consider the first time a region hits one or more exit hyperplanes. The subroutine *Propagate*$_2$ functions like multiple iterations of *Propagate*$_1$, except that at each iteration those trajectories which have crossed an exit hyperplane are not further explored. By the conditions above, this optimization does not compromise soundness —the procedure of HYPERTECH still explores all reachable states.

$Propagate_2(v : \text{location}, U : \text{rectangle})$
$\quad W := U; \text{New-Reached} := \text{New-Frontier} := W_{prev} := T := P := \emptyset;$
$\quad \text{while } W \neq \emptyset \text{ do}$
$\qquad W_{prev} := W;$
$\qquad W := \text{a rectangular overapproximation of } W_{\Delta t, v};$
$\qquad T := \text{a rectangle which contains } inv(v) \cap \left(\bigcup_{0 \leq \Delta\tau \leq \Delta t} U_{\Delta\tau, v}\right);$
$\qquad \text{New-Reached} := \text{New-Reached} \cup \{\{v\} \times T\};$
$\qquad P := \text{the subset of } W \text{ that has crossed } \mathcal{H}_v;$
$\qquad W := W \setminus P;$
$\qquad \text{if } P \neq \emptyset \text{ then New-Frontier} :=$
$\qquad\qquad \text{New-Frontier} \cup \text{Unexplored-Jump-Successors}(v, T) \text{ endif};$
$\quad \text{endwhile};$
$\quad \text{return } (\text{New-Reached}, \text{New-Frontier});$

Fig. 3. Grouping together multiple time step computations

Theorem 2. *Let H be a hybrid automaton satisfying conditions (1) and (2) above, and let $(v, \mathbf{x})$ be a reachable state of H. If the procedure Reachable-States (using subroutine $Propagate_2$) terminates on H, then $(v, \mathbf{x}) \in \bigcup \text{Reached}$.*

4 Three Examples

With the use of interval methods, we obtain *both* a more direct model of the target system (i.e., no rate translation needed) *and* tighter bounds on the sets of reachable states. We substantiate this claim by describing the results of running HYPERTECH on three examples.

4.1 Thermostat With Delay

Consider again the hybrid automaton of Figure 1. We wish to determine the range within which the temperature always lies. The nonlinear dynamics cannot be modeled directly in HyTech. Instead, the dynamics of the temperature x_1 are approximated using rate translation [11]. Using this method, the bounds obtained by HyTech are $0 \leq x_1 \leq 4$. This approximation may be made arbitrarily accurate by splitting each control location and using better bounds on the derivatives in the new locations. By combining rate translation with location splitting, and using a 20-location approximation of the system, HyTech obtains the bounds $0.28 \leq x_1 \leq 3.76$. This 20-location automaton is pictured in Figure 4.

We can run our algorithm directly on the automaton of Figure 1, with a step size of $\Delta t = 0.1$. Initially, $x_1 = 2$, and the automaton is in location on. Our algorithm propagates the values of x_1 according to the differential equation $\dot{x}_1 = -x_1 + 4$, until the interval containing the true value of x_1 entirely crosses the exit condition $x_1 = 3$. At this point, there is a discrete jump to location

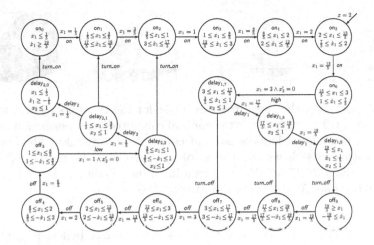

Fig. 4. Rate translation of thermostat automaton, with each location split into five locations

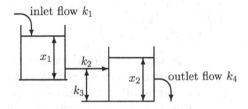

Fig. 5. Two-tank system

delay$_1$. Now our algorithm propagates the interval $[3, 3]$ for one time unit. At the end of one time unit, $x_1 \leq 3.64$, and the automaton jumps to location off. Continuing this process, our algorithm reports that the minimum value of x_1 (which is reached in location delay$_2$) is 0.367. Therefore, using HYPERTECH, the bounds are $0.367 \leq x_1 \leq 3.64$. The bounds found by analytically solving this system are $\frac{1}{e} \leq x_1 \leq 4 - \frac{1}{3}$. Note that $\frac{1}{e} \approx 0.3679$ and $4 - \frac{1}{3} \approx 3.632$. Comparing our results with the analytic solution shows that HYPERTECH computes a close approximation to the actual set of reachable states.

4.2 Two-Tank System

As a second example, we consider the two-tank system of [25] (see Figure 5). The plant consists of two identical interconnected tanks. Into tank 1 flows a stream characterized by the loss parameter k_1.[2] Tank 1's outlet stream, characterized by the loss parameter k_2, flows into tank 2. Tank 1's outlet stream is k_3 meters

---

[2] This loss parameter may be thought of as a friction loss term.

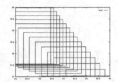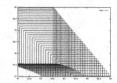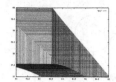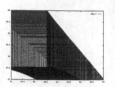

Fig. 6. A portion of the generated rectangles for the two-tank system at times $i\Delta t$, for $i = 0, 1, 2, \ldots$. HyperTech's actual computed overapproximation is the union of all rectangular hulls of pairs of consecutive rectangles. The horizontal (resp. vertical) axis shows the values of x_1 (resp. x_2). From the left: $\Delta t = 5$, running time: 24.27 s.; $\Delta t = 2$, running time: 53.39 s.; $\Delta t = 1$, running time: 98.60 s.; and $\Delta t = 0.5$, running time: 190.64 s.

above tank 2. The outlet stream of tank 2 is characterized by loss parameter k_4. Let x_1 and x_2 denote the heights of the liquid columns in tank 1 and tank 2. Applying Toricelli's law, the dynamics of this system may be seen to be:

$$
\begin{pmatrix} \dot{x}_1 \\ \dot{x}_2 \end{pmatrix} = \begin{cases} \begin{pmatrix} k_1 - k_2\sqrt{x_1 - x_2 + k_3} \\ k_2\sqrt{x_1 - x_2 + k_3} - k_4\sqrt{x_2} \end{pmatrix} & \text{if } x_2 > k_3 \\ \begin{pmatrix} k_1 - k_2\sqrt{x_1} \\ k_2\sqrt{x_1} - k_4\sqrt{x_2} \end{pmatrix} & \text{if } x_2 \leq k_3 \end{cases}
\tag{1}
$$

The dynamical equations change when the liquid level in tank 2 is equal to the height of the connecting pipe. Under this dynamics, the system moves towards an equilibrium point for all $x_i > 0$ and for all $k_i > 0$. For example, for the parameter values $k_2 = k_4 = 1$ $\sqrt{\text{meters}}$ per second, $k_3 = 0.5$ meters, and $k_1 = 0.75$ meters per second, the system moves towards the equilibrium point $x_1 = 0.625\ldots$, $x_2 = 0.563\ldots$. In [25], rate approximation is used to model this dynamical system as a 12-location hybrid automaton; HyTech is then used to overapproximate which states were reachable. With HyperTech, we directly model the system as a hybrid automaton with two states, corresponding to whether $x_2 > k_3$ or not. Further, the analysis is more accurate. For example, HyTech's analysis of the 12-location rate approximation finds that starting from $0.70 \leq x_1 \leq 0.80$ and $0.45 \leq x_2 \leq 0.50$, some states in which both $0.60 \leq x_1 \leq 0.80$ and $0.60 \leq x_2 \leq 0.65$ are reachable, whereas our algorithm shows that these states are unreachable. In Figure 6, we show a part of the overapproximation of the reachable states of the two-tank system, for four different choices of the time step Δt, with the corresponding running times. The running times are obtained on a Sun SPARCstation-20.

4.3   Air-Traffic Conflict Resolution

As a final example, consider an air-traffic conflict resolution system from [26] (see Figure 7). Two aircraft fly towards each other at a fixed altitude and 90 degree relative orientation. When the distance between the aircraft decreases to seven miles, they initiate an avoidance maneuver: each turns 90 degrees to its

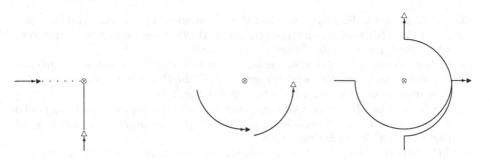

Fig. 7. Aircraft collision avoidance protocol

right, and starts following a half circle. After the half circle is complete, each again turns 90 degrees to its right to continue on the original heading along a straight path.

We model this protocol directly as a three-location hybrid automaton with the original kinematics. In contrast, the protocol would need to be approximated in HyTech in order to be verified. Our model works in a relative coordinate system, so that x_r and y_r give the position of airplane 2 relative to airplane 1, and ψ_r gives the angular orientation of airplane 2 relative to airplane 1. In relative coordinates, the kinematic equations of this system are

$$\dot{x}_r = -v_1 + v_2 \cos \psi_r + \omega_1 y_r , \quad \dot{y}_r = v_2 \sin \psi_r - \omega_1 x_r , \quad \dot{\psi}_r = \omega_1 - \omega_2 , \quad (2)$$

where v_1 (respectively v_2) is the airspeed of airplane 1 (respectively airplane 2) and ω_1 (respectively ω_2) is the angular velocity of airplane 1 (respectively airplane 2). Our automaton has three locations: cruise$_1$, avoid, and cruise$_2$. In location cruise$_1$ the airplanes follow straight-line trajectories, with airspeeds v_1 and v_2 in the range $[.8, 1]$. When the distance between the airplanes decreases to seven miles, the control location changes to avoid. On changing to location avoid, the heading of each aircraft decreases instantaneously by $\frac{\pi}{2}$ radians. In location avoid, $\omega_1 = \omega_2 = 1$ and $v_1 = v_2 = 1$, so that both airplanes follow circular trajectories of the same radius at the same airspeed. When the airplanes have completed their half-circles, the location changes to cruise$_2$. Again the heading of each aircraft decreases instantaneously by $\frac{\pi}{2}$ radians, and the airplanes continue in straight-line trajectories, with airspeeds v_1 and v_2 as in location cruise$_1$. Using this model, we are able to verify in HyperTech that the two airplanes never come within five nautical miles of each other.

References

1. R. Alur, C. Courcoubetis, T.A. Henzinger, and P.-H. Ho. Hybrid automata: an algorithmic approach to the specification and verification of hybrid systems. In *Hybrid Systems I*, LNCS 736, pages 209–229. Springer-Verlag, 1993.

2. B. Bérard and L. Fribourg. Automated verification of a parametric real-time program: the ABR conformance protocol. In *CAV 99: Computer-aided Verification*, LNCS 1633, pages 95–107. Springer-Verlag, 1999.

3. O. Botchkarev and S. Tripakis. Verification of hybrid systems with linear differential inclusions using ellipsoidal approximations. In *HSCC 2000: Hybrid Systems: Computation and Control*, LNCS. Springer-Verlag, 2000.

4. A. Chutinan and B. Krogh. Computing polyhedral approximations to flow pipes for dynamic systems. In *Proceedings of the 37th Conference on Decision and Control*, pages 2089–2094. IEEE Press, 1998.

5. J.C. Corbett. Timing analysis of ADA tasking programs. *IEEE Transactions on Software Engineering*, 22(7):461–483, 1996.

6. T. Dang and O. Maler. Reachability analysis via face lifting. In *HSCC 98: Hybrid Systems: Computation and Control*, LNCS 1386, pages 96–109. Springer-Verlag, 1998.

7. D.L. Dill and H. Wong-Toi. Verification of real-time systems by successive over- and underapproximation. In *CAV 95: Computer-aided Verification*, LNCS 939, pages 409–422. Springer-Verlag, 1995.

8. M.R. Greenstreet and I. Mitchell. Integrating projections. In *HSCC 98: Hybrid Systems: Computation and Control*, LNCS 1386, pages 159–174. Springer-Verlag, 1998.

9. T.A. Henzinger and P.-H. Ho. A note on abstract-interpretation strategies for hybrid automata. In *Hybrid Systems II*, LNCS 999, pages 252–264. Springer-Verlag, 1995.

10. T.A. Henzinger, P.-H. Ho, and H. Wong-Toi. HyTech: a model checker for hybrid systems. *Software Tools for Technology Transfer*, 1:110–122, 1997.

11. T.A. Henzinger, P.-H. Ho, and H. Wong-Toi. Algorithmic analysis of nonlinear hybrid systems. *IEEE Transactions on Automatic Control*, 43(4):540–554, 1998.

12. T.A. Henzinger, P.W. Kopke, A. Puri, and P. Varaiya. What's decidable about hybrid automata? *Journal of Computer and System Sciences*, 57:94–124, 1998.

13. T.A. Henzinger and H. Wong-Toi. Using HyTech to synthesize control parameters for a steam boiler. In *Formal Methods for Industrial Applications: Specifying and Programming the Steam Boiler Control*, LNCS 1165, pages 265–282. Springer-Verlag, 1996.

14. P.-H. Ho. *Automatic Analysis of Hybrid Systems*. PhD thesis, Cornell University, 1995.

15. P.-H. Ho and H. Wong-Toi. Automated analysis of an audio control protocol. In *CAV 95: Computer-aided Verification*, LNCS 939, pages 381–394. Springer-Verlag, 1995.

16. P.-A. Hsiung, F. Wang, , and Y.-S. Kuo. Scheduling system verification. In *TACAS 99: Tools and Algorithms for the Construction and Analysis of Systems*, LNCS 1579, pages 19–33. Springer-Verlag, 1999.

17. R. Klatte, U. Kulisch, M. Neage, D. Ratz, and C. Ullrich. *Pascal-XSC: Language Reference and Examples*. Springer, 1992.

18. O. Knüppel. PROFIL/BIAS: A fast interval library. *Computing*, 53(3–4):277–287, 1994.

19. R. Lohner. Computation of guaranteed enclosures for the solutions of ordinary initial and boundary value problems. In *Computational Ordinary Differential Equations*. Oxford University Press, 1992.

20. R.E. Moore. *Interval Analysis*. Prentice-Hall, 1966.

21. P.J. Mosterman. An overview of hybrid simulation phenomena and their support by simulation packages. In *HSCC 99: Hybrid Systems Computation and Control*, LNCS 1569, pages 165–177. Springer-Verlag, 1999.
22. R. Rihm. Interval methods for initial value problems in ODEs. In *Topics in Validated Computations*. North-Holland, 1994.
23. T. Stauner, O. Müller, and M. Fuchs. Using HYTECH to verify an automotive control system. In *HART 97: Hybrid and Real-time Systems*, LNCS 1201, pages 139–153. Springer-Verlag, 1997.
24. O. Stauning. *Automatic Validation of Numerical Solutions*. PhD thesis, Technical University of Denmark, 1997.
25. O. Stursberg, S. Kowaleski, I. Hoffmann, and J. Preußig. Comparing timed and hybrid automata as approximations of continuous systems. In *Hybrid Systems IV*, LNCS 1273, pages 361–377. Springer-Verlag, 1997.
26. C.J. Tomlin. *Hybrid Control of Air Traffic Management Systems*. PhD thesis, University of California at Berkeley, 1998.
27. T. Villa, H. Wong-Toi, A. Balluchi, J. Preußig, A. Sangiovanni-Vincentelli, and Y. Watanabe. Formal verification of an automotive engine controller in cutoff mode. In *Proceedings of the 37th Conference on Decision and Control*. IEEE Press, 1998.

Robust Undecidability
of Timed and Hybrid Systems*

Thomas A. Henzinger[1] and Jean-François Raskin[1,2]

[1] Department of Electrical Engineering and Computer Sciences
University of California at Berkeley, CA 94720-1770, USA
[2] Département d'Informatique, Faculté des Sciences
Université Libre de Bruxelles, Belgium
{tah,jfr}@eecs.berkeley.edu

Abstract. The algorithmic approach to the analysis of timed and hybrid systems is fundamentally limited by undecidability, of universality in the timed case (where all continuous variables are clocks), and of emptiness in the rectangular case (which includes drifting clocks). Traditional proofs of undecidability encode a single Turing computation by a single timed trajectory. These proofs have nurtured the hope that the introduction of "fuzziness" into timed and hybrid models (in the sense that a system cannot distinguish between trajectories that are sufficiently similar) may lead to decidability. We show that this is not the case, by sharpening both fundamental undecidability results. Besides the obvious blow our results deal to the algorithmic method, they also prove that the standard model of timed and hybrid systems, while not "robust" in its definition of trajectory acceptance (which is affected by tiny perturbations in the timing of events), is quite robust in its mathematical properties: the undecidability barriers are not affected by reasonable perturbations of the model.

1 Introduction

The main limitations of the algorithmic method for analyzing timed and hybrid systems find their precise expression in two well-publicized undecidability results. First, the universality problem for timed automata (does a timed automaton accept all timed words?) is undecidable [AD94]. This implies that timing requirements which are expressible as timed automata cannot be model checked. Consequently, more restrictive subclasses of timing requirements have been studied (e.g., Event-Clock Automata [AFH94], Metric Interval Temporal Logic [AFH96], Event-Clock Logic [RS99]). Second, the emptiness/reachability problem for rectangular automata (does a rectangular automaton accept any timed word, or equivalently, can a rectangular automaton reach a given location?) is undecidable [HKPV95]. While several orthogonal undecidability results are known for hybrid systems, it is the rectangular reachability problem

* This research was supported in part by the DARPA (NASA) grant NAG2-1214, the DARPA (Wright-Patterson AFB) grant F33615-C-98-3614, the ARO MURI grant DAAH-04-96-1-0341, and the NSF CAREER award CCR-9501708.

N. Lynch and B. Krogh (Eds.): HSCC 2000, LNCS 1790, pp. 145–159, 2000.

which best highlights the essential limitations of the algorithmic approach to systems with continuous dynamics. This is because the rectangular automaton model is the minimal generalization of the timed automaton model capable of approximating continuous dynamics (using piecewise linear envelopes). It follows that rectangularity as an abstraction is insufficient for checking invariants of hybrid systems, and further loss of information is necessary (e.g., initialization [HKPV95], discretization [HK97]).

Both central undecidability results have been proved by encoding each computation of some Turing-complete machine model as a trajectory of a timed or hybrid system. The encodings are quite fragile: given a deterministic Turing machine M with empty input, one constructs either a timed automaton that rejects the single trajectory which encodes the halting computation of M (rendering universality undecidable), or a rectangular automaton that accepts that single trajectory (rendering emptiness/reachability undecidable). However, if the specified trajectory is perturbed in the slightest way, it no longer properly encodes the desired Turing computation. This has led researchers to conjecture [Fra99] that undecidability is due to the ability of timed and hybrid automata to differentiate real points in time with infinite precision. Consequently, one might hope that a more realistic, slightly "fuzzy" model of timed and hybrid systems might not suffer from undecidability.[1] In a similar vein, in [GHJ97] it is conjectured that unlike timed automata, robust timed automata, which do not accept or reject individual trajectories but bundles ("tubes") of closely related trajectories, can be complemented.

In this paper, we refute these conjectures. In doing so, we show that the sources of undecidability for timed and hybrid systems are structural, robust, and intrinsic to mixed discrete-continuous dynamics, rather than an artifact of a particular syntax or of the ability to measure time with arbitrary precision. We redo both undecidability proofs by encoding each Turing computation not as a single trajectory but as a trajectory tube of positive diameter. This requires considerable care and constitutes the bulk of this paper. As corollaries we obtain the following results:

Robust timed and rectangular automata Robust   automata   introduce "fuzziness" semantically, by accepting tubes rather than trajectories [GHJ97]. We prove that universality is undecidable for robust timed automata (since emptiness is decidable, it follows that they are not complementable), and that emptiness/reachability is undecidable for robust rectangular automata.

Open rectangular automata Open automata introduce "fuzziness" syntactically, by restricting all guard and differential-inclusion intervals to open

---

[1] Note that "fuzziness," as meant here, is fundamentally distinct from "discretization," which is known to lead to decidability in many cases. Intuitively, fuzziness preserves the density of the time domain, while discretization does not. Mathematically, discretization is performed with respect to a fixed real $\epsilon > 0$ representing finite precision, while fuzziness quantifies over $\epsilon > 0$ existentially.

sets. We prove that emptiness/reachability is undecidable for open rectangular automata. The universality problem for open timed automata is, to our knowledge, still open.

A main impact of these results is, of course, negative: they deal a serious blow to our ability for analyzing timed and hybrid systems automatically, much more so than the previously known results, which rely on questionable, "fragile" modeling assumptions (one trajectory may be accepted even if all slightly perturbed trajectories are rejected, and vice versa). There is, however, also a positive interpretation of our results: they show that the "standard" model for timed and hybrid systems, with its fragile definition of trajectory acceptance, does not give rise to a fragile theory but, on the contrary, is very robust with respect to its mathematical properties (such as decidability versus undecidability). For further decidability/undecidability results about the standard model of hybrid systems, we refer the reader to [AMP95,BT99].

2 Trajectories, Tubes, and Hybrid Automata

In this paper, we consider finite trajectories only. A *trajectory* over an alphabet Σ is an element of the language $(\Sigma \times \mathbb{R}^+)^*$, where $\mathbb{R}^+$ stands for the set of positive reals excluding 0. Thus, a trajectory is a finite sequence of pairs from $\Sigma \times \mathbb{R}^+$. We call the first element of each pair an *event*, and the second element the *time-gap* of the event. The time-gap of an event represents the amount of time that has elapsed since the previous event of the trajectory. For a trajectory τ, we denote its length (i.e., the number of pairs in τ) by $\text{len}(\tau)$, and its projection onto Σ^* (i.e., the sequence of events that results from removing the time-gaps) by $\text{untime}(\tau)$. We assign time-stamps to the events of a trajectory: for the i-th event of τ, the *time-stamp* is defined to be $t_\tau(i) = \sum_{1 \leq j \leq i} \delta_j$, where δ_j is the time-gap associated with the j-th event of τ.

Metrics on trajectories. Let the set of all trajectories be denoted Traj. Assuming that trajectories cannot be generated and recorded with infinite precision, in order to get an estimate of the amount of error in the data that represents a trajectory, we need a metric on Traj. Here we define, as an example, one particular metric d; in [GHJ97], it is shown that all reasonable metrics define the same topology on trajectories. Given two trajectories τ and τ', we define:

- $d(\tau, \tau') = \infty$ if $\text{untime}(\tau) \neq \text{untime}(\tau')$;
- $d(\tau, \tau') = \max\{|t_\tau(i) - t_{\tau'}(i)| : 1 \leq i \leq \text{len}(\tau)\}$ if $\text{untime}(\tau) = \text{untime}(\tau')$.

Thus, only two trajectories with the same length and the same sequence of events have a finite distance, and finite errors may occur only in measuring time. The metric measures the maximal difference in the time-stamps of any two corresponding events: two timed words are close to each other if they have the same events in the same order, and the times at which these events occur are not very different. For instance, for $\tau_1 = (a, 1)(a, 1)(a, 1)$ and $\tau_2 = (a, 0.9)(a, 1.2)(a, 1.2)$, we have $d(\tau_1, \tau_2) = 0.3$.

Given a metric, we use the standard definition of open sets. Formally, for the metric d, a trajectory τ, and a positive real $\epsilon \in \mathbb{R}^+$, define the d-*tube around* τ *of diameter* ϵ to be the set $T(\tau, \epsilon) = \{\tau' : d(\tau, \tau') < \epsilon\}$ of all trajectories at a d-distance less than ϵ from τ. A d-open set O, called a d-*tube*, is any subset of Traj such that for all trajectories $\tau \in O$, there is a positive real $\epsilon \in \mathbb{R}^+$ with $T(\tau, \epsilon) \subseteq O$. Thus, if a d-tube contains a trajectory τ, then it also contains all trajectories in some neighborhood of τ. Let the set of all d-tubes be denoted Tube.

From trajectory languages to tube languages. A *trajectory language* is any subset of Traj; a *tube language* [GHJ97] is any subset of Tube. Every trajectory language L induces a tube language $[L]$, which represents a "fuzzy" rendering of L. In $[L]$ we wish to include a tube iff sufficiently many of its trajectories are contained in L. We define "sufficiently many" as any dense subset, in the topological sense.

For this purpose we review some simple definitions from topology. A set S of trajectories is closed if its complement $S^c = \text{Traj} - S$ is open. The closure $\overline{S}$ of a set S of trajectories is the least closed set containing S, and the interior S^{int} is the greatest open set contained in S. The set S' of trajectories is dense in S iff $S \subseteq \overline{S'}$. Formally, given a trajectory language L, the corresponding tube language is defined as $[L] = \{O \in \text{Tube} : O \subseteq \overline{L}\}$. Thus, a tube O is in $[L]$ if for each trajectory $\tau \in O$ there is a sequence of trajectories with limit τ such that all elements of this sequence are in L. Equivalently, L must be dense in O; that is, for every trajectory $\tau \in O$ and for every positive real $\epsilon \in \mathbb{R}^+$, there is a trajectory $\tau' \in L$ such that $d(\tau, \tau') < \epsilon$. Since the tubes in $[L]$ are closed under subsets and union, the tube language $[L]$ can be identified with the maximal tube in $[L]$, which is the interior $\overline{L}^{int}$ of the closure of L.

We will define the semantics of a robust hybrid automaton with trajectory set L to be the tube set $[L]$. This has the effect that a robust hybrid automaton cannot generate (or accept) a particular trajectory when it refuses to generate (rejects) sufficiently many surrounding trajectories. Neither can the automaton refuse to generate a particular trajectory when it may generate sufficiently many surrounding trajectories.

Timed and rectangular automata. An *interval* has the form (a, b), $[a, b]$, $(a, b]$, or $[a, b)$, where $a \in \mathbb{Q} \cup \{-\infty\}$, $b \in \mathbb{Q} \cup \{\infty\}$, and $a \leq b$ if I is of the form $[a, b]$, and $a < b$ otherwise. We say that the interval I is *open* if it is of the form (a, b), and *closed* if it is of the form $[a, b]$. We write Rect for the set of intervals.

A *rectangular automaton* [HKPV95] is a tuple $A = \langle \Sigma, Q, Q_0, Q_f, C, E, \text{Ev},$ Init, Pre, Reset, Post, Flow$\rangle^2$, where (i) Σ is a finite alphabet of events; (ii) Q is a finite set of locations; (iii) $Q_0 \subseteq Q$ is a set of start locations; (iv) $Q_f \subseteq Q$ is a set of accepting locations; (v) C is a finite set of real-valued variables; (vi)

---

[2] It is often convinient to annotate locations with variable constraints, so-called invariant conditions. Our results extend straight-forwardly to rectangular automata with invariant conditions.

$E \subseteq Q \times Q$ is a finite set of edges; (vii) $\mathsf{Ev} : E \to \Sigma$ is a function that associates with each edge e a letter of the alphabet Σ; (viii) $\mathsf{Init} : Q_0 \to C \to \mathsf{Rect}$ is a function that associates with each start location $q_0 \in Q_0$ and variable $x \in C$ an interval I that contains the possible initial values of this variable when the control of the automaton starts in location q_0; (ix) $\mathsf{Pre} : E \to C \to \mathsf{Rect}$ is a function that associates with each edge e and variable x an interval I such that the value of x must lie in I before crossing the edge e; (x) $\mathsf{Post} : E \to C \to \mathsf{Rect}$ is a function that associates with each edge e and each variable x an interval I such that the value of x must lie in I after crossing the edge e; (xi) $\mathsf{Reset} : E \to 2^C$ is a function that associates with each edge e a subset of variables that are reset when crossing e; if a variable x belongs to the set $\mathsf{Reset}(e)$ then the value, after crossing the edge e, of x is taken nondeterministically from the interval $\mathsf{Post}(e, x)$; (xii) $\mathsf{Flow} : Q \to C \to \mathsf{Rect}$ is a function that associates with each location q and variable x an interval I such that the first derivative of x when the control is in location q lies within I.

Timed automata are a syntactic subset of rectangular automata. A rectangular automaton A is a *timed automaton* [AD94] if the function Flow of A is such that for all locations $q \in Q$, and for all variables $x \in C$, we have $\mathsf{Flow}(q, x) = [1, 1]$; that is, every continuous variable is a clock. The timed automaton A is *open* if all intervals used in the functions Init, Pre, and Post are open. Similarly, a rectangular automaton A is *open* if all intervals used in the functions Init, Pre, Post, and Flow are open.

A rectangular automaton A defines a labeled transition system with an infinite state space S, the infinite set of labels $\mathbb{R}^+ \cup \Sigma$, and the transition relation R. Each transition with label σ correspond to an edge step whose event is $\sigma \in \Sigma$. Each transition with label $\delta \in \mathbb{R}^+$ corresponds to a time step of duration δ. The states and transitions of A are defined as follows. A *state* $(q, \mathbf{x})$ of A consists of a discrete part $q \in Q$ and a continuous part $\mathbf{x} \in \mathbb{R}^n$. The *state space* $S \subset Q \times \mathbb{R}^n$ is the set of all states of A. The state $(q, \mathbf{x})$ is an *initial state* of A if $q \in Q_0$ and $\mathbf{x} \in \mathsf{Init}(q)$[3]. For each edge $e = (q_1, q_2)$ of A, we define the binary relation $\to^e \subset S^2$ by $(q_1, \mathbf{x}) \to^e (q_2, \mathbf{y})$ iff $\mathbf{x} \in \mathsf{Pre}(e)$, $\mathbf{y} \in \mathsf{Post}(e)$, and for every coordinate $i \in \{1, \ldots, n\}$ with $i \notin \mathsf{Reset}(e)$, we have $\mathbf{x}_i = \mathbf{y}_i$. For each event $\sigma \in \Sigma$, we define the *edge-step relation* $\to^\sigma \subset S^2$ by $s_1 \to^\sigma s_2$ iff $s_1 \to^e s_2$ for some edge $e \in E$ with $\mathsf{Ev}(e) = \sigma$. For each positive real $\delta \in \mathbb{R}^+$, we define the binary *time-step relation* $\to^\delta \subset S^2$ by $(q_1, \mathbf{x}) \to^\delta (q_2, \mathbf{y})$ iff $q_1 = q_2$ and $\frac{\mathbf{y} - \mathbf{x}}{\delta} \in \mathsf{Flow}(q_1)$. The transition relation $R \subseteq S \times S$ is defined by $R = \{\to^e | e \in E\} \cup \{\to^\delta | \delta \in \mathbb{R}^+\}$.

Trajectory acceptance and reachable locations. We now define the trajectory language and the reachable locations of a rectangular automaton A. A *run* of the automaton A is a finite path $(q_0, \mathbf{x}_0) \to^{\delta_0} (q_0, \mathbf{y}_0) \to^{\sigma_0} (q_1, \mathbf{x}_1) \to^{\delta_1} (q_1, \mathbf{y}_1) \ldots \to^{\sigma_n} (q_{n+1}, \mathbf{x}_{n+1})$ in the transition system of A that alternates between time steps and edge steps. The run is *initial* if $q_0 \in Q_0$ and $\mathbf{x}_0 \in \mathsf{Init}(q_0)$, and *accepting* if $q_n \in Q_f$. The trajectory $\tau = (\sigma_0, \delta_0)(\sigma_1, \delta_1) \ldots (\sigma_n, \delta_n)$ is *accepted* by the rectangular automaton A if A has an initial and accepting run

---

[3] To simplify notations, we note $\mathbf{x} \in \mathsf{Init}(q)$ instead of $\mathbf{x} \in \mathsf{Init}(q, x)$.

Class of Automata	Emptiness/Reachability	Universality
Timed Automata [AD94]	Decidable	Undecidable
Rectangular Automata [HKPV95]	Undecidable	Undecidable

Fig. 1. Known decidability and undecidability results for timed/rectangular automata.

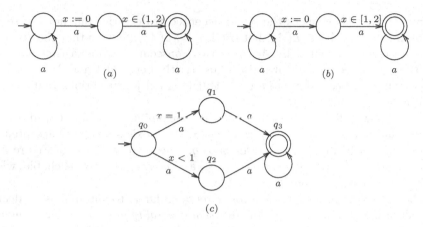

Fig. 2. The timed automata A_1, A_2, and A_3.

$(q_0, \mathbf{x}_0) \rightarrow^{\delta_0} (q_0, \mathbf{y}) \rightarrow^{\sigma_0} (q_1, \mathbf{x}_1) \rightarrow^{\delta_1} \ldots \rightarrow^{\sigma_n} (q_{n+1}, \mathbf{x}_{n+1})$. The trajectory τ *leads to* location q_{n+1}. A location q of A is *reachable* if there exists an trajectory τ accepted by A that leads to q. We denote by $L(A)$ the set of trajectories accepted by A.

The *trajectory-emptiness problem* for a rectangular automaton A is to decide whether or not $L(A)$ is empty. The *trajectory-universality problem* for a rectangular automaton A is to decide whether or not $L(A)$ contains all trajectories over the alphabet Σ. The *location-reachability problem* for a rectangular automaton A is to decide if a given location of A is reachable. Note that the trajectory-emptiness problem for a class of rectangular automaton is decidable iff the location-reachability problem is decidable. The previously known results about these problems are summarized in the table of Figure 1.

Tube acceptance and robustly reachable locations. The rectangular automaton A *accepts* the set $[L(A)]$ of tubes [GHJ97]. The following examples illustrate tube acceptance. First, consider the timed automaton A_1 of Figure 2(a). This automaton accepts all trajectories over the unary alphabet $\{a\}$ which contain two consecutive a events with a time-gap in the open interval $(1, 2)$. This property is invariant under sufficiently small perturbations of the time-stamps. Hence the automaton A_1 accepts precisely those tubes that consist of trajectories in $L(A_1)$, and the maximal accepted tube is $L(A_1)$ itself. In the timed automaton A_2 of Figure 2(b), the open interval $(1, 2)$ is replaced by the closed

Class of Automata	Robust Emptiness/Robust Reachability	Robust Universality
Timed Automata	Decidable	Undecidable
Rectangular Automata	Undecidable	Undecidable

Fig. 3. Decidability results about robust timed and rectangular automata.

interval $[1, 2]$. This changes the set of accepted trajectories but not the set of accepted tubes: $L(A_1) \subset L(A_2)$ but $[L(A_1)] = [L(A_2)]$. Notice that the "boundary trajectories" accepted by A_2, with two consecutive a's at a time-gap of 1 or 2 but no consecutive a's at a time-gap strictly between 1 and 2, are not accepted robustly, because there are arbitrarily small perturbations that are not acceptable.

Let us now define the notion of robust reachability. A location q of a rectangular automaton A is *robustly reachable* if there exists a tube O accepted by A such that each trajectory in O leads to q. The automaton A_3 of Figure 2(c) illustrates this notion: the locations q_0, q_2, and q_3 are robustly reachable, while the location q_1 is not robustly reachable.

The *robust-emptiness problem* for a rectangular automaton A is to decide whether or not $[L(A)]$ is empty. The *robust-universality problem* for a rectangular automaton A is to decide whether or not $[L(A)]$ contains all tubes over Σ. The *robust-reachability problem* for a rectangular automaton A is to decide, given a location q of A, if q is robustly reachable. In the following sections of this paper, we will sharpen the known undecidability results about timed and hybrid systems. We will show that the introduction of fuzziness into timed and hybrid models via the notion of tubes (this fuzziness can be intuitively seen as the semantic removal of equality) does not change the undecidability results. Our results are summarized in the table of Figure 3; only the positive result was previously known [GHJ97].

Some properties of robust timed automata. We recall some results presented in [GHJ97]. We will need these notions to establish our results. The first proposition tells us that when we consider tube acceptance, we can restrict our attention either to closed or to open timed automata.

Proposition 1. *For every timed automaton A, we can construct a timed automaton $\overline{A}$, called the closure of A, whose* Pre, Post, Init *functions use only closed intervals, such that $L(\overline{A}) = \overline{L(A)}$. Furthermore, we can construct an open timed automaton A^{int}, called the interior of A, such that $[L(A)] = [L(A^{int})] = [L(\overline{A})]$.*

The following proposition shows that for open timed automata, tube emptiness coincides with trajectory emptiness.

Proposition 2. *For every open timed automaton A and every trajectory τ, if τ is accepted by A along some path, then there is a positive real $\epsilon \in \mathbb{R}^+$ such that all trajectories in the tube $T(\tau, \epsilon)$ are accepted by A along the same path.*

Before defining the tube complement of a timed automaton, we observe an important property of the trajectory languages that can be defined by timed automata.

Proposition 3. *For every timed automaton A, there is no tube O such that both $L(A)$ and its complement $L(A)^c$ are both dense in O.*

It follows that a tube cannot be accepted by both a timed automaton A and a trajectory complement of A.

For defining the tube complement of a timed automaton A, it is not useful to consider the boolean complement $\mathsf{Tube} - [L(A)]$ of the tube language $[L(A)]$. For $[L(A)]$ is closed under subsets and union. Therefore, unless $[L(A)] = \emptyset$ or $[L(A)] = \mathsf{Tube}$, the boolean complement $\mathsf{Tube} - [L(A)]$ cannot be induced by any trajectory language and, hence, cannot be accepted by any timed automaton. Thus, for every tube language $\mathcal{L} \subseteq \mathsf{Tube}$, we define the *tube complement* of $\mathcal{L}$ to be the set

$$\mathcal{L}^c = \{O \in \mathsf{Tube} : O \cap \bigcup \mathcal{L} = \emptyset\}$$

of tubes that are disjoint from the tubes in $\mathcal{L}$. The following proposition shows that for every timed automaton A, the tube complement $[L(A)]^c$ is induced by the trajectory complement $L(A)^c$; that is, $[L(A)^c] = [L(A)]^c$.

Proposition 4. *If L is a trajectory language and there is no tube O such that both L and L^c are dense in O, then $[L]^c = [L^c]$.*

For two timed automata A and B, we say that B is a *tube complement* of A iff B accepts precisely the tubes that do not intersect any tube accepted by A; that is, $[L(B)] = [L(A)]^c$. From Propositions 3 and 4, it follows that every trajectory complement of a timed automaton is also a tube complement (the converse is generally not true). Since $[L(A)]^c = [L(A^{int})]^c = [L(A^{int})^c]$, in order to construct tube complements, it would suffice to construct trajectory complements of open timed automata.[4] This, however, is not possible, as we show in the next section.

3   The Robust-Universality Problem for Timed Automata

In this section, we show that the halting problem for two-counter machines can be reduced to the robust-universality problem for timed automata. A *two-counter machine* M is a triple $\langle \{b_1, \ldots, b_n\}, C, D \rangle$, where $\{b_1, \ldots, b_n\}$ are n instructions, and C and D are two counters ranging over the natural numbers. Each instruction b_i, $0 \leq i \leq n$, has one of the three possible forms: (i) a conditional jump instruction tests if a counter is 0 and then jumps conditionally to the next instruction; (ii) an increment/decrement instruction increments or decrements the value of one of the two counters and then jumps nondeterministically to one of

---

[4] Similarly, since $[L(A)]^c = [L(\overline{A})]^c = [L(\overline{A})^c]$, it would suffice to construct trajectory complements of closed timed automata. This, however, is known to be impossible [AD94].

two possible next instructions; (iii) a stop instruction puts an end to the machine execution. A *configuration of a two-counter machine* M is a triple $\gamma = \langle i, c, d \rangle$, where i is the program counter indicating the current instruction, and c and d are the values of the counters C and D. A *computation* of M is a finite or infinite sequence $\overline{\gamma} = \gamma_0 \gamma_1 \ldots$ of configurations such that $\gamma_0 = \langle 0, 0, 0 \rangle$, i.e. the first instruction is b_0, and the initial value of the two counters C and D is 0, and for every γ_{i+1} is a M-successor configuration of γ_i, for every $i \geq 0$. If $\overline{\gamma}$ is finite then its last configuration contains a stop instruction. The *halting problem* for a two-counter machine M is to decide whether or not the execution of M has at least one computation that ends in a stop instruction. The problem of deciding if a two-counter machine has a halting computation is undecidable.

Trajectory encoding of a two-counter machine computation. We review how the undecidability of the universality problem for timed automata was established by Alur and Dill [AD94] and explain why their proof does not translate to the robust-universality problem. Given a two-counter machine M, the set $\mathsf{L}_{\mathsf{Traj}}^{\mathsf{Undec}}(M)$ of trajectories is defined as follows: $(\sigma, \delta) \in \mathsf{L}_{\mathsf{Traj}}^{\mathsf{Undec}}(M)$ iff (i) $\sigma = b_{i_0} c^{c_0} d^{d_0} b_{i_1} c^{c_1} d^{d_1} \ldots b_{i_m} c^{c_m} d^{d_m}$ such that $\langle i_0, c_0, d_0 \rangle, \langle i_1, c_1, d_1 \rangle, \ldots \langle i_m, c_m, d_m \rangle$ is a halting computation of M; (ii) for all $j \geq 0$, the time-stamp of b_{i_j} is j; (iii) for all $j \geq 1$, (a) if $c_{j+1} = c_j$, then for every c with time-stamp t in the interval $(j, j+1)$ there is a c with time-stamp $t+1$; (b) if $c_{j+1} = c_j + 1$, then for every c with time-stamp t in the interval $(j+1, j+2)$, except the last one, there is a c with time-stamp $t-1$; (c) if $c_{j+1} = c_j - 1$, then for every c with time-stamp t in the interval $(j, j+1)$, except the last one, there is a c with time-stamp $t+1$; and (iv) the same requirements hold for d's. Then $\mathsf{L}_{\mathsf{Traj}}^{\mathsf{Undec}}(M)$ is nonempty iff M has a halting computation. Furthermore, there exists a timed automaton that accepts exactly the trajectories not in $\mathsf{L}_{\mathsf{Traj}}^{\mathsf{Undec}}(M)$. It follows that the universality problem for timed automata is undecidable.

Note that the i-th configuration is encoded in the interval $[i, i+1)$. To enforce the requirement that the number of c events in two successive configurations is the same, every c in the first interval has a matching c at the exact distance 1, and vice versa. This use of punctuality constraints has the following consequence.

Proposition 5. *Let M be a two-counter machine, there is no tube $O \in$ Tube such that O is dense in $\mathsf{L}_{\mathsf{Traj}}^{\mathsf{Undec}}(M)$; that is, $[\mathsf{L}_{\mathsf{Traj}}^{\mathsf{Undec}}(M)] = \emptyset$.*

This has nurtured some hope that, by removing the possibility to specify punctuality constraints, timed automata might have a decidable robust-universality problem. Unfortunately this is not the case. We next show that we can define a set $\mathsf{L}_{\mathsf{Tube}}^{\mathsf{Undec}}(M)$ of trajectories which forms a tube and encodes halting computations of the given two-counter machine M. Furthermore the tube complement of this tube language can be defined by a robust (open) timed automaton. The undecidability of the robust-universality problem and the nonclosure under complement of robust timed automata will follow.

Tube encoding of a two-counter machine computation. To facilitate the definition of $\mathsf{L}_{\mathsf{Tube}}^{\mathsf{Undec}}(M)$, the undecidable tube language, we first introduce

some new notions. We call an *open (closed) slot* an open (closed) interval of
the real numbers. We define the open (closed) slot between t_1 and t_2 as the set
$\{t \mid t_1 < t < t_2\}$ (respectively, $\{t \mid t_1 \leq t \leq t_2\}$). Given two real numbers t_1 and
t_2 with $t_1 < t_2$, we say that (t_3, t_4) (respectively $[t_3, t_4]$) is the open (closed) slot
generated by t_1 and t_2 if both $t_1 + 1 = t_3$ and $t_2 + 1 = t_4$.

The main idea of $L_{\text{Tube}}^{\text{Undec}}(M)$ is that we encode the configuration i within the
open interval $(i, i+1)$, and the next configuration $i+1$ will be encoded in the open
slot generated by the time of the beginning and the end of configuration i. For
the encoding of the elements of a configuration and their relation with the next
configuration we also use open slots. For instance, we use the triple $\mathsf{B}^{\text{Inst}} \cdot b_{j_i} \cdot \mathsf{E}^{\text{Inst}}$
to encode that b_{j_i} is the instruction executed in the i-th configuration; the letters
B^{Inst} and E^{Inst} are used as delimiters of the instruction, and to generate the slot
for the next instruction. Let us assume that t_1 and t_2 are the time-stamps of B^{Inst}
and E^{Inst}, respectively. Then the encoding of the next instruction has to take place
in the open slot $(t_1 + 1, t_2 + 1)$ generated by the slot for the current instruction.
As we use a dense time domain, this constraint can always be satisfied. We will
proceed in the same way for the encoding of the values of the two counters.
The value of the counters C and D are encoded as follows: if the value of the
counter C is u in configuration i, then the pair $\mathsf{b}^{\text{c}} \cdot \mathsf{e}^{\text{c}}$ is repeated u times in the
encoding of the configuration i. If the counter C is unchanged from configuration
i to configuration $i + 1$, we verify that the $\mathsf{b}^{\text{c}} \cdot \mathsf{e}^{\text{c}}$ sequences in configuration $i + 1$
appear exactly in the open slots defined by the $\mathsf{b}^{\text{c}} \cdot \mathsf{e}^{\text{c}}$ sequences in configuration i.

Having the intuition underlying the language $L_{\text{Tube}}^{\text{Undec}}(M)$, we now define it and
establish that the set of trajectories in $L_{\text{Tube}}^{\text{Undec}}(M)$ correspond to a non empty
set of tubes iff the machine M has a halting computation. The set of events
that we will use in the encoding is the following: (i) B^{Conf} and E^{Conf} are the
delimiters for the beginning and end of the encoding of a configuration; (ii)
B^{Inst} and E^{Inst} are the delimiters for the begin and end of the encoding of the
instruction executed in a configuration; (iii) $b_1, b_2, \ldots, b_n$ are used to represent
the n instructions; (iv) B^{C} and E^{C} are the delimiters for the encoding of the
value of the counter C in a configuration; (v) B^{D} and E^{D}, for the counter D;
(vi) b^{c} and e^{c} are used to encode the value of the counter C; (vii) b^{d} and e^{d},
for D. The trajectories of $L_{\text{Tube}}^{\text{Undec}}(M)$ agree with the following regular expression:
$(\mathsf{B}^{\text{Conf}} \cdot \mathsf{B}^{\text{Inst}} \cdot (b_1 \mid b_2 \mid \ldots \mid b_n) \cdot \mathsf{E}^{\text{Inst}} \cdot \mathsf{B}^{\text{C}} \cdot (\mathsf{b}^{\text{c}} \cdot c \cdot \mathsf{e}^{\text{c}})^* \cdot \mathsf{E}^{\text{C}} \cdot \mathsf{B}^{\text{D}} \cdot (\mathsf{b}^{\text{d}} \cdot d \cdot \mathsf{e}^{\text{d}})^* \cdot \mathsf{E}^{\text{D}} \cdot \mathsf{E}^{\text{Conf}})^*$.
Furthermore, if the configuration i contains the sequence $\mathsf{B}^{\text{Inst}} \cdot b_{j_i} \cdot \mathsf{E}^{\text{Inst}}$, then the
configuration $i + 1$ contains the sequence $\mathsf{B}^{\text{Inst}} \cdot b_{j_{i+1}} \cdot \mathsf{E}^{\text{Inst}}$, where $b_{j_{i+1}}$ is a valid
next instruction of b_{j_i}. The first configuration is encoded in the open interval
$(0, 1)$; that is, if the event B^{Conf} occurs at time t_1 and the event E^{Conf} occurs at
time t_2, then $0 < t_1 < t_2 < 1$. The configuration $i + 1$ is always encoded in the
open slot defined by the configuration i; that is, if the event B^{Conf} of configuration
i occurs at time t_1 and the event E^{Conf} occurs at time t_2, then the encoding of
the configuration $i + 1$ takes place in the open slot $(t_1 + 1, t_2 + 1)$. The encoding
of the instruction executed during the configuration $i + 1$ takes place in the slot
defined by the encoding of the instruction executed in configuration i; that is, if
B^{Inst} and E^{Inst} appear at times t_1 and t_2 in encoding of configuration i, then B^{Inst}

and $\mathsf{E}^{\mathsf{Inst}}$ appear at times t_3 and t_4 in the encoding of configuration $i+1$ with the following (open) real-time constraint: $t_1 + 1 < t_3 < t_4 < t_2 + 1$. We only explain in details the case when the counter C is incremented from configuration i to configuration $i+1$. The other operations are left to the reader. If in configuration i the events B^C and E^C occur at times t_1 and t_2, respectively, then the events B^C and E^C appear for configuration $i+1$ within the open slot $(t_1 + 1, t_2 + 1)$. For each $\mathsf{b}^c \cdot \mathsf{e}^c$ sequence, such that b^c occurs at time t_1 and e^c occurs at time t_2, in the encoding of configuration i, there is exactly one sequence $\mathsf{b}^c \cdot \mathsf{e}^c$ sequence in the encoding of configuration $i+1$ that takes place in the open slot $(t_1 + 1, t_2 + 1)$. Conversely, each $\mathsf{b}^c \cdot \mathsf{e}^c$ that appears in the encoding of the configuration $i+1$, with the exception of the last, must lie in the open slot defined by the $\mathsf{b}^c \cdot \mathsf{e}^c$ sequence of configuration i. This requirement is noted RT_3^c. Finally, the last $\mathsf{b}^c \cdot \mathsf{e}^c$ sequence in the encoding of configuration $i+1$ appears in the slot generated by the two events B^C and E^C if $C = 0$ in configuration i, and appears in the slot generated by the last e^c event and E^C event of configuration i if $C > 0$ in that configuration.

The following proposition is a direct consequence of the use of strict inequalities in the definition of the language $\mathsf{L}_{\mathsf{Tube}}^{\mathsf{Undec}}(M)$.

Proposition 6. *Let M be a two-counter machine, for every trajectory τ_1 that belongs to $\mathsf{L}_{\mathsf{Tube}}^{\mathsf{Undec}}(M)$, there exists a real $\epsilon > 0$ such that for every trajectory τ_2, if $d(\tau_1, \tau_2) < \epsilon$ then $\tau_2 \in \mathsf{L}_{\mathsf{Tube}}^{\mathsf{Undec}}(M)$.*

Corollary 1. *For every two-counter machine M with a halting computation, $[\mathsf{L}_{\mathsf{Tube}}^{\mathsf{Undec}}(M)]$ is a nonempty tube language.*

Corollary 2. *There is no tube O that is dense both in $\mathsf{L}_{\mathsf{Tube}}^{\mathsf{Undec}}(M)$ and in $(\mathsf{L}_{\mathsf{Tube}}^{\mathsf{Undec}}(M))^c$.*

Note also that by Proposition 6 and Corollary 2, we know that the tube semantics of a timed automaton that accepts the complement of the trajectories of $\mathsf{L}_{\mathsf{Tube}}^{\mathsf{Undec}}(M)$, is exactly the complement of the tube language $[\mathsf{L}_{\mathsf{Tube}}^{\mathsf{Undec}}(M)]$. The following lemma shows that it is possible to construct such a timed automaton.

Lemma 1. *There exists a timed automaton A_M that accepts exactly the trajectories that are not in $\mathsf{L}_{\mathsf{Tube}}^{\mathsf{Undec}}(M)$.*

Proof. It is sufficient to show that for each of the requirements defining $\mathsf{L}_{\mathsf{Tube}}^{\mathsf{Undec}}(M)$, we can construct a timed automaton that accepts exactly the trajectories that violate the requirement. The union of these automata is exactly what we are looking for: the timed automaton that accepts the trajectory complement of $\mathsf{L}_{\mathsf{Tube}}^{\mathsf{Undec}}(M)$. Due to the lack of space, we just give here the automaton for the complement of requirement RT_3^c; the other requirements can be found in [HR99]. The timed automata for requirement RT_3^c is shown in Figure 4. This automaton accepts exactly the trajectories which contain two adjacent configurations i and $i+1$ such that (i) the instruction executed in configuration i increments the counter C, that is $b \in I^C$, where I^C is the subset of instructions that increment the counter C; (ii) there is a sequence $\mathsf{b}^c \cdot \mathsf{e}^c$ in configuration i that defines an open slot in configuration $i+1$ which does not contain the sequence $\mathsf{b}^c \cdot \mathsf{e}^c$. □

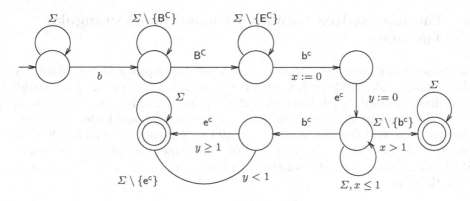

Fig. 4. A timed automaton for the negation of requirement RT_3^c

Combining Lemma 1 and Proposition 4, we obtain the following theorem.

Theorem 1. *For every two-counter machine M, there exists a timed automaton A_M that accepts every tube iff the two-counter machine M has no halting computation.*

Corollary 3. *The robust-universality problem for timed automata is undecidable.*

As the robust-emptiness problem for timed automata is decidable, we obtain the following:

Corollary 4. *There is a tube language definable by a timed automaton whose tube-complement is not definable by a timed automaton.*

From these results we can derive the following result about the trajectory languages of *open* timed automata (already established in [Her98]):

Theorem 2. *There is a trajectory language definable by an open timed automaton which trajectory-complement is not definable by a timed automaton (open or not).*

Proof. By reductio ad absurdum. We have constructed a timed automaton A_M that accepts the complement of the trajectories contained in $L_{Tube}^{Undec}(M)$. This automaton A_M defines a set $L(A_M)$ of trajectories such that $[L(A_M)]$ is exactly the tube complement of $[L_{Tube}^{Undec}(M)]$. By Proposition 1, there exists an open timed automaton, namely, the interior of A_M, denoted A_M^{int}, such that $[L(A_M^{int})] = [L(A_M)] = [L_{Tube}^{Undec}(M)]^c$. By Lemma 4, if we were able to complement the open automaton A_M^{int}, then we could obtain an automaton whose tube semantics would be $[L_{Tube}^{Undec}(M)]$. This, however, is impossible, as the robust-emptiness problem of timed automata is decidable, which would allow us to decide the halting problem for two-counter machines.   □

4   The Robust-Reachability Problem for Rectangular Automata

In this section we investigate undecidable reachability problems and show that they remain undecidable even when we remove equality from the specification formalism. In [HKPV95], it is shown that the formalism of rectangular automata lies at the boundary between decidable hybrid formalisms and undecidable ones. We show here that this boundary stays valid if we do not use equality. For this purpose, we use another tube encoding of two-counter machines computations. With each halting computation $\langle i_0, c_0, d_0 \rangle, \langle i_1, c_1, d_1 \rangle, \ldots, \langle i_n, c_n, d_n \rangle$, we associate the tube

$$(b_{i_j}, t_{(j,0)}), (\mathsf{B}^\mathsf{C}, t_{(j,1)}), (\mathsf{b}^\mathsf{c}, t_{(j,2)}), (\mathsf{e}^\mathsf{c}, t_{(j,3)})(\mathsf{B}^\mathsf{D}, t_{(j,4)})(\mathsf{b}^\mathsf{d}, t_{(j,5)})(\mathsf{e}^\mathsf{d}, t_{(j,6)})$$

with $0 \le j \le n$ and the following timing constraints. We just give the constraints for the encoding of the value of counter C; the same requirements hold for the counter D. Initially the value of the counter C is zero. To encode $C = 0$, we require that if the events $\mathsf{B}^\mathsf{C}, \mathsf{b}^\mathsf{c}$, and e^c are issued at times t_1, t_2, and t_3, then the following constraint is satisfied: $t_1 + \frac{1}{2} < t_2 < t_3 < t_1 + 1$. Let d_1 denote the distance that separates the events B^C and b^c, and let d_2 denote the distance that separates the events B^C and e^c in the encoding of the value of C in configuration i. In the same way, let d_3 and d_4 be those two distances in the encoding of the value of C in configuration $i + 1$. Then we have the following requirements: (a) if C is incremented between i and $i + 1$, then $\frac{d_1}{2} < d_3 < d_4 < \frac{d_2}{2}$; (b) if C is decremented between i and $i+1$, then $2d_1 < d_3 < d_4 < 2d_2$; (c) if C is unchanged between i and $i + 1$, then $d_1 < d_3 < d_4 < d_2$. We denote this trajectory language $\mathsf{L}^{\mathsf{Undec}}_{\mathsf{OpenRect}}(M)$.

Lemma 2. *The trajectory language $\mathsf{L}^{\mathsf{Undec}}_{\mathsf{OpenRect}}(M)$ is definable by an open rectangular automaton A_M.*

Proof. We sketch the proof by giving an open rectangular automaton to increment the counter C. The automaton is given in Figure 6. To see that the automaton checks exactly the desired constraints, we first establish bounds on the values of the variables x and y at times t_0, t_1, t_2, and t_3 represented in Figure 7. The bounds are given in the table of Figure 5. So at time t_3, we have $x \in (d_1, +\infty)$ and $y \in (-\infty, d_2)$. Now let us see the constraints that we obtain on d_3 and d_4. First, by taking into account that $x \in (d_1, +\infty)$ at t_3 and the flow of x in q_5 in included in the interval $(-2, 0)$, we can deduce that $d_3 \in (\frac{d_1}{2}, +\infty)$. Second, by taking into account that $y \in (-\infty, d_2)$ at t_3 and that the flow of y in q_5 is included in the interval $(-\infty, -2)$, we obtain $d_3 \in (-\infty, \frac{d_2}{2})$. As b^c is issued before e^c, we have $d_3 < d_4$, and thus $\frac{d_1}{2} < d_3 < d_4 < \frac{d_2}{2}$, as desired.   □

As a direct consequence of the last lemma, we have the following.

Theorem 3. *The trajectory-emptiness and location-reachability problems for open rectangular automata are undecidable.*

	t_0	t_1	t_2	t_3
$\mathsf{Inf}(x)$	0	d_1	d_1	d_1
$\mathsf{Sup}(x)$	1	$2 \times d_1 + 1$	$d_1 + d_2 + 1$	$+\infty$
$\mathsf{Inf}(y)$	-1	-1	-1	$-\infty$
$\mathsf{Sup}(y)$	0	d_1	d_2	d_2

Fig. 5. Inferior and superior bounds on the values of variables x and y.

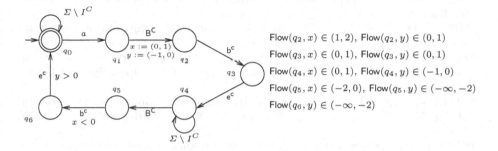

Flow$(q_2, x) \in (1, 2)$, Flow$(q_2, y) \in (0, 1)$

Flow$(q_3, x) \in (0, 1)$, Flow$(q_3, y) \in (0, 1)$

Flow$(q_4, x) \in (0, 1)$, Flow$(q_4, y) \in (-1, 0)$

Flow$(q_5, x) \in (-2, 0)$, Flow$(q_5, y) \in (-\infty, -2)$

Flow$(q_6, y) \in (-\infty, -2)$

Fig. 6. Open rectangular automaton to check incrementation of counter C.

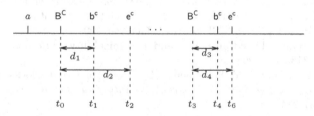

Fig. 7. Two successive encodings of the value of counter C.

The following proposition is a generalization to open rectangular automata of Proposition 2.

Proposition 7. *For every open rectangular automaton A and every trajectory τ, if τ is accepted by A along some path, then there is a positive real $\epsilon \in \mathbb{R}^+$ such that all trajectories in the tube $T(\tau, \epsilon)$ are accepted by A along the same path.*

This proposition implies that tube and trajectory emptiness coincide for open rectangular automata, so we have the following theorem.

Theorem 4. *The robust-emptiness and robust-reachability problems for rectangular automata are undecidable.*

References

[AD94] R. Alur and D.L. Dill. A theory of timed automata. *Theoretical Computer Science*, 126:183–235, 1994.

[AFH94] R. Alur, L. Fix, and T.A. Henzinger. A determinizable class of timed automata, *CAV 94: Computer-aided Verification*, LNCS 818, Springer Verlag, 1–13, 1994.

[AFH96] R. Alur, T. Feder, and T.A. Henzinger. The benefits of relaxing punctuality. *Journal of the ACM*, 43(1):116–146, 1996.

[AMP95] E. Asarin, O. Maler, and A. Pnueli. Reachability analysis of dynamical systems having piecewise-constant derivatives. *Theoretical Computer Science*, 138:65–66, 1995.

[BT99] V. D. Blondel and J. N. Tsitsiklis. A survey of computational complexity results in systems and control. To appear in Automatica, 1999.

[Fra99] M. Franzle. Analysis of Hybrid Systems: An ounce of realism can save an infinity of states, *CSL'99: Computer Science Logic*. LNCS 1683, Springer Verlag, 126–140, 1999.

[GHJ97] V. Gupta, T.A. Henzinger, and R. Jagadeesan. Robust timed automata, *HART 97: Hybrid and Real-time Systems*. LNCS 1201, Springer-Verlag, 331–345, 1997.

[HK97] T.A. Henzinger and P.W. Kopke Discrete-time control for rectangular hybrid automata. *ICALP 97: Automata, Languages, and Programming*. LNCS 1256, Springer-Verlag, 582–593, 1997.

[HKPV95] T.A. Henzinger, P.W. Kopke, A. Puri, and P. Varaiya. What's decidable about hybrid automata? In *27th Annual Symposium on Theory of Computing*, ACM Press, 373–382, 1995.

[HR99] T.A. Henzinger and J.-F. Raskin. *Robust Undecidability of Timed and Hybrid Systems*. Technical Report of the Computer Science Department of the University of California at Berkeley, UCB/CSD-99-1073, October 1999.

[Her98] P. Herrmann. Timed automata and recognizability. *Information Processing Letters*, 65(6):313-318, 1998.

[RS99] J.-F. Raskin and P.-Y. Schobbens. The Logic of Event Clocks: Decidability, Complexity, and Expressiveness. *Journal of Automata, Languages and Combinatorics*, 4(3):247-284, 1999.

Towards a Theory of Stochastic Hybrid Systems

Jianghai Hu, John Lygeros, and Shankar Sastry

Electrical Engineering and Computer Sciences
University of California, Berkeley
Berkeley, CA 94720
{jianghai,lygeros,sastry}@robotics.eecs.berkeley.edu
Tel: (510) 643-2384, Fax: (510) 642-1341

Abstract. In this paper, we present a scheme of stochastic hybrid system which introduces randomness to the deterministic framework of the traditional hybrid systems by allowing the flow inside each invariant set of the discrete state variables to be governed by stochastic differential equation (SDE) rather than the deterministic ones. The notion of embedded Markov chains is proposed for such systems and some illustrative example from high way model is presented. As an important application, these ideas are then applied to the state space discretization of one dimensional SDE to obtain the natural discretized stochastic hybrid system together with its embedded MC. The invariant distribution and exit probability from interval of the MC are studied and it is shown that they converge to their counterparts for the solution process of the original SDE as the discretization step goes to zero. As a result, the discretized stochastic hybrid system provides a useful tool for studying various sample path properties of the SDE.

1 Introduction

In the conventional formulation of hybrid system (See, for example, [6]), there is no place for randomness. Although the deterministic framework captures many characteristics of the real systems in practice, in other cases, the missing flavor of randomness will indeed be a fatal flaw because of the inherent uncertainty in the environment of most real world applications. The idea of introducing stochastic hybrid system is not new. Different researchers have tried to propose different models from their own perspectives. For the most recent and relevant literature, the readers are referred to [1,5,10,2,11,9]. The most important difference lies in where to introduce the randomness.

One obvious choice is to replace the deterministic jumps between discrete states by random jumps governed by some prescribed probabilistic law. Hence the evolution of the discrete states constitutes a time homogeneous Markov chain. The question remained then is when does such jump occur? In [1], the jumps occur every ϵ time, and the effect when $\epsilon \to 0$ is studied. In [10], however, the transitions follow a continuous time Markov process. In both papers, the discrete random transitions are assumed to be independent of the continuous dynamics, therefore the models can actually be better viewed as an extension of

N. Lynch and B. Krogh (Eds.): HSCC 2000, LNCS 1790, pp. 160–173, 2000.

Markov process with some continuous states attached whose evolutions follow state-dependent deterministic differential equations.

Another choice is to replace the deterministic dynamics inside the invariant set of each discrete state by a stochastic differential equation (SDE). Therefore, even if we keep the deterministic discrete transition part, starting from a fixed initial state, different guards can be activated depending on the realization of the solution stochastic process, thus different discrete transitions occur randomly. More general models can be proposed by blending the above two choices.

This paper is organized as following: in Section 2, we will try to give a general definition of stochastic hybrid system based on the second choice mentioned above. An example will be shown in Section 3 together with its analysis. In Section 4, the idea will be applied to a more general problem, in which we will approximate the solution of the SDE in $\mathbb{R}^1$ by the stochastic hybrid automata obtained from state space discretization. And finally we will discuss the special case of gradient system in the last section. The proofs of the theorems are not included due to the limit of space and will appear in subsequent paper.

2   General Definition

Definition 1 (Stochastic Hybrid System). *A stochastic hybrid system (or automata) is a collection* $H = (Q, X, Inv, f, g, G, R)$ *where*

- *Q is a discrete variable taking countably many values in* $\mathbf{Q} = \{q_1, q_2, \cdots\}$;
- *X is a continuous variable taking values in* $\mathbf{X} = \mathbb{R}^N$ *for some* $N \in \mathbb{N}$;
- *$Inv : \mathbf{Q} \to 2^{\mathbf{X}}$ assigns to each* $q \in \mathbf{Q}$ *an invariant open subset of* $\mathbf{X}$;
- *$f, g : \mathbf{Q} \times \mathbf{X} \to T\mathbf{X}$ are vector fields;*
- *$G : E = \mathbf{Q} \times \mathbf{Q} \to 2^{\mathbf{X}}$ assigns to each* $e \in E$ *a guard* $G(e)$ *such that*
  - *For each* $e = (q, q') \in E$, $G(e)$ *is a measurable subset of* $\partial Inv(q)$ *(possibly empty);*
  - *For each* $q \in \mathbf{Q}$, *the family* $\{G(e) : e = (q, q')$ *for some* $q' \in \mathbf{Q}\}$ *is a disjoint partition of* $\partial Inv(q)$.
- *$R : E \times \mathbf{X} \to \mathcal{P}(\mathbf{X})$ assigns to each* $e = (q, q') \in E$ *and* $x \in G(e)$ *a reset probability kernel on* $\mathbf{X}$ *concentrated on* $Inv(q')$. *Here* $\mathcal{P}(\mathbf{X})$ *denote the family of all probability measures on* $\mathbf{X}$. *Furthermore, for any measurable set* $A \subset Inv(q')$, $R(e, x)(A)$ *is a measurable function in* x.

Remark 1. The measurability assumption on R in the preceding definition is made to ensure that the events we encounter later are measurable w.r.t. the underlying σ-field, hence their probabilities make sense.

Definition 2 (Stochastic Execution). *A stochastic process* $(X(t), Q(t)) \in \mathbf{X} \times \mathbf{Q}$ *is called a stochastic execution iff there exists a sequence of stopping times* $\tau_0 = 0 \le \tau_1 \le \tau_2 \le \cdots$ *such that for each* $n \in \mathbb{N}$,

- In each interval $[\tau_n, \tau_{n+1})$, $Q(t) \equiv Q(\tau_n)$ is constant, $X(t)$ is a (continuous) solution to the SDE:

$$dX(t) = f(Q(\tau_n), X(t)) \, dt + g(Q(\tau_n), X(t)) \, dB_t \,,$$

  where B_t is the standard Brownian motion in $\mathbb{R}$;
- $\tau_{n+1} = \inf\{t \geq \tau_n : X(t) \notin Inv(Q(\tau_n))\}$;
- $X(\tau_{n+1}^-) \in G(Q(\tau_n), Q(\tau_{n+1}))$ where $X(\tau_{n+1}^-)$ denotes $\lim_{t \uparrow \tau_{n+1}} X(t)$;
- The probability distribution of $X(\tau_{n+1})$ given $X(\tau_{n+1}^-)$ is governed by the law $R(e_n, X(\tau_{n+1}^-))$, where $e_n = (Q(\tau_n), Q(\tau_{n+1})) \in E$.

Definition 3 (Embedded Markov Process). *In the notation of the previous definition, define $Q_n \triangleq Q(\tau_n)$, $X_n \triangleq X(\tau_n)$. Then $\{(Q_n, X_n), n \geq 0\}$ is called the embedded Markov process for the stochastic execution $(X(t), Q(t))$.*

Under these definitions, for example, a typical stochastic execution starts from (Q_0, X_0) and the continuous state $X(t)$ evolves according to the SDE

$$dX(t) = f(Q_0, X(t)) \, dt + g(Q_0, X(t)) \, dB_t, \quad X(0) = X_0$$

until time τ_1 when $X(t)$ first hits $\partial Inv(Q_0)$. Then depending on the hitting position $X(\tau_1^-)$, (say, $X(\tau_1^-) \in G(e)$ where $e = (Q_0, Q_1)$ for some $Q_1 \in \mathbf{Q}$), the discrete state jumps to $Q(\tau_1) = Q_1$ and the continuous state is reset randomly to $X(\tau_1) = X_1$ according to the conditional probability distribution $R(e, X(\tau_1^-))(\cdot)$ and the same process is repeated with (Q_1, X_1) replacing (Q_0, X_0) and so on.

Lemma 1. $\{(Q_n, X_n)\}$ *defined above is indeed a Markov process with transition probability:*

$$P(Q_{n+1} = q', X_{n+1} \in dx' | Q_n = q, X_n = x) = \int_{y \in G(e)} R(e, y)(dx') \, P(Y_x(\eta) = dy) \,, \tag{1}$$

where $e = (q, q')$, $Y_x(t)$ is the solution to the SDE

$$dY(t) = f(q, Y(t)) \, dt + g(q, Y(t)) \, dB_t, \quad Y(0) = x \,,$$

and $\eta = \inf\{t \geq 0 : Y_x(t) \notin Inv(q)\}$ is the first escape time of $Y_x(t)$ from $Inv(q)$.

Lemma 2. *If the reset kernel $R((q, q'), x) = R(q')$ does not depend on q nor x, then $\{Q_n\}$ itself is a Markov chain (MC) with transition probability $(n \geq 1)$:*

$$P(Q_{n+1} = q' | Q_n = q) = \int_{x \in Inv(q)} P(Y_x(\eta) \in G(e)) R(q)(dx) \,, \tag{2}$$

where e, Y_x, η is defined in the previous lemma. For $n = 0$, the transition probability depends on the initial distribution of $X(0)$.

Remark 2. The condition in Lemma 2 is fairly restrictive and excludes many general stochastic hybrid systems. The point of imposing this condition is to make calculation tractable. Furthermore, as we will see in the later sections, this special class of systems is still rich enough to admit many important applications.

The reason we introduce the embedded MC is that in most cases, it is hard if not impossible to get an explicit expression of the stochastic execution for a stochastic hybrid system. If all we are interested in is the reachability analysis of the discrete states transitions, then $\{Q_n\}$ will capture all the necessary information. This is the case if a subset of the discrete states is defined to be the "bad" states and a controller is designed to minimize the probability of reaching these states within a given time horizon. Or alternatively, some states are defined to be safe and we want to maximize the probability that the execution will remain in these states for as long as possible. At first sight these observation does not seem to be applicable in general, since in most cases, the definition of bad states and safe states involve both the discrete and continuous states. However, by breaking up the corresponding invariant sets and adding more discrete states and trivial reset kernels, we can always reduce the original system to a new one satisfying the above conditions, at least in the case when the support of any reset kernel is contained exclusively in safe or bad set.

3   A Simple Example

To clarify the above concepts, consider the following simple example. Two cars, labeled 1 and 2 with car 2 in the lead, are moving from left to right on a highway (see Figure 1). Due to various random factors such as road condition, wind, and the presence of human operators, the motions of both cars are stochastic. If we absorb all the randomness into the motion of car 1 and ignore the possible occurrence of emergency braking, then the motion of car 2 can be modeled as having a constant speed v_2. Let Δx be the distance between the two cars. Let $d_0 > d_1 > d_2 > d_3 > 0$ be four thresholds. We propose the following hybrid control scheme for car 1 (see the diagram in Figure 2): It consists of 3 discrete states $\{1, 2, 3\}$ corresponding to chasing, keeping and braking respectively.

1. **Chasing**: In this stage, $\Delta x \geq d_2$, and car 1 will try to catch car 2 at speed $v_1 > v_2$. So the perturbed motion of car 1 is governed by $\dot{x}_1 = v_1 + dB_t$, where B_t is a standard 1-D BM;

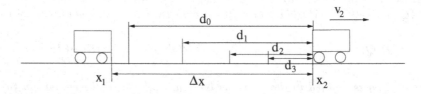

Fig. 1. A two-car platoon on the highway

2. **Keeping:** In this stage $d_3 \leq \Delta x \leq d_1$, and car 1 will try to move at v_2 under the perturbation dB_t;
3. **Braking:** If $\Delta x \leq d_3$, then car 1 will brakes according to some prescribed procedure until $\Delta x = d_0$. For simplicity, we ignore the presence of noise during braking.

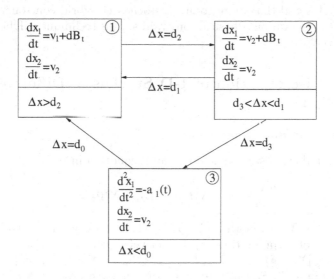

Fig. 2. Diagram for the stochastic hybrid system

The invariant sets and guards for each discrete state are also shown in Figure 2. The reset kernels are trivial, or more precisely, $R(e, x)$ is concentrated at x for any $e = (q, q') \in E$ and any $x \in G(e)$. It is easily seen that H satisfies the condition of Lemma 2. Hence the successive visits to the discrete states $\{Q_n\}$ is a MC. Actually its probability transition matrix is

$$P = \begin{pmatrix} 0 & 1 & 0 \\ p & 0 & 1-p \\ 1 & 0 & 0 \end{pmatrix},$$

where $p = (d_2 - d_3)/(d_1 - d_3)$. The first and third row of P is obvious and the second row follows from ([3]):

Lemma 3. *Let B_t be a standard BM starting from 0. For $a < 0 < b$, define $T_a = \inf\{t \geq 0 : B_t = a\}$, $T_b = \inf\{t \geq 0 : B_t = b\}$. Then $P(T_a < T_b) = \frac{b}{b-a}$ and $E(T_a \wedge T_b) = -ab$. Here $T_a \wedge T_b$ denotes $\min(T_a, T_b)$.*

Calculation shows that the stationary distribution for P is $(\frac{1}{3-p}, \frac{1}{3-p}, \frac{1-p}{3-p})$. Therefore the fraction of time the system spends in each discrete state is proportional to:

$$\left(\frac{ET_1}{3-p}, \frac{ET_2}{3-p}, \frac{(1-p)ET_3}{3-p} \right),$$

where $ET_1 = (d_0 - d_2)/(v_1 - v_2)$, $ET_2 = (d_1 - d_2)(d_2 - d_3)$, $ET_3 = t_3$ are the expected sojourn time in each discrete state respectively.

In practice, we want to maximize the time the stochastic hybrid system spends in the keeping state and minimize the time it spends in the braking state. This can be done by adjusting the thresholds d_0, d_1, d_2, d_3 properly. Sometimes this choice is restricted by other physical constraints. However, we can always use more thresholds and thus more complex stochastic hybrid controller to achieve the goal within the various physical constraints. This technique will be illustrated in the next section.

4   State Discretization of 1-D Stochastic Differential Equation

4.1   Motivation and Definition

Consider the following stochastic differential equation in $\mathbb{R}$:

$$\frac{dX(t)}{dt} = f(X(t)) + dB_t, \quad X(0) = 0, \tag{3}$$

where $f : \mathbb{R} \rightarrow \mathbb{R}$ is smooth and dB_t is white noise with spectral density 1. Define a series of stopping times τ_n inductively as: $\tau_0 = 0$, $\tau_n = \inf\{t \geq \tau_{n-1} : |X(t) - X(\tau_{n-1})| = \delta\}$, $n = 1, 2, 3, \cdots$. Let $S_n = X(\tau_n)$. Then $\{S_n\}$ is a MC taking values in $\delta \cdot \mathbb{Z}$. S_n captures many sample path properties of the solution process $X(t)$, for example, whether $X(t)$ is recurrent. or less obviously, whether $X(t)$ crosses an interval of length less than δ infinitely many times,

Define $\tau_t = \sup_n\{\tau_n : \tau_n \leq t\}$ and let $Y_t = X(\tau_t)$. Then Y_t is piecewise constant with value S_n in time interval $[\tau_n, \tau_{n+1})$. Define $Z(t)$ to be the solution process to the stochastic differential equation:

$$\frac{dZ(t)}{dt} = f(Y_t) + dB_t. \tag{4}$$

Comparing equation (3) and (4) and noticing that during time interval $[\tau_n, \tau_{n+1})$, $|X(t) - Y_t| \leq \delta$ by the definition of τ_n's and f is continuous, we can expect that as $\delta \rightarrow 0$, $Z(t)$ approaches $X(t)$ in distribution, hence $Z(t)$ is a good approximation to $X(t)$ which is often impossible to calculate explicitly. However, it is still difficult to solve equation (4) since Y_t depend on the

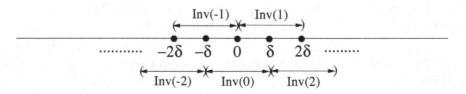

Fig. 3. Discretization of state space

original solution process $X(t)$ through τ_n's and S_n's. So to solve equation (4), theoretically we still have to solve equation (3) first.

One way to get out of this loop is to use the fact that $X(t)$ can be approximated by $Z(t)$, hence τ_n's and S_n's can also be approximated by the corresponding random variables defined from $Z(t)$. This will lead to the discretized stochastic hybrid system (DSHS) defined below.

Definition 4 (Discretized Stochastic Hybrid System). *The discretized stochastic hybrid system for equation (3) is $H = (Q, X, Inv, f, g, G, R)$ where $\mathbf{Q} = \mathbb{Z}$, $\mathbf{X} = \mathbb{R}$, and*
- *$Inv(k) = ((k-1)\delta, (k+1)\delta)$ for any $k \in \mathbf{Q}$;*
- *$f(k, \cdot) = f(k\delta)$, $g(k, \cdot) = 1$ are constant functions;*
- *$G(k, k-1) = \{(k-1)\delta\}$, $G(k, k+1) = \{(k+1)\delta\}$ are singletons and $G(k, l) = \emptyset$ for all other l;*
- *Reset kernels are trivial.*

Since H satisfies the condition of Lemma 2, $\{Q_n\}$ defined as in Section 2 is a MC. By discussion at the beginning of this section, it is expected that $\{Q_n\}$ approximates the MC $\{S_n\}$ defined from the solution $X(t)$ to equation (3). (In the following development, we will use H^δ to stress the dependency of H on the discretization step δ only if necessary).

Obviously the probability transition matrix Q for $\{Q_n\}$ satisfies: $Q_{i,j} = p_i$, if $j = i + 1$; $Q_{i,j} = q_i = 1 - p_i$ if $j = i - 1$ and $Q_{i,j} = 0$ otherwise. Such a chain is called a *death and birth* chain and we will calculate p_k's and q_k's as follows: The solution to the stochastic differential equation $dY(t) = f(k\delta)\,dt + dB_t$ with initial condition $Y(0) = k\delta$ is $Y(t) = k\delta + f(k\delta)t + B_t$, *i.e.* the BM starting from $k\delta$ and with drift $\mu = f(k\delta)$. If we use B_t^μ to denote the BM starting from 0 and with drift μ, then

$$p_k = P(B_t^\mu \text{ reaches } \delta \text{ before it reaches } - \delta). \tag{5}$$

So the problem becomes calculating the exit distribution of B_t^μ from $(-\delta, \delta)$. We will derive the probability in a more general setting. Assume $\mu \neq 0$ since the case when $\mu = 0$ has already been considered in Lemma 3. Let $a < 0 < b$. Denote $T_a = \inf\{t \geq 0 : B_t^\mu = a\}$, $T_b = \inf\{t \geq 0 : B_t^\mu = b\}$.

Lemma 4. *B_t^μ first exits (a, b) from b with probability*

$$P(T_b < T_a) = \frac{e^{-2\mu a} - 1}{e^{-2\mu a} - e^{-2\mu b}}. \tag{6}$$

Therefore by taking $a = -\delta$, $b = \delta$ and $\mu = f(k\delta)$, we have $p_k = \phi[\delta f(k\delta)]$ where ϕ is the monotonically increasing function defined by

$$\phi(x) = \frac{e^{2x} - 1}{e^{2x} - e^{-2x}}, \quad x \neq 0. \tag{7}$$

For a plot of function ϕ, see Figure 4.

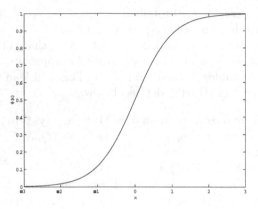

Fig. 4. Plot of ϕ

4.2   Recurrence vs. Stability

Having obtained the probability transition matrix Q, the natural question we will ask ourselves is: what is the relation between the deterministic part of differential equation (3), *i.e.*

$$\frac{dx}{dt} = f(x), \quad x(0) = 0, \tag{8}$$

and the embedded MC $\{Q_n\}$? For (8) we have notions such as equilibrium and various kinds of stability. What are their counterparts for $\{Q_n\}$? Intuitively if equation (8) has a globally stable equilibrium then the sample paths of $\{Q_n\}$ should also be centered around the equilibrium most of the time even if the starting point is far away, thus stability in some probabilistic sense can be expected. It turns out that the notions of recurrence and transience are good candidates for this. Assume MC $\{Q_n\}$ is irreducible, *i.e.* starting from any state, there is a positive probability of jumping to any other state in finite steps.

Definition 5 (Recurrent and Transient MC). *A MC $\{Q_n\}$ on a countable state space S is called recurrent if and only if starting from any state $x \in S$, it will return to x in finite time with probability 1, or more precisely, if and only if*

$$P(T_x < \infty | Q_0 = x) = 1 \quad \forall x \in S,$$

where $T_x \triangleq \inf\{n \geq 1 : Q_n = x\}$. Otherwise $\{Q_n\}$ is called transient.

Definition 6 (Positive Recurrent MC). *A recurrent MC $\{Q_n\}$ on a countable state space S is called positive recurrent if and only if $E[T_x | Q_0 = x] < \infty$ for all $x \in S$.*

An important characteristic of a positive recurrence chain is that its invariant distribution exists and is unique ([3]). In general, positive recurrence implies recurrence, but not the other way around, since symmetric random walk on integer grid $\mathbb{Z}$ is an example of recurrent but not positive recurrent chain.

Now consider the MC $\{Q_n\}$ obtained in subsection 4.1. Obviously it is irreducible. Let $\{Q_n^+\}$ and $\{Q_n^-\}$ be the MC's obtained by observing $\{Q_n\}$ on the subset $\mathbb{N}^+ = \{0, 1, 2, \cdots\}$ and $\mathbb{N}^- = \{0, -1, -2, \cdots\}$ respectively. Both $\{Q_n^+\}$ and $\{Q_n^-\}$ are irreducible. The following lemma justifies our interest in them.

Lemma 5. *$\{Q_n\}$ is (positive) recurrent iff both $\{Q_n^+\}$ and $\{Q_n^-\}$ are (positive) recurrent respectively. Furthermore, if π^+ is the stationary distribution of $\{Q_n^+\}$ on $\mathbb{N}^+$, π^- is the stationary distribution of $\{Q_n^-\}$ on $\mathbb{N}^-$, then $\pi \triangleq \alpha\pi^+ + (1-\alpha)\pi^-$ is the stationary distribution of $\{Q_n\}$ on $\mathbb{Z}$, where*

$$\alpha = \frac{\pi^-(0)p_0}{\pi^-(0)p_0 + \pi^+(0)q_0}.$$

Notice that the transition matrix Q^+ has the property $Q^+(i, j) = 0$ when $|i-j| > 1$, hence it is a death and birth chain. The following lemma is a standard result from probability theory (see [3]):

Lemma 6. *$\{Q_n^+\}$ is recurrent if and only if $\sum_{m=0}^{\infty} \prod_{j=1}^{m} q_j/p_j = \infty$, $\{Q_n^+\}$ is positive recurrent if and only if $\sum_{m=0}^{\infty} \prod_{j=1}^{m} p_{j-1}/q_j < \infty$ (here $p_0 = 1$). In the latter case, the stationary distribution π^+ of $\{Q^+\}$ is:*

$$\pi^+(i) = \prod_{j=1}^{i} \frac{p_{j-1}}{q_j} \bigg/ \sum_{m=0}^{\infty} \prod_{j=1}^{m} \frac{p_{j-1}}{q_j}, \quad i = 0, 1, 2, \cdots$$

Note the products are interpreted as 1 whenever $m = 0$.

Similar argument for $\{Q_n^-\}$ can be established by symmetry. Assembling Lemma 5, Lemma 6, equation (7) together, we get

Theorem 1 (Recurrence of DSHS). *The embedded MC $\{Q_n\}$ of the discretized stochastic hybrid system of (3) is recurrent if and only if*

$$\sum_{m=0}^{\infty} \prod_{j=1}^{m} \frac{1 - \exp[-2\delta f(j\delta)]}{\exp[2\delta f(j\delta)] - 1} = \infty \quad and \quad \sum_{m=0}^{\infty} \prod_{j=-m}^{-1} \frac{\exp[2\delta f(j\delta)] - 1}{1 - \exp[-2\delta f(j\delta)]} = \infty.$$

$$(9)$$

$\{Q_n\}$ is positive recurrent if and only if

$$\sum_{m=0}^{\infty} \prod_{j=1}^{m} \frac{\phi[\delta f((j-1)\delta)]}{1 - \phi[\delta f(j\delta)]} < \infty \quad and \quad \sum_{m=0}^{\infty} \prod_{j=-m}^{-1} \frac{1 - \phi[\delta f((j+1)\delta)]}{\phi[\delta f(j\delta)]} < \infty.$$

$$(10)$$

In the latter case, the stationary distribution π of $\{Q_n\}$ is given by Lemma 5.

4.3   Boundary Between Recurrence and Transience

From Theorem 1, it is evident that whether $\{Q_n\}$ is (positive) recurrent depends only on the "tail" of function f, *i.e.* the asymptotic behavior of $f(x)$ when $x \to \pm\infty$. In general, we have

Lemma 7 (Comparison Lemma). *Suppose $f, g : \mathbb{R} \to \mathbb{R}$ are two smooth vector fields such that*

$$f(x) > g(x), \quad f(-x) < g(-x) \quad \text{for } x \text{ sufficiently large,}$$

Then if $\{Q_n(f)\}$ is (positive) recurrent, so is $\{Q_n(g)\}$. Conversely, if $\{Q_n(g)\}$ is transient, $\{Q_n(f)\}$ is also transient.

Inspired by [3], let us look at f of the form

$$f(x) = \begin{cases} Cx^{-r} & x \geq M \\ -C(-x)^{-r} & x \leq -M \\ \text{do not care} & |x| < M \end{cases} \tag{11}$$

for some constant C and $r > 0$. Note we have deliberately made f to be an odd function outside $(-M, M)$ such that the corresponding MC $\{Q_n^+\}$ and $\{Q_n^-\}$ are mirror image of each other. So by Lemma 5 we need only to consider one of them, say, $\{Q_n^+\}$. If $C \leq 0$, then by the Comparison Lemma and the previous paragraph, $\{Q_n\}$ is recurrent, so we assume $C > 0$ here.

Proposition 1. *Assuming $C > 0$. The DSHS $\{Q_n\}$ corresponding to f in (11) is recurrent if $r > 1$ or if $r = 1$ and $C < 0.5$. $\{Q_n\}$ is transient if $r < 1$ or if $r = 1$ and $C > 0.5$.*

Note the above conclusion is independent of the discretization step δ. Next we will discuss the boundary of positive recurrence. Suppose f is of the form:

$$f(x) = \begin{cases} -Cx^{-r} & x \geq M \\ C(-x)^{-r} & x \leq -M \\ \text{do not care} & |x| < M \end{cases} \tag{12}$$

where C, r are positive constants. A similar argument generates:

Proposition 2. *Assuming $C > 0$. The DSHS $\{Q_n\}$ corresponding to f in (12) is positive recurrent if $r < 1$ or if $r = 1$ and $C > 0.5$. $\{Q_n\}$ is not positive recurrent if $r > 1$ or if $r = 1$ and $C < 0.5$.*

5   DSHS of Gradient System

If equation (8) is a gradient system ([8]) of the form:

$$\frac{dx}{dt} = f(x) = -\nabla V(x) \tag{13}$$

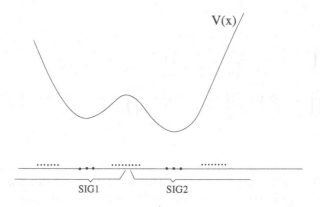

Fig. 5. DSHS for a gradient system

for some $V \in C^2(\mathbb{R})$, then each local minimum of $V(x)$ is an equilibrium of (13) and in the embedded MC $\{Q_n\}$ of the corresponding DSHS, states in the vicinity of each equilibrium constitute an *strongly interacting group* (SIG) in the sense that in any typical execution of $\{Q_n\}$, once the state jumps into an SIG, it will stay inside it for a relatively long period before jumping to another SIG. (See Figure 5). In many applications it is often the case that we want to choose some suitable control so as to make the system evolve inside some desired valleys for as long as possible while avoiding some undesired trap.

Under this setting, the conclusion of Proposition 1 and Proposition 2 in the last subsection translates into: $\{Q_n\}$ is recurrent (transient) if $V(x)$ approaches $-\infty$ slower (faster) than $-\frac{1}{2}\ln(|x|)$ as $|x| \to \infty$ respectively; $\{Q_n\}$ is (not) positive recurrent if $V(x)$ approaches ∞ faster (slower) than $\frac{1}{2}\ln(|x|)$ as $|x| \to \infty$ respectively. Therefore instead of the clear cut boundary between stability and non-stability in the deterministic system, the DSHS have a blurred boundary between positive recurrence and transience, with $V(x)$ growing asymptotically between $-\frac{1}{2}\ln(|x|)$ and $\frac{1}{2}\ln(|x|)$ corresponding to recurrent but not positive recurrent $\{Q_n\}$. In this subsection, we will always assume that $V(x)$ is chosen such that for δ small enough, the corresponding $\{Q_n\}$ is positive recurrent and hence has a stationary distribution π. We will elaborate on the asymptotic behavior of π as $\delta \to 0$ and reveal its relation with $V(x)$.

From Lemma 5 and Lemma 6, π can be written as: $\pi(i) = \alpha\pi^+(i) + (1 - \alpha)\pi^-(i)$ for all $i \in \mathbb{Z}$ with

$$\alpha = \frac{\pi^-(0)\phi[\delta f(0)]}{\pi^-(0)\phi[\delta f(0)] + \pi^+(0)(1 - \phi[\delta f(0)])}$$

and

$$\pi^+(i) = \prod_{j=1}^{i} \frac{\phi[\delta f((j-1)\delta)]}{1 - \phi[\delta f(j\delta)]} \Bigg/ \sum_{m=0}^{\infty} \prod_{j=1}^{m} \frac{\phi[\delta f((j-1)\delta)]}{1 - \phi[\delta f(j\delta)]},$$

$$\pi^-(i) = \prod_{j=-i}^{-1} \frac{1 - \phi[\delta f((j+1)\delta)]}{\phi[\delta f(j\delta)]} \Bigg/ \sum_{m=0}^{\infty} \prod_{j=-m}^{-1} \frac{1 - \phi[\delta f((j+1)\delta)]}{\phi[\delta f(j\delta)]}, \quad \forall i \in \mathbb{Z}.$$

(14)

This messy-looking expression takes an especially simple form as $\delta \to 0$. To reveal this, for each $\delta > 0$ denote π^δ the stationary distribution of $\{Q_n\}$ for the discretized stochastic hybrid system H^δ with discretization step δ. Define function $u^\delta : \mathbb{R} \to \mathbb{R}$ as: $u^\delta(x) = \pi^\delta(k)/\delta$, if $x \in [k\delta, (k+1)\delta)$ for some $k \in \mathbb{Z}$. Then it can be easily checked that u^δ satisfies: $\int_{-\infty}^{\infty} u^\delta(x)\,dx = 1$, and u^δ has roughly the same shape as π^δ. Therefore the discrete distribution π^δ is converted to a continuous density function u^δ. Moreover,

Lemma 8. *Suppose $V(x)$ is chosen such that π^δ exists for $\delta > 0$ small enough. Then*

$$\lim_{\delta \to 0} \frac{u^\delta(y)}{u^\delta(x)} = e^{-2[V(y)-V(x)]} \quad \forall x, y \in \mathbb{R}.$$

We need the following notion to ensure that u^δ converges to a probability density.

Definition 7 (Tightness). *A family $\{u_\alpha, \alpha \in \Lambda\}$ of probability densities indexed by Λ is tight if and only if for each $\epsilon > 0$, there exists an M such that $\int_{-M}^{M} u_\alpha(x)\,dx > 1 - \epsilon$ for all $\alpha \in \Lambda$.*

Theorem 2. *Suppose $V(x)$ is chosen such that π^δ exists for $\delta > 0$ small enough and the resulting $\{u^\delta, \delta > 0\}$ is tight, then $\int_{-\infty}^{\infty} e^{-2V(x)}\,dx < \infty$ and*

$$u^\delta(x) \to u^0(x) \triangleq \frac{e^{-2V(x)}}{\int_{-\infty}^{\infty} e^{-2V(y)}\,dy} \quad as \ \delta \to 0,$$

where the convergence is pointwise.

Shown in Figure 6 are the plots of u^δ for different δ when $V(x) = (x^4 + 20(x-5) + c(x-5)^2)/100$ and $c = 275$. Here we choose $\delta = 40/N$, *i.e.* $[-20, 20]$ is discretized into N subintervals. Notice that the convergence speed is fast: even if the discretization is coarse, the resulting u^δ is still close to the final limit. In Figure 6, the two local minimums are at roughly the same level. By changing the value of c slightly, we can make one valley slightly deeper than the other. However, due to the exponential inverse relation of u^0 to V, this small change will be considerably amplified in u^0.

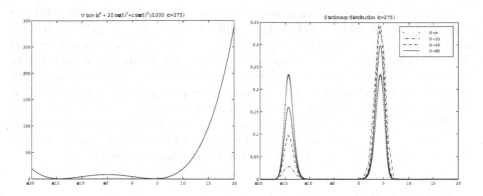

Fig. 6. Left: $V(x)$; Right: u^δ for different $\delta = \frac{40}{N}$

It is expected that the limiting distribution u^0 in Theorem 2 will also be the stationary distribution of the original stochastic differential equation: $dX_t = -\nabla V(X_t)dt + dB_t$ in the sense that if $X(0)$ is distributed as u^0 independently of $\{B_t\}$, then for any $t > 0$, the solution process X_t has the same distribution. We illustrated this in the following example.

Example 1. (**Ornstein-Uhlenbeck process**)   Solution X_t to the SDE $dX_t = \mu X_t + \sigma dB_t$ is called the Ornstein-Uhlenbeck precess ([7]). Consider the case when $\sigma = 1$, $\mu = -a$ for some $a > 0$. Then by Ito formula, $X_t = X_0 e^{-at} + \int_0^t e^{-a(t-s)}dB_s$. If X_0 is Gaussian $N(0,\sigma)$ independently of $\{B_t\}$, then for each $t > 0$, X_t is also Gaussian with mean 0 and variance $\frac{1}{2a} + (\sigma^2 - \frac{1}{2a})e^{-2at}$. Let $\sigma^2 = \frac{1}{2a}$, then we can see that X_t has stationary distribution $N(0, \frac{1}{\sqrt{2a}})$ with density function predicted by Theorem 2.

Next we will discuss the limit behavior of first exit distribution of MC $\{Q_n\}$ from an interval. Consider MC $\{Q_n^+\}$ obtained in subsection 4.2.

Lemma 9. *Suppose $i_1 < i_0 < i_2$ are nonnegative integers. Then the probability that Q_n^+ starting from i_0 hits i_2 first than it hits i_1 is:*

$$\sum_{m=i_1+1}^{i_0-1} \prod_{j=i_1+1}^{m} \frac{q_j}{p_j} \Big/ \sum_{m=i_1+1}^{i_2-1} \prod_{j=i_1+1}^{m} \frac{q_j}{p_j}. \tag{15}$$

Suppose $a, b, c \in \mathbb{R}$ and $a < b < c$. For each $\delta > 0$, define $i_a^\delta = [a/\delta]$, $i_b^\delta = [b/\delta]$, $i_c^\delta = [c/\delta]$. Then for the corresponding embedded MC $\{Q_n\}$, the probability $P_{i_b^\delta}(T_{i_c^\delta} < T_{i_a^\delta})$ can be calculated by Lemma 9. The next theorem characterize the limiting behavior of such probability when $\delta \to 0$.

Theorem 3. *Using the same notation as in the above paragraph. Then as $\delta \to 0$,*

$$P_{i_b^\delta}(T_{i_c^\delta} < T_{i_a^\delta}) \to \frac{\int_a^b e^{-2V(x)}\,dx}{\int_a^c e^{-2V(x)}\,dx}.$$

It can be shown that the above asymptotic expression coincides with the corresponding probability of the original diffusion process (see [4]). Furthermore, under some proper assumptions, the expected escape time from an interval of the embedded MC can be studied as well and can be shown to converge to the corresponding value of the original diffusion process. Therefore the DSHS presents a powerful tool for studying the sample path properties of the SDE, at least when the discretization step is small enough.

The advantage of having closed form formulae for various properties of the stochastic hybrid systems is that it can greatly facilitate the design and evaluation of such systems. These topics will be pursued in future work.

Acknowledgments

This research has been supported by DARPA under grant F33615-98-C-3614, ARO under grant MURI DAAH04-96-1-0341 and by the California PATH project under MOU312.

References

1. E. Altman and V. Gaitsgory. Asymptotic optimization of a nonlinear hybrid system governed by a Markov decision process. *SIAM Journal of Control and Optimization*, 35(6):2070–2085, 1997.
2. Gopal K. Basak, Arnab Bisi, and Mrinal K. Ghosh. Ergodic control of random singular diffusions. In *IEEE Conference on Decision and Control*, pages 2545–2550, Kobe, Japan, 1996.
3. Richard Durrett. *Probability: theory and examples, 2nd edition.* Duxbury Press, 1996.
4. Richard Durrett. *Stochastic calculus: A practical introduction.* CRC Press, 1996.
5. J.A. Filar and V. Gaitsgory. Control of singularly perturbed hybrid stochastic systems. In *IEEE Conference on Decision and Control*, pages 511–516, Kobe, Japan, 1996.
6. John Lygeros. *Hybrid systems: modeling, analysis and control.* preprint, 1999.
7. Bernt Oksendal. *Stochastic Differential Equations, an introduction with application. Fifth edition.* Springer-Verlag, 1998.
8. L. Perko. *Differential equation and dynamical systems, 2nd edition.* Springer-Verlag, 1996.
9. E. Skafidas, R.J. Evans, and I.M. Mareels. Optimal controller switching for stochastic systems. In *IEEE Conference on Decision and Control*, pages 3950–3955, San Diego, CA, 1997.
10. Ching-Chih Tsai. Composite stabilization of singularly perturbed stochastic hybrid systems. *International Journal of Control*, 71(6):1005–1020, 1998.
11. Ching-Chih Tsai and Abraham H. Haddad. Averaging, aggregation and optimal control of singulayly perturbed stochastic hybrid systems. *International Journal of Control*, 68(1):31–50, 1997.

Automatic Compilation of Concurrent Hybrid Factories from Product Assembly Specifications

Eric Klavins

Advanced Technology Laboratories
Department of Electrical Engineering and Computer Science
University of Michigan
1101 Beal Avenue
Ann Arbor, MI 48109-2110, USA
klavins@eecs.umich.edu

Abstract. We address the problem of designing a distributed, hybrid factory given a description of an assembly process and a palette of controllers for basic assembly operations. In particular, we present a method that, starting with a product assembly graph (PAG), allows us to "compile" a factory description, consisting of a geometry and a hybrid, dynamical system representing the motions of robots on that geometry. This method is based on a formalism, which we have described in previous work, that allows us to manage the details of low level, continuous control of robot actuation and high level, logical control of various couplings of robot behaviors. The factory description is intended to be an aid in the design of an actual factory, if not directly implementable itself.

1 Introduction

Large distributed networks of robots and computers form the basis of modern manufacturing systems. These systems should be rapidly reconfigurable, to adjust to design changes in the products they assemble or to changes in the market. Furthermore, they must be easily programmable. These goals, however, are seldom achieved in practice because of the complexity that hundreds of interconnected, concurrently operating robots necessarily incurs. The programming process can be ad hoc and frequently results in a large fraction of the control code being "exception handler code". This cost is felt in terms of expensive programming projects, incompletely understood factory behavior, and a delay in the introduction of new products to the market.

In previous work, [11], [10], we described a formalism for representing and composing concurrent robotic systems which we believe addresses some of the problems in designing distributed, dynamic factories. Specifically, we introduced the notion of a *Threaded Petri Net* (TPN), which combines low level motion control of individual robots or small groups of robots with high level logic control to manage how couplings between robots change over time in the factory. We also introduced a way of composing TPNs to create larger TPNs and demonstrated several properties of TPNs and our composition rules. Although we believe these tools will prove applicable to a broad range of automation settings, our notion of assembly is more immediately inspired by the high flexibility, low volume setting targeted by the "Minifactory" of Rizzi et al. [17], wherein decentralized general

N. Lynch and B. Krogh (Eds.): HSCC 2000, LNCS 1790, pp. 174–187, 2000.

purpose robotic agents accomplish all the factory's parts transport and assembly operations in fluidly choreographed transactions. For example, a complex subassembly task requiring four or six coordinated degrees of freedom can only transpire in such a Minifactory when some subgroup of the decentralized robots "agrees" to collaborate closely in forming the specialized "machine" (the higher degree of freedom coordinated mechanism) suited to the specific task at hand. Of course, that alliance must be temporary, since each of the participating agents is required to play analogous but different roles in other machines, both prior and subsequent to the instantiation of the one in question. The TPN formalism provides tools to frame this problem.

In the present paper, we apply our work on TPNs and composition to the task of *automatically* compiling factory descriptions from a standard representation of a product assembly process called a *product assembly graph* or PAG. The factory description that results consists of: an allocation of robots of various types; a geometrical description of the space that these robots inhabit; and a concurrent hybrid dynamic system, represented by a TPN, which directly corresponds to the robot programs. We use results from our previous work to show that the resulting TPN is live and that it successfully implements the process specified by the PAG input.

It must be stressed that we presuppose an infrastructure of tunable and switchable feedback controllers which our compiler merely "puts together", in a safe and correct way, to realize the assembly process. Such a palette of controllers is relatively easy to build for environments well described by generalized damper dynamics [12], but becomes quite challenging when dynamical dexterity is required. For example, in [2], substantial "hand building" affords deployments of controllers whose domains of attraction explicitly include portions of the forward limit sets of their neighbors. Here, we simply assume that these "dynamical systems details" have been worked out via parameterized families of regulators, and represented in a way that allows us to use them with TPNs (see Section 4). We then focus on the logical coordination and scheduling problems that follow. We have, in fact, built such a palette and a compiler for a simple class of PAGs and simulated the resulting factories. Animations of these factories can be viewed at http://www.eecs.umich.edu/~klavins/mf/.

The paper is organized as follows. In Section 2, we review related research. In Section 3, we review TPNs and our composition method. In Section 4, we introduce mathematical models which represent robots, operations (those in the palette of controllers), and factories. In Section 5, we describe the compilation algorithm in detail and prove that it describes live and correct factories. Finally, in Section 6, we discuss a simple implementation of the compiler.

2   Background and Related Work

The research we report on in this paper draws from several areas: preimage backchaining of motion controllers, autonomous robot assembly, and hybrid discrete/continuous systems including Petri Nets. We review each of these areas as they pertain to the present research.

Preimage backchaining was introduced into the motion planning literature in [14] as a method of sequentially composing motion strategies. In [2] this method was extended to dynamically dexterous robot manipulators in work that serves as the basis of our current research. In [11], we expanded these ideas to include the notion of concurrent composition of behaviors for the case of several robots in a shared workspace based on simple Petri Net composition methods. Similar methods are found in work on the bottom-up synthesis of Petri Nets, especially [13], where simple Petri Nets are combined along paths and invariants of the resulting net are obtained from the constituent nets. In the present paper, we use the properties of our compositional tools design and verify an algorithm that automatically compiles concurrent, hybrid factories.

The approach to assembly in [12], for simple situations, introduces an automatic method for constructing a control law that guides a single robot to assemble a product from its parts based on the notion of an artificial energy landscape wherein the configuration of least energy is the one in which the product is assembled. It is not obvious that this method could be extended to three dimensional systems with orientable parts. In this paper we take the view that the PAG of a product corresponds to a discrete and parallelized version of such a potential function. The individual steps of the assembly may be given by artificial potential field controllers, but the overall logic of the assembly is given by the PAG. This allows us to use multiple robots, as in a high volume factory setting.

Programs such as Archimedes [9] exist which transform the CAD description of a product into a PAG. Little research has been reported concerning translating the PAG directly into a layout and distributed program for a factory, although in the one example we know of, [19], the authors produce elementary conveyer belt layouts. In this paper we introduce a method that we believe will lead to a general procedure for carrying out such a translation.

Hybrid systems combine a discrete state and a continuous state into the same model. A common representation is the *hybrid automaton*, [7]. Many definitions of hybridized Petri Nets, serving various needs, have also been investigated: Continuous and Hybrid Petri Nets [4], Differential Petri Nets [5], and DAE-Petri Nets [1]. The last is most easily seen as an extension of hybrid automata. Our definition of Threaded Petri Net differs from these definitions in several regards. First, we consider a place in a net to be a controlled dynamic system on some subset of the *degrees of freedom* of the system, depending on the marking, and a transition fires when and only when the systems in its preset are in stable equilibrium states. Furthermore, transition firings *redistribute* the degrees of freedom of the system to other dynamic systems in a controlled manner.

3   Definitions and Basic Properties

In this section we introduce the formal ideas that underlie our compiler research. We refer the reader to [10] for the details. We adopt the following definition of a Petri Net, also called a condition/event net, found in [8].

Definition 3.1 *A* **Petri Net** *is a pair (T, P) where T is a finite set of elements called* **transitions** *and $P \subseteq 2^T \times 2^T$ whose elements are called* **places***.*

We use standard Petri Net notation. If $\{\{a_1, ..., a_i\}, \{b_1, ..., b_j\}\} \in P$, we write $[a_1, ..., a_i; b_1, ..., b_j] \in P$. If $p = [a_1, ..., a_i; b_1, ..., b_j]$ then $left(p)$ is the set $\{a_1, ..., a_i\}$ and $right(p)$ is the set $\{b_1, ..., b_j\}$. A **marking** of a net (T, P) is a set $m \subseteq P$. The **flow relation** F of a Petri Net (T, P) is the relation where $(t, p) \in F$ if $t \in left(p)$ and $(p, t) \in F$ if $t \in right(p)$. The **preset** of an element $x \in T \cup P$ is set $\{y \mid y \; F \; x\}$ and is denoted $^{\bullet}x$. The **postset** of x is the set $\{y \mid x \; F \; y\}$ and is denoted $x^{\bullet}$. See [15] for a detailed introduction.

In a graphical representation of a Petri Net, places are represented by circles and transitions by squares. In our research, a place represents a controlled dynamical subsystem decoupled from the entire system in question. Transitions represent discrete changes in the dynamics of subsystems.

3.1   Threaded Petri Nets

Suppose we have a collection of robots $r_1, ..., r_n$ with configuration spaces $\mathcal{C}(r_1)$, ..., $\mathcal{C}(r_n)$ whose continuous state can be given by $\mathbf{x} = (x_1, ..., x_n) \in \mathcal{C}(r_1) \times ... \times \mathcal{C}(r_n)$ and whose global dynamics is simply $\dot{\mathbf{x}} = \mathbf{u}$. The dynamics of components of $\mathbf{x}$ are almost independent of each other. However, robots do interact for short periods of time, as for example during a parts mating operation, so that the dynamics of certain components of $\mathbf{x}$ may occasionally be tightly coupled.

To describe how couplings change and which dynamics are operating on which components of $\mathbf{x}$, we introduce the *Threaded Petri Net*, or TPN. Places correspond to control modes which we will have chosen from a palette of such modes. Thus, for each place p there is a system given by $\dot{\mathbf{y}} = F_p(\mathbf{y})$ where $\mathbf{y}$ is the vector concatenation of l_p vectors (components of $\mathbf{x}$) and F_p is chosen from the palette of controllers that we assume is already constructed. The mode has domain of attraction $\mathcal{D}_p$ and goal set $\mathcal{G}_p$. Formally,

Definition 3.2 *A* **Threaded Petri Net** *(TPN) consists of*

1. *a set T of transitions;*
2. *a set $P \subseteq 2^T \times 2^T$ of places;*
3. *for each $p \in P$, size, dynamics, domain and goal l_p, F_p, $\mathcal{D}_p$ and $\mathcal{G}_p$;*
4. *for each $e \in T$ a bijective function*

$$d_e : \bigcup_{p \in \, ^{\bullet}e} \{p\} \times \{1, ..., l_p\} \to \bigcup_{q \in e^{\bullet}} \{q\} \times \{1, ..., l_q\}$$

called the **redistribution function** *of e;*

subject to the condition that for each $e \in T$,

$$\sum_{p \in \, ^{\bullet}e} l_p = \sum_{q \in e^{\bullet}} l_q$$

(so that it is possible for d_e to be bijective).

Note that the difference between a TPN and a condition/event net is not only the additional information associated with each place. We have also added the redistribution functions, d_e for each $e \in E$, which define what happens to each component of $\mathbf{x}$ as mode changes occur. Graphically, a TPN is depicted as is a simple Petri Net, except that the redistribution functions are shown by curves through the net. See Figure 1 for example.

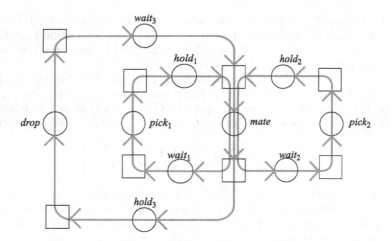

Fig. 1. An example of a Threaded Petri Net which describes the dynamics of three robots in a parts mating procedure.

Definition 3.3 *A **marking** is a pair* (m, f_m) *where* $m \subseteq P$ *and*

$$f_m : \bigcup_{p \in m} \{p\} \times \{1, ..., l_p\} \to \{1, ..., n\}$$

*which specifies which degrees of freedom of the system each mode is operating on. A **legal** marking is one where* f_m *is bijective. We will be concerned only with legal markings in what follows.*

A legal marking (m, f_m) of a TPN says, for each $p \in m$, which components of $\mathbf{x} \; F_p$ is acting on *and* what the dynamics of *each* component of $\mathbf{x}$ are. Thus, we can say how the state of the system is changing given a particular marking (m, f_m). Given $j \in N$, suppose that $f_m^{-1}(j) = (p, i)$. That is, under the marking (m, f_m) the jth component of $\mathbf{x}$ is changing according to the ith component of the mode dynamics of p:

$$\dot{x}_j = \pi_i \circ F_p(x_{f_m(p,1)}, ..., x_{f_m(p,l_p)})$$

where π_i gives the ith component of the function F_p. This is valid until some mode changes, which leads us to a definition of *how* events are triggered.

Definition 3.4 *Let* (m, f_m) *be a legal marking.* $e \in T$ *is* m-**enabled** *with respect to* $\mathbf{x} \in \mathbb{R}^n$ *if*

1. $\bullet e \subseteq m$ *and* $e^\bullet \cap m = \emptyset$;
2. *for each* $p \in \; \bullet e$, $(x_{f_m(p,1)}, ..., x_{f_m(p,l_p)}) \in \mathcal{G}_p$;
3. *for each* $q \in e^\bullet$, $(x_{f_m \circ d_e^{-1}(q,1)}, ..., x_{f_m \circ d_e^{-1}(q,1)}) \in \mathcal{D}_q$.

Notice that condition (1) is just the usual definition of m-enabled for condition/event nets. The second two conditions impose the restriction that the dynamic systems in the preset of the enabled event must be in goal states and the systems in the postset must all be prepared.

A set of events $G \subseteq E$ is called **detached** if whenever e_1 and e_2 are distinct events in G, ${}^{\bullet}e_1 \cap {}^{\bullet}e_2 = e_1{}^{\bullet} \cap e_2{}^{\bullet} = \emptyset$. Suppose we have a marking (m, f_m). The **follower marking** $(m', f_{m'})$ with respect to $G \subseteq E$ is calculated as follows. As with condition/event nets $m' = (m - {}^{\bullet}G) \cup G^{\bullet}$. $f_{m'}$ is the function given by

$$f_{m'}(p, j) = \left\{ \begin{array}{l} f_m(p, j) \;\; if \;\; p \in m - {}^{\bullet}G \\ f_m \circ d_e^{-1}(p, j) \;\; otherwise \end{array} \right\}$$

where e is the single event in $p^{\bullet} \cap G$. We write $(f_m, m) \to^G (f_{m'}, m')$ when $(f_{m'}, m')$ is the follower marking of (f, m) with respect to G. Since legal markings (m, f_m) are such that f_m is bijective, we can be sure that *every* component of **x** is accounted for when the system is in the set of modes given by m. It can be shown that if (f_m, m) is a legal marking and if $(f_m, m) \to^G (f_{m'}, m')$, then $(f_{m'}, m')$ is a legal marking as well.

3.2   Composing Threaded Petri Nets

As mentioned, we intend to compose TPNs into factories. We present a simple type of composition to complete this section. It is based on the idea of a cyclic subprocess, which we call a **gear**, and which we use as the basic building block of our nets. A gear represents the simplest thing a robot in a factory can do, besides remain idle: cycle repeatedly through some set of behaviors.

Definition 3.5 *A* k-**gear** *is a net* (T, P) *where* $T = \{t_0, ..., t_{k-1}\}$ *and* $P = \{[t_i; t_{i+1}] \mid i \in \mathbb{Z}/k\}$. $m \subseteq P$ *is a* **legal marking** *for a* k-*gear if* $|m| = 1$.

(We ignore the dynamics and redistribution functions for now.) A gear for a robot models the program of a single robot. Certain places of a gear must be synchronized with the gears of other robots. Thus, we *compose* gears as follows.

Definition 3.6 *A* **gear net** *is defined recursively:*

1. *A gear is a gear net.*
2. *If* (T, P) *is a gear net and* (S, Q) *is a gear then* $(T \cup S, P \cup Q)$ *is a gear net as long as the following conditions hold:*
   (a) *let* $(T_1, P_1), ..., (T_k, P_k)$ *be the set of gears in* $(T \cup S, P \cup Q)$ *which intersect* (S, Q). *Then* $\bigcap_{i=1}^{k} P_i = \{[a; b]\}$ *and* $\bigcap_{i=1}^{k} T_i = \{a, b\}$ *for some transitions* a *and* b;
   (b) *there exists a transition* $c \in S - T$ *such that* $[c; a] \in Q$.

A **legal marking** *for a gear net is one in which each gear in the net is marked exactly once.*

Since all places in a gear net are of the form $[x; y]$, gear nets are a kind of *marked graph*, a class of nets which have been extensively studied. (See [3], for example.) Conditions (a) and (b) require that gears be added with a "standard interface". We can show the following properties about gear nets.

Theorem 3.1 *(Liveness) Gear nets are deadlock free under legal markings.*

Theorem 3.2 *(Reversibility) Gear nets are reversible given any legal initial marking.*

Thus we are assured that systems we build up from gear nets are live, logically conflict free, and cyclic processes.

4   Representation

Next we describe how to represent the building blocks of factories – products, robots, workspaces and controllers – in a way that is amenable to compilation.

4.1   The Product Assembly Graph

A **product assembly graph** or PAG, is represented as a tree whose leaves represent parts and whose internal nodes represent operations on subtrees which yield subassemblies. For a given set of operations and part types we can define a simple class of PAGs as follows. Suppose that we have part types $part_1, ..., part_k$ and operations $\mathcal{O}_1, ..., \mathcal{O}_j$ where $\mathcal{O}_i$ is an operation which takes m_i subassemblies and produces a single subassembly. Then the class of PAGs is given by:

1. $part_1, ..., part_k$ are all PAGs;
2. for each $i \in \{1, ..., j\}$, if $P_1, ..., P_{m_j}$ are all PAGs then $\mathcal{O}_i(P_1, ..., P_{m_j})$ are PAGs as well.

Clearly, this defines a very simplified class of PAGs. In practice, each operation can take only certain types of subtrees (those representing subassemblies appropriate to the operation), operations are parameterized, and so on. However, we believe that this is a first approximation to the kind of PAGs that we will encounter in practice.

For a given PAG P, we give a unique label to each node P' in P, called $Label(P')$. This identifies the subassembly that is result of the operation.

4.2   Robot Types and Workspaces

We suppose that there is some set of *robot types* at our disposal which we denote by $\mathcal{T} = \{T_1, T_2, ...\}$. Each type T has an "ideal" workspace $\mathcal{W}(T) \subseteq \mathbb{R}^3$ (compact and connected) and a configuration space $\mathcal{C}(T)$. $\mathcal{W}(T)$ describes the geometry of the set of all positions the robot may take – in general, a solid in $\mathbb{R}^3$. $\mathcal{C}(T)$ represents the degrees of freedom of the robot. An example robot type in the Minifactory is the courier, a two degree of freedom planar robot with a workspace that is a rectangular solid $[x_{min}, x_{max}] \times [y_{min}, y_{max}] \times [0, h]$ where the x and y terms represent the limits of movement on a factory platen and h is the height of the robot. The configuration space of a courier is just $\mathbb{R}^2$.

An instantiation of a robot will be denoted by an identifier r with type $Type(r) \in \mathcal{T}$, workspace $\mathcal{W}(r) \simeq \mathcal{W}(Type(r))$, and configuration space $\mathcal{C}(r) = \mathcal{C}(Type(r))$. As we build factories in the compilation procedure defined below, we instantiate new robots and add them to a set R of robot identifiers. We suppose that their ideal workspaces are copies of the ideal workspaces of their types and that for any two distinct instantiated robots r_1 and r_2, we have $\mathcal{W}(r_1) \cap \mathcal{W}(r_2) = \emptyset$. We represent the way in which robots are located with respect to each other by forming an identification (quotient) topology on the union of the workspaces of the robots. This *does not* represent the actual layout of the factory because the resulting geometry may not embed in $\mathbb{R}^3$ without some "stretching". We comment on the layout procedure in Section 5.

A Robot can carry a subassembly, which may be an atomic part or the result of some operation on some number of parts. Which subassembly, if any, a robot is carrying is the discrete state of the robot. It is given by $Label(P)$ for some node P of the PAG that is being compiled. The distinguished label *nopart* will be used to denote the state of a robot not carrying any subassembly.

4.3   Templates for Controllers

In order to use controllers with our assembly compiler, they must be represented in a standard way. Here we describe a template for representing controllers. This template consists of: a description of the robots needed; the index of the robot, called the "carrier", that will hold the result of the operation once the it is complete; a way of combining the workspaces of the robots into a workspace for the operation; a *parts transform pair*; and a control law over the configuration space. The carrier robot and its workspace are used to join the workspaces of controllers as the PAG is traversed during compilation. We have the following definition:

Definition 4.1 *An* **operation template** *is a tuple*

$$\mathcal{O} = (R, j, \sim, \langle \mathbf{a}; \mathbf{b} \rangle, \mathbf{F})$$

where

1. *$R = \langle T_1, ..., T_k \rangle$ is an ordered set of types of robots, with $k = |R|$;*
2. *$j \in \{1, ..., k\}$ is the index of the robot that will carry the result;*
3. *$\sim$ is an equivalence relation. $\mathcal{W} = (\bigcup_{i=1}^{i=k} \mathcal{W}(r_i))/_\sim$ is the resulting workspace;*
4. *$\langle \mathbf{a}; \mathbf{b} \rangle = \langle a_1, ..., a_k; b_1, ..., b_k \rangle$ is the* **parts transform pair** *denoting how the labels of the parts each robot is carrying change as a result of the controller reaching its equilibrium state;*
5. *F is a vector field on $\mathcal{C}$ describing the controlled dynamic system corresponding to the controller with domain $\mathcal{D}_F$ and goal $\mathcal{G}_F$. $\mathcal{C}$ is defined by $\prod_{T \in R} \mathcal{C}(T) - \Delta$ where Δ is the set of configurations which correspond to two robots touching or being in the same place according to $\sim$.*

The operations of interest take some number of subassemblies and perform an operation that produces one new subassembly. Thus, $b_j \neq nopart$ while $b_i = nopart$ for $i \neq j$.

An **instantiation** of a template is an assignment of robot identifiers to R and is written $\mathcal{O}(r_1, ..., r_k)$. The Threaded Petri Net fragment corresponding to this instantiation is denoted $N_\mathcal{O}(r_1, ..., r_k)$ and is depicted, in its general form, in Figure 2.

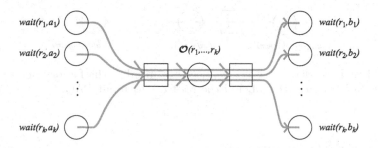

Fig. 2. The Threaded Petri Net associated with the instantiation $\mathcal{O}(r_1, ..., r_k)$

4.4 Factories

We define a factory to be a workspace, a set of robots in the workspace, and a TPN describing the dynamics of the factory. This structure will be built up as the compilation procedure progresses. It will start with a single robot whose task it is to receive the final subassembly from the highest operation in the PAG.

A **factory**, therefore, is a triple $\mathcal{F} = (R, \sim, N)$ where R is a set of robot identifiers, $\sim$ is an equivalence relationship on the union of the workspaces of the robots which describes how the robots are placed in the factory, and N is a TPN which describes the hybrid dynamics of the factory.

5   The Compilation Algorithm

In this section we describe the general form of the compilation procedure for a given class of PAGs. We assume that each operation is already described via a template and that templates for the operations for picking up parts (from parts feeders or trays) and dropping off the final subassembly part are also given. Assume that the type of robot that receives the final subassembly is $OuputType$. The input to the algorithm is a PAG $P = \mathcal{O}(P_1, ..., P_k)$. The function Compile initializes the factory structure with a robot and workspace for the final subassembly $DropOff$ operation and then calls the main function CompileNode.

Compile(P)
    $r \leftarrow Instantiate(OutputType)$
    $R \leftarrow \{r\}$
    $\sim \leftarrow \{(x, x) \mid x \in W(r)\}$
    $N \leftarrow N_0$
    CompileNode(P, r)
End

Here, N is initialized to N_0 which is the Fragment depicted in Figure 3.
The subroutine CompileNode first adds to the factory the robots and workspaces required for the operation $\mathcal{O}$ and then applies itself to each of the subtrees P_1 through P_k. Assume that $P = \mathcal{O}_i(P_1, ..., P_{m_i})$ where$\mathcal{O}_i = (R_i, j_i, \sim_i, \langle \mathbf{a}_i; \mathbf{b}_i \rangle, F_i)$ with $R_i = \langle T_1, ..., T_k \rangle$

Fig. 3. The Threaded Petri Net N_0 used to initialize the factory before compilation. $FinalAssem$ is the label of the root of the input PAG.

```
CompileNode ( P, carrier )
    Allocate robot identifiers r_l where Type(r_l) = T_l for each l ≠ j
    r_j ← carrier
    R ← R ∪ {r_1, ..., r_k}
    ~ ← (~) ∪ (~_i)
    N ← N ∪ N_{O_i}(r_1, ..., r_k)
    For each l ∈ {1, ..., k}
        If a_l ≠ nopart Then
            Choose P ∈ {P_1, ..., P_{m_i}} such that Label(P) = a_i
            CompileNode(P, r_l)
        EndIf
    EndFor
End
```

The CompileNode routine first allocates the new robot identifiers needed for the operation. The factory robots are updated to include these robots as well as the carrier. The equivalence relation is updated as well and becomes a relation over the union of workspaces of all the newly allocated robots as well as the robots that were already in R. Then the TPN that describes the dynamics of the factory is updated to include the fragment for the operation. Finally, for each robot identifier r which, according to the part transition pair $\langle \mathbf{a}; \mathbf{b} \rangle$, should be arriving at the current operation with a part a, CompileNode calls itself on the subtree corresponding to a with r as the new carrier robot. Notice that the recursion eventually bottoms out since the part nodes (leaves) of the PAG have no children.

5.1 Properties of the Resulting Factory

We can show that the factory is a gear net and that its dynamics are correct. We first make use of the following lemma.

Lemma 5.1 Let $(T_1, P_1), ..., (T_k, P_k)$ be gear nets and suppose that for each $i \in \{1, ..., k\}$ we have that $(\{a_i, b_i\}, \{[a : b]\}) \subseteq (T_i, P_i)$ is the intersection of some number of gears in (T_i, P_i). Then, the net obtained by identifying each $(\{a_i, b_i\}, \{[a : b]\})$ is also a gear net.

This result can be used to show that the algorithm above produces gear nets. The proof is inductive on the form of the PAG input. Roughly, we show that PAGs consisting of a single part produce single gears and that assuming that the algorithm compiles gear nets for the subtrees $P_1, ..., P_k$ of the tree $P = \mathcal{O}(P_1, ..., P_k)$, we show that it compiles P correctly into a gear net as well.

Theorem 5.1 *Let N be the TPN resulting from applying the above algorithm to the PAG P. Then N is a gear net.*

Since gear nets are live and reversible and because they are deterministic it is also straightforward to show that under any legal initial conditions (we usually consider the situation where each robot is running $wait(r_i, nopart)$ as the initial marking), that the output robot runs the controller for the $DropOff$ operation infinitely many times in any run. Formally,

Theorem 5.2 *Suppose that $m_0, m_1, \ldots$ is a sequence of markings obtained from a run of the gear net N produced from PAG P. Then there exist infinitely many markings m in the sequence such that $DropOff(r) \in m$ where the robot r is the one instantiated in the initialization routine* Compile.

The workspace that results from compiling a PAG, $\mathcal{W} = (\bigcup_{r \in R} \mathcal{W}(r))/\sim$, does not represent the layout of the factory. In general, $\mathcal{W}$ needs to be "stretched" to be properly embedded in $\mathbb{R}^3$, if it is even possible to do so. At present, we do not have a complete procedure for producing this layout, however, we have an idea of how it will be carried out in practice. Certain workspace types are amenable to stretching in certain directions. For example, the workspace of a planar robot may be extended to be longer or wider but not taller. Thus, there is an allowable family of embeddings $\mathcal{F}$ from $\mathcal{W}$ into $\mathbb{R}^3$ which must be explored. Once one is found, say $f \in \mathcal{F}$, the controllers for the low level operations are composed with f to produce dynamics on the image of f. In the next section, we illustrate this procedure in a simple implementation.

6   The DotFactory: An Example

We have explored the compilation procedure with a simple family of PAGs and a class of "toy" factories called "DotFactories". In the simplest of our investigations, we assume that there is only one part type, $atomic()$, and two operations $mate(\cdot, \cdot)$, $weld(\cdot)$. The robots we consider are all of the same type T_{dot} with workspaces that are copies of the unit interval $[0, 1] \in \mathbb{R}$ and configuration spaces $[0, 1]$ (guidepaths). The physics are simplified: a robot may control its velocity directly $(\dot{x} = u)$; parts move with the robots nearest to them; and part transfers happen instantaneously as long as the robots involved are close together. Robots have width r.

An example template is given next, for the *mate* operation. $mate = (R, j, \sim, \langle \mathbf{a}; \mathbf{b} \rangle, \mathbf{F})$ where

1. $R = \langle T_{dot}, T_{dot}, T_{dot} \rangle$;
2. $j = 3$;
3. $\sim = \{A_1 = A_2 = B_3\}$ where we assume that robot i will have as its workspace the interval $[A_i, B_i] \in \mathbb{R}$.
4. $\langle \mathbf{a}; \mathbf{b} \rangle = \langle LAB_1, LAB_2, nopart; nopart, nopart, LAB_3 \rangle$.
5. $\mathbf{F}$ is a control law over $\mathcal{W}_{mate} = (\bigcup_{i=1}^{3} [A_i, B_i])/\sim$ with $\mathcal{D} = \mathcal{W}_{mate}$ and $\mathcal{G} = B_\epsilon(A_1 + 2r, A_2 + 2r, B_3)$ (a small open ball around the goal point).

We omit a description of the details of F. In the actual implementation, F is derived from a *navigation function* [16] – a method that is quite suitable to the present situation. Similar templates are given for the *weld*, *atomic* and *dropoff* operations.

The input PAG is represented syntactically as in the following example input file:

```
6 parts; // the number of subassemblies
root = sub3; // the finished product
sub3 = mate ( sub2, part3 ); // how to make the subassemblies
sub2 = weld ( sub1 );
sub1 = mate ( part1, part2 );
part1 = atomic();  // these are the actual parts
part2 = atomic();
part3 = atomic()
```

The TPN that is compiled from this PAG describes programs and low level control for six robots in a workspace composed of guidepaths. Since the PAG is a tree, the compiler constructs workspaces that are, topologically, trees as well so that the layout procedure is obvious. The programs for each robot, essentially gears, can be read off directly from this TPN. For example, the gear for the robot, call it r_3, that receives the result of subassembly 1, fixes it to be welded, and then mates it with part 3 is

Loop:
   If $state_3 = nopart$
      Run $\dot{x}_3 = wait$ Until $state_5 = part1 \wedge state_6 = part2$
      Run $\dot{x}_3 = \pi_3 \circ mate(x_5, x_6, x_3)$ Until $x_3 = sub1$
      Break
   If $state_3 = sub1$
      Run $\dot{x}_3 = hold$ Until $state_4 = nopart$
      Run $\dot{x}_3 = \pi_1 \circ weld(x_3, x_4)$ Until $state_3 = sub2$
      Break
   If $state_3 = sub2$
      Run $\dot{x}_3 = wait$ Until $state_1 = nopart \wedge state_2 = part3$
      Run $\dot{x}_3 = \pi_1 \circ mate(x_3, x_2, x_1)$ Until $state_3 = nopart$
      Break
   End Loop

Programs for the other robots are similar. Note that we assume a simple communication system which, in our implementation, is composed of two parts: a *shared memory* where robot i may write its discrete state (the label of the part it is carrying) to memory location i and may read any memory location; and a high speed continuous state sharing link between robots sharing control modes. Because of the distributed nature of the control, the number of continuous states a robot must monitor at any time is less than or equal to the size of the the largest control mode, *independent* of the size of the PAG and the resulting factory. We believe that the method will scale well to significantly larger factories.

Each robot is simulated concurrently at varying operating speeds (chosen randomly) and with varying control speeds. All factories that were compiled performed well under these minor disturbances due to the reactive nature of the low level control method used (borrowed from [16]) and to the robust nature afforded by the gear net structure of the compiled TPN.

We have also investigated robot types with workspaces that are "T-shaped" and shared by another robot. We use the method suggested by Ghrist and Koditschek in [6] for constructing dynamical systems of multiple points on topological graphs. Animations of the factories resulting from several different input PAGs can be viewed at http://www.eecs.umich.edu/~klavins/mf/.

7   Conclusion

We have developed an automatic factory compiler based on our formalism for representing concurrent, hybrid systems. The compiler uses a standard representation of robot workspaces and low level operations and yields a factory geometry, robot task allocation and control programs for each robot. The resulting factory dynamics are shown to be correct using basic properties of our gear net composition method. Our implementation of a simple toy situation suggests that our method yields robust systems and that it scales well.

In the future, we will consider optimizing the compiled net for robot reuse (i.e. reallocating tasks so that one robot alternates between tasks formerly assigned to two robots) and for parallelization of tasks. This leads to TPNs that are not based on gear nets but do have a regular structure, and implies the need for a much more sophisticated layout procedure. The dynamics of the resulting nets must be considered with fairness constraints so that they do not deadlock. We must also address the issues of error recovery and product reworking. We believe that the complexities these issues introduce into our TPNs can be managed by compositional methods similar to those we have already introduced. We are also working on applying these ideas to a factory design tool for a more realistic example which better approaches the Minifactory, mentioned in Section 1 [17].

This research has also lead us is to study the idea of "momentum across transitions" where the dynamical systems corresponding to places are not always controlled to equilibrium states. For example, a robot might toss a ball to another robot which must catch the ball. As the ball approaches the second robot, the transition of that robot into a catching behavior becomes more urgent. We would like to be able to solve this problem not with the explicit use of time as in timed Petri Nets but rather with the intrinsic dynamics of, in this case, a ball in flight. An example of switching between tasks based on urgency can be found in [18] where Rizzi controls a robot to switch between the tasks of bouncing one of two balls on a paddle, effectively juggling them. A systematic approach to this problem may yield factories that are highly dexterous, distributed manipulation systems.

Acknowledgments

The author thanks Professors Bill Rounds and Dan Koditschek for many conversations and advice about this work and Al Rizzi for introducing him to the Minifactory. Eric Klavins is supported in part by the Charles DeVlieg Foundation Fellowship for Manufacturing.

References

1. D. Andreu, J. Pascal, H Pingaud, and R. Valette. Batch process modeling using Petri Nets. In *Proc. of 1994 Intl. Conf. on Systems, man, and Cybernetics*, pages 314–319, October 1994.
2. Robert R. Burridge, Alfred A. Rizzi, and Daniel E. Koditschek. Sequential composition of dynamically dexterous robot behaviors. *International Journal of Robotics Research*, 1998.
3. F. Commoner, A.W. Holt, S. Even, and A. Puneli. Marked directed graphs. *Journal of Computer and System Sciences*, 5:511–523, 1971.
4. R. David and H. Alla. Continuous Petri Nets. In *8th European Workshop on Application and Theory of Petri Nets*, pages 275–294, Saragosse, 1987.
5. I. Demongodin and N. Koussoulas. Differential Petri Nets: Representing continuous systems in a discrete-event world. *IEEE Transactions on Automatic Control*, 43(4):573–579, April 1998.
6. R. Ghrist and D. Koditschek. Safe cooperative robotic motions via dynamics on graphs. In Y. Nakayama, editor, *8th Intl. Symp. on Robotics Research*. Springer Verlag, 1998.
7. T. Henzinger, P.H. Ho, and H. Wong-Toi. HYTECH: A model checker for hybrid systems. *Software Tools for Technology Transfer*, 1:110–122, 1997.
8. Ryszard Janicki. Nets, sequential compositions and concurrency relations. *Theoretical Computer Science*, 29:87–121, 1984.
9. Stephen G. Kaufman et al. The Archimedes 2 mechanical assembly planning system. In *Proceedings of the 1996 IEEE Conference on Robotics and Automation*, pages 3361–3368, 1996.
10. Eric Klavins and Daniel Koditschek. A formalism for the composition of loosely coupled robot behaviors. Technical report no. CSE-TR-412-99, University of Michigan, 1999.
11. Eric Klavins and Daniel Koditschek. A formalism for the composition of concurrent robot behaviors. In *Proceedings of the IEEE Conference on Robotics and Automation*, 2000.
12. Daniel E. Koditschek. An approach to autonomous robot assembly. *Robotica*, 12:137–155, 1994.
13. B. H. Krogh and C. L. Beck. Synthesis of place/transition nets for simulation and control of manufacturing systems. In *4th IFAC/IFORS Symp. Large Scale Systems*, pages 661–666, Zurich, 1986.
14. Tomás Lozano-Perez, Matthew T. Mason, and Russell H. Taylor. Automatic synthesis of fine-motion strategies for robots. *The International Journal for Robotics Research*, 3(1):3–23, 1984.
15. Wolfgang Reisig. *Petri Nets: An Introduction*. Springer Verlag, 1985.
16. Elon Rimon and Daniel E. Koditschek. Exact robot navigation using artificial potential fields. *IEEE Transactions on Robotics and Automation*, 8(5):501–518, October 1992.
17. A. A. Rizzi, J. Gowdy, and R. L. Hollis. Agile assembly architecture: An agent based approach to modular precision assembly systems. In *Proceedings of the 1997 IEEE International Conference on Robotics and Automation*, pages 1511–1516, Albuquerque, NM, April 1997.
18. Alfred A. Rizzi. *Dexterous Robot Manipulation*. PhD thesis, University of Michigan, 1994.
19. Bruce Romney, Cyprien Godard, Michael Goldwasser, and G. Ramkumar. An efficient system for geometric assembly sequence generation and evaluation. In *Proceedings of the 1995 AMSE. Intl. Computers in Engineering Conf.*, pages 699–712, 1995.

A Hybrid Feedback Regulator Approach to Control an Automotive Suspension System*

Xenofon D. Koutsoukos and Panos J. Antsaklis

Department of Electrical Engineering
University of Notre Dame
Notre Dame, IN 46556
{xkoutsou,antsaklis.1}@nd.edu

Abstract. In this paper, we demonstrate a novel hybrid control synthesis approach using an automotive suspension system. Discrete abstractions are used to approximate the continuous dynamics and emphasis is placed on the nondeterministic nature of the abstracting models. The regulator problem for hybrid systems is formulated for safety specifications and algorithms for control design are presented.

1 Introduction

In this paper, a novel systematic methodology for hybrid control synthesis is presented and an example of an automotive suspension system is used to illustrate the approach. The main advantage of the approach is that it provides a convenient general framework for hybrid systems not only for analysis, but more importantly for controller synthesis. Discrete abstractions of the continuous dynamics are studied and the emphasis is placed on the nondeterministic nature of the abstracting models. The notion of quasideterminism is used to characterize discrete abstractions that can be used for control design. The class of systems we are particularly interested in is the class of piecewise-linear systems. Note that the analysis and synthesis algorithms have been implemented using general purpose software, namely Matlab, Simulink, and Stateflow.

Early results of the approach have appeared in [7,6]. The approach has been influenced particularly by [1] where a feedback architecture of a continuous plant with a discrete-event controller is used for hybrid control design. Piecewise-linear systems evolving in discrete-time have been studied in [11,13] and they represent an important class of systems with many practical applications. Recently, the class of piecewise-linear systems has attracted the attention of many researchers, see for example [5,2]. Analysis and synthesis methodologies based on discrete abstractions have been studied extensively in the hybrid system literature [9,8].

The paper is organized as follows. The automotive suspension system is introduced in Section 2. In Section 3, the modeling formalism is briefly outlined. In Section 4, the deterministic nature of the discrete abstractions is discussed and

---

* The partial financial support of the National Science Foundation (ECS95-31485) and the Army Research Office (DAAG55-98-1-0199) is gratefully acknowledged.

N. Lynch and B. Krogh (Eds.): HSCC 2000, LNCS 1790, pp. 188–201, 2000.

algorithms for the computation of the discrete approximations are presented. Finally, the regulator problem for hybrid systems is formulated in Section 5.

2   Automotive Suspension System

This example describes a simplified model of an automotive suspension for an independent wheel. The diagram of Figure 1 illustrates the modeled characteristics. We represent the suspension as a spring/damper system equipped with a compressor and an escape valve. We concentrate only on bounce degrees of freedom, which are represented in the model by the vertical displacement and velocity. The chassis level is raised by pumping air into the system and lowered by opening an escape valve. The suspension influences the bounce according to the equations

$$F = -2k(z + h) - 2c\dot{z} \tag{1}$$
$$m\ddot{z} = F - mg + u \tag{2}$$

where $z, \dot{z}$, and $\ddot{z}$ are the vertical displacement, velocity, and acceleration respectively. The spring and damping rate of the system are represented by the constants k and c. There are two inputs to the model. The first input is the road height h caused by irregularities in the road surface and the second input is the force u caused by the air pressure of the compressor or the escape valve.

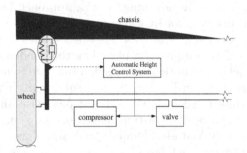

Fig. 1. Automatic height control system

The principal objective in this example is to design an automatic height control system, which increases driving comfort, allows the driver to select the chassis level according to off-road and on-road conditions, and does not violate driving safety. We consider two driving modes for the system, *straight* and *curve*. While in straight driving mode, the driver or a higher level control system in an autonomous vehicle, selects the set-point (sp) for the vertical displacement. The objective of the controller is to guarantee that the vertical displacement remains in a tolerance interval $[sp-lt, sp+ht]$ for any road disturbance from a prescribed bounded set. While in curve mode, the requirement is that the control system does not influence the chassis level, using either the compressor or the escape valve, so not to violate the safety of the system.

In this paper, the design of the controller that selects the action of the compressor and the escape valve is formulated as a hybrid control synthesis problem. A controller is designed based on discrete abstractions of the continuous dynamics using the refinement algorithm presented in Section 5. The controller is responsible for generating the control laws that guarantee that the chassis level will track the set-point within the prescribed tolerance while in straight-driving mode and will suspend the active control while in turning mode. Note that pneumatic suspension system examples have been used in the hybrid system literature to illustrate verification algorithms in a linear hybrid automata setting [4,14,3].

3 Modeling of Hybrid Systems

3.1 Hybrid System Model

We propose to model hybrid systems as *set-dynamical systems* [10]. A *set-dynamical system* (SDS) is denoted as $(X, U, Y; f, g)$ where X is the state set of the system, U is the input set, Y is the output set, $f : X \times U \to X$ is the state transition function, and $g : X \times U \to Y$ is the output function. It is important to distinguish between the controlled and the uncontrolled inputs (disturbances) of an SDS. Furthermore, in the case when the measurements are different than the outputs, a measurement set M and a measurement function m can be included in the system's description.

In order to describe the behavior of a dynamical system, the notion of time must be included in the system's representation and this is accomplished with an index set J equipped with a simple order relation. Assume that the *index set J* is given. Define *index functions* $\alpha : \mathbb{N} \to J$. An index function is said to be *admissible* if $n_1 \leq n_2 \Rightarrow \alpha(n_1) \leq \alpha(n_2)$ (i.e. α is order preserving), and $n_1 \neq n_2 \Rightarrow \alpha(n_1) \neq \alpha(n_2)$ (i.e. α is injective). The state $x \in X$ is associated with an index $j(n)$ meaning the state at time $j(n)$.

A *hybrid dynamical system* (HDS) is defined as an SDS where the constituent sets consist of a continuous and a discrete part. We assume that the continuous part is a subset of a finite dimensional vector space and that the discrete part is finite.

Definition 1. *A hybrid dynamical system is defined by $(X, U, D, Y, M; f, g, m)$ where $X = X_c \times X_d$ is the state set; $U = U_c \times U_d$ is the set of control inputs consisting in general of continuous and discrete controls; $D = D_c \times D_d$ is the set of disturbances; $Y = Y_c \times Y_d$ is the output set; $M = M_c \times M_d$ is the measurement set; $f : X \times U \times D \to X$ is the state transition function; $g : X \times U \times D \to Y$ is the output function; and $m : X \times U \times D \to M$ is the measurement function.*

Presently, we have focused on *piecewise-linear systems* [11,13] to facilitate the development of analysis and synthesis tools. These systems arise when the state set and/or the input set are partitioned into regions described by linear equalities and inequalities and the dynamics at each region are described by linear (or affine) state transitions. Output and measurement maps can be defined also in a similar way. The class of piecewise-linear systems is quite general as it includes linear systems, finite state machines, and their interconnections. They can be used also in many instances as approximations of more general systems.

Control specifications and primary partition Control specifications for hybrid systems can include safety requirements that are usually formulated with respect to a partition of the state space of the system. Consider the state set X of an SDS and define the mapping $\pi : X \to \mathbb{P}(X)$ from X into the power set of X. The mapping π defines an equivalence relation E_π on the set X in the natural way $x_1 \, E_\pi \, x_2$ iff $\pi(x_1) = \pi(x_2)$. The image of the mapping π is called the *quotient space* of X by E_π and is denoted by X/E_π. Adopting this notation we can write $\pi : X \to X/E_\pi$ where π is understood as the *projection* of X onto X/E_π. The mapping π generates a partition of the state set X into the equivalence classes of E_π and will be called *generator*. We assume that the partition defined by π is appropriate for extraction of important information for the system and it will be called the *primary partition*. More specifically, we are interested in the case when $X = \mathbb{R}^n$ and the generator is defined by a set of hyperplanes in $\mathbb{R}^n$. Note that such piecewise-linear regions arise in many applications. Consider the collection $\{h_i\}_{i=1,2,\dots,\ell}$, $h_i : \mathbb{R}^n \to \mathbb{R}$ of real-valued functions of the form $h_i(x) = g_i^T x - w_i$, $i = 1, 2, \dots, \ell$ where $g_i \in \mathbb{R}^n$ and $w_i \in \mathbb{R}$. Let $H_i = \ker(h_i) = \{x \in \mathbb{R}^n : h_i(x) = g_i^T x - w_i = 0\}$ and assume that H_i is an $(n-1)$-dimensional hyperplane ($\nabla h_i(x) = g_i^T \neq 0$). We define the function $\hat{h}_i : \mathbb{R}^n \to \{-1, 0, 1\}$ by

$$\hat{h}_i(x) = \begin{cases} -1 \; if \; h_i(x) < 0 \\ \;\; 0 \;\; if \; h_i(x) = 0 \\ \;\; 1 \;\; if \; h_i(x) > 0 \end{cases} \tag{3}$$

Then, the generator is defined by $\pi(x) = [\hat{h}_1(x), \dots, \hat{h}_\ell(x)]^T$. Although the generator has been defined as $\pi : \mathbb{R}^n \to \{-1, 0, 1\}^\ell$ there is a bijection between $\{-1, 0, 1\}^\ell$ and the quotient set X/E_π (they are the same set).

Measurements and final partition Suppose that at time k we have that $\tilde{y}(k) = \pi(x(k)) \in X/E_\pi$. If it is agreed that the granularity of the partition generated by the mapping π is appropriate for the extraction of useful information regarding the system's behavior, then it is desirable to uniquely determine the state at the next iteration up to its membership on an equivalence class $\tilde{y}(k+1) = \pi(x(k+1)) \in X/E_\pi$. This can be accomplished by considering a finer partition than the partition defined by the generator π to obtain better estimates for the continuous state. This partition will be called the *final partition* and will be determined using the quasideterminism property discussed below. The generator π_F is defined in a similar way as the output function π. Given a partition defined by a finite set of $(n-1)$-dimensional hyperplanes the generator $\pi_F : X \to X/E_{\pi_F}$ separates the state space into a finite number of equivalence classes which correspond to polyhedral regions. The function $z = \pi_F(x)$ can be viewed as a *measurement function* that provides some information about the continuous state. Intuitively, our ability to make decisions to influence the behavior of the system depend on the amount of information contained in the measurement signal.

Example - The automotive suspension system The system contains continuous dynamics due to the spring/damper subsystem and discrete dynamics due to pneumatic part of the suspension. Furthermore, the control specifications contain constraints for both the continuous and discrete variables. For these reasons, the automotive suspension system is modeled as the hybrid dynamical system $(X, U, D, Y, M; f, g, m)$. The state space of the system is $X = X_c \times X_d = \mathbb{R}^2 \times \{straight, curve\}$ representing the displacement and the velocity of the system, and the driving mode. The set of control actions is $U = \{u_0, u_1, u_2\}$ corresponding to the case when the controller is suspended, the compressor is on, and the escape valve is open respectively (the compressor and the valve can not operate simultaneously). The set of exogenous input (that cannot be controlled) is $D = D_c \times D_d = \mathbb{R} \times \{turn, resume\}$ representing the road height and the selection for the driving mode respectively. The output set is $Y = \mathbb{R}$ representing the chassis level. The measurement set is described as the quotient set X/E_{π_F} induced by the final partition π_F that is to be determined in Section 4. The state transition function $f : X \times U \times D \to X$ is described by $x(k + 1) = Ax(k) + Bu(k) + Ed(k)$ where x_1 is the displacement of the chassis, x_2 is the velocity, u is the applied force due to either the compressor or the escape valve, and d is the road height. The parameters of the system A, B, and E are derived from the differential equations (2) by sampling at a prescribed rate T. Finally, the output function is $y(k) = Cx(k)$ where $C = [1, 0]$ and the measurement function $z(k) = \pi_F(x(k))$ returns the membership of the state in one of the equivalences classes of the final partition.

3.2   Control Specifications

Regulatory feedback control of hybrid dynamical systems is based on a representation of the control specifications as a set-dynamical system which is usually called the exosystem. In this paper, we focus on the case when the exosystem is described by a finite automaton. The case when hard time constraints on the transitions of the exosystem are necessary can also be studied in this framework by including clocks in the description of the plant.

Example - The automotive suspension system The control specifications for the automotive suspension system are now described. While in straight driving mode, the driver or a higher level control system in an autonomous vehicle, selects the set point (sp) for the vertical displacement. The objective of the controller is to guarantee that the vertical displacement remains within a tolerance interval $[sp - lt, sp + ht]$ for any road disturbance from a prescribed bounded set. While in curve mode, the requirement is that the control system does not influence the chassis level, using either the compressor or the escape valve, so not to violate the safety of the system. The control specifications can be described formally by the finite automaton shown in Figure 2(i). The state e_0 corresponds to the case the driving mode is straight, where the requirement for the chassis height is to be inside the tolerance interval $[sp - lt, sp + ht]$. The states e_1 corresponds to the case when the driving mode is curve. The input alphabet is $\Sigma = \{turn, resume, \epsilon\}$ where ϵ is a void event.

The primary partition can be derived from the control specifications in a straightforward manner and is described by $h_1(x) = x_1 - (sp + ht)$ and $h_2(x) = x_1 - (sp - lt)$. Then the generator is defined by $\pi(x) = [\hat{h}_1(x), \hat{h}_2(x)]^T$ where the function $\hat{h}_i$ is defined in Equation (3) and it separates the state space into five equivalence classes. For simplicity, we will consider that the safe region is described by the closed interval $[sp - lt, sp + ht]$ and will consider only three regions corresponding to *safe, high, and low* chassis levels as shown in Figure 2(ii).

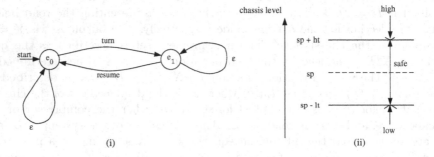

Fig. 2. (i) Exosystem, (ii) Primary Partition

The finite automaton of Figure 2(i) can be represented by the set-dynamical system $(X_e, V_e, Y_e, M_e; f_e, g_e, m_e)$ where $X_e = \{e_0, e_1\}$ is the state set, $V_e = \{turn, resume, \epsilon\}$ is the set of exogenous inputs, $Y_e = \{turn, resume, \epsilon\}$ is the output set (which characterizes part of the exogenous inputs to the plant), and $M_e = X/E_\pi$ is the set of output requests. The state transition function $f_e : X_e \times V_e \rightarrow X_e$ is the state transition of the automaton, the output function $g_e : X_e \times V_e \rightarrow Y_e$ is defined as $g_e(e, v) = v$ for every $e \in X_e$ and $v \in V_e$. Finally, the output request (measurement) function is defined as follows.

$$m_e(e, v) = \begin{cases} safe & \text{for } e = e_0, \ \forall v \in V_e, \\ \tilde{y} \in \{safe, low, high\} & \text{for } e = e_1, \ \forall v \in V_e \end{cases} \tag{4}$$

4 Partition Refinement and Discrete Abstractions

4.1 Motivation

In order to analyze hybrid systems and design control algorithms, it is desirable to induce dynamical systems in finite quotient spaces that preserve the properties of interest and then study the simplified models. Let f be the state transition function of an SDS and assume that the inputs are fixed. Consider the diagram in Figure 3-(a). Intuitively, the map π is used to coarsen the state set of the system. The question that arises is whether the system f can follow this abstraction. This question is concerned with the existence of a mapping $\tilde{f} : X/E_\pi \to X/E_\pi$ that makes the diagram commute. It is shown in [10] that $\tilde{f}$ exists if and only if

$$x_1 E_\pi x_2 \quad \Rightarrow \quad (\pi \circ f)(x_1) = (\pi \circ f)(x_2) \tag{5}$$

(where $\circ$ denotes function composition) and moreover, if (5) is satisfied then $\tilde{f}$ is unique. Note that the above result does not require any structure on the set X or the mappings π and f. Using equivalence relations on the state set X, it is possible to define new dynamical systems in the derived quotient spaces. These systems are called *induced dynamical systems*.

4.2   Quasideterminism

Quasideterminism can be viewed as a desirable property of the partition of the continuous state space. The central characteristic of quasideterministic systems is that only the reachability properties with respect to the control specifications are preserved in the quotient system resulting in more efficient algorithms to partition the state space that are applicable to larger classes of hybrid systems. Quasideterminism is a weaker requirement than the existence of a finite bisimulation. A partition that results in a quasideterminism can be always be computed for piecewise-linear systems, while recent results have shown that finite bisimulations exist only for limited classes of systems [8]. In both approaches an algorithm is used to refine the state space. A bisimulation corresponds to a fixed point of the refinement algorithm. In quasideterminism, we do not require the existence of a fixed point but we stop the refinement at a prescribed fixed iteration. The disadvantage of that is that in this case the quotient system does not completely preserve the reachability properties of the original system, however this is not needed for controller design for an interesting class of problems as this work demonstrates.

Suppose that at time k, $\pi(x(k)) \in X/E_\pi$ is known. In the case when the estimates of the state at time k provide sufficient information to uniquely determine the membership of the state of the induced system at time $k + 1$ on an equivalence class of E_π, the system is said to be quasideterministic. The notion of quasidetermism is illustrated in Figure 3. Although we do not compute an equivalence relation that guarantees the existence of a mapping $\tilde{f}$ that preserves the reachability properties of the original system, we exploit the commutativity of the diagram (c) in Figure 3 in order to analyze the reachability properties with respect to the control specifications. The formal definition for the concept of quasideterminism is given in later in the section.

Denote by $B(X)$ the set of all binary relations on the set X. We can define the poset $(B(X), \leq)$ where the partial order relation $\leq$ on $B(X)$ is defined as $B_1 \leq B_2$ if $(x_1, x_2) \in B_1 \Rightarrow (x_1, x_2) \in B_2$. Let $E(X)$ be the set of all equivalence

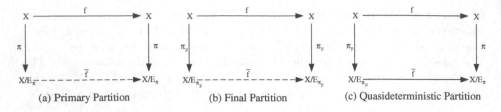

Fig. 3. Quasideterminism and the partitions of the state space

relations on X. We have that $E(X) \subset B(X)$ and $E(X)$ inherits the partial order of $B(X)$. A lattice structure can be developed on the set of all equivalence relations on X (for more details see [10]). The lattice $(E(X), \leq, \wedge, \vee)$ is called the *equivalence lattice*.

Proposition 1. *The set $E_P(X)$ of all equivalence relations on X induced by mappings $\pi : X \to X/E_\pi$ which are defined using finite collections of $(n-1)$-dimensional hyperplanes and thus, they separate the state space X into polyhedral equivalence classes, is a sublattice of the equivalence lattice $E(X)$, and will be called polyhedral equivalence lattice. Furthermore, $E_P(X)$ is not complete.*

Definition 2. *The hybrid system $(X, U, D, Y, M; f, g, m)$ with primary and final partition defined by X/E_π and X/E_{π_F} is quasideterministic with respect to the primary partition if for every region of the final partition $\tilde{z} \in X/E_{\pi_F}$ and for all states $x \in X$ such $\pi_F(x) = \tilde{z}$, there exists unique region of the primary partition $\tilde{y} \in X/E_\pi$ such that $\tilde{y} = \pi(f(x, u, d))$ for every control action $u \in U$ and exogenous input $d \in D$.*

If the hybrid system $(X, U, D, Y, M; f, g, m)$ with primary and final partition defined by X/E_π and X/E_{π_F} is *quasideterministic* with respect to the primary partition π, then it is also quasideterministic if instead of E_{π_F} we use any finer final partition $E_{\pi_q} \leq E_{\pi_F}$. Refinement of the state space partition will terminate if we can guarantee that there is a control policy to satisfy the specifications.

4.3   Partition Refinement

In the following, we present some basic results that will be used in the theoretical analysis of the algorithms for the partition refinement. A *piecewise-linear (PL) subset* [12] of a finite dimensional vector space V is the union of a finite number of sets defined by (finitely many) linear equations $f(x) = a$ and linear inequalities $f(x) > a$. An alternative way to define PL sets which is important for our discussion is the following [12].

Definition 3. *Let $\mathcal{L}$ be the first-order language defined by (i) a set of (countably many) variables $\{x_1, x_2, \ldots\}$, (ii) the connective symbols $\neg$ and $\to$, (iii) the quantifier $\forall$, the parentheses (and) and the comma, (iv) A set of constants $\{r\}$ for each real number r, (v) A set of unary functions $\{r \cdot ()\}$ for each real number, the binary function $+$, (vi) the relational symbols $>$ and $=$.*

Lemma 1. *Every sentence in $\mathcal{L}$ defines a PL set and conversely, every PL subset of $\mathbb{R}^n$ can be defined in this fashion.*

The above lemma is proved in [12]. The conclusion of the lemma is that any set defined using quantifiers can be also defined using only propositional connectives. In order to refine the state space, we define the predecessor operator $pre : \mathbb{P}(X) \to \mathbb{P}(X)$ as

$$pre(P) = \{x | \exists u \in U, \forall d \in D, f(x, u, d) \in P\}. \tag{6}$$

The set $pre(P)$ represents all the states x for which there is a control action that will enforce the state to remain in P for any disturbance d. If the set P is piecewise-linear, then from Lemma 1 it follows that the set $pre(P)$ is also piecewise-linear and can be defined using only propositional connectives.

In the remaining of the paper, we will concentrate on the case the hybrid system is described by

$$(X, U, D, Y, M; f, g, m) \tag{7}$$

with finite input set U, bounded disturbance set D, and transition function given by $x(k+1) = Ax(k) + Bu(k) + Ed(k)$. Similar results can be developed for other classes of piecewise-linear systems.

Initially, assume that the state transition function is given by $x(k + 1) = Ax(k) + Bu(k)$ where $x \in \mathbb{R}^n$ and the input u takes values in a finite set $U \subset \mathbb{R}^m$. For fixed control action $u \in U$ the dynamics of the system are described by the mapping $f_u : \mathbb{R}^n \to \mathbb{R}^n$ with $f_u(x) = Ax + Bu$. We want to compute the set of all the state x that can be driven in P by the control action u by defining the predecessor operator $pre_{f_u}(P) = \{x | f_u(x) = Ax + Bu \in P\}$.

Lemma 2. *Consider the affine function $h(x) = g^T x - w$ and the set $H = \ker(h) = \{x | g^T x - w = 0\}$. Let $H' = \{x | f_u(x) = Ax + Bu \in H\}$ be the set of all $x \in \mathbb{R}^n$ that can be driven in H by application of the affine mapping f_u. Then $H' = \ker(h')$ where $h'(x) = g'^T x - w'$ with $g'^T = g^T A$ and $w' = w - g^T Bu$. In addition, if $Y = int(K')$ is an open halfspace bounded by H', then $f_u(Y) = int(K)$, that is $f_u(Y)$ is an open halfspace bounded by H.*

Next, we define the halfspace $P(g, w) = \{x | g^T x \leq w\}$, $g \neq 0$ and we compute the set of all states that can be driven to P by using the predecessor operator $pre_\exists : \mathbb{P}(X) \to \mathbb{P}(X)$ defined as $pre_\exists(P) = \{x | \exists u \in U, f_u(x) = Ax + Bu \in P\}$.

Lemma 3. *Consider the set $P(g, w) = \{x | g^T x \leq w\}$, $g \neq 0\}$, then $pre_\exists(P) = \{x | g^T Ax \leq w - g^T Bu^*\}$ where u^* is the maximizer of the function $w(u) = w - g^T Bu$ over the set of control actions U.*

Let $f : X \to Y$ be a mapping and consider the sets $D \subset X$ and $E \subset Y$. The *image* of D and the *inverse image* of E under the mapping f are defined by $f(D) = \{f(x) | x \in D\}$, $f^{-1}(E) = \{x | f(x) \in E\}$. It is easily verified that the map $f^{-1} : \mathbb{P}(Y) \to \mathbb{P}(X)$ commutes with unions, intersections, and complements. The operator $pre_{f_u} : \mathbb{P}(X) \to \mathbb{P}(X)$ $(X = \mathbb{R}^n)$ clearly returns the inverse image of P under the mapping f_u for fixed input and therefore commutes with unions, intersection, and complements. The notation pre_{f_u} has been used instead of f_u^{-1} in order to be consistent with the notation when the control action is not fixed. In the case when the input set is finite, the set $pre_\exists(P)$ can be computed for any PL set as the union $\bigcup_{u_i \in U} pre_{f_{u_i}}(P)$.

Next, we consider the case when continuous disturbances are present and we assume that for a fixed discrete control action the description of the system is $x(k+1) = Ax(k) + Bd(k)$ where $x \in \mathbb{R}^n$ and $d \in D \subset \mathbb{R}^m$ a disturbance which takes values in a bounded polyhedron. We define a new predecessor operator $pre_f^d : \mathbb{P}(X) \to \mathbb{P}(X)$ by $pre_f^d(P) = \{x | \forall d \in D, f(x,d) = Ax + Bd \in P\}$. This operator returns all the states which will be in the set P at the next time step for every possible disturbance.

Lemma 4. *Consider the set* $P = P(g, w) = \{x | g^T x \leq w\}$, *then* $pre_f^d(P) = \{x | g^T Ax \leq w - g^T Bd^*\}$ *where* $d^* = argmin_{d \in D}\{-g^T Bd\}$.

The predecessor operator in the case of bounded disturbances commutes with the intersection of halfspaces. Note that this result is a consequence of the equivalence $(\forall x)(\phi(x) \wedge \psi(x)) \leftrightarrow (\forall x)\phi(x) \wedge (\forall x)\psi(x)$ in predicate logic.

In the following, we consider the system $x(k+1) = Ax(k) + Bu(k) + Ed(k)$ where the disturbance d takes values in a bounded polyhedral set D and the control input u takes values in a finite set U and the polyhedral set $P = \{x | g_1^T x \leq w_1 \wedge \cdots \wedge g_p^T \leq w_p\}$. Then by using the results of this section we have that

$$pre(P) = \{x | \exists u \in U, \forall d \in D, f(x, u, d) = Ax + Bu + Ed \in P\}$$

$$= \bigcup_{u_i \in U} pre_{f_{u_i}}^d(P)$$

$$= \bigcup_{u_i \in U} \{x | g_1^T Ax \leq w_1 - g_1^T Bu_i - g_1^T Ed_1^* \wedge \cdots \wedge g_p^T Ax \leq w_p - g_p^T Bu_i - g_p^T Ed_p^*\}$$

where $d_i^* = argmax_{d \in D}\{-g_i^T Bd\}$. Next, consider the hyperplanes $h_i'(x) = g_1^T Ax - (w_1 - g_1^T Bu_i - g_1^T Bd_1^*)$, $i = 1, \ldots, p$ and the partition $\pi' \in E_P(X)$ defined by those hyperplanes using Equation (3).

Proposition 2. *The hybrid system (7) with primary and final partition defined by* X/E_π *and* $\inf(E_\pi, E_{\pi'})$ *respectively is quasideterministic with respect to the primary partition.*

The implication of the above proposition is that for every state, every control action, and every disturbance the membership of the state at the next time step to an equivalence class of the primary partition can be uniquely determined from the current region of the final partition. Given a fixed time window repetitive applications of the predecessor operator can take into consideration more than one time steps. At this point it is possible to construct a discrete-event system based on the final partition π_F and extend supervisory control techniques in order to exploit the information that is preserved in the discrete abstraction due to quasideterminism. However, we continue with our analysis of specific control problems for which we can formulate conditions for the existence of control policies that guarantee that the specifications are satisfied.

4.4   Safety

In the following, we focus on the safety problem and we describe algorithms for the refinement of the state space partition that result in quasideterministic systems. Given a set of safe states described by the piecewise-linear set $P \subset \mathbb{R}^n$ and an initial condition $x_0 = x(0) \in P$, we say that the system is *safe* if $x(k) \in P$ for every k. The system is safe with respect to the set P if

$$P \subseteq pre(P) = \{x | \exists u \in U, \forall d \in D, f(x, u, d) \in P\}. \tag{8}$$

The validity of equation (8) can be tested using the representation of $pre(P)$ without quantifiers. Since the set $pre(P)$ is piecewise-linear but not polyhedral, the development of efficient algorithms that test if the equation (8) holds is necessary and is a topic of current research. A simple algorithm to perform this test consists of representing the complement of $pre(P)$ as the union of polyhedra $Q = [pre(P)]^c = \bigcup_{i=1,...,\rho} Q_i$ and then, testing if $P \cap Q_i = \emptyset$ for every $i = 1, ..., \rho$ using linear programming techniques. A simple way to express Q as the union of polyhedra is to consider all the inequalities that define Q pairwise and eliminate all the pairs that correspond to parallel hyperplanes.

Proposition 3. *Given the polyhedral set of safe states P and the hybrid system (7), if $P \cap Q_i = \emptyset, i = 1, ... \rho$ where $Q = [pre(P)]^c = \bigcup_{i=1,...,\rho} Q_i$, then there exists control policy that guarantees that the system is safe.*

Example - Automotive Suspension System The automotive suspension system is safe if the chassis level is inside the interval $[sp - lt, sp + ht]$ while in straight driving mode. Our approach for the design of the controller is that given the desired-set point and therefore the primary partition, a final partition can be constructed and the conditions of Proposition 3 can be tested in an autonomous manner. If there exists a control policy that guarantees that the system is safe, then a controller that implements such a policy can be designed based on the discrete abstraction induced by X/E_{π_F}. The same approach can be used also off-line to characterize all the set-points for which there exists a control policy that guarantees safety.

In order to construct the final partition, we translate the control specification from the output space to the input space to obtain the set $P_1 = \{(x_1, x_2) | sp - lt \leq x_1 \leq sp + ht\}$. Clearly, the set P_1 is unbounded in the state space $\mathbb{R}^2$. From Lemma 2 it follows that the set $pre(P_1)$ is bounded by hyperplanes that in general intersect with P_1 and therefore, it is not possible that $P_1 \subseteq pre(P_1)$. The practical implication of this observation is that if the chassis level is very close to the boundary of the set P_1, then if the chassis vertical velocity is large and directed towards the unsafe region, there will be no finite control input that will guarantee safety. In order to proceed with the controller design we have to determine a bounded approximation of the set P_1 by taking into consideration realistic bounds for the chassis vertical velocity. The final partition can be determined using the partition refinement algorithms described above. The primary and final partition for typical values of the system parameters are shown in Figure 4 where it can be seen that $P \subset pre(P)$.

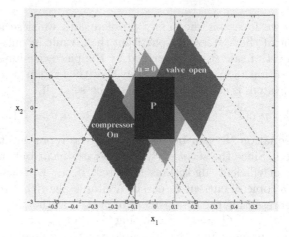

Fig. 4. Final partition

5   Hybrid System Regulator

In this section, the regulator problem for hybrid systems is formulated. In general, a regulator requests certain types of outputs from the plant so that these are attained in the presence of disturbances. The desired outputs are characterized by a regulation condition and they can be described as the outputs of another SDS, called the *exosystem*. The plant and the exosystem are linked by a controller to form a regulator as shown in the Figure 5(i). A feedback controller can be designed to regulate the system. The main characteristic of the controller is that it contains a copy of the exosystem in accordance to the "internal model principle".

In the following, we consider the safety problem and we describe how a controller can be designed based on the discrete abstraction induced by the final partition. The state of the controller correspond to the regions of the final partition and the current state $x_c = \pi_F(x)$ can be determined by filtering the plant measurements using the inequalities that define the equivalence classes of the final partition. The controller can be described by the SDS $\mathcal{C} = (X_c, Y \times M, U; f_c, g_c)$ where X_c is the state set of the controller; $Y \times M$ is the input set of the controller consisting of pairs describing the output request and that actual plant output every time instant; U is the output set representing the control actions; $f_c : X_c \times (Y \times M) \to X_c$ is the state transition function for the controller; and $g_c : X_c \times (Y \times M) \to U$ is the output function given by $u = g_c(x_c, (m_e(x_e, v), \pi_F(x)))$. Since for some states there exist more than one control inputs that can be applied for safety, there are several ways to implement the output function of the controller. For example, the output function can defined by

$$
u = \begin{cases}
u_0 & \text{if } x \in P_0 = pre^d_{f_{u_0}} \\
u_1 & \text{if } x \in P_1 = pre^d_{f_{u_1}} \setminus P_0 \\
\vdots & \vdots \\
u_N & \text{if } x \in P_N = pre^d_{f_{u_N}} \setminus P_{N-1}
\end{cases} \tag{9}
$$

Example - Automotive Suspension System The controller for the automotive suspension system is shown in Figure 5(ii). For the straight driving mode the controller is represented as a finite automaton with three different states corresponding to the regions of the final partition for the set P and output function defined by (9). For the curve driving mode, the controller consists of one state with constant output function $u = 0$. The controller communicates with the plant and the exosystem in a synchronous manner.

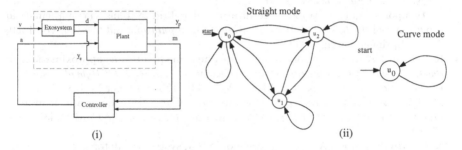

(i)   (ii)

Fig. 5. (i) Hybrid system regulator, (ii) Controller

Remark A problem related to safety is to examine if there exists a control policy that will drive the state of the system to a prescribed region. For example, since at the end of a curve the chassis level may not be inside the interval $[sp - lt, sp + ht]$, it is required that as soon as the system is in *straight* mode the chassis level must be driven to the safety region by using either the compressor or the valve. This is a reachability specification that can be also studied in the framework presented in the paper. The final partition can be constructed by repetitive applications of the predecessor operator. For the termination of the partition refinement algorithm, the reachability specifications should be characterized by bounds on the time for the state to reach the desired region.

6   Conclusions

A novel hybrid control synthesis approach is demonstrated using an automotive suspension system. Controller design is based on quasideterministic discrete abstractions of the continuous dynamics. The regulator problem for hybrid systems is formulated for safety specifications and algorithms for control design are presented. Although a second-order system was used the approach, the methodologies and the algorithms described are applicable to more complex systems. The approach has been validated with simulations using Matlab, Simulink, and Stateflow but simulation results are omitted due to length limitations. An important point is that the above approach is potentially implementable on-line for real-time control. Note that due to space limitations, detailed descriptions of the technical results were omitted, but they can be obtained by contacting the authors.

References

1. P. Antsaklis, J. Stiver, and M. Lemmon. Hybrid system modeling and autonomous control systems. In R. L. Grossman, A. Nerode, A. P. Ravn, and H. Rischel, eds., *Hybrid Systems*, Vol. 736, *LNCS*, 366–392. Springer-Verlag, 1993.
2. A. Bemporad and M. Morari. Control of systems integrating logic, dynamics, and constraints. *Automatica*, 35(3):407–427, 1999.
3. A. Bemporad and M. Morari. Verification of hybrid systems via mathematical programming. In *HSCC 99: Hybrid Systems—Computation and Control*, Vol. 1569, *LNCS*. Springer-Verlag, 1999.
4. R. Fehnker. Automotive control revised - linear inequalities as approximations of reachable sets. In T. Henzinger and S. Sastry, eds., *HSCC 98: Hybrid Systems—Computation and Control*, Vol. 1386, *LNCS*, 110–125. Springer-Verlag, 1998.
5. M. Johansson and A. Rantzer. Computation of piecewise quadratic Lyapunov functions for hybrid systems. *IEEE Transactions on Automatic Control*, 43(1):31–45, 1998.
6. X. Koutsoukos and P. Antsaklis. Design of hybrid system regulators. In *Proceedings of the 38th IEEE Conference on Decision and Control*, 3990–3995, Phoenix, AZ, Dec. 1999.
7. X. Koutsoukos and P. Antsaklis. Hybrid control of a robotic manufacturing system. In *Proceedings of the 7th IEEE Mediterranean Confereence on Control and Automation*, 144–159, Haifa, Israel, June 1999.
8. G. Lafferriere, G. Pappas, and S. Sastry. Reachability analysis of hybrid systems using bisimulations. In *Proceedings of the 37th IEEE Conference on Decision and Control*, 1623–1628, Tampa, FL, 1998.
9. J. Raisch and S. O'Young. Discrete approximation and supervisory control of continuous systems. *IEEE Transactions on Automatic Control*, 43(4):568–573, 1998.
10. M. Sain. *Introduction to Algebraic System Theory*. Academic Press, 1981.
11. E. Sontag. Nonlinear regulation: The piecewise linear approach. *IEEE Transactions on Automatic Control*, 26(2):346–358, 1981.
12. E. Sontag. Remarks on piecewise-linear algebra. *Pacific Journal of Mathematics*, 92(1):183–210, 1982.
13. E. Sontag. Interconnected automata and linear systems: A theoretical framework in discrete-time. In R. Alur, T. Henzinger, and E. Sontag, eds., *Hybrid Systems III, Verification and Control*, Vol. 1066, *LNCS*, 436–448. Springer-Verlag, 1996.
14. T. Stauner, O. Muller, and M. Fuchs. Using HYTECH to verify an automotive control system. In O. Maler, ed., *Hybrid and Real-Time Systems*, Vol. 1201, *LNCS*, 139–153. Springer-Verlag, 1997.

Ellipsoidal Techniques for Reachability Analysis*

Alexander B. Kurzhanski ** and Pravin Varaiya

ERL, EECS
University of California at Berkeley
195M Cory Hall
Berkeley, CA, 94720-1770
{kurzhans,varaiya}@eecs.berkeley.edu

Abstract. This report describes the calculation of the reach sets and tubes for linear control systems with time-varying coefficients and hard bounds on the controls through tight external and internal ellipsoidal approximations. These approximating tubes touch the reach tubes from outside and inside respectively at *every point* of their boundary so that the surface of the reach tube is totally covered by curves that belong to the approximating tubes. The proposed approximation scheme induces a very small computational burden compared with other methods of reach set calculation.

In particular such approximations may be expressed through ordinary differential equations with coefficients given in explicit analytical form. This yields exact parametric representation of reach tubes through families of external and internal ellipsoidal tubes. The proposed techniques, combined with calculation of external and internal approximations for intersections of ellipsoids, provide an approach to reachability problems for hybrid systems.

Introduction

Recent activities to promote advanced automation of real-time processes have motivated new interest in the problem of reachability for controlled systems. This is also related to the problem of verification of hybrid systems [4]. Effective and implementable solutions to these problems must incorporate procedures for calculating reach sets and reach tubes for continuous-time systems [13]. Another demand for effectively performing such calculations comes from interval analysis in scientific computation. [11].

Among methods for reachability analysis are those based on ellipsoidal techniques, (see, for example [2], [3], [6]). Publications in this area were mostly concentrated on deriving a single equation that would produce a sub-optimal (with respect to volume) ellipsoidal approximation to the exact reach set.

* Research supported by National Science Foundation Grant ECS 9725148
** Corresponding author.

N. Lynch and B. Krogh (Eds.): HSCC 2000, LNCS 1790, pp. 202–214, 2000.
© Springer-Verlag Berlin Heidelberg 2000

However, it turns that ellipsoidal methods allow *exact representations* of the reach sets and tubes for linear systems through *parametrized families of both external and internal ellipsoids* (see [6]). But to ensure effective calculation, an important open question is how to effectively single out such families of tightest ellipsoidal approximations to the reach tube that would touch its surface or the surface of its neighborhood at *every point*, (both from inside and outside !) and would thus *totally cover* this tube. A crucial point in organizing the calculation is to indicate such a parametrized variety of curves along which the procedure could be realized recurrently in time, without having to calculate the solution "afresh" for every new instant of time. A positive answer to the latter problem is given in this presentation for both external and internal approximations. It removes an unnecessary computational burden present in other methods and also opens new routes for deriving adequate numerical error estimates and new methods for systems other than those treated here [16], [12], [14]. The suggested approach is particularly relevant for hybrid systems since it allows further propagation to systems with resets. [1] An application of the proposed techniques to the verification of hybrid systems is given in paper [1].

In this paper we deal with reach tubes for control systems with linear dynamics and hard bounds on the control. We study the following question : given a reach tube (or its $\epsilon-$ neighborhood) and a smooth curve that runs along its surface, do there exist ellipsoid-valued external (internal) tubes that would contain (be contained in) the reach tube and touch the reach tube precisely along the given curve? The answer to this question is positive. However the properties of the respective ellipsoidal tubes do depend strongly on the given curve. The "good" situation is when the given curve may be realized *as a trajectory of the original control system.* [2] The required ellipsoidal tubes are then generated by ellipsoid-valued maps which satisfy the semigroup property and thus generate some generalized dynamical systems. Moreover, the approximating tubes are tight in the sense that there exists no other ellipsoidal tube that could be squeezed in between the approximation and the reach tube (for both external and internal ellipsoids). Lastly, the parameters of the ellipsoidal approximations are described by fairly simple ordinary differential equations. The paper also indicates the properties of the basic equations (18), (24) that allow them to be used correctly, without misunderstanding. Thus, it may be shown that when given is *any smooth curve* on the surface of the reach tube, which is not itself a system trajectory, there again exists ellipsoidal tubes that touch the reach sets along this curve. But now the respective ellipsoidal-valued maps may not satisfy the semigroup property and their evolution in time is not described by equations as simple as in the "good" case. The calculations then cannot be realized recur-

---

[1] These questions as well as the internal representations given here were not discussed in book [6].

[2] This happens when the given curve (a system trajectory) develops along the points of support for hyperplanes generated by vectors that are realized as the motions of the linear system adjoint to the homogeneous part of the control system under investigation.

sively. They require procedures that have to memorize additional items and are therefore computationally heavier than in the "good" case. A simplification of the computational procedure in this general case to the level of the "good" case results in *non-tight* approximations(!).

1   The Reachability Problem

Consider the linear system

$$\dot{x} = A(t)x + B(t)u, \quad t_0 \le t \le t_1, \tag{1}$$

where $x \in \mathbb{R}^n$ is the state and $u \in \mathbb{R}^m$ is the control. The matrices $A(t), B(t)$ are continuous and the system is *completely controllable* (see [9]). The control $u = u(t)$ is any measurable function restricted by hard bounds $u(t) \in \mathcal{P}(t)$, for almost all t, where $\mathcal{P}(t)$ is a nondegenerate ellipsoid continuous in t, namely, $\mathcal{P}(t) = \mathcal{E}(q(t), Q(t))$, and

$$\mathcal{E}(q(t), Q(t)) = \{u : (u - q(t), Q^{-1}(t)(u - q(t)) \le 1\}, \tag{2}$$

with $q(t) \in \mathbb{R}^m$ (the center of the ellipsoid) and positive definite matrix function $Q(t) \in \mathbb{R}^{m \times m}$ (the matrix of the ellipsoid) continuous in t. The *support function* of the ellipsoid is

$$\rho(l|\mathcal{E}(q(t), Q(t))) = \max\{(l, x)|x \in \mathcal{E}(q(t), Q(t)\} = (l, q(t)) + (l, Q(t)l)^{1/2}.$$

The continuity of $Q(t)$ means that its support function $\rho(l|Q(t))$ is continuous in t uniformly in l with $(l, l) \le 1$.

Definition 11 *Given position $\{t_0, x^0\}$, the **reach set** (or "attainability domain") $\mathcal{X}(\tau, t_0, x^0)$ at time $\tau > t_0$ from this position is the set*

$$\mathcal{X}[\tau] = \mathcal{X}(\tau, t_0, x^0) = \{x[\tau]\}$$

*of all states $x[\tau] = x(\tau, t_0, x^0)$ reachable at time τ by system (1), with $x(t_0) = x^0$, through all possible controls u that satisfy the constraint (2). The set-valued function $\tau \mapsto \mathcal{X}[\tau] = \mathcal{X}(\tau, t_0, x^0)$ is known as the **reach tube**.*
*The reach set $\mathcal{X}(\tau, t_0, \mathcal{X}^0)$ (at time τ, **from set** $\mathcal{X}^0 = \mathcal{X}(t_0)$) is the union*

$$\mathcal{X}(\tau, t_0, X^0) = \cup\{\mathcal{X}(\tau, t_0, x^0)|x^0 \in \mathcal{X}^0\}.$$

*The set-valued function $\tau \mapsto \mathcal{X}[\tau] = \mathcal{X}(\tau, t_0, \mathcal{X}_0)$ is known as the **reach tube** from set $\mathcal{X}^0$.*

The following properties may be checked directly.

Lemma 1. *The set-valued map $\mathcal{X}(t, t_0, \mathcal{X}^0)$ satisfies the semigroup property*

$$\mathcal{X}(t, t_0, \mathcal{X}^0) = \mathcal{X}(t, \tau, \mathcal{X}(\tau, t_0, \mathcal{X}^0)). \tag{3}$$

In the sequel it is assumed that $\mathcal{X}^0 = \mathcal{E}(x^0, X^0)$ is an ellipsoid. It is worth noting that the set $\mathcal{X}[\tau]$ may also be treated as the cut $\mathcal{X}[\tau] = \mathcal{X}(\tau, t_0, \mathcal{E}(x^0, X^0))$ of the solution tube $\mathcal{X}(\cdot) = \{\mathcal{X}[t] : t \geq t_0\}$ to the differential inclusion

$$\dot{x} \in A(t)x + \mathcal{E}(B(t)q(t), B(t)Q(t)B'(t)), \quad t \geq t_0, \quad x^0 \in \mathcal{E}(x^0, X^0). \quad (4)$$

A standard calculation using convex analysis indicates the following (see, for example[6]).

Lemma 2. *The support function*

$$\rho(l|\mathcal{X}(t, t_0, \mathcal{E}(x^0, X^0))) = (l, x^\star(t)) + (l, X(t, t_0)X^0 X'(t, t_0)l)^{1/2} + \quad (5)$$

$$+ \int_{t_0}^{t} (l, X(t, s)B(s)Q(s)B'(s)X'(t, s)l)^{1/2} ds.$$

Here $X(t, s)$ is the transition matrix for the homogeneous system (1),

$$\partial X(t, s)/\partial t = A(t)X(t, s), \quad X(s, s) = I, \quad \dot{x}^\star = A(t)x^\star + B(t)q(t), \quad x^\star(t_0) = x^0,$$

where I is the identity matrix. For a time-invariant system $A(t) = A = const$, and $X(t, s) = \exp(A(t - s))$. The last representation leads to the next result.

Lemma 3. *The reach set* $\mathcal{X}[t] = \mathcal{X}(t, t_0, \mathcal{E}(x^0, X^0))$ *is a convex compact set in* $\mathbb{R}^n$ *that evolves continuously in t.*

Points on the boundary of the reach set $\mathcal{X}[t]$ have an important characterization. Consider a point x^* on the boundary $\partial\mathcal{X}[\tau]$ of the reach set $\mathcal{X}[\tau] = \mathcal{X}(\tau, t_0, \mathcal{E}(x^0, X^0))$.[3] Then there exists a related *support vector* l^* such that

$$(l^*, x^*) = \rho(l^*|\mathcal{X}[\tau]). \quad (6)$$

The control $u = u^*(t)$ and the initial state $x(t_0) = x^{*0} \in \mathcal{E}(x^0, X^0)$ which transfer system (1) from state $x(t_0) = x^{*0}$ to $x(\tau) = x^*$ is specified by the well-known "maximum principle" (see details in [9]). However, the calculation of the reach sets directly from these relations, especially in large dimensions, is cumbersome. Among the effective methods for these problems are those that rely on ellipsoidal techniques, as given in [6].

Remark 1.1 Due to the controllability assumption we will further assume, without loss of generality, that $B(t) = I$. To return to the case $B(t) \neq I$ it suffices in the sequel to substitute everywhere $Q(t)$ by $B(t)Q(t)B'(t)$. However, in the last case, for computational purposes it may be useful to start the approximation process at time $t = t_0 + \delta, \delta > 0$, to have $W(t_0 + \delta, t_0) > 0$.

---

[3] The boundary $\partial\mathcal{X}[\tau]$ of set $\mathcal{X}[\tau]$ may be here defined as the set $\partial\mathcal{X}[\tau] = \mathcal{X}[\tau] \setminus int\mathcal{X}[\tau]$. Under the controllability assumption, set $\mathcal{X}[\tau]$ has a non-void interior $int\mathcal{X}[\tau] \neq \emptyset$ for $\tau > t_0$.

2 Ellipsoidal Approximation of Reach Sets

Although the initial set $\mathcal{E}(x^0, X^0))$ and the control set $\mathcal{E}(q(t), Q(t))$ are ellipsoids, the reach set $\mathcal{X}[t] = \mathcal{X}(t, t_0, \mathcal{E}(x^0, X^0))$ will *not* generally be an ellipsoid. As indicated in [6], the reachability set $\mathcal{X}[t]$ may be approximated both externally and internally by ellipsoids $\mathcal{E}_-$ and $\mathcal{E}_+$, with $\mathcal{E}_- \subseteq \mathcal{X}[t] \subseteq \mathcal{E}_+$. The approximations are said to be *tight* if for any ellipsoid $\mathcal{E}$ the inclusion $\mathcal{X}[t] \subseteq \mathcal{E} \subseteq \mathcal{E}_+$ implies $\mathcal{E} = \mathcal{E}_+$, while inclusion $\mathcal{E}_- \subseteq \mathcal{E} \subseteq \mathcal{X}[t]$ implies $\mathcal{E} = \mathcal{E}_-$. Here we shall deal with both tight *external and internal* approximations.

Problem 2.1. Given a vector function $l^*(t)$, $(l^*, l^*) = 1$, continuously differentiable in t, find external and internal ellipsoids $\mathcal{E}_-^*[t] \subseteq \mathcal{X}[t] \subseteq \mathcal{E}_+^*[t]$ such that **for all $t \geq t_0$, the equalities**

$$\rho(l^*(t)|\mathcal{X}[t]) = \rho(l^*(t)|\mathcal{E}_+[t]) = \rho(l^*(t)|\mathcal{E}_-[t]) = (l^*(t), x^*(t)), \qquad (7)$$

hold, so that the supporting hyperplane for $\mathcal{X}[t]$ generated by $l^*(t)$, namely, the plane $(x - x^*(t), l^*(t)) = 0$ that touches $\mathcal{X}[t]$ at point $x^*(t)$, is also a supporting hyperplane for $\mathcal{E}_+^*[t], \mathcal{E}_-^*[t]$ and touch them at the same point.

The solutions to this problem are given within the following statements.

Theorem 21 *With $l(t) = l^*(t)$ given, the solution to Problem 2.1(external) is an ellipsoid $\mathcal{E}_+[t] = \mathcal{E}(x^*(t), X_+^*[t])$, where*

$$X_+^*[t] = \left(\int_{t_0}^t p_t^*(s)ds + p_0^*(t) \right)$$

$$\left(\int_{t_0}^t (p_t^*(s))^{-1} X(t, s)Q(s)X'(t, s)ds + p_0^{*-1}(t)X(t, t_0)X^0 X'(t, t_0) \right), \qquad (8)$$

and

$$p_t^*(s) = (l^*(t), X(t, s)Q(s)X'(t, s)l^*(t))^{1/2}, \qquad (9)$$

$$p_0^*(t) = (l^*(t_0), X(t, t_0)X^0 X'(t, t_0)l^*(t_0))^{1/2}.$$

This result follows from [6], [7]. Since the calculations have to be made for all t, the parametrizing functions $p_t(s)$, $s \in [t_0, t]$, $p_0(t)$ must depend on t. Note therefore that the result requires the evaluation of the integrals in (8) for each time t and vector l. If the computation burden for each evaluation of (8) is $C_n t$, and we estimate the reach tube via (8) for T values of time t and L values of l, the total computational burden would be $C_n TL$.

In other words, relations (8), (9) need to be solved "afresh" for each t. It may be more convenient for computational purposes to have them given in the form of recurrence relations. As indicated further, in the next Section, this could be done by selecting function $l^*(t)$ of Problem 2.1 in an appropriate way.

A similar result is available for internal approximations.

Theorem 22 *With $l = l^*(t)$ given, the solution to Problem 2.1 (internal) is an ellipsoid $\mathcal{E}(x_-(t), X_-(t))$, where*

$$X_-(t) = \qquad (10)$$

$$= X(t,t_0)\Big(Q_0^{1/2}S'_{0t}(t_0) + \int_{t_0}^t X(t_0,\tau)Q^{1/2}(\tau)S'_t(\tau)d\tau\Big)'$$

$$\Big(S_{0t}(t_0)Q_0^{1/2} + \int_{t_0}^t S_t(\tau)Q^{1/2}(\tau)X'(t_0,\tau)\Big)X'(t,t_0). \qquad (11)$$

with $S_0, S_t(\tau)$ satisfying relations

$$S_t(\tau)Q^{1/2}(\tau)X'(t,\tau)l^*(t) = \lambda_t(\tau)S_{0t}Q_0^{1/2}X'(t,t_0)l^*(t), \qquad (12)$$

and $S'_{0t}S_{0t} = I; S'_t(\tau)S_t(\tau) \equiv I$ for all $t \geq t_0, \tau \in [t_0,t]$, where

$$\lambda_t(\tau) = (l^*(t), X(t,\tau)Q(\tau)X'(t,\tau)l^*(t))^{1/2}(l^*(t), X(t,t_0)Q_0X'(t,t_0)l^*(t))^{1/2}. \qquad (13)$$

The parametrizing functions are *orthogonal matrix-valued functions* $S_t(\tau), S_{0t}$. They too are dependent on t, so that the calculations have to be done "afresh" for each t as in the "external" case. Thus, the computaion in general is not recursive. To ease the computational burden we look for recurrence relations.

3   Recurrence Relations

There is a special selection of functions $l^*(t)$ that lead to recurrence relations.

Assumption 31 *The function $l^*(t)$ is of the form, $l^*(t) = X(t_0,t)l$, with $l \in \mathbb{R}^n$ given. For the time-invariant case $l^*(t) = e^{-A'(t-t_0)}l$.* [4]

Then $p_t^*(s), p_0^*(t), X_+^*[t]$ of (9), (8) transform into

$$p_t^*(s) = (l, X(t_0,s)Q(s)X'(t_0,s)l)^{1/2} = p^*(s); \quad p_0^*(t) = (l, X^0l)^{1/2} = p_0^*, \qquad (14)$$

and

$$X_+^*[t] = X(t,t_0)X_+(t)X'(t,t_0), \quad X_+[t] = \Big(\int_{t_0}^t p^*(s)ds + p_0^*\Big)\Psi(t), \qquad (15)$$

where

$$\Psi(t) = \qquad (16)$$

$$= \int_{t_0}^t (l, X(t_0,s)Q(s)X'(t_0,s)l)^{-1/2}X(t_0,s)Q(s)X'(t_0,s)ds + (l, X^0l)^{-1/2}X^0.$$

In this particular case $p_t^*(s)$ does not depend on t ($p_{t'}^*(s) = p_{t''}^*(s)$ for $t' \neq t''$) and the lower index t may be dropped.

---

[4]   Under this Assumption the vector $l^*(t)$ is the solution to equation

$$\dot{l}^* = -A'(t)l^*, \quad l^*(t_0) = l,$$

which is the adjoint to the homogeneous part of equation (1).

Direct differentiation of $X_+[t]$ yields

$$\dot{X}_+[t] = \pi^*(t)X_+[t] + \pi^{*-1}(t)X(t_0,t)Q(t)X'(t_0,t), \quad X_+[t_0] = X^0, \qquad (17)$$

where

$$\pi^*(t) = p^*(t)\left(\int_{t_0}^t p^*(s)ds + p_0^*\right)^{-1}.$$

Calculating

$$(l, X_+[t]l) = \left(\int_{t_0}^t p^*(s)ds + p_0^*\right)(l, \Psi(t)l) = (\int_{t_0}^t p^*(s)ds + p_0^*)^2,$$

one may observe that

$$\pi^*(t) = (l, X(t_0,t)Q(t)X'(t_0,t)l)^{1/2}(l, X_+[t]l)^{-1/2}. \qquad (18)$$

In order to pass to the matrix function $X_+^*[t]$ we note that

$$\dot{X}_+^*[t] = A(t)X(t,t_0)X_+[t]X'(t,t_0) + X(t,t_0)X_+[t]X'(t,t_0)A(t)$$
$$+X(t,t_0)\dot{X}_+[t]X'(t,t_0).$$

After a substitution from (16) this gives

$$\dot{X}_+^* = A(t)X_+^* + X_+^* A'(t) + \pi^*(t)X_+^* + \pi^{*-1}(t)Q(t), \quad X^*(t_0) = X^0. \qquad (19)$$

We summarize these results as follows.

Theorem 31 *Under Assumption 3.1 the solution to Problem 2.1(external) is given by the ellipsoid $\mathcal{E}_+^*[t] = \mathcal{E}(x_+(t), X_+^*[t])$, where $x_+(t) = x^*(t)$ and $X_+^*[t]$ is a solution to equations (18), (16).*

Since the set $X_+^*[t]$ depends on vector l, we denote $X_+^*[t] = X_+^*[t]_l$.

Theorem 32 *For any $t \geq t_0$ the reach set $\mathcal{X}[t]$ may be described as*

$$\mathcal{X}[t] = \cap\{\mathcal{E}(x_+(t), X_+^*[t]_l)\}|\, l : (l,l) = 1\}. \qquad (20)$$

This is a direct consequence of Theorems 3.1.

Thus, if $l^*(t)$ satisfies Assumption 3.1, the complexity of computing a tight, external ellipsoidal approximation to the reach set for all t, is the same as computing the solution to the differential equation (18). If L values of l and T values of t are evaluated, the computational burden is $C_n TL$.

For the general(non-recursive) case, the relation corresponding to (18) is far more complicated and is actually a functional-differential equation which requires recalculations for each t. If however (18) is still used for the general case, the inclusion $\mathcal{X}[t] \subset \mathcal{E}_+[t]$ remains true but the tightness property is lost.

Throughout the previous discussion we have observed that under Assumption 3.1 the tight external ellipsoidal approximation $\mathcal{E}(x^*, X_+^*(t))$ is governed by the simple ordinary differential equations (18). Moreover, in this case the points $x^*(t)$ of support for the hyperplanes generated by vector $l(t)$ run along *a system trajectory* of (1) which is generated by a control that satisfies the maximum principle.

Similar facts are also true for internal approximations. We now again select function $l^*(t)$ to satisfy Assumption 3.1. Then substituting $l^*(t)$ in (12),(13), we observe that the relations for calculating $S_t(\tau), \lambda_t(\tau)$ transform into

$$S_t(\tau)Q^{1/2}(\tau)X'(t_0,\tau)l = \lambda_t(\tau)S_{0t}Q_0^{1/2}l; \quad S_0'S_0 = I; S'(\tau)S(\tau) \equiv I \quad (21)$$

and

$$\lambda_t(\tau) = (l, X(t_0,\tau)Q(\tau)X'(t_0,\tau)l)^{1/2}/(l, Q_0 l)^{1/2}. \quad (22)$$

Here the known functions used for calculating $S_t(\tau), \lambda_t(\tau)$ *do not depend on t.* Therefore, the unknown functions $S_t(\tau), \lambda_t(\tau)$ *do not depend on t* either, no matter what is the interval $[t_0, t]$. The lower indices t in S_{0t}, S_t, λ_t may be dropped. Differentiating (10) in view of the last remark, we come to

$$\dot{X}_- = A(t)X_- + X_-A'(t) + \dot{Q}_*'Q_* + Q_*'\dot{Q}_*, \quad (23)$$

where

$$Q_*(t) = S_0 Q_0^{1/2} X'(t,t_0) + \int_{t_0}^t S(\tau)Q^{1/2}(\tau)X'(t,\tau)d\tau,$$

$$\dot{Q}_*(t) = S(t)Q^{1/2}(t), Q_*(t_0) = S_0 Q_0.$$

Using the notation

$$H(t) = Q_*^{-1}(t)S(t)Q^{1/2}(t) = Q_*^{-1}(t)\dot{Q}_*(t), \quad (24)$$

we further come to equation

$$\dot{X}_- = A(t)X_- + X_-A'(t) + H'(t)X_-(t) + X_-(t)H(t), \quad X(t_0) = Q_0. \quad (25)$$

and also observe that the center $x_-(t) = x_+(t) = x^*(t)$. This leads to the following theorem.

Theorem 33 *Under Assumption 3.1 the solution to Problem 3.1 (internal) is given by ellipsoid $\mathcal{E}(x_-(t), X_-(t))$ where $X_-(t)$ is given by equations (24), (23), and the functions $S(t), \lambda(t)$ involved in the calculation of $H(t)$ satisfy together with S_0 the relations (20), (21), where the lower indices t in S_{0t}, S_t, λ_t are to be dropped.*

Function $H(t) = Q_*^{-1}(t)S(t)Q^{1/2}(t)$ in (23) may be also expressed through equation

$$\dot{Q}_* = Q_*A'(t) + S(t)Q^{1/2}(t), \quad Q_*(t_0) = S_0 Q_0^{1/2}. \quad (26)$$

This gives the result

Lemma 4. *The ellipsoid* $\mathcal{E}(x_-(t), X_-(t))$ *of Theorem 3.3 given by equations (23)-(25) depends on the selection of the orthogonal matrix function $S(t)$ and for any such $S(t)$ the inclusion*

$$\mathcal{E}(x_-(t), X_-(t)) \subseteq \mathcal{X}[t], \quad t \geq t_0, \tag{27}$$

is true with equalities (7)(internal) attained under conditions (20), (21). The following relation is true

$$\mathcal{X}[t] = cl\{\cup\{\mathcal{E}(x_-(t), X_-[t]_l)\} \mid l : (l, l) = 1\}\}.$$

where clY stands for the closure of set Y.

The boundary of $\mathcal{X}[t]$ is thus described as a function of a *finite-dimensional parameter* $l \in \mathbb{R}^n$.

Let us now suppose that function $l(t)$ of Problem 2.1 (internal) is *any* continuous curve on the surface of $\mathcal{X}[t]$. Then one has to use formula (10), keeping in mind that $S_{0t}, S_t(\tau)$ do depend on t. After a differentiation of (10) in t, one may observe that (25) transforms into

$$\dot{X}_- = A(t)X_- + X_- A'(t) + H'(t)X_-(t) + X_-(t)H(t) + \Phi(t, \cdot), \quad X_(t_0) = Q_0. \tag{28}$$

where $\Phi(t, \cdot)$ is a functional of $S_t(\tau), S_{0t}$. The calculations are then far more cumbersome than under Assumption 3.1. If in this general case we still use the simpler equation (24), then the inclusion (26) will still be true, but the property of tightness will be lost. Note that under Assumption 3.1 the term $\Phi(t, \cdot)$ disappears.

4   The Reach Tube

The results of the previous Sections may be thus summarized as follows. Suppose Assumption 3.1 is fulfilled. then the points $x^*(t)$ of support for vector $l^*(t) = X'(t, t_0)l$, $l \in \mathbb{R}^n$, namely, those for which the equalities

$$(l^*(t), x^*(t)) = \rho(l^*(t) \mid \mathcal{X}[t]) = \rho(l^*(t) \mid \mathcal{E}(x^*(t), X_+^*[t])) \tag{29}$$

are true **for all** $t \geq t_0$, may be reached from initial state

$$x^{*0} = x^*(t_0) = \frac{X^0 l}{(l, X^0 l)^{1/2}} + x^0. \tag{30}$$

and from a *trajectory* $x^*(t)$ that satisfies the following "maximum relation":

$$(l^*(t), x^*(t)) = \max\{(l^*(t), x) \mid x \in \mathcal{E}(x^*, X_+^*[t])\}, \tag{31}$$

which is attained at

$$x^*(t) = x^*(t) + X_+^*[t]l^*(t)(l^*(t), X_+^*[t]l^*(t))^{-1/2}, \tag{32}$$

where $X_+^*[t] = X[t]$ is the solution to equations (18), (17).

For $B(t) \equiv I$ and $Q(t)$ nondegenerate the same trajectory (31) may be attained through *internal ellipsoids* with

$$x^*(t) = x^*(t) + X_-[t]l^*(t)(l^*(t), X_-[t]l^*(t))^{-1/2}, \qquad (33)$$

where $X_-[t]$ is a solution to (24), (23). The same property holds if $Q(t)$ is nondegenerate and the system (1) is controllable.

Denoting $x^*(t) = x[t, l]$, we thus come to a two-parameter surface $x[t, l]$ that defines the boundary $\partial \mathcal{X}$ of the reachability tube $\mathcal{X} = \cup\{X[t],\ t \geq t_0\}$. With $t = t'$ fixed and $l \in \mathcal{S}$ varying,($\mathcal{S}$ is a unit sphere), the vector $x[t', l]$ runs along the boundary $\partial \mathcal{X}[t']$. On the other hand, with $l = l'$ fixed and with t varying, the vector $x[t, l']$ moves along one of the trajectories $x^*(t)$ that touch the reachability set $\mathcal{X}[t]$ according to (7). Then

$$\cup\{x[t, l] | l \in \mathcal{S}\} = \partial \mathcal{X}[t], \quad \cup\{x[t, l] | l \in \mathcal{S},\ t \geq t_0\} = \partial \mathcal{X}$$

Remark 4.1. The possibility of using both external and internal representations is important for treating hybrid dynamics for systems that allow resets. Thus, if for example set $\mathcal{X}^0 = \cap \mathcal{E}_i^0$, one may introduce approximations of type $\mathcal{E}_-^0 \subseteq \mathcal{X}^0 \subseteq \mathcal{E}_+^0$ to start the calculations of the reach set $\mathcal{X}[t]$. On the other hand, if for some $t' > t_0$ we have $\mathcal{X}[t'] \cap \mathcal{E}_M$, where $\mathcal{E}_M$ stands for a given guard, we may introduce approximations of type

$$\mathcal{E}_-^*[t'] \subseteq \cup\{\mathcal{E}(x_-(t'), X_-(t')) | (l, l) \leq 1\} \cap \mathcal{E}_M$$
$$\subseteq \cap\{\mathcal{E}(x_+(t'), X_+(t')) | (l, l) \leq 1\} \cap \mathcal{E}_M \subseteq \mathcal{E}_+^*[t']$$

for the resets and proceed for $t \geq t'$ with the procedures of Sections 2-4.

5 An Example

Taking system

$$\dot{x}_1 = x_2, \quad \dot{x}_2 = u,$$

$$x_1(0) = x_1^0, x_2(0) = x_2^0,\ |u| \leq \mu,\ \mu > 0,\ X^0 = \{x : (x, x) \leq \epsilon^2\}.$$

and omitting the calculations, we indicate the external and internal ellipsoidal approximations of the respective reach set $\mathcal{X}[t] = \mathcal{X}(t, 0, X^0)$.

Here the "good" curves of Assumption 3.1 have the form of straight lines: $l^*(t) = \exp(-A't)l$ or $l_1^* = l_1, l_2^* = l_2 - t l_1$. They are shown in fig.1 for $\epsilon > 0$. The external and internal approximations that touch the reach set $\mathcal{X}[t]$ along these lines are shown in fig.2 and fig.3 for $\epsilon = 0$ and in fig.4 for $\epsilon > 0$.

6   Conclusion

This paper specifies and studies the behavior of the tight external and internal ellipsoidal approximations of reach sets and reach tubes for linear time-variant

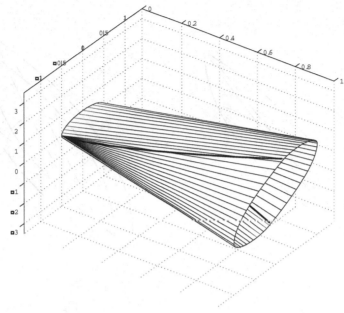

Fig.1

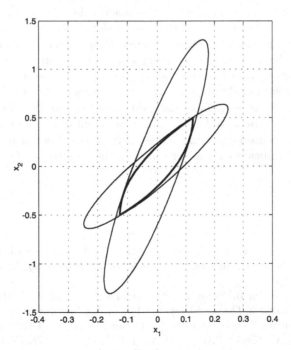

Fig.2

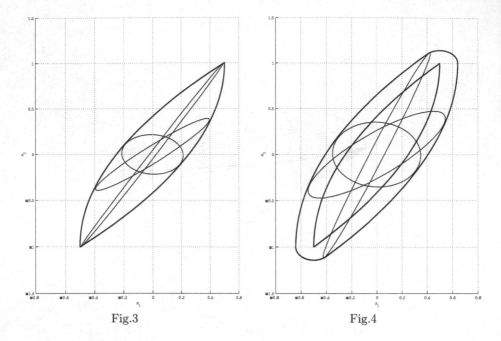

Fig.3   Fig.4

control systems. It shows that equations (18), (24) with appropriately chosen parametrizing functions $\pi(t), S(t)$ generate two family of tight (external and internal) ellipsoidal aproximations to the reach tube $\mathcal{X}[t]$ which touch it along a certain family of "good" curves that *cover the whole tube*. It gives analytical representations that allow to achieve a substantial reduction of the computation burden for calculating these sets as compared to direct methods and thus gives effective techniques for calculating the reach tubes in a compact recursive form. The analytical relations developed in this paper open routes to the investigation of precise error estimates in ellipsoidal approximations for problems of evolution, estimation and control as well as to the development of new computational tools for classes of systems more complicated than those treated in this paper. In particular, they indicate convenient tools for the treatment of hybrid dynamics (see [1]).

References

1. Botchkarev O., Tripakis S., Verification of Hybrid Systems with Linear Differential Inclusions using Ellipsoidal Approximations. *Proceedings of this Conference*, Pittsburg, 2000.
2. Boyd S., El Ghaoui L., Feron E., Balakrishnan V., *Linear Matrix Inequalities in System and Control Theory*, SIAM, Studies in Applied Mathematics, 1994.
3. Chernousko F.L., *State Estimation for Dynamic Systems*, CRC Press, 1994.
4. Henzinger T.A., Kopke P.W., Puri A. and Varaiya P., What's Decidable about Hybrid Automata ? *Proc. 27-th STOC*, pp.373 - 382, 1995.

5. Kurzhanski A.B., Filippova T.F., On the Theory of Trajectory Tubes: a Mathematical Formalism for Uncertain Dynamics, Viability and Control, in: *Advances in Nonlinear Dynamics and Control*, ser. PSCT 17, pp.122 - 188, Birkhäuser, Boston, 1993.
6. Kurzhanski A. B., Vályi I. *Ellipsoidal Calculus for Estimation and Control*, Birkhäuser, Boston, ser. SCFA, 1997.
7. Kurzhanski A. B., Varaiya P., Ellipsoidal techniques for reachability analysis. To appear.
8. Kurzhanski A. B., Varaiya P., Ellipsoidal techniques for reachability analysis. Internal approximations. To appear.
9. Lee E.B., Marcus L., *Foundations of Optimal Control Theory*, Wiley, NY,, 1967.
10. Leitmann G., Optimality and reachability via feedback controls. In :*Dynamic Systems and Mycrophysics*, Blaquiere A., Leitmann G. eds.,1982.
11. Lempio F., Veliov V., Discrete Approximations of Differential Inclusions, *Bayreuther Mathematische Schriften*, Heft 54, pp. 149 - 232, 1998.
12. Pappas G.L., Sastry S., Straightening out differential inclusions. *System and Control Letters*, 35(2), pp.79-85, Sept.1998.
13. Puri A., Borkar V. and Varaiya P., ϵ - Approximations of Differential Inclusions, in: R.Alur, T.A.Henzinger, and E.D.Sonntag eds., *Hybrid Systems*,pp. 109 - 123, LNCS 1201, Springer, 1996.
14. Puri A., Varaiya P., Decidability of hybrid systems with rectangular inclusions. In D.Dill ed., *Proc. CAV - 94, LNCS 1066*, Springer, 1996.
15. Rockafellar, R. T., *Convex Analysis*, 2-nd ed., Princeton University Press, 1999.
16. Varaiya P. Reach Set Computation Using Optimal Control, in *Proc.of KIT Workshop on Verification of Hybrid Systems*, Verimag, Grenoble, 1998.

Uniform Reachability Algorithms

Gerardo Lafferriere[1] and Chris Miller[2],[*]

[1] Department of Mathematical Sciences
Portland State University, P.O. Box 751, Portland, OR 97207
`gerardo@mth.pdx.edu`
[2] Department of Mathematics
The Ohio State University, 231 W. 18th Avenue, Columbus, OH 43210
`miller@math.ohio-state.edu`

Abstract. We introduce the notion of a parametrized family $(H_p)_{p \in P}$ of hybrid systems, and consider questions of reachability in the systems H_p as the parameter p ranges over P. Under the assumption of a uniform (as p ranges over P) finite bound on the number of discrete transitions associated to the individual systems H_p, the notion of reachability is first-order (in the sense of mathematical logic) and uniform in the parameter p. Techniques from logic can then be used to analyze computational questions associated to the family of systems.

This paper is concerned with uniform verification of reachability properties for parametrized families of hybrid systems.

The central reachability question for a hybrid system (no matter how this is defined) is to determine, given two states of the system, whether there is a trajectory which takes the system from one state to the other. Ideally, one has an algorithm which takes as input pairs of states (x, y) and computes whether there exists such a a trajectory; see, for example, [1,5]. Tools from mathematical logic can be useful in these investigations.

More generally, one can consider reachability questions for families of hybrid systems that are linked up in some reasonable fashion, and hope that one can find algorithms that work uniformly as one varies the systems under consideration. Most of the work in this paper goes into making precise statements of these loose notions.

Here is an outline of this paper. In Section 1, we extend the definition of hybrid system given in [4,5] to that of a *parametrized family* of hybrid systems. Section 2 contains some relevant material from model-theoretic definability theory. We present the main results in Section 3, followed by some examples and applications in Section 4.

1   Families of Hybrid Systems

We begin with an informal discussion. The intuitive notion of a parametrized family of hybrid systems is fairly clear (once we have a clear notion of hybrid

[*] Research supported by NSF grant DMS-9896225

N. Lynch and B. Krogh (Eds.): HSCC 2000, LNCS 1790, pp. 215–228, 2000.

system, that is). If we think of a hybrid system as some sort of black box, then a parametrized family $(H_p)_{p \in P}$ of hybrid systems is a black box H with a console $P = (P_1, \ldots, P_l)$ of dials $P_1, \ldots, P_l$ such that each setting $p = (p_1, \ldots, p_l)$ of the dials yields a hybrid system H_p. The dials control the various relevant sets, perhaps the vector fields involved, perhaps even the entire state spaces. Naturally, we would like the resulting systems to vary in some sensible way with respect to the settings of the dials.

Similarly, the intuitive notion of uniform decidability (or computability, or whatever) of a family of hybrid systems is easy to describe (again, once we have a clear notion of decidability or computability of a hybrid system). For H as above, we should have algorithms, working uniformly over all settings p, for answering various questions about the systems H_p. For example, we should have some computable function Φ_H such that: (a) given a setting p and states x, y of H_p, we have $\Phi_H(p, x, y) = 0$ if and only if y is reachable from x in the system H_p; and (b) for a given pair (x, y) of possible states x, y, we can compute the set of all p such that x, y are states of H_p and $\Phi_H(p, x, y) = 0$. Further variations easily come to mind, but we should not digress too far at this point.

We now begin to make these intuitive notions precise. First we carefully parametrize all data involved in the definition of a hybrid system, as given in [4] or [5]. Now, any set of hybrid systems can be made into a parametrized family: just index it set-theoretically by some suitable ordinal. This is a rather useless approach, of course. We incorporate in our definition a certain amount of desirable uniformity. This is unavoidably rather tedious notationally, and we advise the reader to keep the informal discussion above in mind throughout. We stress that several other ways of doing this easily come to mind; some less complicated, some more complicated. We have chosen an approach somewhere in the middle.

Of crucial importance in this paper are the notions of *parametrized families* of sets and maps. Let X, Y be sets, $A \subseteq X \times Y$ and $x \in X$; then A_x denotes the *fiber* of A over x, that is, the set $\{ y \in Y : (x, y) \in A \}$. (One can define similarly the fiber of A over $y \in Y$, but we will not introduce notation for this.) The *(first) projection* $\pi(A)$ of A is the set of all $x \in X$ such that $A_x \neq \emptyset$. Given a map $f : A \to Z$ (with Z some set) and $x \in X$, let $f(x, \cdot) : A_x \to Z$ denote the map $y \mapsto f(x, y) : A_x \to Z$. Let $B \subseteq X$ and consider the indexed families $(A_x)_{x \in B}$ and $(f(x, \cdot) : A_x \to Z)_{x \in B}$. The former is called a *parametrized (by B) family* of subsets of Y, while the latter is called a parametrized family of maps. (After identifying a map with its graph, a parametrized family of maps is just a special kind of parametrized family of sets.) Of particular interest is the case $B = \pi(A)$.

Let M be a set and $m, n \in \mathbb{N}$. We identify $M^m \times M^n$ with M^{m+n} whenever convenient. (Regard M^0 as the one-point space $\{\emptyset\}$, and functions $f : M^0 \to M$ as the corresponding constant $f(\emptyset)$.) Hence, given $A \subseteq M^{m+n}$ and $x \in M^m$, $A_x \subseteq M^n$ denotes the fiber of A over x, and (unless stated otherwise) $\pi(A)$ denotes the projection of A on the first m coordinates. (We rely on context to indicate when subscripts indicate taking fibers, and when they are used as indices.) Moreover, for $i = 1, \ldots, m$, we let π_i denote the projection on the i-th coordinate. (There will be times when, in order to avoid ambiguity, we shall have to abandon some of this notation, and just state things in words.)

Examples. (a) Consider the set

$$\{ (a,b,c,d,e,f,x,y) \in \mathbb{R}^8 : ax^2 + bxy + cy^2 + dx + ey + f = 0 \}.$$

As the tuple (a,b,c,d,e,f) varies over $\mathbb{R}^6$, we obtain the family of all conic sections in the plane. (b) Given a set $E \subseteq \mathbb{R}$, let $\mathrm{GL}_n(E)$ denote the set of all invertible $n \times n$ matrices with entries from E. Then

$$(x \mapsto Ax : \mathbb{R}^n \to \mathbb{R}^n)_{A \in \mathrm{GL}_n(E)}$$

is a parametrized family of maps. Note that a map $A \in \mathrm{GL}_n(E)$ can be identified with (or coded up as) as a point in $\mathbb{R}^{n^2}$, since the action of the map is determined by its coefficients.

Definition 1. *Let M be a set. A parametrized family of hybrid systems on M is a 5-tuple $H = (M, S, I, \Gamma, F)$ where $M, S, I,$ and Γ are sets, and F is a map, with the properties indicated below.*

- *The set M is a Hausdorff, second countable, (sufficiently) differentiable manifold.*
- *The graph space S is a nonempty subset of*

$$\{(x, y, y_i, y_j) : x \in \mathbb{R}^n, y \in \mathbb{R}^m, \quad 1 \le i,j \le m, \quad y_1 < \ldots < y_m\} \subseteq \mathbb{R}^{n+m+2}.$$

  The parameter space P is the projection of S on the first $n+m$ coordinates. The projection of P on the last m coordinates is denoted by Q.
- *$I \subseteq \mathbb{R}^{n+m} \times M^m$, and its projection on the first $n+m$ coordinates is equal to P.*
- *$\Gamma \subseteq \mathbb{R}^{n+m+2} \times M^2$, and its projection on the first $n+m+2$ coordinates is S. For a fixed $z = (p, p_{n+i}, p_{n+j}) \in S$, where $1 \le i,j \le m$, we require that $\pi_1(\Gamma_z) \subseteq \pi_i(I_p)$ and $\pi_2(\Gamma_z) \subseteq \pi_j(I_p)$.*
- *The map $F : P \times M \to (TM)^m$ is such that for each $p \in P$, each component $F_i(p, \cdot) : M \to TM$ of $F(p, \cdot)$ is a complete (i.e. trajectories are defined for all time) vector field on M. Here, TM denotes the tangent bundle of M.*

The *(parametrized) flow* of the map $F : P \times M \to (TM)^m$ associated to H is the function $\phi : P \times \mathbb{R}^m \times M^m \to M^m$ defined by $\phi(p, t, x) = y$ if and only if for each $i = 1, \ldots, m$, the integral curve of $F_i(p, \cdot)$ with initial condition x_i passes through y_i at time t_i.

Each element of Q is an ordered list of vertices or *locations* for a single hybrid system. The continuous components of hybrid trajectories lie in subsets of M. For a fixed parameter $p \in P$ and the corresponding list of vertices $q = (p_{n+1}, \ldots, p_{n+m}) = (q_1, \ldots, q_m)$ the projection $\pi_i(I_p)$ is nonempty and is referred to as the *invariant set* at location q_i. The projection of S on the last two coordinates is the set E of *edges*. For each $z = (p, p_{n+i}, p_{n+j}) \in S$, Γ_z defines a relation on M which induces discrete transitions (see below). The sets $R(e) = \{p_{n+i}\} \times \pi_1(\Gamma_z)$ and $G(e) = \{p_{n+j}\} \times \pi_2(\Gamma_z)$ are respectively the *reset* and *guard* associated to the edge $e = (p_{n+i}, p_{n+j})$.

Note that our definitions allow a parametrized variation of the discrete locations (but not their number).

Additional parametrized features could be added to the definition in an obvious way. For example, one may wish to specify some distinguished initial and final sets for trajectories $(C_o, C_f \subseteq \mathbb{R}^{n+m+1} \times M$, with suitable projections).

While the present definition is concise, in special cases it may help intuition to have parameters separated into groups depending on which entities they parametrize: initial conditions, vector fields, invariant sets, and so on.

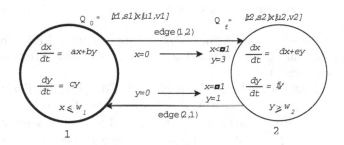

Fig. 1. Parametrized family of hybrid systems

Example. Figure 1 provides a schematic representation of a family of hybrid systems. This is encoded as follows, where a through f, r_i, s_i, p_i, q_i, and u_i all vary over appropriate sets of real numbers:

- $M = \mathbb{R}^2$
- $P = \{(a, b, c, d, e, f, r_1, r_2, s_1, s_2, u_1, u_2, v_1, v_2, w_1, w_2, 1, 2)\}$
- $S = \{(p, i, j) : p \in P,\ i, j = 1, 2,\ i \neq j\}$
- $C_o = \{(p, 1) : p \in P\} \times [r_1, s_1] \times [u_1, v_1]$
- $C_f = \{(p, 2) : p \in P\} \times [r_2, s_2] \times [u_2, v_2]$
- $I = \{(p, (x_1, y_1), (x_2, y_2)) : x_1 \leq w_1, y_2 \geq w_2\}$
- $\Gamma = \{(p, 1, 2, (x_1, y_1), (x_2, y_2)) : x_1 = 0, x_2 < -1, y_2 = 3\}$ $\cup$
  $\{(p, 2, 1, (x_1, y_1), (x_2, y_2)) : x_1 = -1, y = 1, y_2 = 0\}$
- $F(p, x) = (A_1 x, A_2 x)$, where A_i are upper triangular matrices with entries a, b, c, and d, e, f respectively.

Some restrictions on the parameters are needed to satisfy the inclusion requirements (for example, $w_1 \geq 0$, $w_2 \leq 0$). In this example, parameters affect the vector fields and the initial and final sets, but do not influence the locations or the relation Γ.

For each fixed parameter $p \in P$, H determines a hybrid system H_p similar to those introduced in [4,5]. Let $X_D = \{p_{n+1}, \dots, p_{n+m}\}$ be the set of discrete locations. Put $H_p = (X, X_o, X_f, \text{Funct}, \text{Edge}, \text{Inv}, \text{Rel})$, where:

- $X = X_D \times M$ is the state space.
- $X_o = \{(q, x) \in X : \exists\, 1 \leq i \leq m, \text{with } q = p_{n+i}, \text{ and } (p, q, x) \in C_o\}$ is the set of initial states. The final states X_f are defined similarly.
- $\text{Funct} : X \to TM$ is defined by $\text{Funct}(p_{n+i}, x) = F_i(p, x)$ for each $1 \leq i \leq m$. For each $q \in X_D$, $\text{Funct}(q, \cdot)$ defines a vector field on M.
- $\text{Edge} = S_p$ is the set of edges along which discrete transitions occur.
- $\text{Inv} : X_D \longrightarrow 2^M$ assigns to each location the set $\text{Inv}(p_{n+i}) := \pi_i(I_p)$.
- $\text{Rel} = \{(p_{n+i}, x, p_{n+j}, y) : z = (p, p_{n+i}, p_{n+j}) \in S, (x, y) \in \Gamma_z\}$ defines discrete transitions.

(For convenience, we omitted notation for the dependence of the new objects on p.)

For each edge $e = (q, r)$ define $\text{Rel}(e) = \{(q, x, r, y) \in \text{Rel}\}$. The systems in [4,5] are a special case corresponding to $\text{Rel}(e) = G(e) \times R(e)$ (guard times reset as defined earlier).

Consider the (single) hybrid system $K = (X, \text{Edge}, \text{Inv}, \text{Rel}, \text{Funct})$. An element $((q, x), e, (r, y)) \in X \times E \times X$ is a *discrete transition (along e)*, denoted by $(q, x) \xrightarrow{e} (r, y)$, if $e = (q, r)$ and $(q, x, r, y) \in \text{Rel}(e)$. Fix some $\tau \notin E$. An element $((q, x), \tau, (r, y)) \in X \times \{\tau\} \times X$ is a *continuous transition*, denoted by $(q, x) \xrightarrow{\tau} (r, y)$, if $q = r$ and there exist $\delta \geq 0$ and a differentiable curve $\gamma : [0, \delta] \longrightarrow \text{Inv}(q)$ satisfying $\gamma' = \text{Funct}(q, \gamma)$, $\gamma(0) = x$, and $\gamma(\delta) = y$ (that is, there is an arc of a trajectory of the vector field $F(q, \cdot)$ connecting x to y within Inv). Note that if $(q, x) \xrightarrow{\tau} (q, y)$ and $(q, y) \xrightarrow{\tau} (q, z)$, then $(q, x) \xrightarrow{\tau} (q, z)$. Given $a, b \in X$, we say that b *is reachable from a (in K)* if there exist $k \in \mathbb{N}$, $\sigma_1, \dots, \sigma_k \in E \cup \{\tau\}$ and states $a_0, \dots, a_k \in X$ such that $a = a_0$, $b = a_k$ and $a_{i-1} \xrightarrow{\sigma_i} a_i$ for $i = 1, \dots, k$.

When we wish to emphasize the parameter set P we use the notation $(H_p)_{p \in P}$ to denote the parametrized family H.

Given a family $H = (H_p)_{p \in P}$ of hybrid systems, we define the *reachability set of H*, denoted by $\text{Reach}(H)$, to be the set of all (p, x, y) such that $p \in P$ and y is reachable from x in H_p. The *reachability problem for H* is just the set membership question for $\text{Reach}(H)$.

Important Note. From now on, we will restrict our study to systems in which, for each e, $\text{Rel}(e)$ is of a special form, namely, a finite union of subsets of $\text{Rel}(e)$, each a cartesian product $A \times B \subseteq (\{q\} \times M) \times (\{r\} \times M)$, where $e = (q, r)$. That is, we assume that for each edge e there is an integer $n(e)$ and sets $A_i(e)$, $B_i(e)$ as above such that $\text{Rel}(e) = \cup_{i=1}^{n(e)} (A_i(e) \times B_i(e))$. The results of [4] on bisimulations extend to this case with minor modifications. For such systems, since concatenations of continuous transitions collapse, there is an integer N_K such that, if b is reachable from a in K, then we may take $k \leq N_K$. Moreover, for a parametrized family $(H_p)_{p \in P}$ we assume that the set $\{n_p(e) : e \in E, p \in P\}$ is bounded above (i.e. the numbers $n_p(e)$ are bounded uniformly in p).

2   Definability Theory[1]

We require some notions from model theory (a branch of mathematical logic) which, in its most general form, is the study of classes of models of theories in given languages, and the relationships between syntax and semantics. At its root, there are quite a few important—but rather tedious—technical definitions, creating pitfalls for the unwary outsider. To make matters worse, the subject has undergone something of a revolution in the last decade or so, resulting in changes of terminology, as well as entire points of view. Many of these recent changes have not found their way into standard texts. But there is a fairly small fragment of model theory that often suffices for *applications* to other subjects, especially for explaining and applying model-theoretic results: what can be called (first-order) definability theory. We present in this section a brief introduction to the subject. There are two equivalent approaches—informally, the top-down and bottom-up—each more useful than the other at times.

We provide here neither history nor a comprehensive treatment of basic results (and, for ease of exposition, we still gloss over some minor technicalities). Rather, our goal is to equip the reader with the basic technology necessary in order to understand how some current developments in model theory can be applied to hybrid systems. We also recast and clarify some material from [4,5]. The reader interested in historical context, original sources, detailed statements of results, proofs, and so on, may begin by consulting [15,16,17] for information.

2.1   The Top-Down Approach

In this scenario, we are interested in some particular class of sets that are (or that we hope are) closed under first-order definability; we make this notion precise.

Let M be a nonempty set. A *structure on* M is a sequence $\mathfrak{M} = (\mathfrak{M}_n)_{n \in \mathbb{N}}$ such that for each $n \in \mathbb{N}$:

- (S1) $M^n \in \mathfrak{M}_n$ and $\mathfrak{M}_n$ is a boolean algebra of subsets of M^n (that is, $\mathfrak{M}_n$ is closed under taking complements and finite unions).
- (S2) $\{ (x_1, \ldots, x_n) \in M^n : x_i = x_j \} \in \mathfrak{M}_n$, $1 \leq i < j \leq n$.
- (S3) If $A \in \mathfrak{M}_n$, then $A \times M$, $M \times A \in \mathfrak{M}_{n+1}$.
- (S4) If $A \in \mathfrak{M}_{n+1}$, then $\pi(A) \in \mathfrak{M}_n$.

We say that a set $A \subseteq M^n$ is *definable in* $\mathfrak{M}$, or that $\mathfrak{M}$ *defines* A, if $A \in \mathfrak{M}_n$. If no ambient space M^n is mentioned, then "definable set" (in $\mathfrak{M}$) means "definable subset of M^n, for some $n \in \mathbb{N}$". A map $f : A \to M^n$, $A \subseteq M^m$, is definable if its graph $\{ (x, f(x)) : x \in A \} \subseteq M^{m+n}$ is definable. Whenever a particular structure $\mathfrak{M}$ is under consideration, we just say "definable".

The use of the word "definable" comes from a connection to first-order logic. Note the correspondence between the set-theoretic operations of complementation, union, intersection and projection, and the logical operations of negation,

---

[1] This section is partially based on a lecture given by the second author at HSCC'99 (Berg en Dal).

disjunction, conjunction and existential quantification. Closure under these logical operations is a very strong condition; see Appendices A and B of [17] for some examples of how this can be exploited.

It is crucial to understand that definability is always taken with respect to some particular structure. Whenever we have more than one structure under consideration, we must take care to avoid ambiguities.

Let $m, n \in \mathbb{N}$, $A \subseteq M^{m+n}$ and $B \subseteq M^m$. Then the parametrized family $(A_x)_{x \in B}$ is a *definable family* (of subsets of M^n) if A and B are definable. If, moreover, $f : A \to M^p$ is a map, then the parametrized family $(f(x, \cdot) : A_x \to M^p)_{x \in B}$ is called a definable family of maps if f and B are definable. (The definability of A follows from that of f, since A is the projection on the first $m + n$ coordinates of the graph of f.) Note that we can code up any finite collection of definable families as a single definable family (in the same structure).

There is a natural partial order on the class of all structures on M. Given structures $\mathfrak{M} = (\mathfrak{M}_n)$ and $\mathfrak{M}' = (\mathfrak{M}'_n)$ on M we put $\mathfrak{M} \subseteq \mathfrak{M}'$ if $\mathfrak{M}_n \subseteq \mathfrak{M}'_n$ for all $n \in \mathbb{N}$. If $\mathfrak{M} \subseteq \mathfrak{M}'$, then we say that (a) $\mathfrak{M}$ is a *reduct* of $\mathfrak{M}'$; (b) $\mathfrak{M}'$ is an *expansion* of $\mathfrak{M}$; or (c) $\mathfrak{M}'$ *expands* $\mathfrak{M}$.

Clearly, M has a largest structure on it: For each $n \in \mathbb{N}$, just let $\mathfrak{M}_n$ be the collection of all subsets of M^n. This is not a very interesting structure, but its existence is occasionally useful for theoretical purposes. There is also a smallest structure on M (also not very interesting). Usually, we are interested in structures that come equipped with some extra basic information.

Let $S_\beta \subseteq M^{n(\beta)}$ be sets (β in some index set J) and $f_\alpha : M^{n(\alpha)} \to M$ be functions (α in some index set I). A *structure on* $(M, (S_\beta), (f_\alpha))$ is a structure $\mathfrak{M}$ on M such that each f_α and each S_β is definable in $\mathfrak{M}$. Equivalently, we say that $\mathfrak{M}$ is an *expansion* of $(M, (S_\beta), (f_\alpha))$.

Given $S \subseteq M$, we say that A is S-*definable* (in $\mathfrak{M}$), or *definable with parameters from* S, if A is definable in $(\mathfrak{M}, (c)_{c \in S})$, that is, in the expansion of $\mathfrak{M}$ by constants for each $c \in S$. In the case $S = M$, we say that A is *parametrically definable*. Note that "definable" and "$\emptyset$-definable" mean the same thing. The distinction between "definable" and "parametrically definable" is often extremely important in model-theoretic statements and arguments.

In some branches of model theory, it has become more customary to use "definable" to mean "parametrically definable" (it's more convenient for analytic and geometric purposes). But when computation is at issue, this is irksome: We don't want to be involved with computing, say, arbitrary real numbers, and it doesn't seem to make sense to talk about decision procedures (or algorithms) that range over uncountable collections of sets. When consulting the literature, one must determine in which sense "definable" is being used; when using the notion, one must take care to use it consistently. [2]

Examples

− *Semilinear sets.* Let K be a subfield of $\mathbb{R}$ and V be a K-linear subspace of $\mathbb{R}$. For each $n \in \mathbb{N}$, let $\mathfrak{M}_n$ be the collection of all finite unions of sets of the

---

[2] This has been a problem in some earlier papers on hybrid systems.

form

$$\{\, x \in V^n : f_1(x) = \cdots = f_k(x) = 0, \ g_1(x) < 0, \ldots, g_l(x) < 0 \,\}$$

where each f_i and each g_j are affine K-linear maps $V^n \to V$. (If $K = \mathbb{Q}$, then we can take the coefficients of the maps to be integers.) It's routine to check that each $\mathfrak{M}_n$ satisfies (S1)–(S3). Verifying (S4) takes a bit more work, but it's not difficult; see e.g. pages 25–27 of [16].

– *Semialgebraic sets.* Let R be a real-closed ordered field ($\mathbb{R}$, for example). For each $n \in \mathbb{N}$, let $\mathfrak{M}_n$ be the collection of all finite unions of sets

$$\{\, x \in R^n : f(x) = 0, \ g_1(x) < 0, \ldots, g_l(x) < 0 \,\}$$

where f and each g_j are n-variable polynomial functions with coefficients from R. (If R is the field of real algebraic numbers, then we can take the coefficients to be integers.) It's again routine to check that each collection satisfies (S1)–(S3). That (S4) holds is due to A. Tarski [13]; for an interesting alternate proof, due to S. Łojasiewicz, see Ch. 2 of [16].

– *Subexponential sets.* For each $n \in \mathbb{N}$, let $\mathfrak{M}_n$ be the collection of all projections on the first n variables of sets $\{\, (x,y) \in \mathbb{R}^{n+k} : F(x,y) = 0 \,\}$, where $k \in \mathbb{N}$ and $F : \mathbb{R}^{n+k} \to \mathbb{R}$ is a function from the ring

$$\mathbb{Z}[x_1, \ldots, x_n, y_1, \ldots, y_k, e^{x_1}, \ldots, e^{x_n}, e^{y_1}, \ldots, e^{y_k}].$$

In this case, verifying (S2)–(S4), along with showing that these collections are closed under finite intersections, is the routine part. The (rather hard) work of showing that they are closed under complementation (hence also under finite unions) is due to A. Wilkie [18].

– *Finitely (or globally) subanalytic sets.* For each $n \in \mathbb{N}$, let $\mathfrak{M}_n$ be the collection of all subsets of $\mathbb{R}^n$ whose image under the map

$$(x_1, \ldots, x_n) \mapsto \left(\frac{x_1}{\sqrt{1+x_1^2}}, \ldots, \frac{x_n}{\sqrt{1+x_n^2}} \right) : \mathbb{R}^n \to (-1,1)^n$$

is subanalytic. Here, again, closure under complementation is the hard part. The result is essentially due to A. Gabrielov [2]; see also L. van den Dries [14]. For applications of subanalytic geometry in control theory, see e.g. [12].

2.2   The Bottom-up Approach

In this case, we are given the set M together with some functions on, and subsets of, various cartesian products, and we close off under definability.

For each $n \in \mathbb{N}$, let $\mathcal{P}_n$ be a (possibly empty) collection of subsets of M^n and $\mathcal{F}_n$ be a (possibly empty) collection of functions $M^n \to M$. Elements of $\mathcal{P}_n$ and $\mathcal{F}_n$ are sometimes called, respectively, *primitive* relations and functions, or just *primitives*. We now regard M as being equipped with these relations and functions, that is, we consider the *structure* $(\, M, (\mathcal{P}_n), (\mathcal{F}_n) \,)$ as an algebraic object. (We shall see that our use of the word "structure" for two formally different objects causes no trouble when we are concerned with definability.)

For each $n \in \mathbb{N}$, let $\mathcal{T}_n$ be the smallest set of functions on M^n such that: (a) $\mathcal{T}_n$ contains the coordinate projections $\pi_i : M^n \to M$ for $i = 1, \ldots, n$; and (b) for all $m \in \mathbb{N}$, $F \in \mathcal{F}_m$, and $f_1, \ldots, f_m \in \mathcal{T}_n$, we have $F \circ (f_1, \ldots, f_m) \in \mathcal{T}_n$.

We construct collections $\mathfrak{M}_{n,k}$ of subsets of M^n by induction on $k \in \mathbb{N}$. For $k = 0$ and $n \in \mathbb{N}$, let $\mathfrak{M}_{n,0}$ be the boolean algebra generated by the collection of all sets of the following forms:

$$\{ x \in M^n : f(x) = g(x) \}, \quad f, g \in \mathcal{T}_n$$
$$\{ x \in M^n : (f_1(x), \ldots, f_m(x)) \in P \}, \quad f \in \mathcal{T}_n, \ P \in \mathcal{P}_m, \ m \in \mathbb{N}$$

Assume that the stage k collections have been constructed. For $n \in \mathbb{N}$, let $\mathfrak{M}_{n,k+1}$ be the boolean algebra of subsets of M^n generated by $\mathfrak{M}_{n,k} \cup \{ \pi(A) : A \in \mathfrak{M}_{n+1,k} \}$.

For $n \in \mathfrak{N}$, put $\mathfrak{M}_n := \bigcup_{k \in \mathbb{N}} \mathfrak{M}_{n,k}$. It's easy to see that $(\mathfrak{M}_n)_{n \in \mathbb{N}}$ is the smallest structure, in the top-down sense, on $(M, (\mathcal{P}_n), (\mathcal{F}_n))$; so we just denote it by $(M, (\mathcal{P}_n), (\mathcal{F}_n))$ and call it the *structure on M generated by* $(\mathcal{P}_n), (\mathcal{F}_n)$. (Often, for convenience, we just list the primitives.) And, of course, we say that $A \subseteq M^n$ is definable in $(M, (\mathcal{P}_n), (\mathcal{F}_n))$ if $A \in \mathfrak{M}_n$.

Let $\mathfrak{M} = (\mathfrak{M}_n)$ be a structure on M in the top-down sense. Clearly, a set A is definable in $\mathfrak{M}$ in the top-down sense if and only if A is definable in $(M, (\mathfrak{M}_n))$ in the bottom-up sense.

Examples

- The sets definable in $(\mathbb{Q}, <, +, -, 0, 1)$ are the (rational) semilinear sets. Here, the symbol $-$ denotes the function $x \mapsto -x : \mathbb{Q} \to \mathbb{Q}$. The function $-$ and the constant 0 are definable in $(\mathbb{Q}, <, +, 1)$, but often we include them as primitives for convenience. On the other hand, 1 is not definable in $(\mathbb{Q}, <, +)$.
- If R is the set of all real algebraic numbers, then the sets definable in $(R, +, \cdot)$ are the (algebraic) semialgebraic sets. Each of $<, -, 0$ and 1 are definable:

$$x = 1 \Leftrightarrow \forall y [xy = y], \quad x < y \Leftrightarrow \exists z [z^2 (y - x) = 1] .$$

- The sets definable in $(\mathbb{R}, +, \cdot, (r)_{r \in \mathbb{R}})$ are the (real) semialgebraic sets.
- The sets definable in $(\mathbb{R}, +, e^x)$ are the subexponential sets. (Multiplication is definable from addition and exponentiation: For $a, b, c \in \mathbb{R}$, we have $ab = c$ if and only if there exist $x, y, z \in \mathbb{R}$ such that $e^x = a$, $e^y = b$, $e^z = c$, and $x + y = z$.)
- A set is finitely subanalytic if and only if it is definable in $(\mathbb{R}, +, \cdot, (f))$, where f ranges over all real-analytic functions $f : [-1, 1]^n \to \mathbb{R}$, n ranging over $\mathbb{N}$; see [14].

All of the examples given so far have the property that they have some explicit, fairly simple, top-down form arising visibly from some nice collection of primitives; it's probably fair to say that this is the exception rather than the rule.

For example, the structure $(\mathbb{R}, +, \cdot, \mathbb{Z})$ is extremely complicated, involving even set-theoretic independence issues; see, for example, Ch. V of [3]. So although the generating primitives are familiar, fairly simple (at least, seemingly so) objects, the descriptions of the definable sets become increasingly complicated logically.

The central issue in definability theory is the attempt to understand the definable sets generated from given sets of primitives.

2.3   O-minimality

Let $(M, <)$ be a dense linearly ordered set without endpoints, and let $\mathfrak{M}$ be an expansion of $(M, <)$. Note that every finite union of points and open intervals (with endpoints from $M \cup \{\pm\infty\}$) contained in M is parametrically definable in $\mathfrak{M}$. The structure $\mathfrak{M}$ is *o-minimal* (short for order-minimal) if every *parametrically* definable subset of M is a finite union of points and open intervals. For $M = \mathbb{R}$, this is the same as saying that every parametrically definable subset of $\mathbb{R}$ has finitely many connected components. For expansions of $(\mathbb{R}, <, +, 1)$, this is the same as requiring only that every *definable* subset of $\mathbb{R}$ have finitely many connected components;[3] see [8] for a proof.

Clearly, o-minimality is preserved downward: If $(M, <) \subseteq \mathfrak{M}' \subseteq \mathfrak{M}$ and $\mathfrak{M}$ is o-minimal, then so is $\mathfrak{M}'$.

Examples

– It's easy to check that a semilinear subset of $\mathbb{Q}$ is a finite union of points and intervals, so $(\mathbb{Q}, <, +)$ is o-minimal. On the other hand, $(\mathbb{Q}, <, +, \cdot)$ is not o-minimal: The definable set $\{x \in \mathbb{Q} : x^2 < 2\}$ is not a finite union of points and open intervals (with rational endpoints). It's not known if there are any *proper*—that is, strictly larger—o-minimal expansions of $(\mathbb{Q}, <, +)$.
– If R is a real-closed field, then $(R, +, \cdot)$ is o-minimal, since a semialgebraic subset of R is a finite union of points and open intervals.
– The structure on $\mathbb{R}$ consisting of all finitely subanalytic sets is o-minimal; see [14].
– Let $\mathfrak{R}$ be an o-minimal expansion of $(\mathbb{R}, <, +)$. Then the expansion of $\mathfrak{R}$ by exponentiation is o-minimal (hence so is the expansion of $\mathfrak{R}$ by multiplication); see Y. Peterzil *et al.* [10] and P. Speissegger [11].) In particular, $(\mathbb{R}, +, e^x)$ is o-minimal.

As a counterpoint to the last item above, o-minimal expansions of $(\mathbb{R}, <, +)$ that do not define multiplication are exceptional (in the sense that they have rather special properties) as are o-minimal expansions of $(\mathbb{R}, +, \cdot)$ that do not define exponentiation; see [9] for information.

O-minimal structures have so many nice properties that it takes pages to describe (let alone prove) them; we only touch on the subject here.

---

[3] This resolves some of the "definable versus parametrically definable" problems in some earlier papers on o-minimal hybrid systems.

Proposition 1 (Uniform Finiteness). *Let* $\mathfrak{M} = (M, <, \ldots)$ *be o-minimal and* $A \subseteq M^{m+n}$ *be parametrically definable. Then there exists* $N \in \mathbb{N}$ *such that for all* $x \in M^m$, *if* A_x *is finite, then* A_x *contains less than* N *elements.*

Proposition 2 (Definable Choice). *Let* $\mathfrak{M}$ *be an o-minimal expansion of a dense linearly ordered group* $(M, <, +, 0, 1)$ *with a distiguished element* $1 > 0$. *Let* $\emptyset \neq A \subseteq M^{m+n}$ *be definable. Then there is a definable map* $f : \pi(A) \rightarrow M^n$ *such that* $(x, f(x)) \in A$ *for all* $x \in \pi(A)$.

3 Main Results

Let $H = (H_p)_{p \in P}$ be a family of hybrid systems on M, with parametrized flow ϕ, where the manifold M is assumed to be contained in some cartesian product $\mathbb{R}^K$. Let $\mathfrak{R}(H)$ denote the structure $(\mathbb{R}, +, \cdot, I, \Gamma, \phi,)$. Throughout this section, "definable" means " definable in $\mathfrak{R}(H)$".

Proposition 3. Reach(H) *is definable, as is the set* Reach$^*(H)$, *consisting of all pairs of states* (x, y) *for which there exists* $p \in P$ *such that* y *is reachable from* x *in* H_p.

Proof. This is immediate since: (a) the parametrized flow is definable; (b) there exists $N \in \mathbb{N}$ such that for all $(p, x, y) \in$ Reach(H), y is reachable from x in H_p via a (hybrid) trajectory of length $\leq N$ (c) the state spaces are uniformly definable, and (d) Reach$^*(H)$ is the projection of Reach(H) on the last two variables.   □

Corollary 1. *If there is an algorithm for deciding membership questions for sets definable in* $\mathfrak{R}(H)$, *then the reachability problem for* H *is decidable.*

The first-order theory of real-closed fields (in the language of ordered rings) is decidable [13]; hence:

Corollary 2. *If* I, Γ *and* ϕ *are definable in* $(\mathbb{R}, +, \cdot)$, *then the reachability problem for* H *is decidable.*

Proposition 4 (Parameter Selection). *If* $\mathfrak{R}(H)$ *is o-minimal, then there is a definable map* $\Psi :$ Reach$^*(H) \rightarrow P$ *such that for all* $(x, y) \in$ Reach$^*(H)$, y *is reachable from* x *in* $H_{\Psi(x,y)}$.

Proof. Apply Definable Choice.   □

As mentioned in Section 1, we may include (parametrized) initial and final sets C_o, C_f in families of hybrid systems, and study reachability between them. Note that the set of all (p, x, y) such that $x \in (C_o)_p$, $y \in (C_f)_p$, and y is reachable from x in H_p is definable in $(\mathfrak{R}(H), C_o, C_f)$, the expansion of $\mathfrak{R}(H)$ by C_o and C_f.

Bisimulations and o-minimality. Suppose that $\mathfrak{R}(H)$ is equipped with initial and final states, and is o-minimal. Then, for each $p \in P$, each structure $\mathfrak{R}(H_p)$ is also o-minimal. By Theorem 5.3 of [4], each system H_p admits a finite bisimulation. But more is true: An examination of the proof—together with Uniform Finiteness—shows that the bisimulations associated to the H_p are obtained uniformly in p (including a uniform finite bound on the number of iterations needed in the bisimulation algorithm). As of this writing, we do not know of any applications of this observation that are not obtained more directly by standard o-minimal arguments, so we do not give the precise (rather technical) statement here. Further investigation of possible applications is in order.

Time-Abstraction and o-minimality. The time-abstract view of hybrid systems ignores all analytic-geometric properties of the continuous part of the hybrid trajectories, but some of the most powerful and striking applications of o-minimality are to the study of these properties; see e.g. [17].

4   Applications

We illustrate the results of the previous section by some examples, obtained by uniformizing some results from [5].

Throughout this section, "semialgebraic" means "definable in $(\mathbb{R}, +, \cdot)$" (arbitrary real constants *not* included).

Consider the parametrized family of hybrid systems $(H_p)_{p \in P}$ defined as follows. Let U be the set of m-tuples of $k \times k$ nilpotent matrices with real entries (a $k \times k$ matrix A is nilpotent if $A^k = 0$). Identify U with a semialgebraic subset of $(\mathbb{R}^{k^2})^m$. Put $H = (M, S, I, \Gamma, F)$ where each object is specified below.

- $S = \{(x, (1, 2, \dots, m), i, j) : x \in V, 1 \le i, j \le m\}$ where V is any semialgebraic subset of U. In particular, we use a fixed set of locations $(1, 2, \dots, m)$ and $P = V \times \{(1, 2, \dots, m)\}$.
- $M = \mathbb{R}^k$
- $I = P \times M^m$. That is, I is identified with a subset of $\mathbb{R}^{mk^2 + m + km}$.
- Γ is a semialgebraic set satisfying the conditions of Definition 1.
- For each $p = (a_{11}(1), \dots, a_{kk}(1), \dots, a_{11}(m), \dots, a_{kk}(m), (1, 2, \dots, m)) \in P$ and $x \in M$ define $F(p, x) = (A(1)x, \dots, A(m)x)$. So, each component of $F(p, \cdot)$ is a linear vector field with a nilpotent matrix.

Since the flows of the vector fields defined above consist of polynomial functions in t with semialgebraic coefficients, such flows are themselves semialgebraic.

By Proposition 3, there is a semialgebraic function Φ such that for any choice of nilpotent matrices $A_1, \dots, A_m$ satisfying $(A_1, \dots, A_m) \in V$ and any two points x, y in $\{1, \dots, m\} \times \mathbb{R}^k$, y is reachable from x in the corresponding hybrid system if and only if $\Phi(p, x, y) = 0$ (where $p = (a_{11}(1), \dots, a_{kk}(1), \dots, a_{11}(m), \dots, a_{kk}(m), (1, 2, \dots, m)))$.

As another example, consider a finite set $\mathcal{D}$ of diagonal matrices with rational entries and let $\mathcal{A}$ be a semialgebraic subset of $GL_k(\mathbb{R})$. Let all data be as in the previous example except that we replace V with the set of (diagonalizable) matrices of the form $T\Lambda T^{-1}$, where $T \in \mathcal{A}$ and $\Lambda \in \mathcal{D}$. In this case Reach(H) is definable in $(\mathbb{R}, +, \cdot, \exp)$. Moreover, it was shown in [5] that Reach(H) is in fact semialgebraic. Also, a kind of converse is true: If Reach(H) is semialgebraic, and the matrices involved are diagonalizable, then, up to scaling, the set $\mathcal{D}$ must be finite. Using results from [6,7], this is not difficult to prove, but it would take us too far afield to do it here.

The set Reach(H) may be definable in an o-minimal structure even if the parametrized flow is not. Consider a system as above but where the matrices are similar to one with purely imaginary eigenvalues (with rational imaginary part) and of a special real Jordan form (having 2×2 blocks). Then the flows are complete and periodic and so the system is not o-minimal. On the other hand, using the calculations in §5.3 of [5] one can show that Reach(H) is, in fact, semialgebraic.

Conclusion

In this paper, we introduced parametrized families of hybrid systems and obtained some reachability results uniformly in parameters.

While some decidability results are available for special classes of systems, they all rely on the decidability of the theory of real closed fields. New interesting results may be obtained by finding classes of formulas decidable within the theory of the field of reals with exponentiation.

References

1. R. Alur and D. Dill, *A theory of timed automata*, Theoret. Comput. Sci. **126** (1994), 183–235.
2. A. Gabrielov, *Projections of semi-analytic sets*, Funct. Anal. Appl. **2** (1968), 282–291.
3. A. Kechris, *Classical Descriptive Set Theory*, Springer-Verlag, 1995.
4. G. Lafferriere, G. Pappas, and S. Sastry, *O-minimal hybrid systems*, Math. Control Signals Systems, to appear.
5. G. Lafferriere, G. Pappas, and S. Yovine, *A new class of decidable hybrid systems*, Hybrid Systems : Computation and Control (F. Vaandrager and J. van Schuppen, eds.), Lecture Notes in Comput. Sci., vol. 1569, Springer-Verlag, Berlin, 1999, pp. 137–151.
6. C. Miller, *Expansions of the real field with power functions*, Ann. Pure Appl. Logic **68** (1994), 79–94.
7. C. Miller, *Exponentiation is hard to avoid*, Proc. Amer. Math. Soc. **122** (1994), 257–259.
8. C. Miller and P. Speissegger, *Expansions of the real line by open sets: o-minimality and open cores*, Fund. Math. **162** (1999), 193–208.
9. C. Miller and S. Starchenko, *A growth dichotomy for o-minimal expansions of ordered groups*, Trans. Amer. Math. Soc. **350** (1998), 3505–3521.

10. Y. Peterzil, P. Speissegger, and S. Starchenko, *Adding multiplication to an o-minimal expansion of the additive group of real numbers*, Logic Colloquium '98, to appear.
11. P. Speissegger, *The Pfaffian closure of an o-minimal structure*, J. Reine Angew. Math. **508** (1999), 189–211.
12. H. Sussmann, *Subanalytic sets and feedback control*, J. Differential Equations **31** (1979), 31–52.
13. A. Tarski, *A decision method for elementary algebra and geometry*, second ed., University of California Press, 1951.
14. L. van den Dries, *A generalization of the Tarski-Seidenberg theorem, and some nondefinability results*, Bull. Amer. Math. Soc. (N.S) **15** (1986), 189–193.
15. L. van den Dries, *o-Minimal structures*, Logic: From Foundations to Applications, Oxford Sci. Publ., Oxford University Press, New York, 1996, pp. 137–185.
16. L. van den Dries, *Tame Topology and O-minimal Structures*, London Math. Soc. Lecture Note Ser., vol. 248, Cambridge University Press, 1998.
17. L. van den Dries and C. Miller, *Geometric categories and o-minimal structures*, Duke Math. J. **84** (1996), 497–540.
18. A. Wilkie, *Model completeness results for expansions of the ordered field of real numbers by restricted Pfaffian functions and the exponential function*, J. Amer. Math. Soc. **9** (1996), 1051–1094.

On the Existence of Solutions
to Controlled Hybrid Automata

Michael Lemmon

University of Notre Dame
Dept. of Elect. Eng., Notre Dame, IN 46556, USA
lemmon@maddog.ee.nd.edu

Abstract. This paper studies the existence of solutions to a class of
hybrid automata in which the underlying continuous dynamics are represented by inhomogeneous linear time-invariant systems whose inputs
are *controls* that can be determined by the user. The principal result of
the paper is a procedure that searches for global periodic non-terminating
solutions of systems having a single cycle.

1 Introduction

A *controlled hybrid automaton* is a hybrid automaton [Alu93] [Lyn96] whose underlying continuous-state dynamics are modeled as inhomogeneous differential
equations. In particular, we restrict our attention to continuous-dynamics represented by linear time-invariant (LTI) systems of the form $\dot{x}(t) = Ax(t) + Bu(t)$
where $A \in \Re^{n \times n}$, $x(t) \in \Re^n$, $B \in \Re^n$, and $u(t) \in \Re$. The scalar $u(t)$ is the *control input* at time $t \in \Re$ and it is selected by the system designer. In this paper,
we further restrict our attention to systems with only a single cycle. This paper
presents preliminary work examining conditions under which non-chattering and
non-terminating solutions exist for the controlled hybrid automaton. The principal result is a gradient-following algorithm that provides a systematic means
of searching for global periodic non-terminating solutions of systems with single
cycles.

The remainder of the paper is organized as follows. Section 2 defines the
controlled hybrid automaton and defines the sense in which a hybrid trajectory
satisfies such a system. Section 3 outlines conditions for the existence of local
non-chattering solutions. Section 4 outlines conditions for global periodic non-
terminating system trajectories. Final remarks are found in section 5.

2 Controlled Hybrid Automata

A *controlled hybrid automaton* is a labeled digraph characterized by the 4-tuple
(N, A, ℓ_N, ℓ_A). N is a set of *nodes* in the directed graph (represented graphically
as open circles). The set of nodes is usually taken as a subset of the positive
integers. $A \subset N \times N$ is a set of *directed arcs* between nodes. The arc (i, j) from
node i to node j is graphically represented as an arrow that starts at node i and

N. Lynch and B. Krogh (Eds.): HSCC 2000, LNCS 1790, pp. 229–242, 2000.
© Springer-Verlag Berlin Heidelberg 2000

terminates at node j. The ordered pair (N, A) is the *finite automaton* associated with the hybrid system. The map $\ell_N : N \times \Re^{n \times n} \times \Re^n$ associates a pair of real vectors with the node. In particular, the label $\ell_N(i) = (A_i, B_i)$ associates a real matrix $A_i \in \Re^{n \times n}$ and a real matrix (vector) $B_i \in \Re^n$ with the ith node. Associated with node i is the following inhomogeneous differential equation,

$$\dot{x}(t) = A_i x(t) + B_i u(t) \tag{1}$$

where $x(t) \in \Re^n$ and $u(t) \in \Re$. Equation 1 is called the *modal equation* of the ith node. The map $\ell_A : A \to \mathcal{P}(\Re^n)$ maps an arc $a_1 \in A$ onto a collection of vectors in $\Re^n$. In particular, if arc a_1 is labeled as

$$\ell_A(a_1) = \{v_{11}, v_{12}, \cdots, v_{1p_1}\}$$

then we can associate with a_1 a special subset $\Gamma(a_1) \subset \Re^n$ that is called the *guard* of the arc. The *guard* is defined to be the convex hull of the points in the collection $\ell_A(a_1)$. By the standard representation theorems for convex sets, we therefore know that $\Gamma(a_1)$ can be characterized as

$$\Gamma(a_1) = \left\{ x = \sum_{i=1}^{p_1} \lambda_{1i} v_{1i} \ : \ \sum_{i=1}^{p_1} \lambda_{1i} = 1 \ , \ \lambda_{1i} \geq 0 \ , \ v_{1i} \in \ell_A(a_1) \right\}$$

From the above equation it should be clear that the vectors in $\ell_A(a_1)$ are the extreme points (vertices) for convex polytope $\Gamma(a_1)$.

Remark: In this paper we've adopted the convention of representing guards as convex combinations of vertices, rather than as feasible regions bounded by hypersurfaces.

A *controlled hybrid trajectory* $z : \Re \to X \times N \times U$ is a function mapping a real number $\tau \in \Re$ onto the ordered triple $(x(\tau), i(\tau), u(\tau))$ where $x(\tau) \in X \subset \Re^n$ is called the the *continuous state*, $i(\tau) \in N$ is called the *discrete state*, and $u(\tau) \in U \subset \Re$ is the *control*. It is assumed that X is a closed connected subset of $\Re^n$ and it is assumed that U is a compact subset of $\Re$.

A time instant $\tau \in \Re$ is said to be *regular* if z is continuous at τ. (In this case, we assume that N is equipped with a discrete metric $d(i, j) = 1$ if $i \neq j$ and is zero if $i = j$). If τ is not a regular point, then it is called a *switching instant*. Controlled hybrid trajectories with a finite number of switching instants in any closed time interval are said to be *non-chattering*. A controlled trajectory with an infinite number of switching instants is said to be *non-terminating*. The trajectory is said to be *local* if its maximum interval of existence has the form $[\tau_0, \tau_0 + T)$ and T is finite. The trajectory is said to be *global* if its maximum interval of existence of $[\tau_0, \infty)$.

A controlled hybrid trajectory $z : [\tau_0, \tau_0 + T) \to X \times N \times U$ is said to *satisfy* the controlled hybrid automaton (N, A, ℓ_N, ℓ_A) with initial condition $x_0 \in X$ and $i_0 \in N$ at time $\tau_0 \in \Re$ if and only if

- $x(\tau_0) = x_0$, $i(\tau_0) = i_0$, and $u(\tau_0) \in U$.
- For all closed intervals $[\tau_a, \tau_b]$ containing *no* switching instant, there exists a $j \in N$, an absolutely continuous [Aub84] trajectory $x : [\tau_a, \tau_b] \to X$, and a measurable control $u : [\tau_a, \tau_b] \to U$ such that $i(\tau) = j$ and $\dot{x}(\tau) = A_j x(\tau) + B_j u(\tau)$ for all $\tau \in [\tau_a, \tau_b]$.
- At any switching instant, $\tau_s \in \Re$, there exists a j and k in N such that $(j, k) \in A$, $\lim_{\tau \to \tau_s^-} i(\tau) = k$, and $x(\tau_s) \in \Gamma((j, k))$.

Such trajectories are also said to be *solutions* of the hybrid automaton. A system that can generate non-chattering solutions will be said to be *non-Zeno*. A system that can generate non-terminating solutions will be said to be *deadlock-free*.

Remark: Note that switching can occur anywhere within the guard set.

Consider a controlled trajectory z defined over $[\tau_0, \tau_0 + T)$ with discrete state trajectory $i : [\tau_0, \tau_0 + T) \to N$. The sequence of discrete states associated with i can be denoted by the string $\sigma \in N^*$ where N^* is the Kleene closure of N. We refer to σ as the trajectory's *event sequence*. By the pumping lemma [Dav83] , we know any finite length event sequence can be decomposed as $\sigma = usv$ such that the event sequence $us^n v$ (for any positive n) is accepted by the *finite automaton* (N, A) associated with our system. This means that the sequence s represents a *cycle* of events. If there exist trajectories such that the *hybrid automaton* can execute this cycle repeatedly, then we say that the hybrid automaton is deadlock-free with respect to s. A key issue in the study of hybrid automata (whether or not they are controlled) concerns the deadlock-freedom of such systems. This issue is, in essence, a question concerns the existence of global non-terminating solutions to hybrid automata.

3   Local Non-chattering Solutions

Figure 1 shows a cyclic controlled hybrid automaton. Assume that the initial continuous state at time τ_0 is x_0 and that the initial discrete state is $i_0 = 1$. In this section, we briefly examine conditions ensuring the existence of a $T > 0$ such that there exists a controlled hybrid trajectory z over the interval $[\tau_0, \tau_0 + T)$ that is a solution to the controlled hybrid automaton. In this section, we consider two distinct cases. The first case occurs when x_0 is not in $\Gamma((1, 2))$. The second case occurs when $x_0 \in \Gamma((1, 2))$.

The following results are a routine application of *viability theory* [Aub84] and are presented here for the sake of completeness. See [Aub84] for a precise statement of the definitions and theorems cited below.

Let's assume that $x_0 \notin \Gamma((1, 2))$. Since the guards are closed sets, this means that x_0 belongs to an open set so we can enclose x_0 in an open neighborhood $B_\epsilon(x_0)$ that is contained completely within the complement of the two guards. Over this neighborhood we can define a set valued mapping $F : X \to \mathcal{P}(\Re^n)$ that takes the value

$$F(x) = \{A_1 x + B_1 u \ , \ u \in U\} \tag{2}$$

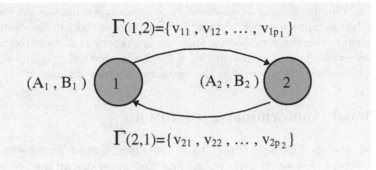

Fig. 1. cyclic controlled hybrid automata

at x. Since U is compact, we know that $F(x)$ will be an upper semi-continuous (USC) convex-valued map and we can use the USC Existence Theorem to infer that there exists a $\delta > 0$ and an absolutely continuous $x : [\tau_0, \tau_0 + \delta) \to X$ that satisfies the differential inclusion $\dot{x} \in F(x)$. By the Measurable Selection Theorem we can infer the existence of a measurable u over the same interval. Since there are no switching instants over this time interval, we know that $i(\tau) = 1$ for all $\tau \in [\tau_0, \tau_0 + \delta)$. This particular case, therefore, has a hybrid trajectory satisfying the system.

The other major case of interest occurs when $x_0 \in \Gamma((1,2))$. Let's first consider the case when $x_0 \in \Gamma((1,2))$ and $x_0 \notin \Gamma((2,1))$. We consider the set valued map, $F(x)$, of equation 2. Since x_0 may lie on the boundary of $\Gamma((1,2))$, we cannot enclose x_0 in an open neighborhood over which F is defined. However, F is upper semicontinuous and provided we can ensure that F satisfies the *tangential condition* [Aub84], then we can use the Viability Theorem to ensure the existence of of an absolutely continuous solution to the differential inclusion $\dot{x} \in F(x)$ that is viable in $\Gamma((1,2))$. Finally, the Measurable Selection Theorem ensures the existence of the desired Lebesgue measurable control, u.

Now let's consider the case when $x \in K = \Gamma((1,2)) \cap \Gamma((2,1))$ Consider a set valued map, F, that takes the value

$$F(x) = \{A_i x + B_i u \ : \ u \in U, i = \{1,2\}\}$$

at point $x_0 \in K$. Since K is compact, then x_0 may lie on the boundary of K and cannot be enclosed in an open neighborhood over which F is defined. Morever, F may not be upper semicontinuous over K. Therefore we cannot use the USC Existence Theorem to establish the existence of local trajectories. However, the convex hull $\overline{co}(F(x))$ of $F(x)$ is clearly convex valued and Lipschitzean on K. Moreover, if we can ensure that the tangential condition holds, then the Viability Theorem ensures the existence of an absolutely continuous solution to $\dot{x} \in \overline{co}(F(x))$ that is viable in K. The Relaxation Theorem can then be used to infer the existence of absolutely continuous solutions to the original differential inclusion $\dot{x} \in F(x)$. As before, an application of the Measurable Selection Theorem ensures the existence of a Lebesgue measurable u for this selected trajectory x. Finally, since x is absolutely continuous, we know that each closed interval has a finite number of switching instants in which the discrete state changes value thereby establishing that the trajectory is non-chattering. We've therefore established the existence of a local non-chattering solution provided the tangential condition found in the viability thoerem is satisfied.

4   Global Nonterminating Solutions

A global nonterminating solution to a controlled hybrid automaton is a hybrid trajectory that exists over $[\tau_0, \infty)$ and that generates an infinite number of switching instants. This section studies the existence of global non-terminating periodic trajectories for the cyclic controlled hybrid automaton in figure 1. This hybrid automaton consists of two nodes (1 and 2) and two arcs. The ith node is labeled with the system (A_i, B_i) and arcs $(1,2)$ and $(2,1)$ are labeled with vertex collections $\mathbf{V}_1 = \{v_{11}, v_{12}, \cdots, v_{1p_1}\}$ and $\mathbf{V}_2 = \{v_{21}, v_{22}, \cdots, v_{2p_2}\}$, respectively. The ith guard associated with the arc entering the ith node denoted as $\Gamma_i = \overline{\mathrm{co}}(\mathbf{V}_i)$.

Consider one of the modal systems ($i = 1$ or 2)

$$\dot{x}(t) = A_i x(t) + B_i u(t) \tag{3}$$

where $x(t) \in \Re^n$, $A_i \in \Re^{n \times n}$, $B_i \in \Re^n$, and $u(t) \in \Re$. We say a state $v \in \Re^n$ is *reachable* from a state $w \in \Re^n$ if there exists a time $T > 0$ and a measurable control $u : [\tau_0, \tau_0+T] \to U$ such that the controlled trajectory $x : [\tau_0, \tau_0+T] \to X$ satisfies equation 3 with $x(\tau_0) = w$ and $x(\tau_0 + T) = v$. The set of all points from which v is reachable is called the *preset* of v and will be denoted as pre(v). The preset of a subset $\Gamma \subset X$ is denoted as pre(Γ) and is defined by the equation pre(Γ) = $\bigcup_{v \in \Gamma}$ pre(v). A necessary and sufficient condition [Ant97] for w to lie in the preset of v is that there exist a $T > 0$ such that

$$e^{A_i T} w - v \in \mathcal{R}(\mathcal{C}_i) \tag{4}$$

where

$$\mathcal{C}_i = \begin{bmatrix} B_i & A_i B_i & A_i^2 B_i & \cdots & A_i^{n-1} B_i \end{bmatrix} \tag{5}$$

is called the *controllability matrix* for the ith modal system. The range space of $\mathcal{C}_i$ is denoted as $\mathcal{R}(\mathcal{C}_i)$ and we assume it has a dimension of r_i. In the following discussion, $\mathbf{E}_i$ is a matrix of dimension $n \times r_i$ ($i = 1, 2$) whose columns are standard basis vectors for the subspace $\mathcal{R}(\mathcal{C}_i)$.

Remark: Note that this reachability condition applies when the control u can be unbounded (as is the case in so-called impulsive controls).

Remark: Note that the term *reachability* is used in a somewhat different sense that what is found in traditional algorithmic verification [Alu95]. Traditional hybrid automata have homoegeneous modal equations and as a result pre(v) for a fixed transition time T consists of a single point. In view of equation

4, it is apparent that the introduction of the control extends the preset of v to a set formed from affine varieties of the controllability subspace.

We assume an initial condition $x_0 \in X$ and $i_0 = 1$. The question to be answered is whether there exists a control u and a pair of switching times T_1 and T_2 such that a hybrid trajectory z is a solution of the system over the interval $[\tau_0, \infty)$ and such that z generates an infinite sequence of switching instants

$$\tau_0, \tau_{11}, \tau_{21}, \tau_{21}, \tau_{22}, \cdots, \tau_{ij}, \cdots$$

where τ_{ij} is the jth switching instant out of mode i, $\tau_{2j} - \tau_{1j} = T_1$, and $\tau_{1,j+1} - \tau_{2j} = T_2$. In other words, the hybrid trajectory z is nonterminating and periodic in time.

By our definition of a solution to a controlled hybrid automaton, we know that the continuous state at each switching instant must lie in the appropriate guard set. In other words $x(\tau_{ij}) \in \Gamma_i$ for all j and $i = 1, 2$. Since the guards are convex polytopes, the switching instants $x(\tau_{ij})$ can be represented as convex combination of the form

$$x(\tau_{1j}) = \sum_{i=1}^{p_1} \lambda_{1i} v_{1i}$$

$$x(\tau_{2j}) = \sum_{i=1}^{p_2} \lambda_{2i} v_{2i}$$

where $\lambda_{ij} \geq 0$ for $i = 1, 2$ and all j and where $\sum_{i=1}^{p_j} \lambda_{ij} = 1$ for $j = 1, 2$. Therefore if we are to have a nonterminating behavior, we know that $x(\tau_{2j})$ must be reachable from $x(\tau_{1j})$ in time T_1 and $x(\tau_{1,j+1})$ is reachable from $x(\tau_{2j})$ in time T_2. From equation 4, this condition is satisfied if there exist vectors $\overline{\beta}_1 = [\beta_{11}, \beta_{12}, \cdots, \beta_{1r_1}]^T$ and $\overline{\beta}_2 = [\beta_{21}, \beta_{22}, \cdots, \beta_{2r_2}]^T$ such that

$$0 = \sum_{i=1}^{r_1} \beta_{1i} e_{1i} + e^{A_1 T_1} \sum_{i=1}^{p_1} \lambda_{1i} v_{1i} - \sum_{i=1}^{p_2} \lambda_{2i} v_{2i} \tag{6}$$

$$0 = \sum_{i=1}^{r_2} \beta_{2i} e_{2i} - \sum_{i=1}^{p_1} \lambda_{1i} v_{1i} + e^{A_2 T_2} \sum_{i=1}^{p_2} \lambda_{2i} v_{2i} \tag{7}$$

$$1 = \sum_{i=1}^{p_1} \lambda_{1i} \tag{8}$$

$$0 \geq \lambda_{1i} \;, (i = 1, \ldots, p_1) \tag{9}$$

$$1 = \sum_{i=1}^{p_2} \lambda_{2i} \tag{10}$$

$$0 \geq \lambda_{2i} \;, (i = 1, \ldots, p_2) \tag{11}$$

We reframe equations 6, 7, 8, and 10 as the matrix vector equation

$$\mathbf{c} = \mathbf{S}\overline{\eta} \tag{12}$$

$$
\begin{bmatrix} \mathbf{0}_{2n \times 1} \\ 1 \\ 1 \end{bmatrix} = \left[\begin{array}{cc|cc} \mathbf{E}_1 & \mathbf{0}_{n \times r_2} & e^{A_1 T_1} \mathbf{V}_1 & -\mathbf{V}_2 \\ \mathbf{0}_{n \times r_1} & \mathbf{E}_2 & -\mathbf{V}_1 & e^{A_2 T_2} \mathbf{V}_2 \\ \hline \mathbf{0}_{1 \times r_1} & \mathbf{0}_{1 \times r_2} & \mathbf{1}_{1 \times p_1} & \mathbf{0}_{1 \times p_2} \\ \mathbf{0}_{1 \times r_1} & \mathbf{0}_{1 \times r_2} & \mathbf{0}_{1 \times p_1} & \mathbf{1}_{1 \times p_2} \end{array} \right] \begin{bmatrix} \overline{\beta}_1 \\ \overline{\beta}_2 \\ \overline{\lambda}_1 \\ \overline{\lambda}_2 \end{bmatrix} \tag{13}
$$

$$
= \left[\mathbf{G}^T \,|\, \mathbf{F}^T \right] \begin{bmatrix} \mathbf{z} \\ \mathbf{y} \end{bmatrix} \tag{14}
$$

where $\mathbf{V}_1 = [v_{11}, v_{12}, \cdots, v_{1p_1}]$, $\mathbf{V}_2 = [v_{21}, v_{22}, \cdots, v_{2p_2}]$ (matrices whose columns are the guard vertices) , $\mathbf{c} = [\mathbf{0}_{n \times 1} \mid \mathbf{1}_{2 \times 1}]^T$, $\overline{\eta} = [\mathbf{z}^T, \mathbf{y}^T]^T$, $\mathbf{z} = [\beta_1, \cdots, \beta_r]^T$, and $\mathbf{y} = [\overline{\lambda}_1^T, \overline{\lambda}_2^T]^T$.

Remark: The vectors $\overline{\lambda}_1$, $\overline{\lambda}_2$, $\overline{\beta}_1$, and $\overline{\beta}_2$ satisfying equations 6 to 11 characterize affine spaces which are mutually reachable from each other. Note that these solutions provide an explicit characterization of mutually reachable presets in terms of the vertices of the guards. This explicit representation of the presets of the system is the reason why the guards were represented as convex combinations of vertices.

By the theorem of the alternative [Baz93], a necessary and sufficient condition for equations 6 to 11 to have a non-negative solution is that there exist no vector $\mathbf{x}$ such that

$$
\mathbf{G}\mathbf{x} \leq 0 \ , \quad \mathbf{F}\mathbf{x} = 0 \ , \quad \mathbf{c}^T \mathbf{x} > 0 \tag{15}
$$

The solution to equation 15 can be checked by solving the associated linear program

$$
\begin{aligned}
\textbf{maximize}: \ & \mathbf{c}^T \mathbf{x} \\
\textbf{subject to}: \ & \mathbf{G}\mathbf{x} \leq 0 \\
& \mathbf{F}\mathbf{x} = 0
\end{aligned} \tag{16}
$$

Solutions to the above problem have a special form due to the fact that $\mathbf{G}\mathbf{x}$ forms a polytopic cone whose apex is at the origin. Figure 2 shows the possible situations that can occur with this linear program. The figure shows that solutions to this linear program are either unbounded and positive or bounded and equal to zero. If the solution is $\mathbf{x} = 0$, then the alternative problem in equation 15 has no solution since $\mathbf{c}^T \mathbf{x} = 0$. This means that equation 12 has a non-negative solution and we can infer that for the fixed time T that the guard Γ is reachable from $\overline{co}(\mathbf{W})$. If an unbounded solution occurs then equation 12 has no non-negative solutions and we can infer that for the given T, the guard Γ is not directly reachable from $\overline{co}(\mathbf{W})$.

Remark: A feasible solution at $\mathbf{x} = 0$ implies that the specified cycle exists between the two guard sets and an unbounded solution implies that a cycle does not exist with the specified transition times T_1 and T_2. Note that the existence of an unbounded solution does not imply that the guards don't support a recurrent cycle, for there may be other transition times for which the cycle exists and it may be possible that the guards support a cycle in which the transitions are not necessarily periodic.

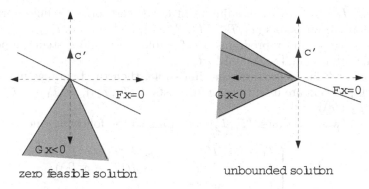

zero feasible solution   unbounded solution

The alternative problem 's linear program must have a solution
lying in the nullspace of F will either be unbounded or lie at
the apex of the cone formed by the equation G x < 0.

Fig. 2. The Alternative Problem's Linear Program

If the linear program in equation 16 returns an unbounded solution, then
it may be possible to adjust the transition times T_1 and T_2 to force a solution
at $\mathbf{x} = 0$. From duality theory [Baz93], we know that if the primal problem in
equation 16 is unbounded, then its dual is infeasible. The infeasibility of the dual
can be readily checked by examining the Lagrange multipliers associated with
the inequality constraints of the primal problem. These Lagrange multipliers are
generated by any primal-dual linear programming algorithm. They represent
part of the solution to the dual problem and are used to help assess how close
a linear programming algorithm is to being finished. If these multipliers are
negative, then the dual is infeasible and we can immediately conclude that the
guard does not support a cycle at the specified transition times.

The preceding observation suggests a simple heuristic method for adjusting
the times T_1 and T_2 in order to force the dual problem to be feasible. Let ν_k denote the kth Lagrange multiplier associated with the linear program's inequality
constraints. We define a performance measure associated with a specific pair of
times (T_1, T_2) as

$$J((T_1, T_2)) = \min_k \nu_k \tag{17}$$

This measure identifies the smallest Lagrange multiplier and uses it as a measure
of how close the dual problem is to being feasible. The obvious strategy is to
perturb the current transition times T_1 and T_2, observe the change in J and then
select a new set of times that will increase J. We continue in this manner until
J becomes positive.

This idea was tested using the following, very simple, search strategy. First
initialize the search by selecting a set of times T_1 and T_2. The search is then
executed by the following steps.

1. Perturb (T_1, T_2) by a small adjustment $\delta > 0$ and solve the linear program (equation 16) for points (T_1, T_2), $(T_1, T_2 + \delta)$, and $(T_1 + \delta, T_2)$, .
2. If any of these linear programs are feasible, then the system supports a periodic solution and we're finished.
3. If all of these linear programs are infeasible, then the Lagrange multipliers for each problem are used to compute costs $J((T_1, T_2))$, $J((T_1 + \delta, T_2))$, and $J((T_1, T_2 + \delta))$.
4. Select a new set of times, (T_1', T_2') according to the following rule,

$$T_1' = \begin{cases} T_1 + \delta, & \text{if } J((T_1, T_2)) < J((T_1 + \delta, T_2)) \\ T_1 - \delta, & \text{if } J((T_1, T_2)) > J((T_1 + \delta, T_2)) \\ T_1, & \text{otherwise} \end{cases}$$

$$T_2' = \begin{cases} T_2 + \delta, & \text{if } J((T_1, T_2)) < J((T_1, T_2 + \delta)) \\ T_2 - \delta, & \text{if } J(T_1, T_2)) > J((T_1, T_2 + \delta)) \\ T_2, & \text{otherwise} \end{cases}$$

5. Set $T_1 = T_1'$, $T_2 = T_2'$ and return to step 1.

What this algorithm does is attempt to solve a nonlinear optimization problem using a gradient-following strategy. The preceding steps describe the master algorithm that uses the results of the linear program in equation 16 to select a set of better times.

Remark: In the procedure we've chosen, of course, there are no guarantees that this search will terminate as it is currently unclear how the times, T_1 and T_2 are related to the problem's Lagrange multipliers. Nonetheless, this search program provides what seems to be a very pragmatic method for testing for the existence of global solutions and if it does terminate, then we know for certain that the cycle is live.

The following example illustrates the proposed search algorithm. Consider the cyclic hybrid automaton shown in figure 1 where the nodes are labeled as

$$\ell_N(1) = \left(\begin{bmatrix} 0 & 4 \\ 1/4 & 0 \end{bmatrix}, \begin{bmatrix} 4 \\ 1 \end{bmatrix} \right)$$

$$\ell_N(2) = \left(\begin{bmatrix} 0 & -10 \\ -1/10 & 0 \end{bmatrix}, \begin{bmatrix} 10 \\ 1 \end{bmatrix} \right)$$

The arcs are labeled with vertex collections

$$\mathbf{V_1} = \left\{ \begin{bmatrix} -2 \\ 0 \end{bmatrix}, \begin{bmatrix} -3 \\ 0 \end{bmatrix}, \begin{bmatrix} -3 \\ 1 \end{bmatrix} \right\}$$

$$\mathbf{V_2} = \left\{ \begin{bmatrix} 2 \\ 0 \end{bmatrix}, \begin{bmatrix} 3 \\ 0 \end{bmatrix}, \begin{bmatrix} 3 \\ 1 \end{bmatrix} \right\}$$

A MatLab script was written to implement the master program given above and this script was used to search for a global non-terminating solution of our hybrid automaton. The lefthand plot in figure 3 illustrates the results of this search. The x-axis shows the times T_1 and T_2 whereas the y-axis shows the value of the

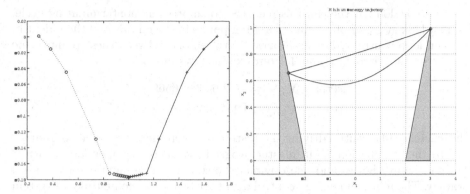

Fig. 3. Performance Measure J versus Switching Intervals and Simulation Results

cost $J((T_1, T_2))$. The search starts with $T_1 = T_2 = 1$. This starting point is in the middle of the x-axis. The intermediate values of J computed by the master algorithm are shown by the solid and dotted line trajectories (the dotted line for T_1 and the solid line for T_2). We see that the master algorithm computes a monotone sequence of times in which T_2 is decreasing and T_1 is increasing. After a finite number of iterations the master program has identified the times $T_2 = 0.29$ and $T_1 = 1.7$ as points whose linear programs have non-negative Lagrange multipliers. These points, therefore, are feasible and characterize a global non-terminating periodic cycle for this system.

The master algorithm allows us to assert that a global periodic solution to this system exists. The intermediate results of the algorithm also allow us to characterize the switching sets and we can actually identify some of the control strategies, u, that enforce this periodic solution. This additional information is contained in all non-negative solution vectors $\overline{\lambda}_1$ and $\overline{\lambda}_2$ satisfying our system $\mathbf{S}\overline{\eta} = \mathbf{c}$. The set of all solutions can be parameterized as $\overline{\eta} \in \overline{\eta}_{p0} + \text{null}(\mathbf{S})$ where η_{p0} is a particular solution to the inhomogeneous equation $\mathbf{c} = \mathbf{S}\overline{\eta}$. Note that this implies that the mutually reachable sets in the guards are affine sets. It was our parametrization of the guard as a convex combination of vertices that allowed us to obtain such a simple and explicit representation of these sets. For the example above, we can readily identify these sets in which the particular solution is $\overline{\eta}_{p0} = \begin{bmatrix} 0 & -1.94 & 1.39 & -1.10 & 0.71 & 0 & 0.048 & 0.95 \end{bmatrix}^T$ and the null space of $\mathbf{S}$ is spanned by the columns of the matrix

$$N = \begin{bmatrix} -0.9237 & 0 \\ 0.1380 & -0.4162 \\ 0.1820 & -0.3445 \\ -0.2321 & 0.1920 \\ 0.0501 & 0.1525 \\ -0.1536 & -0.6431 \\ 0.1144 & 0.4389 \\ 0.0393 & 0.2042 \end{bmatrix}$$

Let's now look at the controls required to enforce the nonterminating cycle. We first look at a pair of specific switching points in the guards and then identify an open loop control enforcing a periodic trajectory between these points. One specific set of switching points for our example system is

$$x(\tau_1) = \mathbf{V}_1\bar{\lambda}_1 = \begin{bmatrix} -2.6569 & 0.6569 \end{bmatrix}$$
$$x(\tau_2) = \mathbf{V}_2\bar{\lambda}_2 = \begin{bmatrix} 3 & 0.9903 \end{bmatrix}$$

The existence of a control driving the system between these two points is guaranteed by the termination of our master program. What is this open loop control? We have many choices and one obvious choice is the minimum energy control strategy. The minimum energy control $u(t)$ that transfers the first modal system from the initial state $x(\tau_1)$ to target state $x(\tau_2)$ satisfies the condition

$$x(\tau_2) - e^{A_1 T_1} x(\tau_1) \in \mathcal{R}(\mathcal{C}_1)$$

is given by

$$u_1(t) = B_1^T e^{A_1^T (T_1 - t)} \eta_1$$

where η_1 is the solution of the equation

$$W_1(0, T_1)\eta_1 = x(\tau_2) - e^{A_1 T_1} x(\tau_1)$$

$W_1(0, T_1)$ is the controllability Gramian of (A_1, B_1). For the system at hand the solution is

$$u_1(t) = \frac{2e^{T_1}}{e^{2T_2} - 1} \beta_1 e^{-t} = 0.3349e^{-t}$$

Similarly the minimum energy solution for the second mode is

$$u_2(t) = \frac{2e^{-T_2}}{1 - e^{-2T_2}} \beta_2 e^t = -0.9758e^t$$

The hybrid automaton's trajectory with this minimum-energy control is shown in the righthand plot in figure 3. This figure shows the state space for our system. The two triangular regions in this plot represent the guards. For the specific choice of points $x(\tau_1)$ and $x(\tau_2)$, we use the control $u(t)$ identified above to compute the state trajectory between these points. The solid line in figure 3 shows the resulting controlled trajectory.

It is, of course, possible to obtain other controls realizing this cycle. For instance, an "impulsive" control strategy can be employed, in which we impulsively drive the system state along an affine variety of the controllable subspace and then allow the system to relax into the guard. (In other words we let $u(t)$ be an impulse function of specified magnitude). The lefthand plot of figure 4 illustrates the state trajectory generated by this control law. As in figure 3, we are looking at the system's state space. The triangular regions represent the guards and the solid lines denote the state trajectory generated by the impulsive control.

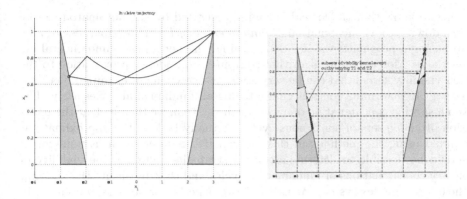

Fig. 4. Impulsive simulation results and sweeping out the viability kernel

For a specific pair of times, our master algorithm identifies all states in the guards that are mutually reachable from each other. If we were able to identify a *range* of feasible switching times, then it should be possible to identify larger subsets of the guard that are mutually reachable from each other. The complete set of states in the guards that are mutually reachable from each other (under any control strategy) is sometimes referred to as the *viability kernel*. Our algorithm, therefore, provides a means of approximating the viability kernel. Note that this is an under approximation to the viable set (as opposed to the over-approximation computed by model checking algorithms [Alu95]). For our specific example, we were able to identify a range of times over which periodic solutions could be guaranteed. This range was computed to be. $1.7 < T_1 < 3.75$ and $.1059 < T_2 < .29$. The set of points swept out by these various times is shown in the righthand plot of figure 4. We've compared this set to the actual viability kernel for this system and the specified bounds appear to provide a close approximation to the actual viability kernel.

Remark: The failure of the master program to find any feasible solution does not guarantee that a global solution doesn't exist. How quickly we find a feasible solution clearly depends upon the type of search strategy the master program uses and depends on our initial guess.

Remark: Our approach focuses on identifying *periodic* global solutions and obviously it may be possible that this is overly restrictive. For instance, it may be possible that only chaotic trajectories exist between the two guards, or that a more complex periodic behavior exists between the two guards.

Remark: The preceding discussion focused on establishing non-terminating solutions to a rather simple hybrid automaton. This problem was chosen as a *canonical problem* in the sense that its solution may provide a foundation upon which to establish the existence of global solutions to more complex systems. How might this be done? This is the topic of another paper, but we can speculate on a possible strategy based on prior results on the role of cycles in hybrid automata [He98] [Zhi98]. Essentially, the argument runs as follows. From the pumping lemma, we know that the logical behavior generated by any automaton can be broken down into a concatenation of *fundamental cycles*. This paper, essentially, is proposing a pragmatic way for determining whether a given fundamental cycle is viable. Moreover, our algorithm computes an under-approximation to the cycle's viability kernel that can be very good (as shown in our example). Let's assume we can determine controls guaranteeing all fundamental cycles are viable. Given a specific concatenation of cycles in the system, we then look at the intersection of viable sets of contiguous cycles (actually look at the approximations computed using the methods in this paper). If this intersection is non-empty, it should be possible to determine control strategies enforcing the viability of arbitrary concatenations of fundamental cycles and thereby ensure the viability of the entire complex system. As noted above, whether or not this approach will work is still under study.

5   Conclusions

Controlled hybrid automata are automata in which a user-determined input control signal can be used to help supervise overall system behavior. In this paper, we assumed the modal systems were linear and time invariant with polytopic guards formed from the convex combination of vertices. This paper studied the existence of solutions to this class of hybrid system. A routine application of Viability theory was used to characterize the existence of local trajectories. This paper presented a necessary and sufficient condition for the existence of a global periodic non-terminating trajectory with specified switching intervals. This result was used to propose a gradient following search strategy for determining a set of switching intervals ensuring a global nonterminating trajectory. The proposed method also provides an under-approximation of the cycle's viability kernel that could be used in extending this work to more complex switching systems. A distinguishing feature of this study is the explicit use of the open loop control signal $u(t)$ as a means of enforcing a cycle's viability.

This work is preliminary in that there are still a number of open questions that need to be answered. There is uncertainty over the performance of the proposed search algorithm. It should be noted, however, that such gradient following heuristics often work extremely well on real-life problems, so this approach may still be a pragmatic approach to hybrid system verification. Another open issue concerns the conservatism imposed by confining our search to periodic non-terminating solutions. While this might appear to be very restrictive on the surface, it must be realized that the proposed approach can actually identify a *set* of periodic solutions and that other non-periodic solutions might be seen as limiting points of this set. Another interesting issue brought up by this paper is the explicit use of control. Traditional analyses of hybrid systems assume no control and the verification process can be seen as a "take it or leave it" analysis that provides little guidance on determining how "close" a system is to being viable. The use of control advocated in this paper may provide the system designer with a more sophisticated approach to verification in which control becomes a necessary component in system design. Finally, this paper has focused on hybrid systems containing only one cycle. This simple problem is viewed as a necessary starting place for the analysis of more complex hybrid systems and the details of this later analysis will be the subject of future papers.

References

Alu93.   R. Alur, C. Courcoubetis, T.A. Henzinger, and P.-H. Po, " Hybrid Automata: an Algorithmic Approach to the Specification and Verification of Hybrid Systems", Robert L. Grossman, Anil Nerode, Anders P. Ravn, and Hans Rischel, editors, *Hybrid Systems*, Lecture Notes in Computer Science, vol. 736, Springer-Verlag, pp. 209-229, 1993.

Alu95.   R. Alur, C. Courcoubetis, Halbwachs, T.A. Henzinger, P-H Ho , X Nicollin, Olivero, J. Sifjakis, and S. Yovine, "The Algorithmic Analysis of Hybrid Systems", *Theoretical Computer Science* Vol. 138:, pp. 3-34, 1995

Ant97.   P.J. Antsaklis, A.N. Michel, *Linear Systems*, MacGraw-Hill, 1997.

Aub84.   J.P. Aubin and A. Cellina, *Differential Inclusions*, Springer Verlag, 1984.

Baz93.   M. Bazaraa, H.D. Sherali, and C.M. Shetty, *Nonlinear Programming: theory and algorithms*, 2nd Edition, John-Wiley, 1993.

Dav83.   M.D. Davis and E.J. Weyuker, *Computability, Complexity, and Languages*, Academic Press, 1983.

He98.   K.X. He and M.D. Lemmon, Lyapunov stability of continuous valued systems under the supervision of discrete event transition systems, in *Proceedings of Hybrid Systems: Control and Computation*, lecture notes in computer science Vol. 1386, Springer Verlag, 1998.

Lyn96.   N. Lynch, R. Segala, F. Vaandrager, and H.B. Weinberg, "Hybrid I/O Automata", In R. Alur, T. Henzinger, and E. Sontag, editors, *Hybrid Systems III: Verification and Control* (DIMACS/SYCON Workshop on Verification and Control of Hybrid Systems), New Brunswick, New Jersey, October 1995), volume 1066 of Lecture Notes in Computer Science, pages 496-510. Springer-Verlag 1996.

Zhi98.   P. Zhivoglyadov and R.H. Middleton, On Stability in Hybrid Systems, Proceedings off 37th IEEE Conference on Decision and Control, Tampa Florida, USA, pp. 3687-3692, December 1998.

Nonlinear Stabilization
by Hybrid Quantized Feedback[*]

Daniel Liberzon

Dept. of Elect. Eng., Yale University
New Haven, CT 06520-8267 U.S.A.
daniel.liberzon@yale.edu

Abstract. This paper is concerned with global asymptotic stabilization
of continuous-time control systems by means of quantized feedback. For
linear systems, a hybrid control strategy for dealing with this problem
was recently proposed by Roger Brockett and the author. The solution
is based on making discrete on-line adjustments to the sensitivity of
the quantizer. In the present paper we extend this method to a class of
nonlinear systems.

1 Introduction

We study the problem of stabilizing a control system with quantized state feedback. This problem consists in designing a stabilizing control law which, instead
of using the measurements of the system's state x directly, is only allowed to
depend on the quantized measurements $q_\Delta(x)$. Here q_Δ is a piecewise constant
function with a finite set of values, called a *quantizer* (see Figure 1).

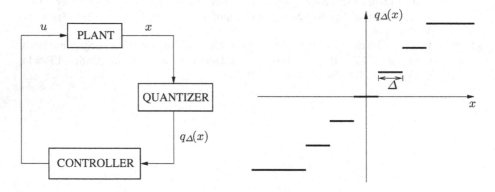

Fig. 1. Quantized state feedback

Given a positive real number Δ and a positive integer M, we denote by $\mathcal{I}$
the set $\{k \in \mathbb{Z} : -M \leq k \leq M\}$, and define the function $q_\Delta : \mathbb{R} \to \{k\Delta : k \in \mathcal{I}\}$

[*] This research was supported in part by AFOSR, DARPA, and NSF.

N. Lynch and B. Krogh (Eds.): HSCC 2000, LNCS 1790, pp. 243–257, 2000.

by the formula

$$q_\Delta(x) = \begin{cases} k\Delta & \text{if } (k-1/2)\Delta \leq x < (k+1/2)\Delta \text{ where } k \in \mathcal{I} \\ M\Delta & \text{if } x \geq (M+1/2)\Delta \\ -M\Delta & \text{if } x < -(M+1/2)\Delta \end{cases} \qquad (1)$$

We will call Δ the *sensitivity* of the quantizer and M the *saturation constant*. When $x \in \mathbb{R}^n$, the quantized vector $q_\Delta(x)$ is defined componentwise using the function (1). Geometrically, $\mathbb{R}^n$ is thereby divided into a finite number of rectilinear regions called *quantization blocks*, each corresponding to a fixed value of q_Δ.

Quantizers of the kind described above are commonly used to model finite-precision effects which arise, for example, when the state measurements to be used for feedback are processed by a digital computer or transmitted via a digital communication channel. In such situations, the sensitivity and the number of values of the quantizer are given *a priori* and cannot be changed by the control designer. This places significant constraints on what can be accomplished by using feedback. In particular, we see from (1) that $q_\Delta(x) = 0$ for all x in a sufficiently small neighborhood of the origin, hence asymptotic stabilization is impossible (unless there exists an open-loop control law that drives all the states in this neighborhood to the origin). See, e.g., [3,5,7,14,20,21,22] for more information regarding quantization issues.

However, numerical quantization in computer-controlled systems is just one example of a situation where the problem of developing techniques for quantized feedback stabilization is relevant. There are many other cases in which a function of the form (1) can be used to represent information that is available to the controller. For instance, imagine having a sensor that determines whether the temperature of a certain object is "normal", "too high", or "too low". Such a sensor can be modeled by a quantizer with saturation constant $M = 1$ (of course, a higher value of M can also be used if one demands more information from the sensor). Since it is reasonable to assume that one is allowed to adjust the threshold settings from time to time, the sensitivity Δ is not necessarily fixed in this case. As another example, a video or photographic camera with zooming capability can be described by a quantizer of the type considered here, with M determined by the number of pixels. By zooming in or zooming out, one effectively varies the sensitivity of such a quantizer.

Following [2], we take the approach that it is possible to change the sensitivity Δ (but not the saturation constant M) of the quantizer on the basis of available quantized measurements. The problem then is to design a quantized feedback law that yields asymptotic stability of the equilibrium $x = 0$. The above discussion suggests that this problem, besides being of theoretical interest, is also quite meaningful in many practical applications. We will assume that the given system evolves in continuous time. The values of Δ, on the other hand, will belong to a discrete set and will be updated at discrete instants of time (these events will be triggered by the values of a suitable Lyapunov function). The closed-loop system will therefore be a hybrid system, with continuous state x and discrete state Δ.

Another logical variable z (for "zoom") will also be used. The systems obtained in this paper fall into the general framework for hybrid systems presented in [1] (see [2] for details). As in [2], all feedback control laws will be constructed explicitly, and solutions to all differential equations are well defined, with the understanding that they are to be interpreted in the sense of Filippov [8] if necessary.

Control policies based on the above idea of making discrete on-line adjustments to the sensitivity Δ of the quantizer will be referred to as *hybrid quantized feedback control policies*. Our reasons for adopting a hybrid control approach, rather than varying Δ continuously, are threefold. First, in specific situations there may be some constraints on how many values Δ is allowed to take and how frequently it can be changed. Thus a discrete adjustment policy for Δ is more natural and easier to implement than a continuous one. Second, the analysis of hybrid systems obtained in this way is usually much more tractable than that of systems resulting from continuous adjustments of Δ. In fact, we will see that a method based on computation of invariant regions defined by level sets of a Lyapunov function provides a very simple and effective tool for studying the behavior of the closed-loop system. Finally, a discrete adjustment policy is more robust with respect to time delays, which constitute another important issue to consider in the present "limited information feedback" setting (cf. [22]).

The hybrid control approach to the quantized feedback stabilization problem was first introduced by Roger Brockett and the author in the recent paper [2]. That paper deals with linear control systems. It is shown there that if a linear system can be stabilized by a linear feedback law, then it can also be stabilized by a hybrid quantized feedback control policy. One strategy proposed in [2] for achieving global asymptotic stability consists of two stages. First, since the initial state is unknown, we "zoom out" by increasing Δ until the state of the system can be adequately measured. Second, we "zoom in" by decreasing Δ in such a way as to drive the state to 0. (The discrete "zoom" variable z equals 1 in the first case and -1 in the second case.)

The goal of this paper is to extend the above method to nonlinear systems. It can be shown via a linearization argument that by using the approach of [2] one can obtain local asymptotic stability for a nonlinear system, provided that the corresponding linearized system is stabilizable (see [11]). Here we are concerned with the problem of achieving global or at least semi-global asymptotic stability. Working with a given nonlinear system directly, one gains an advantage even if only local asymptotic stability is sought, because the linearization of a stabilizable nonlinear system may fail to be stabilizable. We will demonstrate that the techniques developed in [2] can be extended in a natural way to those nonlinear systems that are "externally stabilizable" in a certain sense to be made precise below. For linear systems, external stabilizability follows automatically from the usual (internal) stabilizability, but for nonlinear systems this leads to a nontrivial problem which is a subject of ongoing research. We thus reveal an interesting interplay between the problem of quantized feedback stabilization, the theory of hybrid systems, and recent advances in nonlinear control design.

The rest of the paper is structured as follows. In Section 2 we review a result from [2] concerning linear systems. We provide its complete proof (slightly modified from the original version). This method is then applied to a class of nonlinear systems in Section 3, where the main results of this paper are obtained. The ideas behind the control strategy are essentially the same as in the linear case; however, the analysis is inherently nonlinear and involves several concepts from the modern nonlinear control literature. Section 4 discusses these concepts and relevant developments, as well as the results presented here and their possible extensions.

2   Linear Systems

Given a quantizer q_Δ, by its *saturation region* we will mean the union of those quantization blocks that are infinite, i.e., the set of vectors $x \in \mathbb{R}^n$ with at least one component exceeding $(M - 1/2)\Delta$ in magnitude. When x is such that the quantizer does not saturate, the quantization error satisfies the bound

$$\|x - q_\Delta(x)\| \le \Delta\sqrt{n}/2 \tag{2}$$

where $\| \cdot \|$ stands for the standard Euclidean norm. Observe that, according to the above terminology, x belongs to the saturation region of q_Δ whenever at least one of the components of the vector $q_\Delta(x)$ equals $\pm M\Delta$, even though this does not automatically imply that the inequality (2) fails to hold.

Consider the linear control system

$$\dot{x} = Ax + Bu, \qquad x \in \mathbb{R}^n, \ u \in \mathbb{R}^m. \tag{3}$$

Suppose that it is *stabilizable*, which means that there exists a matrix K such that the eigenvalues of $A + BK$ have negative real parts. Since the linear feedback law $u = Kx$ cannot be implemented, it seems logical to try the quantized feedback law $u = Kq_\Delta(x)$. As shown in [2], one can define a hybrid quantized feedback control policy based on this feedback law to render $x = 0$ a globally asymptotically stable equilibrium of the continuous part of the closed-loop system.

Theorem 1. [2] *There exists a hybrid quantized feedback control policy that makes the system* (3) *globally asymptotically stable.*

Proof. The control law will take the form

$$u(q_\Delta(x), z) = \begin{cases} 0 & \text{if } z = 1 \\ Kq_\Delta(x) & \text{if } z = -1 \end{cases}$$

To define a desired control policy, we need to describe the evolution of Δ and z. Before proceeding to do that, we make some observations regarding the behavior of the system

$$\dot{x} = Ax + BKq_\Delta(x)$$

which can also be written as

$$\dot{x} = (A + BK)x - BK(x - q_\Delta(x)). \tag{4}$$

We will let $\lambda_{min}(\cdot)$ and $\lambda_{max}(\cdot)$ denote the smallest and the largest eigenvalue of a symmetric matrix, respectively. Recall that by the standard Lyapunov stability theory there exist positive definite symmetric matrices Q and D such that $(A + BK)^T Q + Q(A + BK) = -D$. Whenever the inequality (2) holds, the derivative of $x^T Q x$ along the solutions of (4) is given by

$$\frac{d}{dt} x^T Q x = -x^T D x - 2x^T Q B K (x - q_\Delta(x))$$

$$\leq -\lambda_{min}(D)\|x\|^2 + 2\|x\|\|QBK\|\Delta\sqrt{n}/2$$

$$= -\|x\|(\lambda_{min}(D)\|x\| - \|QBK\|\Delta\sqrt{n})$$

The last expression is negative outside the ball $\{x : \|x\| \leq \Theta\Delta\sqrt{n}\}$, where

$$\Theta := \|QBK\|/\lambda_{min}(D).$$

In other words, if q_Δ does not saturate, we have

$$\|x\| > \Theta\Delta\sqrt{n} \Rightarrow \frac{d}{dt} x^T Q x < 0. \tag{5}$$

In what follows we will use the simple facts that the radius of the ball inscribed in an ellipsoid of the form $\{x : x^T Q x \leq \gamma^2\}$ equals $\gamma/\sqrt{\lambda_{max}(Q)}$ and the radius of the ball circumscribed about the same ellipsoid equals $\gamma/\sqrt{\lambda_{min}(Q)}$. Fix an arbitrary $\epsilon > 0$. Define the *scaling factor* Ω by the formula

$$\Omega := \left((\Theta\sqrt{n} + \epsilon)\sqrt{\frac{\lambda_{max}(Q)}{\lambda_{min}(Q)}} + \sqrt{n}\right)\sqrt{\frac{\lambda_{max}(Q)}{\lambda_{min}(Q)}}\left(M - \frac{1}{2}\right)^{-1} \tag{6}$$

and take the saturation constant M of the quantizer q_Δ to be large enough so that we have $\Omega < 1$.

We now describe the "zooming-out" stage of the control strategy ($z = 1$). Set the control to 0 and choose an arbitrary $\Delta(0) > 0$. Then increase Δ in a piecewise constant fashion, fast enough to dominate the rate of growth of $\|e^{At}\|$. For example, one can fix a positive number τ and let $\Delta(t) = \Delta(0)$ for $t \in [0, \tau)$, $\Delta(t) = e^{2\|A\|\tau}\Delta(0)$ for $t \in [\tau, 2\tau)$, $\Delta(t) = e^{2\|A\|2\tau}\Delta(0)$ for $t \in [2\tau, 3\tau)$, and so on. Clearly, there will be a time $t \geq 0$ such that

$$\|x(t)\| \leq \Delta(t)\left(\left(M - \frac{1}{2}\right)\sqrt{\frac{\lambda_{min}(Q)}{\lambda_{max}(Q)}} - \sqrt{n}\right)$$

hence

$$\|q_{\Delta(t)}(x(t))\| \leq \Delta(t)\left(\left(M - \frac{1}{2}\right)\sqrt{\frac{\lambda_{min}(Q)}{\lambda_{max}(Q)}} - \frac{\sqrt{n}}{2}\right)$$

by virtue of (2). We can thus pick a time $t_0 > 0$ such that

$$\|q_{\Delta(t_0)}(x(t_0))\| \leq \Delta(t_0)\left(\left(M - \frac{1}{2}\right)\sqrt{\frac{\lambda_{min}(Q)}{\lambda_{max}(Q)}} - \frac{\sqrt{n}}{2}\right)$$

which implies that

$$\|x(t_0)\| \leq \Delta(t_0)\left(M - \frac{1}{2}\right)\sqrt{\frac{\lambda_{min}(Q)}{\lambda_{max}(Q)}}. \tag{7}$$

This inequality guarantees, in particular, that at time t_0 the quantizer does not saturate. It is important to note that the time instant t_0 was determined on the basis of the quantized measurements only.

Next, we come to the "zooming-in" stage ($z = -1$). Starting at $t = t_0$, we let $u = Kq_\Delta(x)$. We keep Δ equal to $\Delta(t_0)$ until a later time to be specified below. It follows from (7) that $x(t_0)$ belongs to the ellipsoid

$$\left\{x : x^T Q x \leq (\Delta(t_0))^2 \left(M - \frac{1}{2}\right)^2 \lambda_{min}(Q)\right\}. \tag{8}$$

Since $\Omega < 1$, it is not difficult to see that $(M-1/2)\sqrt{\lambda_{min}(Q)} > \Theta\sqrt{n}\sqrt{\lambda_{max}(Q)}$. From this and (5) we conclude that x will not leave the ellipsoid (8) for as long as $\Delta = \Delta(t_0)$, hence the quantizer will not saturate. Moreover, x will approach the smaller ellipsoid

$$\{x : x^T Q x \leq (\Delta(t_0))^2 \Theta^2 n \lambda_{max}(Q)\}$$

(it might even happen that $x(t_0)$ already belongs to this ellipsoid). Thus we can pick a time $t_1 > t_0$ such that

$$\|q_{\Delta(t_0)}(x(t_1))\| \leq \Delta(t_0)(\Theta\sqrt{n} + \epsilon)\sqrt{\frac{\lambda_{max}(Q)}{\lambda_{min}(Q)}} + \frac{\sqrt{n}}{2}$$

which implies that

$$\|x(t_1)\| \leq \Delta(t_0)\left((\Theta\sqrt{n} + \epsilon)\sqrt{\frac{\lambda_{max}(Q)}{\lambda_{min}(Q)}} + \sqrt{n}\right).$$

Therefore, $x(t_1)$ belongs to the ellipsoid

$$\left\{x : x^T Q x \leq (\Delta(t_0))^2 \left((\Theta\sqrt{n} + \epsilon)\sqrt{\frac{\lambda_{max}(Q)}{\lambda_{min}(Q)}} + \sqrt{n}\right)^2 \lambda_{max}(Q)\right\}. \tag{9}$$

Again, note that the time instant t_1 was selected using only the quantized measurements.

The basic idea that allows us to achieve asymptotic stability is to decrease Δ by means of multiplying it by the scaling factor Ω. Namely, we let $\Delta(t_1) :=$

$\Omega\Delta(t_0)$. Using (6), it is straightforward to verify that the ellipsoid (9) is identical to the one defined by (8) with $\Omega\Delta(t_0)$ in place of $\Delta(t_0)$. This means that we can continue the analysis for $t \geq t_1$ as before. (The fact that the scaling is performed at $t = t_1$ is not crucial: since all the ellipsoids considered here are invariant regions for the closed-loop system, Δ could also be scaled at an arbitrary time $t > t_1$.) Thus there exists a time $t_2 > t_1$, which can be determined from the quantized measurements, such that $x(t_2)$ belongs to the ellipsoid defined by (9) with $\Delta(t_1)$ in place of $\Delta(t_0)$. When $t = t_2$ (or at an arbitrary time $t > t_2$) we set $\Delta := \Omega\Delta(t_1)$. Repeating this procedure, we obtain the desired control policy. Indeed, stability of the equilibrium $x = 0$ in the sense of Lyapunov follows directly from the adjustment policy for Δ (note that the amount by which Δ needs to be increased initially is proportional to $\|x(0)\|$). Moreover, we have $\Delta(t) \to 0$ as $t \to \infty$, and by the above analysis the same is true for $x(t)$. Figure 2 illustrates the proof. □

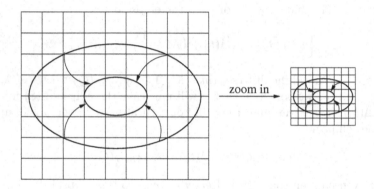

Fig. 2. Stabilization by hybrid quantized feedback

In the preceding, Δ takes a countable set of values which is not assumed to be fixed in advance. In some situations Δ may be restricted to take values in some given set S. It is not hard to see that the proposed method, suitably modified, will still work in this case, provided that the set S satisfies the following properties:

1. S contains a sequence $\Delta_{11}, \Delta_{21}, \ldots$ that increases to ∞.
2. Each Δ_{i1} from this sequence belongs to a sequence $\Delta_{i1}, \Delta_{i2}, \ldots$ in S that decreases to 0 and is such that we have $\Omega \leq \Delta_{i,j+1}/\Delta_{ij}$ for each j.

If the set of possible values for Δ is finite rather than countable, we can only obtain practical stability and not global asymptotic stability.

The above proof reflects just one among several different quantized feedback control strategies for linear systems presented in [2]. That paper treats the topics of achieving exponential convergence, quantized output feedback stabilization, quantized feedback stabilization under sampling, and stabilization using quantizers with small saturation constants. Some of these questions are further pursued in [11]. We have chosen the particular approach described here because it appears to be the most suitable one for generalization to nonlinear systems.

3   Nonlinear Systems

We now turn to nonlinear control systems of the form

$$\dot{x} = f(x, u), \qquad x \in \mathbb{R}^n, \ u \in \mathbb{R}^m. \tag{10}$$

In this section, all vector fields and control laws are assumed to be sufficiently regular (e.g., smooth). Given a feedback law $u = k(x)$, we consider the system

$$\dot{x} = f(x, k(x + e)) \tag{11}$$

where e is a measurement disturbance input, later to be associated with the quantization error. For the purposes of this paper, we will say that the system (11) is *input-to-state stable* (ISS) with respect to e if there exists a positive definite radially unbounded smooth function $V : \mathbb{R}^n \to \mathbb{R}$ such that for some class $\mathcal{K}_\infty$ functions[1] α_1, α_2 and ρ, for all $x \neq 0$, and for all e we have

$$\alpha_1(\|x\|) \leq V(x) \leq \alpha_2(\|x\|) \tag{12}$$

and

$$\|x\| \geq \rho(\|e\|) \Rightarrow \nabla V(x) f(x, k(x + e)) < 0. \tag{13}$$

According to the results of [18], this is equivalent to the original definition of ISS given by Sontag in [15].

In this section we assume that the given system (10) has the property that there exists a feedback law $u = k(x)$ which makes the system (11) input-to-state stable with respect to e. This is the property which we referred to as "external stabilizability" in the Introduction. We also assume that the functions α_1, α_2 and ρ satisfy the following condition:

$$(\alpha_2^{-1} \circ \alpha_1)'(0) > 0 \quad \text{and} \quad \rho'(0) < \infty \tag{14}$$

(there is no loss of generality in requiring that the above derivatives exist, if one allows the possibility that $\rho'(0)$ might equal ∞). We postpone a close examination of these assumptions until Section 4. Everything that follows remains valid if one replaces a static control law $u = k(x)$ by a dynamic control law.

As before, let q_Δ be a quantizer with sensitivity Δ and saturation constant M. The problem under consideration is to find a quantized state feedback law that makes the system (10) asymptotically stable. The idea that we propose is to use the above control law k, which results in the closed-loop system

$$\dot{x} = f(x, k(q_\Delta(x))) = f(x, k(x - (x - q_\Delta(x)))). \tag{15}$$

---

[1] A function $\alpha : [0, \infty) \to [0, \infty)$ is said to be of class $\mathcal{K}_\infty$ if it is continuous, strictly increasing, and such that $\alpha(0) = 0$ and $\alpha(r) \to \infty$ as $r \to \infty$.

If q_Δ does not saturate, then the inequality (2) holds, and we deduce from (13) that the derivative of V along solutions of the system (15) satisfies

$$\|x\| > \rho(\Delta\sqrt{n}/2) \Rightarrow \dot{V} < 0. \tag{16}$$

This fact will be used in the sequel in much the same way as the formula (5) has been used in the previous section.

Fix a positive number ϵ, and define the functions

$$\gamma_1(\Delta) := \alpha_1^{-1} \circ \alpha_2 \circ \rho(\Delta(\sqrt{n}/2 + \epsilon)) + \Delta\sqrt{n}$$

and

$$\gamma_2(\Delta) := \alpha_2^{-1} \circ \alpha_1((M - 1/2)\Delta).$$

Suppose for the moment that an upper bound on the initial state is known: $\|x(0)\| \le E_0$, where E_0 is a positive number. A desired hybrid quantized feedback control policy can then be described as follows. Choose an arbitrary $\Delta(0) > 0$. In view of (14), an elementary argument shows that if M was taken to be large enough, then we have

$$\gamma_2(\Delta) > \gamma_1(\Delta) \qquad \forall \Delta \in (0, \Delta(0)] \tag{17}$$

and furthermore

$$\gamma_2(\Delta(0)) \ge E_0. \tag{18}$$

It follows from (12) and (18) that $x(0)$ belongs to the region

$$\{x : V(x) \le \alpha_1((M - 1/2)\Delta(0))\}. \tag{19}$$

Using (16) and the inequality $\gamma_2(\Delta(0)) > \gamma_1(\Delta(0))$, we conclude that x will not leave the region (19) for as long as $\Delta = \Delta(0)$, and the quantizer will not saturate. Moreover, x must approach the smaller region

$$\{x : V(x) \le \alpha_2 \circ \rho(\Delta(0)\sqrt{n}/2)\}.$$

Thus we can pick a time $t_1 > 0$ such that

$$\|q_{\Delta(0)}(x(t_1))\| \le \alpha_1^{-1} \circ \alpha_2 \circ \rho(\Delta(0)\sqrt{n}/2 + \Delta(0)\epsilon) + \Delta(0)\sqrt{n}/2$$

hence

$$\|x(t_1)\| \le \alpha_1^{-1} \circ \alpha_2 \circ \rho(\Delta(0)\sqrt{n}/2 + \Delta(0)\epsilon) + \Delta(0)\sqrt{n} = \gamma_1(\Delta(0)).$$

Therefore, $x(t_1)$ belongs to the region

$$\{x : V(x) \le \alpha_2 \circ \gamma_1(\Delta(0))\}. \tag{20}$$

When $t = t_1$, set $\Delta := \gamma_2^{-1} \circ \gamma_1(\Delta(0))$. It is not hard to check that the region (20) is the same as the one given by (19) with $\Delta(0)$ replaced by this new value of

Δ. This means that we can continue the same analysis for $t \geq t_1$. Repeating this procedure, we have $\Delta(t) \to 0$ as $t \to \infty$ because of (17), and asymptotic stability follows. The above argument is very similar to the one employed at the "zooming-in" stage in the proof of Theorem 1.

Now suppose that there exists a value of M for which

$$\gamma_2(\Delta) > \gamma_1(\Delta) \qquad \forall \Delta > 0. \tag{21}$$

In this case, an *a priori* bound on the initial state is not necessary, because we can apply a "zooming-out" procedure. Namely, suppose that the unforced system

$$\dot{x} = f(x, 0) \tag{22}$$

is *forward complete*, meaning that for every initial state $x(0)$ its solution, which we denote by $\xi(x(0), t)$, is defined for all $t \geq 0$. Then we can set $u = 0$ and increase Δ fast enough to dominate the rate of growth of $\|x(t)\|$. For example, take M large enough so that the function $\chi(r) := \gamma_2(r) - r\sqrt{n}$ is of class $\mathcal{K}_\infty$. Fix a positive number τ and let $\Delta(t) = \chi^{-1}(\max_{\|x(0)\| \leq \tau} \|\xi(x(0), \tau)\|)$ for $t \in [0, \tau)$, $\Delta(t) = \chi^{-1}(\max_{\|x(0)\| \leq 2\tau} \|\xi(x(0), 2\tau)\|)$ for $t \in [\tau, 2\tau)$, and so on. Then there will be a time t such that

$$\|x(t)\| \leq \gamma_2(\Delta(t)) - \Delta(t)\sqrt{n}$$

hence

$$\|q_{\Delta(t)}(x(t))\| \leq \gamma_2(\Delta(t)) - \Delta(t)\sqrt{n}/2$$

by (2). Thus we can pick a time $t_0 > 0$ such that

$$\|q_{\Delta(t_0)}(x(t_0))\| \leq \gamma_2(\Delta(t_0)) - \Delta(t_0)\sqrt{n}/2$$

hence

$$\|x(t_0)\| \leq \gamma_2(\Delta(t_0)).$$

This implies that $x(t_0)$ belongs to the region (19), and from this point on the analysis can be continued exactly as before. We have thus proved the following theorem.

Theorem 2. *Suppose that the system (10) is input-to-state stabilizable with respect to measurement disturbances, in the sense defined above. Suppose also that the condition (14) holds. Then for each number $E_0 > 0$ there exists a hybrid quantized feedback control policy that makes the system (10) asymptotically stable, with domain of attraction containing all initial states $x(0)$ such that $\|x(0)\| \leq E_0$ (semiglobal asymptotic stability). In addition, if the system (22) is forward complete and for some M the inequality (21) holds, then there exists a hybrid quantized feedback control policy that makes the system (10) globally asymptotically stable.*

If the condition (14) does not hold, then the inequality (17) cannot be satisfied by choosing a sufficiently large M. However, for any given numbers $\Delta(0) > \delta > 0$ there exists a positive integer M such that we have

$$\gamma_2(\Delta) > \gamma_1(\Delta) \qquad \forall \Delta \in (\delta, \Delta(0)].$$

It is not difficult to see that in this case the above procedure gives semiglobal practical stability, in the sense that all initial states whose norm satisfies a known bound are driven to the region given by (20) with δ in place of $\Delta(0)$. As explained in the next section, such a property can actually be achieved without imposing the input-to-state stabilizability assumption.

We also point out that, in view of [19, Lemma I.2], every asymptotically stabilizing feedback law is automatically input-to-state stabilizing with respect to the measurement error e locally, i.e., for sufficiently small values of x and e. This leads at once to local versions of our results.

4 Discussion of Results

The concept of input-to-state stability (ISS) captures the property that bounded inputs lead to bounded states, and inputs converging to zero produce states that converge to zero. This concept was introduced by Sontag in [15]. In the same paper he proved that if an affine system of the form

$$\dot{x} = f(x) + G(x)u$$

is asymptotically stabilizable by a feedback law $u = k_0(x)$, then one can always find a feedback law $u = k(x)$ that makes the system

$$\dot{x} = f(x) + G(x)(k(x) + d)$$

ISS with respect to an actuator disturbance d. However, there might not exist a feedback law that makes the system

$$\dot{x} = f(x) + G(x)k(x + e)$$

ISS with respect to a measurement disturbance e, as was shown by way of counterexamples in [9] and later in [6]. Of course, for linear systems with linear feedback laws all three concepts (asymptotic stabilizability, input-to-state stabilizability with respect to actuator errors, and input-to-state stabilizability with respect to measurement errors) are equivalent.

Thus the problem of finding control laws that achieve ISS with respect to measurement disturbances for the system (11) is a nontrivial one, even for affine systems. This problem has recently attracted considerable attention in the literature (see [6,10,12]). In particular, it was shown in [10] that the class of systems that admit such control laws includes single-input plants in strict feedback form. As pointed out in [16], it also includes systems that admit globally Lipschitz control laws achieving ISS with respect to actuator disturbances, although this condition is quite restrictive. In the paper [12] small-gain techniques are applied to handle certain classes of systems with unknown parameters and unmodeled dynamics.

In Section 3 we showed how to achieve semiglobal or global asymptotic stability for nonlinear systems by means of quantized state feedback. The assumptions that were needed to prove the result included the existence of a feedback law achieving ISS with respect to measurement disturbances. We also imposed the condition (14) which characterizes the behavior near zero of the functions α_1, α_2 and ρ that appear in the definition of ISS. To obtain global asymptotic stability, we required that there exist a saturation constant M for which the inequality (21) holds. This depends on the relative rate of growth of the above functions at infinity. We now give a simple example of a system for which all of these hypotheses are satisfied.

EXAMPLE. Consider the following system, which is a simplified version of the system treated in the example on page 811 in [12]:

$$\dot{x} = x^3 + xu, \qquad x, u \in \mathbb{R}.$$

In [12] it is shown how to construct a feedback law k such that the closed-loop system

$$\dot{x} = x^3 + xk(x + e)$$

is ISS with respect to e. It follows from the analysis of [12] that the inequalities (12) and (13) hold with $V(x) = x^2/2$, so one can take $\alpha_1(r) = \alpha_2(r) = r^2/2$, and for ρ one can take any linear function $\rho(r) = cr$ with $c > 1$. We have $(\alpha_2^{-1} \circ \alpha_1)'(0) = 1$ and $\rho'(0) = c$, so (14) is obviously valid. Moreover, (21) holds for every $M > c(\sqrt{n}/2 + \epsilon) + \sqrt{n} + 1/2$.

We conclude the paper with a discussion of how one can achieve semiglobal practical stability for a much more general class of nonlinear systems than the one considered in the previous section. The idea is based on the work reported by Sontag in [17] on designing stabilizing control laws that are robust with respect to *small* measurement errors. Suppose that the system (10) is *globally asymptotically controllable to the origin*, in the sense that every initial state $x(0)$ can be driven to 0 by a bounded control. Every system that is stabilizable by continuous feedback certainly belongs to this category. The following statement is a direct consequence of [17, Theorem 1]: there is a class $\mathcal{K}_\infty$ function Γ with the property that for any numbers $E_0 > \epsilon > 0$ there exist positive numbers $\delta(\epsilon, E_0)$, $\kappa(\epsilon, E_0)$ and $T(\epsilon, E_0)$ and a feedback law $u = k(x)$ such that for every measurement disturbance e with $\|e(t)\| \leq \kappa\delta \ \forall t \geq 0$ each solution of the system (11) with $\|x(0)\| \leq E_0$ satisfies

$$\|x(t)\| \leq \Gamma(E_0) \qquad \forall t \geq 0 \tag{23}$$

and

$$\|x(t)\| \leq \epsilon \qquad \forall t \geq T. \tag{24}$$

The feedback law k will in general be discontinuous. The solutions of (11) are then to be interpreted in the "closed-loop system sampling" sense as defined in

[4], with sampling period δ. Actually, the statement remains valid if the sampling period varies within the bounds $\underline{\delta} < \overline{\delta}$, provided that the inequalities

$$\overline{\delta} \leq \delta \tag{25}$$

and

$$\|e(t)\| \leq k\underline{\delta} \tag{26}$$

are satisfied. The proof of this result relies on the existence of a continuous (but in general not necessarily smooth) control Lyapunov function for a given asymptotically controllable system. The feedback law is defined in terms of all sampling schedules that are sufficiently fast, as expressed by the inequality (25). On the other hand, robustness with respect to small measurement errors is only assured if the sampling is not too fast, as expressed by the inequality (26). This modification is not necessary, however, if the system possesses a *smooth* control Lyapunov function (see [13]). The need for sampling disappears altogether if the stabilizing feedback law k is continuous, in which case the solutions of the closed-loop system can be interpreted in the classical sense.

The above result suggests the following control strategy. Suppose that we know an upper bound on the initial state: $\|x(0)\| \leq E_0$. Fix an arbitrary $\epsilon \in (0, E_0)$. Choose Δ small enough so that

$$\Delta\sqrt{n}/2 \leq \kappa\delta \tag{27}$$

and take M large enough to have

$$\Gamma(E_0) \leq (M - 1/2)\Delta. \tag{28}$$

If we now take the above control law k and let $u = k(q_\Delta(x))$, then the quantizer will never saturate, and $x(T)$ will be in a ball of radius ϵ around the origin. We arrive at the following result.

Proposition 3. *Suppose that the system (10) is globally asymptotically controllable to the origin. Then there is a class $\mathcal{K}_\infty$ function Γ with the following property: for any numbers $E_0 > \epsilon > 0$ there exists a quantized feedback control law such that, whenever $\|x(0)\| \leq E_0$, the solution of the closed-loop system satisfies the estimates (23) and (24) for some $T > 0$ (semiglobal practical stability).*

One might also attempt to use a hybrid quantized feedback control policy, as in the previous sections, to drive x into a ball around 0 of a smaller radius $\epsilon' < \epsilon$. Namely, one can try to "zoom in", i.e., replace Δ at time T by a smaller value for which the inequality (28) holds with ϵ in place of E_0. The difficulty here is that both κ and δ in the inequality (27) depend on ϵ and E_0, so it is in general not guaranteed that this inequality will still be satisfied with $\kappa(\epsilon', \epsilon)$ and $\delta(\epsilon', \epsilon)$. Thus we see that the task of estimating the size of the smallest possible attractor requires a careful examination of the findings of [17], which is an interesting topic for further research.

Acknowledgments

The author is grateful to Roger Brockett, Eduardo Sontag, and Bo Hu for helpful discussions.

References

1. R. W. Brockett, Hybrid models for motion control systems, in *Essays on Control: Perspectives in the Theory and Its Applications* (H. Trentelman and J. C. Willems, eds.), Birkhäuser, Boston, 1993, pp. 29–53.
2. R. W. Brockett, D. Liberzon, Quantized feedback stabilization of linear systems, *IEEE Trans. Automat. Control*, Jul 2000, to appear.
3. J.-H. Chou, S.-H. Chen, I.-R. Horng, Robust stability bound on linear time-varying uncertainties for linear digital control systems under finite wordlength effects, *JSME Internat. J. Series C*, vol. 39, 1996, pp. 767–771.
4. F. H. Clarke, Yu. S. Ledyaev, E. D. Sontag, A. I. Subbotin, Asymptotic controllability implies feedback stabilization, *IEEE Trans. Automat. Control*, vol. 42, 1997, pp. 1394–1407.
5. D. F. Delchamps, Stabilizing a linear system with quantized state feedback, *IEEE Trans. Automat. Control*, vol. 35, 1990, pp. 916–924.
6. N. C. S. Fah, Input-to-state stability with respect to measurement disturbances for one-dimensional systems, *ESAIM J. Control, Optimization and Calculus of Variations*, vol. 4, 1999, pp. 99–122.
7. X. Feng, K. A. Loparo, Active probing for information in control systems with quantized state measurements: a minimum entropy approach, *IEEE Trans. Automat. Control*, vol. 42, 1997, pp. 216–238.
8. A. F. Filippov, *Differential Equations with Discontinuous Righthand Sides*, Kluwer, Dordrecht, 1988.
9. R. Freeman, Global internal stabilizability does not imply global external stabilizability for small sensor disturbances, *IEEE Trans. Automat. Control*, vol. 40, 1995, pp. 2119–2122.
10. R. Freeman, P. V. Kokotovic, Global robustness of nonlinear systems to state measurement disturbances, in *Proc. 32nd Conf. on Decision and Control*, 1993, pp. 1507–1512.
11. B. Hu, Z. Feng, A. N. Michel, Quantized sampled-data feedback stabilization for linear and nonlinear control systems, in *Proc. 38th Conf. on Decision and Control*, 1999, pp. 4392–4397.
12. Z.-P. Jiang, I. Mareels, D. Hill, Robust control of uncertain nonlinear systems via measurement feedback, *IEEE Trans. Automat. Control*, vol. 44, 1999, pp. 807–812.
13. Yu. S. Ledyaev, E. D. Sontag, A Lyapunov characterization of robust stabilization, *J. Nonlinear Analysis*, vol. 37, 1999, pp. 813–840.
14. J. Raisch, Control of continuous plants by symbolic output feedback, in *Hybrid Systems II* (P. Antsaklis et al., eds.), Springer-Verlag, Berlin, 1995, pp. 370–390.
15. E. D. Sontag, Smooth stabilization implies coprime factorization, *IEEE Trans. Automat. Control*, vol. 34, 1989, pp. 435–443.
16. E. D. Sontag, Input/output and state-space stability, in *New Trends in Systems Theory* (G. Conte et al., eds.), Birkhäuser, Boston, 1991, pp. 684–691.
17. E. D. Sontag, Feedback insensitivity to small measurement errors, in *Proc. 38th Conf. on Decision and Control*, 1999, pp. 2661–2666.

18. E. D. Sontag, Y. Wang, On characterizations of the input-to-state stability property, *Systems Control Lett.*, vol. 24, 1995, pp. 351–359.
19. E. D. Sontag, Y. Wang, New characterizations of input to state stability, *IEEE Trans. Automat. Control*, vol. 41, 1996, pp. 1283-1294.
20. J. Sur, B. E. Paden, State observer for linear time-invariant systems with quantized output, *ASME J. Dynamic Systems, Measurement, and Control*, vol. 120, 1998, pp. 423–426.
21. W. S. Wong, R. W. Brockett, Systems with finite communication bandwidth constraints I: State estimation problems, *IEEE Trans. Automat. Control*, vol. 42, 1997, pp. 1294–1299.
22. W. S. Wong, R. W. Brockett, Systems with finite communication bandwidth constraints II: Stabilization with limited information feedback, *IEEE Trans. Automat. Control*, vol. 44, 1999, pp. 1049–1053.

Diagnosis of Quantised Systems by Means of Timed Discrete-Event Representations

Jan Lunze

Technische Universität Hamburg–Harburg, Arbeitsbereich Regelungstechnik, Eissendorfer Str. 40, D–21071 Hamburg, Germany, Lunze@tu-harburg.de

Abstract. The paper deals with the diagnosis of continuous–variable or hybrid systems whose state can be measured only by means of a quantiser. Hence, the on–line information used in the diagnosis is given by the sequences of input and output events. The paper describes how the quantised system can be represented by a semi–Markov process and how the diagnostic problem can be solved by using this timed discrete–event representation. A specific result is obtained if the model is does not include probabilistic information about the event occurrence. The diagnostic method is illustrated by considering a numerical example which concerns a part of a batch process. The results show that the temporal information included in the semi–Markov process is crucial for fault diagnosis of discrete–event systems.

1 Introduction

Diagnosis of quantised systems. This paper is concerned with the diagnosis of dynamical systems with discrete inputs and outputs. As shown in Fig. 1, the system under consideration is a continuous–variable continuous–time system that can be described by the state–space model

$$\dot{x} = f(x(t), u(t), f), \quad x(0) = x_0 .$$ (1)

The system behaviour depends on the fault $f \in \mathcal{F}$ where $f_0 \in \mathcal{F}$ symbolises the faultless system.

The system state x is accessible only through a quantiser, which generates an event whenever the state changes its qualitative value $[x]$. Hence, the system output is a timed event sequence

$$E_t(0...t_h) = (E_0, T_0; \ E_1, T_1; \ E_2, T_2; \ ...; E_H, T_H)$$ (2)

in which E_k denotes the name and T_k the occurrence time of the k–th event. H is the number of events that the quantised system generates in the time interval $[0, t_h]$.

The injector associates with a given input event sequence

$$V(0...t_h) = (v_0, v_1, v_2, ..., v_H)$$

N. Lynch and B. Krogh (Eds.): HSCC 2000, LNCS 1790, pp. 258–271, 2000.

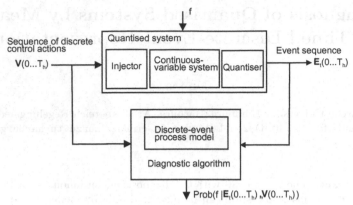

Fig. 1. Diagnosis of quantised systems

the input $u(t)$ of the continuous–variable system defined by

$$u(t) = u^{v_k} \quad \text{for} \quad t_k \leq t < t_{k+1} . \tag{3}$$

The system consisting of the continuous–variable system, the quantiser and the injector is called the *quantised system*.

The **diagnostic problem** is given as follows:

Given: Model M of the quantised system
        Input and output event sequences $E_t(0...t_h)$ and $V(0...t_h)$
Find: Fault f

The development of a diagnostic method that solves this problem consists of two major steps:

1. **Modelling:** A model M of the quantised system has to be found that is simple enough to be used in the diagnostic algorithm. Here, M is a semi–Markov process. In Sect. 5 it will be shown how this model can be set up for the quantised system.
2. **Diagnosis:** A diagnostic algorithm has to be found for determining the fault f for given input and output event sequences. In Sect. 6 such a diagnostic algorithm will be elaborated for the semi–Markov process M.

After presenting the method for modelling hybrid systems by semi–Markov processes and for diagnosing quantised systems by using the semi–Markov process, specific results are obtained for timed nondeterministic representations that do not include probabilistic information about the quantised system. A comparison of the different results show that the temporal information about the quantised system is crucial for diagnosis.

Relevant literature. Results along this line of research have been obtained in two fields. The modelling problem for quantised systems has been investigated, for example, in [2], [3], [10] or [13]. As an extension to these results, the model proposed in [4] is used here. It describes the *timed* event sequences and, thus, includes more information about the quantised system. On the other hand, diagnosing quantised systems by means of a discrete–event representation has been investigated in [1], [7], [8], [11] or [12]. This paper extends these methods to timed discrete–event representations. The proof of the diagnostic algorithm can be found in [5].

2   Example: Diagnosis of a Batch Process

The class of diagnostic problems considered in this paper is illustrated by the batch process depicted in Fig. 2. The tank system is a hybrid systems because the dynamic properties are switched depending on the current liquid levels. The dashed lines mark liquid levels, which are measured by sensors that indicate only if the level is higher or lower than its position. These sensors act as quantisers.

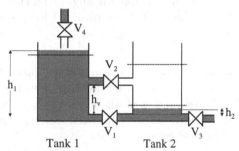

Fig. 2. Example of a batch process

The following operation from a batch process is considered. At $t = 0$ the liquid level in Tank 1 (left) is "high" (i.e. higher than the dashed line) and the level in Tank 2 is "low". The aim is to bring the level in the right tank above the upper dashed line. To do this, the Valves V_1, V_2 and V_4 are opened and Valve V_3 closed.

The only on–line information is obtained from the qualitative sensors positioned at the dashed lines in Fig. 2. Consequently, the behaviour of the system is considered in the partitioned state space depicted in Fig. 3. If the trajectory of the system crosses one of the partition borders, an event is generated. Figure 3 shows as two examples the events e_{34} and e_{43}.

The fault set $\mathcal{F} = \{f_0, f_1, f_2, f_3, f_4\}$ is considered where f_1, f_2 and f_4 denote the situation that the Valve V_1, V_2 or V_4 is not opened, respectively, and f_3 describes that Valve V_3 is not closed. The diagnostic problem is to find the fault as quickly as possible after the control input, which opens the valves V_1, V_2 and V_4, has been applied.

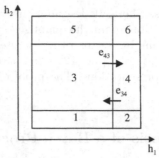

Fig. 3. Partition of the state space of the two tanks (x_1 = level of Tank 1; x_2 = level of Tank 2)

3   Quantised Continuous–Variable Systems

The quantised system consists of the continuous–variable system (1), the quantiser and the injector. It is assumed that for any initial state x_0 and input $u(t)$ (1) has a unique solution, which will be considered for the time interval $[0, t_h]$ and denoted by $x_{[0,t_h]}$.

Quantisation of the state space. The quantiser maps the state space IR^n onto a finite set $\mathcal{N}_x = \{0, 1, 2, ..., N\}$ of qualitative values and, thus, introduces a partition of IR^n into a finite number of disjoint sets $\mathcal{Q}_x(i)$, where $\mathcal{Q}_x(i)$ denotes the set of states $x \in \mathsf{IR}^n$ with the same qualitative value i. The mapping invoked by the quantiser is symbolised by the symbol $[.]$:

$$[x] = i \iff x \in \mathcal{Q}_x(i) \ . \tag{4}$$

The sets $\mathcal{Q}_x(i)$ $(i = 1, ..., N)$ are assumed to be bounded while $\mathcal{Q}_x(0)$ is the unbounded "remaining" subset of IR^n: $\mathcal{Q}_x(0) = \mathsf{IR}^n \setminus \left(\cup_{i=1}^N \mathcal{Q}_x(i) \right)$. For the bounded sets, $\delta\mathcal{Q}_x(i)$ denotes the hull of $\mathcal{Q}_x(i)$. Figure 3 illustrates the state quantisation for the example batch process.

The quantised system is said to generate the **event** e_{ij} at time t_k if

$$[x(t_k + \delta t)] = i \quad \text{and} \quad [x(t_k - \delta t)] = j \tag{5}$$

hold for some $i, j \in \mathcal{N}_x, i \neq j$ for arbitrarily small $\delta t > 0$. In this way, the timed event sequence (2) is obtained. The relation between a continuous state trajectory $x_{[0,t_h]}$ and the event sequence E_t is given by the quantiser, which is symbolised by

$$E_t(0...t_h) = \mathrm{Quant}_t \left(x_{[0,t_h]} \right) \ . \tag{6}$$

Injector. The injector associates with the discrete input value $[u(t)] = v$ ($v \in \mathcal{N}_u$) a quantitative value $u^v \in \mathcal{U} \subseteq \mathsf{IR}^m$. It is assumed that the input u changes its value simultaneously with the qualitative state $[x]$. In this way, a given input event sequence V is transformed into the input function (3).

4   The Modelling Problem

For the diagnosis, a model has to be used which generates for every given initial event e_0 and input event sequence V the event sequence $E_t(0...t_h)$ for all faults $f \in \mathcal{F}$. Such a model is available if (1) is combined with the quantiser and the injector. However, this model includes continuous–variable and discrete–event parts. For diagnosis, a more compact model has to be found.

Nondeterminism of the discrete–event behaviour. An inherent problem of this modelling task results from the from the fact that these event sequences are not unique [6]. The nondeterminism of the discrete–event behaviour means that the quantised system may generate one of a set of different event sequences E_t and it is impossible to select the true sequence in advance. The reason for this is given by the fact that the initial state x_0 of the system (1) is unknown. After the first event e_0 has been observed at time t_0, the state of the system is known to lie in a subset $\delta \mathcal{Q}(e_0)$ of the corresponding partition border . If, for notational convenience, t_0 is assumed to be zero, the occurrence of the event e_0 restricts x_0 to be in the set $\delta \mathcal{Q}(e_0)$ at that time instant. Depending on x_0 the system may produce one event sequence of the set

$$\mathcal{S}_t(e_0, V(0...t_h), f) = \{ E_t = \text{Quant}_t(x_{[0,t_h]}) \mid \text{Eqns. (1), (3) hold for some}$$
$$x_0 \in \delta \mathcal{Q}_x(e_0) \} . \tag{7}$$

Moreover, the temporal distance of the events may vary considerably.

Stochastic properties of the quantised system. A compact representation of the nondeterministic behaviour of the quantised system can be obtained by a statistical evaluation. It is assumed that the initial state x_0 of the continuous–variable system (1) is uniformly distributed over the set $\delta \mathcal{Q}(e_0)$. Then the event sequence E_t is a random sequence with $E_t \in \mathcal{S}_t(e_0, V, f)$. The probability that the event e has occurred before or at time t is denoted by

$$V_e(e, t, f) = \sum_k \text{Prob}\,(E_k = e,\, T_k \le t \mid F = f) . \tag{8}$$

Figure 4 shows the statistical properties of the quantised tank system. The strips depict the probability $V_e(e, t, f_1)$ in grey scale. The strips are shown only for the time interval in which $\frac{dV_e}{dt} > 0$ holds, because the event e may occur exactly in this time interval. The darker the strip is the more probable is the occurrence of the event until the corresponding time instant. The numbers on the right margin show the final probability values.

Modelling aim. The model, which will be used for diagnosis, should describe the relation between the initial event e_0, the qualitative input sequence $V(0...t_h)$ and the event sequence $E_t(0...t_h)$ for all faults $f \in \mathcal{F}$. Since the behaviour is

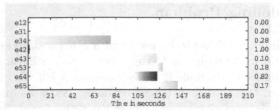

Fig. 4. Graphical representation of the statistical properties of the tank system for fault f_1 and initial event e_{42}

nondeterministic, a nondeterministic model has to be used which generates a set $\mathcal{M}_t(e_0, \boldsymbol{V}, f)$ of event sequences. The modelling aim is to find a representation such that the relation

$$\mathcal{M}_t(e_0, \boldsymbol{V}(0...t_h), f) \supseteq \mathcal{S}_t(e_0, \boldsymbol{V}(0..t_h), f) \tag{9}$$

holds for all e_0, $\boldsymbol{V}$, f and t_h. That is, the model is *complete* in the sense that it generates all event sequences that the quantised system may generate.

5   Representation of Quantised Systems by Semi–Markov Process

5.1   Brief Introduction to Semi–Markov Processes

In a semi–Markov process $M_T(\mathcal{Z}, \mathcal{V}, f_T, z_0)$, $\mathcal{Z}$ is the set of states, $\mathcal{V}$ the set of input values (input alphabet), f_T the probability density function and z_0 the initial state. The semi–Markov process changes its state Z instantaneously at the time instances T_k $(k = 1, 2, ..., H)$ so that the process can be described by the sequence

$$\boldsymbol{Z}_t(0...t_h) = (Z_0, T_0; Z_1, T_1; Z_2, T_2; ...; Z_H, T_H) , \tag{10}$$

which means that the process assumes state Z_k in the time interval $[T_k, T_{k+1})$.

As the state of the semi–Markov process cannot be unambiguously predicted, the process is described by the state probability $p_t(z, t) = \text{Prob}\,(Z(t) = z)$. It is assumed that the input $V(t)$ of the semi–Markov process changes simultaneously with the state such that $V(t) = V_k$ for $T_k \leq t < T_{k+1}$.

In order to determine $p_t(z, t)$ for $t > 0$, the transition relation between any pair of states $z, \tilde{z} \in \mathcal{Z}, z \neq \tilde{z}$ has to be described. This is done in terms of the sojourn time $\tau = T_k - T_{k-1}$ by the probability distribution

$$F_{z\tilde{z}}(\tau, v) = \text{Prob}\,(Z_1 = z, T_1 \leq \tau \,|\, Z_0 = \tilde{z}, T_0 = 0, V_0 = v) \text{ for } z \neq \tilde{z}. \tag{11}$$

of the semi–Markov process, which is assumed to be homogeneous.

The semi–Markov process can generate any trajectory (10) for which any pair $(z_{k+1}, t_{k+1}; z_k, t_k)$ of successive states can occur with non–vanishing probability. Hence, the set of trajectories with length H that the semi–Markov process generates when starting in the initial state z_0 under the influence of the input sequence $\boldsymbol{V}(0...t_h)$ is given by

$$\mathcal{M}_t(z_0, \boldsymbol{V}(0...t_h)) = \{(z_0, 0;\ z_1, t_1;\ ...;\ z_H, t_h) \,|\, f_{z_{k+1}z_k}(t_{k+1} - t_k, v_k) > 0$$
$$\text{holds for } k = 0, 1, ..., H - 1\} \tag{12}$$

with

$$f_{z\tilde{z}}(\tau, v) = \frac{d}{d\tau} F_{z\tilde{z}}(\tau, v) \ .$$

5.2   Representation of the Quantised System by a Semi–Markov Process

This section extends the results of [4] for describing quantised systems by a semi–Markov process to systems with inputs and faults. Lemma 1 states how the probability density f_T of the semi–Markov process has to be chosen for a given quantised system in order to satisfy the modelling aim (9).

The semi–Markov process $M_T(\mathcal{E}, \mathcal{V}, \mathcal{F}, f_T, e_0)$ is now used with $\mathcal{Z} = \mathcal{E}$, where $\mathcal{E}$ is the set of all events that the quantised system may generate. Thus, all relations of the preceeding section can be written with e replacing z. The fault f considered occurs as new argument in all functions that have been introduced for the semi–Markov process, particularly in the probability density

$$f_T : \mathcal{E} \times \mathcal{E} \times \mathsf{IR}^+ \times \mathcal{V} \times \mathcal{F} \longrightarrow [0, 1]$$

which will be referred to by the abbreviation

$$f_{e_{k+1}e_k}(\tau, v, f) = f_T(e_{k+1}, e_k, \tau, v, f) \ .$$

In order to satisfy the modelling aim (9) the probability density function of the semi–Markov model has to be chosen according to the relation

$$\boxed{\begin{array}{l} \text{Timed Abstraction:} \\[2mm] f_{e\tilde{e}}(\tau, v, f) = \dfrac{d}{d\tau}\text{Prob}\left(E_1 = e, T_1 \leq \tau \mid E_0 = \tilde{e}, T_0 = 0, V_0 = v, F = f\right) \\[2mm] \hfill \text{for } e \neq \tilde{e} \ . \end{array}} \tag{13}$$

On the right–hand side, a pair $(\tilde{e}, e)$ of succeeding events is considered and the probability of its occurrence determined by means of (1), (3) for given v and f.

Lemma 1 *The semi–Markov process $M_T(\mathcal{E}, \mathcal{V}, \mathcal{F}, f_T, e_0)$ satisfies the modelling aim (9) if the probability density f_T satisfies (13).*

This lemma follows from Theorem 1 in [4] if the model is considered for fixed $v \in \mathcal{V}$ and $f \in \mathcal{F}$. The set $\mathcal{M}_t(e_0, \boldsymbol{V}(0...t_h), f)$ generated by the semi–Markov process is given by

$$\mathcal{M}_t(e_0, \boldsymbol{V}(0...t_h), f) = \{(e_0, 0;\ e_1, t_1;\ ...;\ e_H, t_H) \,|\, f_{e_{k+1}e_k}(t_{k+1} - t_k, v_k, f) > 0$$
$$\text{holds for } k = 0, 1, ..., H - 1\} \ .$$

6   Diagnosis of the Quantised System

6.1   The Principle of Consistency–Based Diagnosis

The diagnostic problem can be posed as the following question:

Can the quantised system generate the event sequence $E_t(0...t_h)$ if it has obtained the input sequence $V(0...t_h)$, i.e. does the relation

$$E_t(0...t_h) \in S_t(e_0, V(0...t_h), f) \tag{14}$$

hold for some $f \in \mathcal{F}$?

Note that e_0 on the right–hand side of (14) is the first element of E_t on the left–hand side. The diagnostic result is denoted by $p(f, t_h)$ as follows:

$$\begin{aligned} p(f, t_h) > 0 \text{ if } & E_t(0...t_h) \in S_t(e_0, V(0...t_h), f) \\ p(f, t_h) = 0 \text{ else .} & \end{aligned} \tag{15}$$

$p(f, t_h) > 0$ says that the observed behaviour over the time horizon $[0, t_h]$ is *consistent* with the quantised system and $p(f, t_h) = 0$ means that the fault f cannot have occurred.

6.2   Diagnosis of Semi–Markov Processes

The diagnostic problem is first solved for the model $M_T(\mathcal{E}, \mathcal{V}, \mathcal{F}, f_T, e_0)$. The result is denoted by

$$p_M(f, t_h) = \text{Prob} \left(f \mid E_t(0...t_h), V(0...t_h) \right) , \tag{16}$$

so that the relation

$$\begin{aligned} p_M(f, t_h) > 0 \text{ if } & E_t(0...t_h) \in M_t(e_0, V(0...t_h), f) \\ p_M(f, t_h) = 0 \text{ else} & \end{aligned} \tag{17}$$

holds.

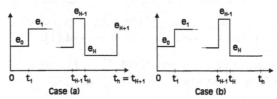

Fig. 5. State sequence of the semi–Markov process

The solution will be described with the symbols defined in Figure 5. The input and state sequences are considered for the closed time interval $[0, t_h]$. It is assumed that H is the number of events that occurred in the open time interval

$[0, t_h)$. The event e_H was generated at time t_H. v_H is the input to the system for $t \geq t_H$.

For the solution of the diagnostic problem two cases have to be distinguished:

Case (a): At time t_h the $(H+1)$-st event e_{H+1} occurs, i.e. $t_{H+1} = t_h$
Case (b): There is no event occurring at time t_h, i.e. $t_{H+1} > t_h$.

Since the fault may occur at any time $t \leq 0$ it may or may not influence the initial event e_0, which is assumed to occur at time $t_0 = 0$. Hence,

$$p_M(f, 0) = \frac{1}{n_F} \quad \text{for all } f \in \mathcal{F} \tag{18}$$

is used where n_F denotes the number of faults considered.

The diagnostic result is obtained for time t_h by first determining an auxiliary function p_a:

$$\begin{aligned} \text{Case (a): } p_a(f, t_h) &= f_{e_{H+1}e_H}(t_h - t_H, v_H, f)\, p_M(f, t_H) \\ \text{Case (b): } p_a(f, t_h) &= F_{e_H}(t_h - t_H, v_H, f)\, p_M(f, t_H) \end{aligned} \tag{19}$$

with

$$F_{\tilde{e}e}(\tau, v, f) = \int_0^\tau f_{\tilde{e}e}(\tau, v, f)\, d\tau$$

$$F_e(\tau, v, f) = 1 - \sum_{\tilde{e} \in \mathcal{E}, \tilde{e} \neq e} F_{\tilde{e}e}(\tau, v, f). \tag{20}$$

Second, the diagnostic result is obtained from

$$p_M(f, t_h) = \frac{p_a(f, t_h)}{\sum_{f \in \mathcal{F}} p_a(f, t_h)} \tag{21}$$

provided that

$$\sum_{f \in \mathcal{F}} p_a(f, t_h) > 0 \tag{22}$$

holds.

Theorem 1 $p_M(f, t_h)$ *obtained by (21) describes the probability (16) that the output sequence $\boldsymbol{E}_t(0...t_h)$ has been generated for the input sequence $\boldsymbol{V}(0...t_h)$ by the semi–Markov process with fault f.*

If (22) is violated for some time t_h the event sequence $\boldsymbol{E}_t$ is inconsistent with the semi–Markov process for all $f \in \mathcal{F}$. The proof is given in [5].

6.3   Diagnosis of Quantised Systems

The algorithm described in Sect. 6 is now used to solve the diagnostic problem for the quantised system. This can be done due to the following theorem [5]:

Theorem 2 *Assume that the diagnostic algorithm (18) – (21) has been applied to the semi–Markov process M_T for a given input sequence $V(0...t_h)$ and event sequence $E_t(0...t_h)$. The relation*

$$p_M(f, t_h) = 0 \text{ implies } p(f, t_h) = 0 \qquad (23)$$

holds if and only if the model is complete and, thus, the requirement (9) is satisfied.

Theorem 2 shows that the semi–Markov process, which is obtained from the abstraction operation described in (13) can be used for diagnosing the quantised system. Equation (23) yields the following corollary.

Corollary 1 *The diagnostic algorithm (18) – (21) which is applied to the semi–Markov process that satisfies the relation (13) yields the following results:*

- **Fault detection:** *If $p_M(f_0, t_h) = 0$ holds (where f_0 symbolises the faultless system), then some fault has occurred in the quantised system.*
- **Fault identification:** *If $p_M(f, t_h) = 0$ holds, the quantised system has not been effected by the fault f.*

If $p_M(f, t_h) \neq 0$ holds, the fault f may have occurred. Fault identification by consistency–based diagnosis means to *exclude* those faults that, according to the information available, is known not to have occurred.

6.4   Diagnosis by Means of a Nondeterministic Representation of the Quantised System

In this section, the diagnosis will be considered under the assumption that the probabilistic information included in the probability density function f_T of the semi–Markov process is not available. Then the timed description provides only time intervals

$$\mathcal{T}_{e\tilde{e}}(v, f) = [t_{\min e\tilde{e}}(v, f), t_{\max e\tilde{e}}(v, f)]$$

with the upper and lower bounds $t_{\max e\tilde{e}}$ and $t_{\min e\tilde{e}}$ of the time that passes after the event $\tilde{e}$ before the successor event e is generated by the quantised system. This result is interesting for three reasons. First, if the model (1) of the continuous–variable system is not available, experiments made with the quantised system can bring about the information required to determine the time interval $\mathcal{T}_{e\tilde{e}}$, whereas these experiments may be insufficient to provide the probabilistic information included in f_T. Second, if the diagnostic results obtained for the semi–Markov process and this nondeterministic representation are compared it becomes obvious whether the probabilistic or the temporal information included in the semi–Markov process is of more importance for the efficiency of the diagnosis. Third, in discrete–event systems theory models are used which describe the temporal distance of events by time intervals, for example, time–labelled Petri nets. In [13] a method is described for obtaining such a model for quantised systems. Such models can be used in the following diagnostic algorithm.

If the semi–Markov process is given, the borders of the time interval $\mathcal{T}_{e\tilde{e}}$ can be determined as follows:

$$t_{\mathrm{mine}\tilde{e}}(v, f) = \min_{\tau} f_{e\tilde{e}}(\tau, v, f) \neq 0$$
$$t_{\mathrm{maxe}\tilde{e}}(v, f) = \max_{\tau} f_{e\tilde{e}}(\tau, v, f) \neq 0 \ .$$

However, it can be determined also experimentally by measuring the time that passes between the events $\tilde{e}$ and e if the quantised system with input v and fault f has different initial states x_0.

The diagnostic result $p_M(f, t_h)$ is no longer the probability of the fault occurrence, but it only shows whether the fault f can be diagnosed until time t_h $(p_M(f, t_h) = 1)$ or not $(p_M(f, t_h) = 0)$. Therefore, the following modifications have to be made for the diagnostic algorithm. The initial values are

$$p_M(f, 0) = 1 \quad \text{for all } f \in \mathcal{F} \tag{24}$$

because no fault can be excluded without any on–line information about the quantised system. The auxiliary function p_a says whether the quantised system subject to fault f and input v_H can generate the event e_{H+1} after the sojourn time $t_h - t_H$:

$$\text{Case (a): } p_a(f, t_h) = \begin{cases} 1 \text{ if } p_M(f, t_H) = 1 \text{ and } t_h - t_H \in \mathcal{T}_{e_{H+1}e_H}(v_H, f) \\ 0 \text{ else} \end{cases}$$
$$\text{Case (b): } p_a(f, t_h) = \begin{cases} 1 \text{ if } p_M(f, t_H) = 1 \text{ and } t_h - t_H \in \mathcal{T}_{e_H}(v_H, f) \\ 0 \text{ else} \end{cases} \tag{25}$$

$\mathcal{T}_{e_H}$ describes which time may pass until the quantised system generates the successor event of e_H:

$$\mathcal{T}_{e_H}(v, f) = \{t \mid \exists \bar{t} > t, e : \bar{t} \in \mathcal{T}_{ee_H}\} \ .$$

The diagnostic result is obtained from

$$p_M(f, t_h) = p_a(f, t_h) \tag{26}$$

provided that

$$\sum_{f \in \mathcal{F}} p_a(f, t_h) > 0 \tag{27}$$

holds.

7   Example: Diagnostic Results for the Batch Process

The diagnostic algorithm is now applied to the batch process. Figure 6 compares the event sequences that the tank system generates for the different faults with

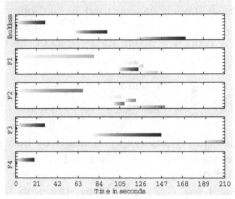

Fig. 6. Comparison of the discrete–event behaviour for different faults

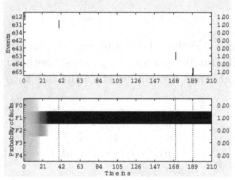

Fig. 7. Discrete–event behaviour of the process subject to fault f_1 and diagnostic result

initial event e_{42}. Note that the event sequence $(e_{42}, e_{34}, e_{53}, e_{65})$ may be generated by the faultless system as well as by the system with the faults f_1, f_2 or f_3. Therefore, fault diagnosis is possible only if the temporal distance of the events are taken into account.

In the upper part of Fig. 7 the discrete–event behaviour of the tank system is presented for fault f_1. The dashes show at which time the events e_{12}, e_{31}, e_{53}, and e_{65} occur. These time instants are marked in the lower part of the figure by dotted lines. The lower part shows the diagnostic result, where the probability $p_M(f, t_h)$ is depicted in grey scale. Obviously, the fault f_1 is uniquely detected after about 30sec. That is, $p_M(f_1, t_h) = 1$ holds for $t_h > 30$, which is also indicated at the right margin of the figure. Note that the diagnosis is finished before the second event occurs.

The figures show how the fault probabilities change over time. In practical applications, a threshold will be used and a fault is announced only if its probability exceeds this threshold. This, however, includes some heuristics concerning the threshold level, which is not the subject of this paper.

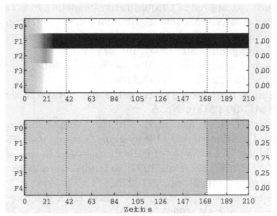

Fig. 8. Comparison of the diagnostic result obtained by means of the semi–Markov model (top) and the untimed stochastic automaton (bottom)

If the nondeterministic model without probabilistic information is used in the diagnosis, the result is the same with the only difference that all stripes in Fig. 7 are black rather than grey. Consequently, the results of fault identification are the same. The additional probabilistic information included in the semi–Markov process makes it possible to distinguish all fault, which cannot be excluded, concerning the degree of certainty with which they exist in the quantised system. This degree of certainty is described by the different grey levels in the figure.

Figure 8 shows a comparison with the diagnostic result obtained by using an untimed description of the quantised system. Obviously, the result with the timed model is much better. This demonstrates the fact that the temporal information included in the semi–Markov model is the key information for fault identification.

Conclusions

The paper has presented a method for diagnosing quantised continuous–variable systems. The method is based on a timed discrete–event representation of the quantised system by means of a semi–Markov process. It has been shown how the probability density function of the semi–Markov process can be obtained from the quantised system and how this model can be used for diagnosis. The diagnostic algorithm is very simple. It includes only some multiplications to be carried out in each recursion step. The simplicity of the algorithm is based on the simplification of the model, which has been introduced by the timed abstraction.

References

1. G. Lichtenberg, A. Steele: "An approach to fault diagnosis using parallel qualitative observers", *Workshop on Discrete Event Systems*, Edinburgh 1996, pp. 290–295.
2. J. Lunze: A Petri-net approach to qualitative modeling of continuous dynamical systems, Systems Analysis, Modelling, Simulation, **9** (1992), pp 88–111.
3. J. Lunze, "Qualitative modelling of linear dynamical systems with quantized state measurements", *automatica* **30** (1994), pp. 417–431.
4. J. Lunze, "A timed discrete–event abstraction of continuous–variable systems", *Intern. J. Control* **72** (1999), pp. 1147-1164.
5. J. Lunze, "Diagnosis fo quantised systems based on a timed discrete–event model", *IEEE Trans.* **SMC-30** (2000), No. 5.
6. J. Lunze, B. Nixdorf, B., J. Schröder, "On the nondeterminism of discrete–event representations of continuous–variable systems," *automatica* **35** (1999), 395–408.
7. J. Lunze, F. Schiller, "An example of fault diagnosis by means of probabilistic logic reasoning", *SAFEPROCESS*, Hull 1997; extended version to appear in *Control Engineering Practice* **7** (1999), pp. 271–278.
8. J. Lunze; J. Schröder: Process diagnosis based on a discrete–event description, *Automatisierungstechnik* **47** (1999), 358–365.
9. J. Lunze; T. Serbesow: Logikbasierte Prozeßdiagnose unter Berücksichtigung der Prozeßdynamik, *Messen, Steuern, Regeln* **34** (1991), 163–165 und 253–257.
10. J. Raisch, S. O'Young, "A totally ordered set of discrete abstractions for a given hybrid or continuous system", In: P. Antsaklis, W. Kohn, A. Nerode, S. Sastry, Eds., *Hybrid Systems IV*, Lecture Notes in Computer Science, vol. 1273, pp. 342–360 Berlin: Springer–Verlag, 1997.
11. M. Sampath, R. Sengupta, S. Lafurtune, K. Sinnamohideen, D. Teneketzis, "Diagnosability of discrete event systems", *IEEE Trans.*, vol. AC–40, pp. 1555-1575, 1995.
12. V.S. Srinivasan, M.A. Jafari, "Fault detection/monitoring using timed Petri nets", *IEEE Trans.*, vol. SMC–23, 1993.
13. O. Stursberg; S. Kowalewski; S. Engell: Generating timed discrete models, *2–nd MATHMOD*, Vienna 1997, pp. 203–207.

Existence and Stability of Limit Cycles in Switched Single Server Flow Networks Modelled as Hybrid Dynamical Systems [*]

Alexey S. Matveev[1] and Andrey V. Savkin[2]

[1] Department of Mathematics and Mechanics, St.Petersburg State University,
Bybliotechnaya 2, Petrodvoretz, 198904, St.Petersburg, Russia
almat@am1540.spb.edu
[2] Department of Electrical and Electronic Engineering,
the University of Western Australia, Nedlands, WA 6907, Australia
savkin@ee.uwa.edu.au

Abstract. The paper deals with the qualitative analysis of the so-called switched flow networks. Such networks are used to model various communication, computer, and flexible manufacturing systems. We prove that for any deterministic network from a specific class, there exists a finite number of limit cycles attracting all the trajectories. Furthermore, we determine this number.

1   Introduction

The paper considers hybrid dynamical systems that are called switched flow networks. Special classes of such networks were introduced in [7] to model flexible manufacturing/assembly/disassembly systems. These networks are also useful to model various computer and communication systems, especially those with time-sharing schemes. Other examples concern batch processes, chemical kinetics, and biotechnological processes.

As is known, even very simple flow networks of the second order can exhibit a chaotic, irregular, unpredictable behavior [1,2]. Such a behavior is unacceptable for most of real systems. A typical synthesis problem (see [2,10,11,3]) is to find a feedback switching policy that ensures a regular, predictable behavior of the network. Dealing with this problem involves qualitative analysis of the dynamics of the close-loop system. Up to now samples of such investigations [1,2,10,3] were mainly confined to specific two-dimensional systems. The main idea underlying the theoretical analysis was reduction to iterated maps of an interval into itself.

The network studied in this paper consists of buffers (nodes) connected with links (edges). We refer to the content of buffers as "work"; it will be convenient to think of work as a fluid, and a buffer as a tank. (In applications, "work" may represent a continuous approximation of a discrete flow of jobs in a computer system or parts in a manufacturing system, etc.) Work arrives from outside the system at fixed rates at certain buffers. The network is processed by a single

---

[*] This work was supported by the Australian Research Council

N. Lynch and B. Krogh (Eds.): HSCC 2000, LNCS 1790, pp. 272–282, 2000.

server, which is able to deal with only one buffer at any moment. The server removes work from a selected buffer and delivers it at fixed rates along the edges departing from this buffer. The location of the server is a discrete control variable determined by a feedback policy.

We consider quite general networks of arbitrary dimension. More precisely, we assume that the network may have an arbitrary number of nodes and any node may have an arbitrary number of edges both departing from and arriving at it. Nevertheless, an edge coming from inside the system and one coming from outside it cannot arrive at a common node. Furthermore, we suppose that the network contains neither cycles nor impasses, i.e., for any node, there exists an edge both arriving at and departing from it. We consider a deterministic network; more precisely, the rates at which work is transferred along the edges are assumed to be constant and fixed. This model generalizes in particular those from [1,3], where the case of three buffers with no edges between them and certain specific control policies was studied.

We show that, depending on the system parameters, either 1) the total amount of work in the buffers converges to infinity in course of time for any switching policy, or 2) no policy can keep the system working for a long time, so far as infinitely many buffer changes accumulate at the vicinity of a finite time instant, or 3) a scaled total amount of work in the buffers remains constant whatever control policy be adopted. (Underscore that the statement 2) concerns the fluid model of the network. At the same time, accumulation of buffer changes signals that the conditions under which the continuous (fluid) approximation can be employed to model the real (discrete) network are violated. So the conclusion in question certainly cannot be directly extended on the real-life discrete prototype of the model at hand.) The further consideration is focused on the case 3). We study a natural switching strategy that extends the so-called Clear-the-Largest-Buffer-Level [7] one. Our main result is that the close-loop system exhibits a periodic behavior almost always, i.e., whenever the tuple p of its parameters lies outside a certain set E of the zero measure. More precisely, there exists a finite number of limit cycles each being locally asymptotically stable, and any trajectory converges to some of them. Furthermore we count these cycles and discuss phenomena that occur if $p \in E$.

To obtain criteria for existence of self-excited oscillations or limit cycles is an old and challenging problem of the classic qualitative theory of differential equations whose origins may be traced back to the work of Poincaré and Lyapunov (see e.g. [6]). Few constructive results are known for nonlinear systems of order higher than 2. It is even harder to study stability of limit cycles. Our result shows that constructive criteria for existence and global stability of limit cycles can be proved for quite general switched flow networks. This appears to be surprising and gives us a hope that it is possible to develop a qualitative theory of some classes of hybrid dynamical systems, which will be even more constructive than the classic qualitative theory of differential equations.

The ideas underlying the proofs of the results presented are related to the general theory developed in [4,8,5,9].

2   Single Server Flow Networks

Consider an oriented graph with the set of the nodes

$$\widehat{G} := \{g_1, \ldots, g_L, g_{L+1} = \infty\}.$$

The edge departing from g_i and arriving at g_j is denoted by (g_i, g_j). (There is no more than one such edge.) The special node ∞ is interpreted as the exterior of the system. Correspondingly, any edge of the form (∞, g_i) and (g_i, ∞) (where $i = 1, \ldots, L$) is regarded as coming from outside and going outside the system, respectively.

Assumption 1 *The graph satisfies the following properties:*

— *If (∞, g_i) $(i = 1, \ldots, L)$ is an edge, there is no other edge arriving at the same node g_i.*
— *The graph contains no cycles. (In particular, (g, g) is not an edge for any $g \in \widehat{G}$.)*
— *For any node g_j $(j = 1, \ldots, L)$, there is an edge arriving at g_j, as well as that departing from g_j.*

Associated with each node

$$g \in G := \{g_1, \ldots, g_L\}$$

is a buffer (or tank). Its content is called "work" and interpreted as fluid. The work arrives to the system continuously along the edges of the form (∞, g) at a constant rate $\rho_g > 0$. There also is a server (or machine), which serves buffers. At any time, the server is able to deal with only one buffer. While so doing with a specific buffer g, the server removes work at a constant rate $s_g > 0$ and delivers it along the edges departing from g. The distribution of the work flow among the edges is in a given proportion. In other words, the server sends work along the edge (g, g') at a constant rate $s_g \rho(g, g')$, where $\rho(g, g') > 0$ and

$$\sum_{g' \in G(g)} \rho(g, g') = 1 \qquad \forall g \in G. \tag{1}$$

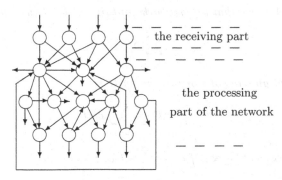

Fig. 1. A flow network.

Here
$$G(g) := \left\{ g' \in \widehat{G} : (g, g') \text{ is an edge} \right\}.$$

The location of the server is a control variable, which is chosen in accordance with a prescribed feedback control policy. We assume that the server switches between buffers instantaneously.

Depending on the system's parameters, certain dynamical properties can or cannot be ensured by choice of a switching policy. To specify this statement, we introduce some notations. Put

$$G_r(g) := G(g) \setminus \{\infty\}, \qquad S_{-1} := \{g \in G : G_r(g) = \emptyset\}.$$

Introduce also the sets $S_{-2}, S_{-3}, \ldots$ by setting iteratively for $i = -1, -2, \ldots$,

$$S_{i-1} := \{g \in G : G_r(g) \subset \mathfrak{S}_i := S_i \cup S_{i+1} \cup \ldots \cup S_{-1} \quad \text{and} \quad g \overline{\in} \mathfrak{S}_i\}. \quad (2)$$

As can be easily shown, the sets S_i are pair-wise disjoint and there exists an integer N such that $S_{-i} = \emptyset$ for all $i > N$, $S_{-1} \neq \emptyset, \ldots, S_{-N} \neq \emptyset$. Furthermore

$$G = S_{-1} \cup S_{-2} \cup \ldots \cup S_{-N}.$$

Next, we define a number $\delta_g > 0$ for any node $g \in G$. We first put $\delta_g := s_g^{-1}$ for all $g \in S_{-1}$. Suppose that the number δ_g has been defined for all $g \in \mathfrak{S}_i$. Then we put

$$\delta_g := s_g^{-1} + \sum_{g' \in G_r(g)} \delta_{g'} \rho(g, g') \qquad \forall g \in S_{-i-1}. \quad (3)$$

(In the sum on the right, the multiplier $\delta_{g'}$ is already defined in view of (2) and the induction hypothesis.) Finally, we introduce the set of the nodes at which work arrives from outside the system

$$\mathfrak{R} := \{g \in G : (\infty, g) \text{ is an edge}\}. \quad (4)$$

Denote by x_g the content of the buffer g.

Lemma 1. *Assume that the network contains at least two buffers. Suppose also that*

$$\sum_{g \in \mathfrak{R}} \delta_g \rho_g > 1. \quad (5)$$

Then the total amount of work in the system

$$w(t) := \sum_{g \in G} x_g(t)$$

converges to ∞ *as* $t \to \infty$. *If on the contrary*

$$\sum_{g \in \mathfrak{R}} \delta_g \rho_g < 1, \quad (6)$$

infinitely many buffer changes accumulate at the vicinity of a finite time $t_* \geq 0$. *These assertions are true irrespective of what control policy be adopted.**

Thus no control policy can make the system even dissipative (in the sense that $\lim\sup_{t\to\infty} w(t) < \infty$) if (5) holds. If on the contrary (6) is true, no control policy can keep the system working for a long time. (We underscore once more that Lemma 1 concerns the fluid model of the network. At the same time, accumulation of buffer changes means that while dealing with a specific buffer, the server processes an amount of work that tends to zero. This apparently contradicts the conditions under which the continuous fluid-like "work" can be considered as a proper model of a discrete and quantified flow of jobs in a computer system or parts in a manufacturing one etc. So in the case (6), the conclusions of the lemma cannot be directly applied to the real-life prototype of the model at hand if this prototype is discrete in its nature.) Further we consider the case where

$$\sum_{g\in\mathfrak{R}} \delta_g \rho_g = 1. \tag{7}$$

From now on this relation is assumed to be valid. It is easy to see that then the scaled total amount of work in the system

$$\sigma := \sum_{g\in G} \delta_g x_g \tag{8}$$

remains constant in the course of time.

3   A Switching Control Policy

In this paper, the network is regarded as composed of *receiving and processing* parts. The first one is constituted by the nodes from the set (4), the second one $\mathfrak{P}$ consists of the rest of the nodes (except for the "exterior" one ∞) $\mathfrak{P} := G\backslash\mathfrak{R}$. We assume that the server processes these parts separately on the base of the Clear-the-Largest-Buffer-Level policy [7]. Thus its work splits into consecutive *sessions* of serving either the receiving or the processing parts, respectively. More precisely, we consider the following switching policy.

SP1 *The server starts with the receiving part of the network.*

SP2 *This part is served on the basis of the Clear-the-Largest-Buffer-Level strategy* [7]. *This means that the server switches when the current buffer is emptied and to a buffer* $g \in \mathfrak{R}$ *with the largest (over* $\mathfrak{R}$ *) scaled content* $\zeta_g := c_g x_g$. *(Here* $c_g > 0$ *is a given scaled coefficient.) Likewise, the server starts*

---

* We assume, however, that the server is working constantly, i.e., there are no periods when it is standing idle. Note also that $t_* \leq \bar{t}$, where the time $\bar{t} = \bar{t}[w(0)]$ is independent of the switching policy. Any policy that makes the system working for the longest possible time $\tilde{t}$ clears up the network in the sense that $w(t) \to 0$ as $t \to \tilde{t}-0$.

to process the part in question with a buffer $g \in \mathfrak{R}$ having the largest value of ζ_g. **

SP3 *The server ends to deal with the receiving part when it has changed buffers $(k-1)$ times (within the current session) and has emptied the kth buffer $g \in \mathfrak{R}$. (Here k is the number of the buffers in the receiving part $\mathfrak{R}$.)*

SP4 *After this, the server enters the processing part of the network and serves it on the basis of a similar strategy. More precisely, the server switches when the current buffer $g \in \mathfrak{P} = G \setminus \mathfrak{R}$ is emptied and to a buffer $g \in \mathfrak{P}$ with the largest (over $\mathfrak{P}$) scaled content $\zeta_g := c_g x_g$. Likewise, it starts with a buffer $g \in \mathfrak{P}$ having the largest value of ζ_g.*

SP5 *The server deals with the processing part of the network until it becomes empty $x_g = 0 \; \forall g \in \mathfrak{P}$.* *** *Then it returns to the receiving part of the network.*

In some cases, the server can be switched to an empty buffer in accordance with the above policy. (Such a case may occur only at the first service session and for special initial data.) Then the next server switching is implemented immediately. Thus the server can make several instantaneous buffer changes until it reaches a nonempty one.

The state of the system is described by a pair (x, q) consisting of the "continuous" x and "discrete" q components. Here $x = \{x_g\}_{g \in G}$ and

$$q \in Q := \big\{(g, i)\big\}_{g \in G, i = 1, 2, \ldots}.$$

Being in the state (g, i), where either $g \in \mathfrak{R}$ or $g \in \mathfrak{P}$, means that the server is dealing with the buffer g and this buffer is ith in the current session of serving the receiving or the processing parts of the network, respectively.

The evolution of the system is described by the following logic-differential equations:

$$\textbf{if} \quad q = (g, i) \quad \textbf{then} \quad \begin{cases} \dot{x}_{g'} = \rho_{g'} \text{ whenever } g' \in \mathfrak{R} \text{ and } g' \neq g \\ \dot{x}_g = \begin{cases} -s_g & \text{if } g \bar{\in} \mathfrak{R} \\ \rho_g - s_g & \text{if } g \in \mathfrak{R} \end{cases} \\ \dot{x}_{g'} = s_g \rho(g, g') \text{ if } g' \in G_r(g) \\ \dot{x}_{g'} = 0 \text{ otherwise} \end{cases} \quad (9)$$

$$\textbf{if} \quad q(t) = (g, i), g \in G, i = 1, 2, \ldots \quad \textbf{and} \quad x_g(t) = 0 \quad \textbf{then}$$

** If the largest scaled content is attained at several buffers, there is a variety of candidate buffers to be switched to. Though there is no reason to prefer any of them, one can do so by specifying the control policy. We, however, consider all the possible decisions. Therefore in the event in question, several continuations of a given trajectory are taken into account.

*** Note that a given buffer $g \in \mathfrak{p}$ can be visited several times during one session.

$$q(t+0) := \begin{cases} (g', i+1) \text{ if } g \in \mathfrak{R} \text{ and } i < k & \left. \begin{array}{l} \text{where } g' \in \mathfrak{R} \\ \text{is such that} \\ \zeta_{g'}(t) \geq \zeta_{g''}(t) \\ \forall g'' \in \mathfrak{R} \end{array} \right. \\ (g', 1) \quad \text{ if } g \in \mathfrak{P} \text{ and } \sum_{g'' \in \mathfrak{P}} x_{g''} \leq 0 \\ (g', i+1) \text{ if } g \in \mathfrak{P} \text{ and } \sum_{g'' \in \mathfrak{P}} x_{g''} > 0 & \left. \begin{array}{l} \text{where } g' \in \mathfrak{P} \\ \text{is such that} \\ \zeta_{g'}(t) \geq \zeta_{g''}(t) \\ \forall g'' \in \mathfrak{P} \end{array} \right. \\ (g', 1) \quad \text{ if } g \in \mathfrak{R} \text{ and } i = k \end{cases}$$

$q(0) - (g, 1)$ where $g \in \mathfrak{R}$ is such that $\zeta_g(0) \geq \zeta_{g'}(0) \ \forall g' \in \mathfrak{R}$.

Except for the events specified, the discrete state $q(t)$ keeps its value in course of time.

Strictly speaking, the second formula holds only if the buffer corresponding to $q(t+0)$ is not empty. Otherwise, several buffer changes are performed instantaneously at the time t and the formula for $q(t+0)$ must be modified. We omit the details so far as, on the one hand, they are apparent and, on the other hand, the event in question is not typical: it may occur only at the first service session and for initial data from a set of the zero measure.

Any pair of functions $[x(\cdot), q(\cdot)]$ with $x(\cdot)$ absolute continuous and $q(\cdot)$ piecewise constant and left-continuous that satisfy the above equations is called a *trajectory*. A given initial data may give rise to several trajectories since the buffer g with the largest scaled content ζ_g is not determined uniquely in certain cases. A simple analysis shows that any trajectory can be extended on an infinite time interval.[†] (From now on, we consider trajectories defined on such an interval.) Furthermore the times of discrete state transitions do not accumulate and, being put in ascending order, form an infinite sequence $\{t_n\}_{n=1}^{\infty}$ such that $t_n \to \infty$ as $n \to \infty$. Supplemented by the term $t_0 := 0$, this sequence is called the *switching time sequence* of the trajectory.

We assume that (7) holds and consider trajectories with $\sigma(0) = 1$, where the quantity σ is given by (8). The system is studied in the invariant domain

$$K := \{(x, q) : q \in Q, \quad x_g \geq 0 \ \forall g, \quad \sigma = 1\}. \tag{10}$$

4   Asymptotic Behavior of the System

For $x = \{x_g\}_{g \in G}$ ($x_g \in \mathbb{R}$), we put $\|x\| := \sum_{g \in G} |x_g|$. The symbol **mes** stands for the Lebesgue measure. We start the section with several definitions from [5].

[†] We recall that the case (7) is considered.

Definition 1. *Let* $[x(t), q(t)]$ *be a periodic trajectory,* $T > 0$ *be its minimal period, and let* $\{t_k\}$ *be its switching time sequence. An integer* s *is said to be the order of this periodic trajectory if*

$$t_s \leq T < t_{s+1}.$$

It is easy to see that $t_{ls+j} = lT + t_j$ for $l = 0, 1, \ldots$ and $j = 1, 2, \ldots$.

Definition 2. *Let* $\mathsf{t}_p = [x_p(\cdot), q_p(\cdot)]$ *be a periodic trajectory,* $\{t_k\}$ *be its switching time sequence, and let* s *be its order. Furthermore, let* $\mathsf{t} = [x(\cdot), q(\cdot)]$ *be another trajectory, and let* $\{\hat{t}_k\}$ *be its switching time sequence. Then* t *is said to converge to* t_p *as* $t \to \infty$ *if there exists an integer* $N \geq 0$ *such that*

$$q_p(t_k) = q(\hat{t}_{k+N}) \quad \forall k = 0, 1, 2, 3, \ldots,$$

$$\left. \begin{array}{l} \displaystyle\lim_{i \to +\infty} x(\hat{t}_{is+N+j}) = x_p(t_j), \\[2mm] \displaystyle\lim_{i \to +\infty} (\hat{t}_{is+N+j+1} - \hat{t}_{is+N+j}) = t_{j+1} - t_j \end{array} \right\} \quad \forall j = 1, \ldots, s.$$

It can be shown (see [5] for details) that then there exists a sequence $\{\tau_i\} \subset (0, +\infty)$ such that $\tau_{i+1} - \tau_i \to T$ as $i \to \infty$ and, for any $\lambda > 0$,

$$\left. \begin{array}{l} \max\left\{\|x(t + \tau_i) - x_p(t)\| : t \in [0, \lambda]\right\} \to 0 \\[2mm] \mathbf{mes}\left\{t \in [0, \lambda] : q(t + \tau_i) \neq q_p(t)\right\} \to 0 \end{array} \right\} \quad \text{as} \quad i \to \infty.$$

This in particular means that the continuous components $x(t)$ and $x_p(\theta)$ of the trajectories t and t_p, respectively, come close not only for the selected time instants $t = \hat{t}_{is+N+j}$ and $\theta = t_j$, as was stated in Definition 2.

Let t converge to t_p as $t \to \infty$. Then it evidently converges to any trajectory that is a shift $\mathsf{t}_p^{(\tau)}(t) := \mathsf{t}_p(t + \tau)$ ($\tau = \mathrm{const} > 0$) of t_p in time.

Definition 3. *A periodic trajectory* $\mathsf{t}_p = [x_p(\cdot), q_p(\cdot)]$ *lying in the invariant domain* (10) *is said to be* locally asymptotically stable *in* K *if for some* $\varepsilon > 0$, *any trajectory* $\mathsf{t} = [x(\cdot), q(\cdot)]$ *such that* $\|x(0) - x_p(0)\| < \varepsilon$ *and* $q(0) = q_p(0)$ *converges to* t_p *as* $t \to \infty$.

Let a periodic trajectory t_p be locally asymptotically stable. Then so clearly is any trajectory that is a shift $\mathsf{t}_p^{(\tau)}(t) := \mathsf{t}_p(t + \tau)$ ($\tau = \mathrm{const} > 0$) of t_p in time.

Definition 4. *A* limit cycle *is a class* $\mathcal{LC}$ *of periodic trajectories such that, along with any trajectory* t, *it contains all the trajectories that are shifts of* t *and one of any two trajectories from* $\mathcal{LC}$ *is a shift of the other. A limit cycle* $\mathcal{LC}$ *is said to* lie in K *if any trajectory constituting it lies in* K.

All the periodic trajectories constituting a given cycle evidently have a common order, which is called the *order of the cycle*.

Definition 5. *A trajectory* t *is said to* converge to a limit cycle $\mathfrak{LC}$ *if it converges to any periodic trajectory constituting* $\mathfrak{LC}$. *A limit cycle lying in* K *is said to be* locally asymptotically stable in K *if so is any periodic trajectory from it.*

As follows from the foregoing remarks, it suffices to verify any of these properties for only one periodic trajectory from the cycle.

Let us revert to the network in question. The tuple of its parameters

$$p := \left[\{s_g\}_{g \in G} \,,\, \{\rho_g\}_{g \in \mathfrak{R}} \,,\, \{\rho(g, g')\}_{g \in G, g' \in G(g)} \right] \tag{11}$$

(the scaling coefficients c_g are not included since they are regarded as related to the switching policy) belongs to the set

$$\mathcal{P} := \left\{ \begin{matrix} p : s_g > 0 \,\forall g \in G, & \rho_g > 0 \,\forall g \in \mathfrak{R}, \\ \rho(g, g') > 0 \,\forall g \in G, g' \in G(g), & \text{(1) and (7) hold} \end{matrix} \right\}. \tag{12}$$

In a natural way, this set can be regarded as an analytical manifold of dimension $m + k - 1$. Here m is the number of the pairs (g, g') such that $g \in G$ and $g' \in G(g)$, and k is the number of the buffers in the receiving part $\mathfrak{R}$.

Theorem 1. *Assume that the processing part of the system is not empty and* $c_g := \rho_g^{-1} \,\forall g \in \mathfrak{R}$. *(This scaling means that while dealing with the receiving part, the server switches to the buffer with the longest period of being unserved. This is true since the beginning of the second service session.) Suppose that Assumption 1 and relations (1), (7) hold. Consider the control policy* **SP1–SP5** *and denote by* k *the number of the buffers in the receiving part* $\mathfrak{R}$.

Then the parameter manifold (12) contains a subset E *of the zero Lebesgue measure such that whenever the tuple (11) of the parameters lies outside* E, *the following statements hold:*

1. *There exist limit cycles lying in the invariant domain (10).*
2. *Their number equals* $k! := 1 \times 2 \times \cdots \times k$.
3. *Each of these cycles is locally asymptotically stable in this domain.*
4. *Any trajectory lying in it converges to one of the above limit cycles.*

Thus "almost all" systems from the class under consideration exhibit a regular and predictable behavior.

From now on, the hypotheses of Theorem 1 are assumed to hold. For given tuples of parameters $p \in \mathcal{P}$ and $\{c_g\}_{g \in G}$, either the statements 1—4 are true or the domain (10) contains infinitely many limit cycles, as well as a continuum of trajectories that converge to no limit cycle. More precisely, the second case occurs if and only if there exists a periodic trajectory for which the largest scaled content is attained at several buffers at a moment when the server switches to a buffer from the processing part. (Such event never occurs along periodic trajectories at times when the server switches to a buffer from the receiving part.) At this moment, the trajectory splits into a number of continuations. It can be shown that any of them can be chosen periodic. Moreover, they can be chosen to have a common period T and so that these trajectories are the same (up to a shift of time) during any session of serving the receiving part. The restrictions $\chi_1, \ldots, \chi_r$ of these trajectories on $[0, T]$ can be clearly combined arbitrarily $\chi_{i(1)}, \chi_{i(2)}, \chi_{i(3)}, \ldots$ in course of time to form a new trajectory. Corresponding to a periodic sequence $\chi_{i(1)}, \chi_{i(2)}, \chi_{i(3)}, \ldots$ is a periodic trajectory whose period is multiple of T. Countably many periodic trajectories can evidently be obtained so. If the above sequence is not periodic, the trajectory converges to no limit cycle. There obviously is a continuum of such trajectories.

It will follow from the proof of Theorem 1 that its statement is related to the policy **SP1–SP3** of serving the receiving part of the network much more than to that **SP4,SP5** of dealing with the processing one. More precisely, this statement remains true under various alterations of the second policy. For example, it can be replaced by the following one. The server first serves the buffers from S_{-N}, then from S_{-N+1}, and so on, up to serving S_{-1}, and then returns to the receiving part of the network. Each of the sets S_{-i} is processed on the basis of the Clear-the-Largest-Buffer-Level policy. In other words, the server switches when the current buffer $g \in S_{-i}$ is emptied and to a buffer $g \in S_{-i}$ with the largest (over S_{-i}) scaled content. Likewise, it starts with a buffer $g \in S_{-i}$ having the largest value of this content at the moment. The server deals with the layer S_{-i} until it becomes empty. (The advantage of this policy is that it excludes multiple passing through a buffer within a given session.)

References

1. C. Chase, J. Serrano, and P. Ramadge. Periodicity and chaos from switched flow systems: Contrasting examples of discretely controlled continuous systems. *IEEE Transactions on Automatic Control*, 38(1):70–83, 1993.
2. C. Horn and P.J. Ramadge. A topological analysis of a family of dynamical systems with nonstandard chaotic and periodic behavior. *International Journal of Control*, 67(6):979–1020, 1997.
3. Z. Li, C.B. Soh, and X. Xu. Stability of hybrid dynamic systems. In *Proceedings of the 2-nd Asian Control Conference*, pages 105–108, Seoul, Korea, July 1997.
4. A.S. Matveev and A.V. Savkin. Reduction and decomposition of differential automata: Theory and applications. In T.A. Henzinger and S. Sastry, editors, *Hybrid Systems: Computation and Control*. Springer-Verlag, Berlin, 1998.
5. A.S. Matveev and A.V. Savkin. *Qualitative Theory of Hybrid Dynamical Systems*. Birkhauser, Boston, 1999.
6. V.V. Nemytskii and V.V. Stepanov. *Qualitative theory of differential equations*. Princeton University Press, Princeton, New Jersey, 1960.
7. J.R. Perkins and P.R. Kumar. Stable, distributed, real-time scheduling of flexible manufacturing/assembly/disassembly systems. *IEEE Transactions on Automatic Control*, 34(2):139–148, 1989.
8. A.V. Savkin. A hybrid dynamical system of order n with all trajectories converging to $(n-1)!$ limit cycles. In *Proceedings of the 14th IFAC World Congress*, Beijing, China, July 1999.
9. A.V. Savkin and A.S. Matveev. Cyclic linear differential automata: A simple class of hybrid dynamical systems. *Automatica*, 36(5), 2000.

10. T. Ushio, H. Ueda, and K. Hirai. Controlling chaos in a switched arrival system. *Systems and Control Letters*, 26:335–339, 1995.
11. T. Ushio, H. Ueda, and K. Hirai. Stabilization of periodic orbits in switched arrival systems with n buffers. In *Proceedings of the 35th IEEE Conference on Decision and Control*, pages 1213–1214, Kobe, Japan, December 1996.

Hybrid Systems Diagnosis

Sheila McIlraith[1], Gautam Biswas[2], Dan Clancy[3], and Vineet Gupta[3]

[1] Knowledge Systems Lab, Stanford University, Stanford, CA 94305
[2] Computer Science Department, Vanderbilt University, Nashville, TN 37212
[3] Caelum Research Corporation, NASA Ames Research Center, Moffett Field, CA 94035

Abstract. This paper reports on an on-going project to investigate techniques to diagnose complex dynamical systems that are modeled as hybrid systems. In particular, we examine continuous systems with embedded supervisory controllers that experience abrupt, partial or full failure of component devices. We cast the diagnosis problem as a model selection problem. To reduce the space of potential models under consideration, we exploit techniques from qualitative reasoning to conjecture an initial set of qualitative candidate diagnoses, which induce a smaller set of models. We refine these diagnoses using parameter estimation and model fitting techniques. As a motivating case study, we have examined the problem of diagnosing NASA's Sprint AERCam, a small spherical robotic camera unit with 12 thrusters that enable both linear and rotational motion.

1   Introduction

The objective of our project has been to investigate how to diagnose hybrid systems – complex dynamical systems whose behavior is modeled as a hybrid system. Hybrid models comprise both discrete and continuous behavior. They are typically represented as a sequence of piecewise continuous behaviors interleaved with discrete transitions (e.g., [7]). Each period of continuous behavior represents a so-called *mode* of the system. For example, in the case of NASA's Sprint AERCam, modes might include *translate_X-axis*, *rotate_X-axis*, *translate_Y-axis*, etc. [1]. In the case of an Airbus fly-by-wire system, modes might include *take-off*, *landing*, *climbing*, and *cruise*. Mode transitions generally result in changes to the set of equations governing the continuous behavior of the system, as well as to the state vector that initializes that behavior in the new mode. Discrete transitions that dictate mode switching are modeled by finite state automata, temporal logics, switching functions, or some other transition system, while continuous behavior within a mode is modeled by, e.g., ordinary differential equations (ODEs) or differential and algebraic equations (DAEs).

The problem we address in this paper is how to diagnose such hybrid systems. For the purposes of this paper, we consider the class of hybrid systems that are continuous systems with an embedded supervisory controller, but whose hybrid models contain no autonomous jumps. I.e., all nominal transitions between system modes are induced by a controller action, none are induced by the system state and model [7]. The class of systems we consider can be modeled as a composition of a set of component subsystems, each of which is itself a hybrid system. We assume that the system operation is being tracked by a monitoring and observer system (e.g., [19]) that ensures that the system behavior predicted by the model does not deviate significantly from the observed

N. Lynch and B. Krogh (Eds.): HSCC 2000, LNCS 1790, pp. 282–295, 2000.

behavior in normal system operation. When observations occur outside this range, the behavior is deemed to be aberrant and diagnosis is initiated. In this paper, we consider faults whose onset is abrupt, and which result in partial or complete degradation of component behavior. The general problem we wish to address can be stated as follows: *Given a hybrid model of system behavior, a history of executed controller actions, a history of observations, including observations of aberrant behavior relative to the model, isolate the fault that is the cause for the aberrant behavior.* Diagnosis is done online in conjunction with the continued operation of the system. Hence, we divide our diagnosis task into two stages, initial conjecturing of candidate diagnosis and subsequent refinement and tracking to select the most likely diagnoses.

In this paper we conceive the diagnosis problem as a model selection problem. The task is to find a mathematical model and associated parameter values that best fit the system data. These models dictate the components of the system that have malfunctioned, their mode of failure, the estimated time of failure and any additional parameters that further characterize the failure. To address this diagnosis problem, we propose to exploit AI techniques for qualitative diagnosis of continuous systems to generate an initial set of qualitative candidate diagnoses and associated models, thus drastically reducing the number of potential models for our system. This is followed by parameter estimation and model fitting techniques to select the most likely mode and system parameters for candidate models of system behavior, given both past and subsequent observations of system behavior and controller actions. The main contributions of the paper are: 1) formulation of the hybrid diagnosis problem; 2) the exploitation of techniques for qualitative diagnosis of continuous systems to reduce the diagnosis search space; and 3) the use of parameter estimation and data fitting techniques for evaluation and comparison of candidate diagnoses.

In Section 2 we provide a brief description of NASA's Sprint AERCam, which we have used as a motivating example and which we will use to illustrate certain concepts in this paper. In Section 3 we present a formal characterization of the class of hybrid systems we study and the diagnosis problem they present. In Section 4 we describe our approach to hybrid diagnosis and the algorithms we use to achieve hybrid diagnosis. The generation of initial candidate qualitative diagnoses is described in Section 4.1, and the subsequent quantitative fitting and tracking of candidate diagnoses and their models is described in Section 4.2. In the final two sections, we briefly discuss related work and summarize our contributions.

2   Motivating Example: The AERCam

We are using NASA's Sprint AERCam and a simulation of system dynamics and the controller written in Hybrid CC (HCC) as a testbed for this work. We describe the dynamic model of the AERCam system briefly, a more detailed description of the model and simulation appear in [1].

The AERCam is a small spherical robotic camera unit, with 12 thrusters that allow both linear and rotational motion (Fig. 1). For the purposes of this model, we assume the sphere is uniform, and the fuel that powers the movement is in the center of the sphere. The fuel depletes as the thrusters fire.

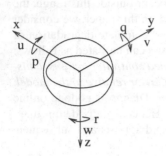

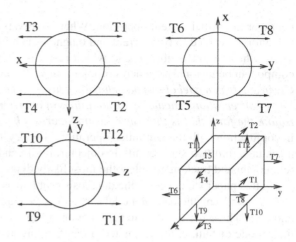

The Body frame of reference and the directions of velocities (u,v,w) are the components of the translation velocity, while (p,q,r) are components of the angular velocity.

Three views of the AERCam, showing the thrusters, and showing all the thrusters together in the cube circumscribing the AERCam.

Fig. 1. The AERCam axes and thrusters

The dynamics of the AERCam are described in the AERCam body frame of reference. The translation velocity of this frame with respect to the shuttle inertial frame of reference is 0. However, its orientation is the same as the orientation of the AERCam, thus its orientation with respect to the shuttle reference frame changes as the AERCam rotates (i.e., it is not an inertial frame). The twelve thrusters are aligned so that there are four along each major axis in the AERCam body frame. For modeling purposes, we assume the positions of the thrusters are on the centers of the edges of a cube circumscribing the AERCam. Thus, for example, thrusters T_1, T_2, T_3, T_4 are parallel to the x-axis and are used for translation along the x-axis or rotation around the y-axis. I.e., firing thrusters T_1 and T_2 results in translation along the positive x-axis, and firing thrusters T_1 and T_4 results in a negative rotation around the y-axis. AERCam operations are simplified by limiting them to either translation or rotation. Thrusters are either on or off, therefore, the control actions are discrete. In a normal mode of operation, only two thrusters are on at any time.

2.1   AERCam Dynamics

A simplified model of the AERCam dynamics based on Newtonian laws is derived using an inertial frame of reference fixed to the space shuttle. The AERCam position in this frame is defined as the triple (x, y, z). Let $\vec{V}$ be the velocity in the AERCam body frame, with its vector components given by (u, v, w). The frame rotates with respect to the inertial reference frame with velocity $\omega = (p, q, r)$, the angular velocity of the AERCam. The rotating body frame implies an additional Coriolis force acting upon the AERCam. We assume uniform rotational velocity since in the normal mode of operation, the AERCam does not translate and rotate at the same time [2, pg. 130]. Similar equations can be derived for the rotational dynamics [1].

$$d(m\,\vec{V})/dt = \vec{F} - 2m(\vec{V} \times \vec{\omega}) \quad \text{Newton's Law}$$

$$\vec{V}\,dm/dt + md(\vec{V})/dt = \vec{F} - 2m(\vec{\omega} \times \vec{V})$$

The resultant equation for each coordinate:

$$du/dt = F_x/m - 2(qw - vr) - (u/m) * dm/dt$$

$$dv/dt = F_y/m - 2(ru - pw) - (v/m) * dm/dt$$

$$dw/dt = F_z/m - 2(pv - qu) - (w/m) * dm/dt$$

2.2   Position Control Mode of the AERCam

In the position control mode, the AERCam is directed to go to a specified position and point the camera in a particular direction. Assume the AERCam is at position A and directed to go to position B. In the first phase, the AERCam rotates to get one set of thrusters pointed towards B. These are then fired, and the AERCam cruises towards B. Upon reaching a position close to B, it fires thrusters to converge to B, and then rotates to point the camera in the desired direction.

To facilitate the illustration of the diagnosis problem, we use a simple trapezoidal controller, which we explain in two dimensions. Suppose the task is to travel along the x-axis for some distance, then along the y-axis. Such manoeuvres are needed for navigating in the space shuttle. In order to do this, the AERCam fires its x thrusters for some time. Upon reaching the desired velocity, these are switched off. When the AERCam has reached a position close to the desired x position, the reverse thrusters are switched on, and the AERCam is brought to a halt — the velocity graph is a trapezium. The process is analogous for the y direction.

3   Problem Formulation

In this section we provide our formulation of the hybrid diagnosis problem.

Definition 1 (Hybrid System). A hybrid system is a 5-tuple $\langle \mathcal{M}, X, \mathcal{F}, \Sigma, \phi \rangle$, where

- $\mathcal{M}$, finite set of system modes $(\mu_1, \ldots, \mu_k)$.
- $X \subseteq R^n$, continuous state variables. $x(t)$ is the continuous behavior at time t.
- $\mathcal{F}$, finite set of functions $\{f_{\mu_1}, \ldots, f_{\mu_k}\}$, and associated parameter values θ such that for each mode, μ_i, $f_{\mu_i}(t, \theta, x(t)) : R \times R \times X \to X$ defines the continuous behavior of the system in μ_i.[1]
- Σ, finite set of actions $(\sigma_1, \ldots, \sigma_l)$, which transition the system between modes.
- ϕ, transition function which maps an action, mode and system state vector into a new mode and initial state vector, i.e., $\phi : \Sigma \times \mathcal{M} \times X \to \mathcal{M} \times X$.

To define the hybrid diagnosis problem, we augment Definition 1 as follows.

---

[1] Parameter value ranges may be associated with θ.

Definition 2 (Diagnosable Hybrid System). A diagnosable hybrid system, $\langle \mathcal{M}, X, \mathcal{F}, \Sigma, \phi, COMPS \rangle$ is a hybrid system comprised of m potentially malfunctioning components $COMPS = (c_1, \ldots, c_m)$ where

- For each $\mu \in \mathcal{M}$, μ includes a designation of whether each $c_i \in COMPS$ is operating normally, or abnormally, i.e., $(\neg)ab(c_i)$.
- We assume that transitions to fault modes are achieved by exogenous actions. Hence, $\Sigma = \Sigma_c \cup \Sigma_e$, where
  - Σ_c is a finite set of controller actions, and
  - Σ_e is a finite set of exogenous actions.
- $\mathcal{A}$, the controller action history, the sequence of time-indexed controller actions performed.
- $X_{obs} \subseteq X$, continuous state variables that are observable. $x_{obs}(t)$ is the observations at time t.
- $\mathcal{O}$, the observation history, the sequence of time-indexed observations.

For notational convenience, μ_F denotes a faulty mode, i.e., a mode for which at least one $c_i \in COMPS$ is $ab(c_i)$ in μ_F. θ_F denotes the parameters associated with f_{μ_F}.

In the case of the AERCam example, the potentially malfunctioning components are the 12 thrusters, and a mode μ includes the behavior mode (e.g., translate-x, translate-y, rotate-x, etc.) and $(\neg)ab(T_i)$, $i = 1, \ldots, 12$, for each thruster. The continuous state vector includes the x, y, z position of the AERCam, velocity and acceleration. The parameter values, θ associated with each f_μ are the percentage degradation of each of the thrusters.

Definition 3 (Model). A model, Mod of a diagnosable hybrid systems is a time-indexed mode sequence and associated parameter values $([\mu_1, \ldots, \mu_m], [\theta_1, \ldots, \theta_m])$.

Notice that each model of the system, $(\boldsymbol{\mu}, \boldsymbol{\theta})$ induces a corresponding time-indexed piecewise continuous sequence of functions $[f_{\mu_1}, \ldots, f_{\mu_m}]$ dictating system behavior.

In this paper we make several simplifying assumptions regarding our diagnosis task. In particular, we make a single-time fault assumption. We assume that our systems do not experience multiple sequential faults. Further, we assume that faults are abrupt, resulting in partial or full degradation of component behavior. We cast the hybrid diagnosis task as the problem of finding the most likely model for the observation history, $P(Mod \mid \mathcal{O})$. I.e, the sequence of modes and parameter values $(\boldsymbol{\mu}, \boldsymbol{\theta})$ that best fit the observations over time. Under normal operation, the model of the system Mod_{normal} is fully dictated by the sequence of controller actions $\mathcal{A}$ and the nominal parameter values, θ. Once again, we assume that the system operation is being tracked by a monitoring and observer system (e.g., [19]) that ensures that the system behavior predicted by the model does not deviate significantly from the observed behavior in normal system operation. When observations occur outside this range, the behavior is deemed to be aberrant and diagnosis is initiated. Given a diagnosable hybrid system $\langle \mathcal{M}, X, \mathcal{F}, \Sigma, \phi, COMPS \rangle$, a controller action history, $\mathcal{A}$ and a history of observations, $\mathcal{O}$ which includes observations of aberrant behavior, the **hybrid diagnosis task** is to determine what components are faulty, what fault mode caused the aberrant behavior, when it occurred, and what the values of the parameters associated with the fault mode are. In the AERCam system, a diagnosis might be that thruster T_1 experienced a blockage fault of 50%, at time t_i.

Once Mod_{normal} has been rejected, we must find a new most likely model from among the potentially exponential (in $COMPS$) number of mode sequences, occurring within a large but bounded time range. We propose to exploit previous research on temporal causal graphs for qualitative diagnosis of continuous systems [18], to compute a set of candidate qualitative diagnoses that are consistent with our system, in order to identify a preliminary subset of candidate models, whose likelihood can be estimated.

Definition 4 (D-tuple). A D-tuple is a 4-tuple $\langle C, \mu_F, t_F, \theta_F \rangle$, where μ_F is a fault mode, t_F is the time the fault mode commenced, θ_F is the parameter values associated with the fault mode behavior, and C is the set of failed (abnormal) components in μ_F.

Definition 5 (Candidate Qualitative Diagnosis). Given a diagnosable hybrid system with model $Mod = (\mu, \theta)$ an action history $\mathcal{A}$, and a history of observations, $\mathcal{O}$ which includes observations of aberrant behavior, D-tuple $\langle C, \mu_F, t_F, \theta_F \rangle$ is a candidate qualitative diagnosis iff there exists a range of parameter values $\theta_F = [\theta_l, \theta_u]$, and time range $t_F = [t_l, t_u]$ such that the occurrence of fault mode μ_F with parameter values θ_F in time range t_F is consistent with $\mathcal{O}$, $\mathcal{A}$ and Mod.

Hence, a candidate qualitative diagnosis stipulates a fault mode, including one or more faulty components. It also stipulates a lower and upper bound, $[t_l, t_u]$, on the time the fault mode occurred. This range generally corresponds to the start times of the controller induced modes preceding and following the fault, or up to the point the fault was detected. This candidate diagnosis induces an associated *candidate model*, $Mod_C = ([\mu_1, \ldots, \mu_i, \mu_F, \mu'_{i+1}, \ldots, \mu'_m], [\theta_1, \ldots, \theta_i, \theta_F, \theta'_{i+1}, \ldots, \theta'_m])$ corresponding to Mod with the fault mode μ_F and θ_F inserted at t_F. Every subsequent mode, $\mu_{i+1}, \ldots, \mu_m$, has $ab(c_i), c_i \in C$ enforced, and every subsequent set of parameters has the parameters associated with faulty components C enforced. Computing candidate qualitative diagnoses is discussed in Section 4.1.

Since each candidate qualitative diagnosis only conjectured ranges for the time of the fault mode, t_F and parameter values associated with the fault mode, θ_F, the associated candidate models are underconstrained. In Section 4.2, we discuss methods for estimating unique values for t_F and θ_F and for estimating a posterior probability for each of the candidate models, Mod_C, given $\mathcal{O}$.

Definition 6 (Candidate Diagnosis). Given a diagnosable hybrid system, a history of controller actions $\mathcal{A}$, and a history of observations $\mathcal{O}$, D-tuple $\langle C, \mu_F, t_F, \theta_F \rangle$ with associated model Mod_C is a candidate diagnosis for the hybrid system, iff $P(Mod_C \mid \mathcal{O}) > \alpha$, for defined threshold value $\alpha \in [0, 1]$.

4   Diagnosing Hybrid Systems

In this section we discuss one method for computing hybrid diagnoses. In Section 4.1 we discuss a technique for generating candidate qualitative diagnoses, and their associated candidate models. In Section 4.2 we discuss techniques for model fitting and for model (and hence diagnosis) comparison. In particular we discuss techniques for estimating the parameters of the candidate models, and the likelihood of the models, and for continued monitoring and refinement of the candidate models as the system continues to operate and observations continue to be made.

We illustrate these techniques with the following simple AERCam example. Consider the scenario depicted in Fig. 2. In the first accelerate phase, the AERCam is being powered by thrusters $T1$ and $T2$. Assume that at some point in this phase, a sudden leak in the $T2$ thruster causes an abrupt change in its output. As a consequence, the AERCam starts veering to the right of the desired trajectory, as illustrated by the left-most dotted lines in Fig. 2. (The other dotted lines represent other potential candidate diagnoses consistent with the point of detection of the failure.) Soon after this occurs, the supervisory controller commands the AERCam to turn off Thrusters $T1$ and $T2$ with the objective of getting the AERCam to cruise in a straight line. In the faulty situation, the AERCam has some residual angular velocity about the z-axis, so it continues to rotate in the cruise mode. Then the controller turns on thrusters $T3$ and $T4$, to decelerate the AERCam with the objective of bringing it to a halt. Again, this objective is not entirely achieved in the the faulty situation. Next, thrusters $T5$ and $T6$ are switched on, to move the AERCam in the y direction. However, since the AERCam is not in the desired orientation after the failure, the position error due to faulty thruster $T2$ accumulates causing a greater and greater deviation from the desired trajectory of the system. The position of the AERCam is being continuously sensed, filtered for noise and monitored. At some point within the y translation the trajectory exceeds the error bound, i.e., $P(Mod_{normal} < \alpha)$ and is flagged by the monitoring system as aberrant relative to Mod_{normal}. At this point, the diagnosis task begins.

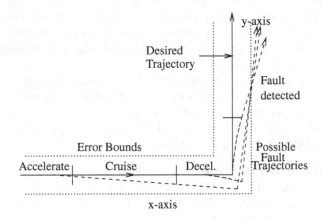

Fig. 2. Possible fault trajectories of AERCam (simplified for illustration purposes).

4.1   Qualitative Candidate Generation

Given the current system model $Mod = (\boldsymbol{\mu}, \boldsymbol{\theta})$ (commonly Mod_{normal}), a history of controller actions $\mathcal{A}$, and a history of observations $\mathcal{O}$ including one or more observa-

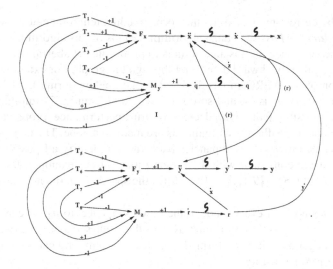

Fig. 3. A subset of the temporal causal graph showing the relations between Thrusters $T1 - T8$ and the x and y positions of the AERCam.

tions of aberrant behavior, we wish to generate a set of *candidate qualitative diagnoses* $\langle C, \mu_F, t_F, \theta_F \rangle$, and associated *candidate models* as described in Definition 5. To do so, we extend techniques for generating qualitative diagnoses of continuous dynamic systems to deal with hybrid systems with multiple modes. The model and propagation mechanism, as applied to continuous systems diagnosis, is described in [18].

In the case of our AERCam example, the action history $\mathcal{A}$ is [(on($T1$), on($T2$)), (off($T1$), off($T2$)), (on($T3$), on($T4$)), (off($T3$), off($T4$), on($T5$), on($T6$)), (off($T5$), off($T6$))]; the model, Mod_{normal} is the time-indexed sequence [(*accelerate_x*, $\neg ab(T1-T12), \theta$), (*cruise_x*, $\neg ab(T1-T12), \theta$), (*decelerate_x*, $\neg ab(T1-T12), \theta$), (*accelerate_y*, $\neg ab(T1-T12), \theta$), (*cruise_y*, $\neg ab(T1-T12), \theta$)], where θ is a vector of length 12 all of whose entries are 0 (percent degradation in thrusters).

To generate candidate qualitative diagnoses we construct an abstract model of the dynamic system behavior, Mod_{normal} as a temporal causal graph. A part of the temporal causal graph for the AERCam dynamics is shown in Fig. 3. The graph expresses directed cause-effect relations between component parameters and the system state variables. Links between variables are labeled as: (i) +1, implying direct proportionality, (ii) −1, implying inverse proportionality, and (iii) $\int$, implying an integrating relation. An integrating relation introduces a temporal delay in that a change on the cause side of the relation affects the derivative of the variable on the effect side. This adds temporal characteristics to the relations between variables. Some edges are labeled by variables, implying the sign of the variable in the particular situation defines the nature of the relationship. The candidate generation algorithm is invoked for every initial instance of an aberrant observation. The aberrant observation plus the controller action history $\mathcal{A}$ are input to a backward propagation algorithm that operates on the temporal causal graph. The algorithm operates backwards from the last mode in the mode sequence of Mod:

Step 1 For the current mode, extract the corresponding temporal causal graph model, and apply the *Identify Possible Faults* algorithm. Details of this algorithm are presented in [18], but the key aspect of this algorithm is to propagate the aberrant observation expressed as a $\pm$ value, backward depth-first through the graph. For example, given that the y−position of the AERCam has deviated − (i.e., below normal), backward propagation implies $d(y)/dt$ is −, and so on, till we get T_5^- and T_6^-, implying thrusters $T5$ and $T6$ are possibly faulty with decreased thrust performance. Propagation along a path can terminate if conflicting assignments are made to a node. The goal is to systematically propagate observed discrepancies backward to identify all possible candidate hypotheses that are consistent with the observations. In our example, the component parameters, $COMPS = \{T1, \ldots, T12\}$ form the space of candidate faults.

Step 2 Repeat Step 1 for every mode in the mode sequence, to μ_1. The system model needs to be substituted as the algorithm traverses the mode sequence backwards. Therefore, back propagation will be performed on a different temporal causal graph for each mode in the controller history[2].

The output of this step is a set of qualitative diagnoses $\langle C, \mu_F, t_F, \theta_F \rangle$, each with an associated candidate model, as described in Section 3. Returning to our AERCam example, three qualitative candidate diagnoses are generated. The first candidate diagnosis is that $T2$ failed in the x acceleration phase. The time of the fault mode transition is $[t_1, t_2]$, and the parameters associated with the failure – the percentage degradation of the component is in the range $[0, 100]$. So the first candidate qualitative diagnosis is $\langle T2, (accelerate\_x, ab(T2), \neg ab(T1, T3 - T12), \theta_F), [t_1, t_2], [0, 100] \rangle$. The candidate model simply has $(accelerate\_x, ab(T2), \neg ab(T1), \neg ab(T3 - T12))$ inserted after the mode $(accelerate\_x, \neg ab(T1 - T12))$, and $ab(T2)$ enforced in every subsequent mode. The second candidate qualitative diagnosis is that $T4$ failed in the deceleration phase of x translation, i.e., $\langle T4, (decelerate\_x, ab(T4), \neg ab(T1 - T3, T5 - T12), \theta_F), [t_3, t_4], [0, 100] \rangle$. The third candidate is that $T6$ failed during y acceleration, i.e., $\langle T6, (accelerate\_y, ab(T6), \neg ab(T1 - T5, T7 - T12), \theta_F), [t_4, t_D], [0, 100] \rangle$, where t_D is the time of detection of the aberrant behavior. In each case θ_F is a vector of length 12 with every entry equal to 0 (percentage degradation), except the entries corresponding to the faulty thrusters, C which will have the range $[0, 100]$.

4.2   Model Fitting and Comparison

Given the candidate qualitative diagnoses and their associated candidate models, the next phase of the diagnosis process is quantitative refinement of the qualitative candidate diagnoses and their associated models through parameter estimation and data fitting, followed by tracking of the fit of subsequent observations to the candidate models. The goal is to at least provide a probabilistic ranking of the plausible candidates, if not a unique model (and hence diagnosis).

As observed in the previous section, the model associated with the candidate qualitative diagnosis, Mod_C is underconstrained. Both the time of the fault mode occurrence, t_F and the parameters associated with the faulty behavior θ_F are represented as ranges

---

[2] We may cut off back-propagation along the mode sequence beyond a time limit.

and must be estimated. Further, the candidate qualitative diagnoses were generated from initial observations of aberrant behavior, and their consistency can be further evaluated by monitoring the qualitative transients associated with each candidate. The refinement process is performed by a set of *trackers* [21], one for each candidate diagnosis and associated model. Each tracker comprises both a *qualitative transient analysis* component and a quantitative *model estimation*, component. The two components operate in parallel as described below.

Qualitative Transient Analysis

The qualitative transient analysis component performs a further qualitative analysis of the consistency of candidate qualitative diagnoses based on monitoring of higher-order transients whose manifestation is seen over a longer period of time. If the transients of a candidate qualitative diagnosis do not remain consistent with subsequent observations, the candidate diagnosis will be eliminated and the *model estimation* component informed. The technique we employ is derived from techniques for qualitative monitoring of continuous systems. Details of the algorithm appear in [18].

Model Estimation

The purpose of the model estimation component is to perform quantitative model fitting, i.e., to provide a quantitative estimate of the parameters of the models and to assign a probability to each of the candidate models (and hence candidate diagnoses), given the noisy observed data. In particular, given a candidate model, Mod_C the model estimation component uses parameter estimation techniques to estimate both the time at which the failure occurred, t_F, and the value for the parameters, θ_F, associated with the conjectured failure mode. In this paper we discuss two alternate approaches to our time and parameter estimation problem. The first approach is based on Expectation Maximization (EM) (e.g., [8]), an iterative technique that converges to an optimal value for t_F and θ_F simultaneously. The second approach we consider employs General Likelihood Ratio (GLR) techniques (e.g., [5]) to estimate the time of failure t_F, and then uses the observations obtained after the failure to estimate the fault parameters, θ_F, by a least squares method. As described in Section 3, the outcome of both approaches is a unique value for t_F and θ_F and a measure of the likelihood of Mod_C given the observations. The proposed approaches to model fitting have trade-offs and we are currently assessing the efficacy of these and other alternative approaches through experimentation.

EM-Based Approach The Expectation Maximization (EM) algorithm (e.g., [8]) provides a technique for finding the maximum-likelihood estimate of the parameters of an underlying distribution from a given set of data, when that data is incomplete or has missing values. The parameter estimation problem we address in this paper is a variant of the motion segmentation problem described in [24]. Here, we define the basic algorithm and the intuition behind our approach. (See [8] for more details.)

The time of failure, $t_F = [t_l, t_u]$ of our candidate qualitative diagnosis dictates the mode in which the failure is conjectured to have occurred. Let us call this mode μ_i. The behavior of our hybrid system in mode μ_i is described by the continuous function f_{μ_i}, with *known* parameters θ_i. At some (to be estimated) time point t_F within the predicted time period of μ_i, we have conjectured that the system experienced a fault which transitions it into mode μ_F. The behavior of our hybrid system in mode μ_F is described by the continuous function f_{μ_F}, with *unknown* parameters, θ_F. We also have a set of data points $\mathcal{O}' = [x_{obs}(t_l), \ldots, x_{obs}(t_u)] \subseteq \mathcal{O}$, which either reflect the behavior of the system under f_{μ_i} or under f_{μ_F}.

Given all this information, our task is to find 1) values for parameters θ_F, and 2) an assignment of the data points $\mathcal{O}'$ to either μ_i or μ_F so that we maximize the fit of the data to the two functions. The assignment of data points will in turn tell us the value of t_F. EM provides an iterative algorithm which converges to provide a maximum-likelihood estimate for θ_F given $\mathcal{O}'$, i.e., roughly we are calculating the likelihood of θ,
$$L(\theta) = P(\mathcal{O}' \mid \theta_F, Mod_C).$$

The basic EM algorithm comprises two steps: an Expectation Step (E Step), and a Maximization Step (M Step) [24]:
- Select an initial (random) value for θ_F.
- Iterate until convergence:

  - E Step: assign data points to either $f_{\mu_i}(\theta_i)$ or $f_{\mu_F}(\theta_F)$, which ever fits it best.

  - M Step: re-estimate θ_F using the data points assigned to $f_{\mu_F}(\theta_F)$.

The assignment of data points to μ_i and μ_F provides an estimate for t_F. We may exploit the fact that the assignment of data points is temporally correlated with all points before t_f belonging to μ_i, and all points after t_f belonging to μ_f. We may also exploit the fact that data points at the beginning of the interval will belong to μ_i, while those at the end will belong to μ_F. These task-specific qualities help our algorithm converge more quickly.

EM provides a rich algorithm for maximum-likelihood parameter estimation when we don't know the value of t_F. In some hybrid diagnosis applications, depending upon the sensors in our system, and the level of noise in the sensors, we may be able to develop monitoring techniques that will help isolate a reasonable value for t_F, minimizing the need for iteration in EM. In such cases, an alternative to the EM-based approach is to first estimate t_F using the Generalized Likelihood Ratio (GLR) method [5], followed by parameter estimation of θ_F.

GLR + Least Squares Approach Here, we divide the parameter estimation problem into two parts: (i) estimate the time of failure, t_F, using the Generalized Likelihood Ratio (GLR) method, and (ii) apply a standard least squares method for parameter estimation. The intuition is that solving the problem in two parts simplifies the estimation process, and very likely mitigates the numerical convergence problems that arise in dealing with complex higher-order models.

The GLR method for detecting abrupt changes in continuous signals is described in [5]. We have applied it to fault transients analysis in complex fluid thermal systems [16]. Here we provide an overview of the method for the single parameter case. Assume that the signal under scrutiny is a time-indexed sequence of random variables $y(k)$, with probability density function, $p_{\theta_i}(y)$ in desired mode μ_i, and $p_{\theta_F}(y)$ in fault mode μ_F. y is either contained in x_{obs} or computed from x_{obs}. We assume that a fault causes an abrupt change in $y(k)$. In the case of the AERCam, y captures the difference between the observed and expected values of the, e.g., acceleration, as predicted by the model.

The central quantity in the change detection algorithm is the cumulative sum of the log-likelihood ratio for a window of observations between times m and n,

$$S_m^n(\theta_F) = \sum_{k=m}^n ln\frac{p_{\theta_F}(y(k))}{p_{\theta_i}(y(k))}.$$

Again, this ratio is a function of two unknowns: t_F and θ_F. The common statistical solution is to use maximum likelihood estimates for these two parameters, resulting in a double maximization:

$$g_n = \max_{1 \le m \le n} \sup_{\theta_F} S_m^n(\theta_F).$$

If we assume that probability density functions, $p_{\theta_i}(y)$ and $p_{\theta_F}(y)$ are Gaussian, then g_n reduces to:

$$g_n = \frac{1}{2\sigma_i^2} \max_{1 \le m \le n} \frac{1}{n-m+1}\left[\sum_{k=m}^n (y(k) - \omega_i)\right]^2,$$

where ω_i and σ_i^2 are the mean and variance for $p_{\theta_i}(y)$, respectively.

When processing a sequence of samples, the point of abrupt change, t_F, is computed from $min\{n : g_n \ge h\}$, where h is an appropriately defined threshold. Hence, the smaller the value of h, the more sensitive the function to change, and unfortunately to false alarms, so h must be set carefully.

Once t_F is estimated, data points observed after t_F, are used to estimate the parameter, θ_F for a hypothesized fault using regression techniques. In the case of the AERCam, the position vector of the AERCam is modeled as a set of quadratic functions in terms of the thruster force. These functions contain one unknown, θ_F, the parameter that corresponds to the degree of degradation in the faulty thruster. The least squares estimate for θ_F is computed, and the the measure of fit of the candidate model to the observed data used to estimated the probability of the candidate model (and hence, diagnosis).

Model Comparison

From the model estimation component, each tracker computes the likelihood of its model Mod_C, and hence of the associated candidate diagnosis $\langle C, \mu_F, t_F, \theta_F \rangle$, as a measure of fit of the observations to the model. As new data $x_{obs}(t)$ are observed, θ_F and t_F, are adjusted and $P(Mod_C \mid x_{obs}(t))$ computed. If the likelihood of Mod_C falls below a predefined acceptable likelihood threshold, α, then its tracker is terminated, and the associated candidate diagnosis $\langle C, \mu_F, t_F, \theta_F \rangle$ removed from the list of candidate diagnoses. Tracking terminates when a unique diagnosis is obtained, or when the diagnoses are sufficiently discriminated to determine suitable controller actions.

5   Related Work

The specific problem of diagnosing hybrid systems has received little attention to date, although there is much related work. Within the AI community, there has been a great deal of research on diagnosing static systems (e.g., [14]), while much less on diagnosing discrete dynamical systems (e.g., [17,25]), and qualitative representations of continuous systems (e.g.,. [18]). Within the FDI community, the largest proportion of research has focused on diagnosing continuous systems (e.g., [13,11]). The most common model-based approaches use observer schemes(e.g., [12,20]), where the goal is to design residual generators based on observed discrepancies, such that individual residuals are sensitive to a particular subset of faults. There is also complementary work by Basseville [4], using model-based statistical processing techniques for early fault detection and residual identification. [18] perform residual generation and analysis task in a qualitative framework to address some of the computational issues that arise in handling the complex dynamics that occur in fault transients, with some preliminary work on building multiple observers for hybrid systems [19]. Diagnosis of discrete-event systems has also been studied within the FDI community (e.g, [22,15]). Fabre et al. [10] have employed stochastic Petri nets based on a Hidden Markov Model probabilistic scheme for alarm analysis. Unfortunately, it is not clear how to systematically derive such representations from the physical system models that we work with.

6   Summary

In this paper we addressed the problem of diagnosing hybrid systems. The main contributions of the paper are 1) formulation of the hybrid diagnosis problem as model selection; 2) the exploitation of techniques for qualitative diagnosis of continuous systems to reduce the diagnosis search space; and 3) the use of parameter estimation and data fitting techniques for evaluation and comparison of candidate diagnoses. This work continues with experimental analysis of the proposed techniques, and a more formal characterization of our approach in terms of Bayesian model selection.

Acknowledgements

This work was funded in part by NASA grant NAG 21337. The first author would like to thank David Fleet for useful discussion relating to this work.

References

1. L. Alenius and V. Gupta. Modeling an AERCam: A case study in modeling with concurrent constraint languages. In *Proceedings of the CP'97 Workshop on Modeling and Computation in the Concurrent Constraint Languages*, 1998.
2. V. I. Arnold. *Mathematical Methods of Classical Mechanics*. Springer Verlag, 1978.
3. P. Baroni, G. Lamperti, P. Pogliano and M. Zanella Diagnosis of large active systems *Artificial Intelligence*, 110(1):135–183, 1999.
4. M. Basseville. On-board component fault detection and isolation using a statistical local approach. *Automatica,* vol. 34, no. 11, 1998.
5. M. Basseville and I.V. Nikiforov. *Detection of Abrupt Changes: Theory and Applications*. Prentice Hall, Englewood Cliffs, NJ, 1993.

6. J. A. Blimes. A gentle tutorial of the EM algorithm and its application to parameter estimation for gaussian mixture and hidden markov models. Technical Report TR-97-021, International Computer Science Institute (ICSI) and Computer Science Division, Dept. of Electrical Engineering and Computer Science, U.C. Berkeley, 1998.

7. M. Branicky. *Studies in Hybrid Systems: Modeling, Analysis, and Control.* PhD thesis, Department of Electrical Engineering and Computer Science, Massachusetts Institute of Technology, 1995.

8. A. P. Dempster, N. M. Laird, and D. B. Rubin. Maximum likelihood from incomplete data. *Journal of the Royal Statistical Society Ser. B,* 39:1–38, 1977.

9. B. Etkin and L. D. Reid. *Dynamics of Flight:Stability and Control.* John Wiley and Sons, 1995.

10. E. Fabre, A. Aghasaryan, A. Benveniste, R. Boubour and C. Jard. Fault detection and diagnosis in distributed systems: an approach by partially stochastic Petri nets. *Journal of Discrete Event Dynamic Systems,* vol. 8, no. 2, pp. 203-231, 1998.

11. P.M. Frank. Fault diagnosis in dynamic systems using analytic and knowledge-based redundancy: a survey and some new results. *Automatica,* vol. 26, pp. 459-474, 1990.

12. E.A. Garcia and P.M. Frank. Deterministic nonlinear observer-based approaches to fault diagnosis: a survey. *Control Engineering Practice,* 5(5):663–670, 1999.

13. J.J. Gertler. *Fault Detection and Diagnosis in Engineering Systems.* Marcel Dekker, New York, 1988.

14. W. Hamscher, L. Console and J. de Kleer *Readings in Model-based Diagnosis.* Morgan Kaufmann, 1992.

15. J. Lunze. A timed discrete-event abstraction of continuous-variable systems. *Intl. Jour. of Control,* vol. 72, no. 13, pp. 1147-1164, 1999.

16. E.J. Manders, P.J. Mosterman, and G. Biswas. Signal to symbol transformation techniques for robust diagnosis in transcend. In *10th Int. Workshop on Principles of Diagnosis,* pp. 155–165, 1999.

17. S. McIlraith. Explanatory diagnosis: Conjecturing actions to explain observations. In *Proceedings of the Sixth International Conference on Principles of Knowledge Representation and Reasoning (KR'98),* pp. 167–177, 1998.

18. P. Mosterman and G. Biswas. Diagnosis of continuous valued systems in transient operating regions. *IEEE Transactions on Systems, Man, and Cybernetics,* 1999. vol. 29, no. 6, pp. 554-565, 1999.

19. P. Mosterman and G. Biswas. Building hybrid observers for complex dynamic systems using model abstractions. In *International Workshop on Hybrid Systems: Computation and Control,* Nijmegen, Netherlands, March 1999.

20. R.J. Patton and J. Chen. Observer-based fault detection and isolation: robustness and applications. *Control Engineering Practice,* 5(5):671–682, 1997.

21. B. Rinner and B. Kuipers. Monitoring piecewise continuous behavior by refining trackers and models. In *Hybrid Systems and AI: Modeling, Analysis and Control of Discrete + Continuous Systems,* AAAI Technical Report SS-99-05, pp. 164–169, 1999.

22. M. Sampath, R. Sengupta, S. Lafortune, K. Sinnamohideen and D. Teneketzis. Failure diagnosis using discrete-event models. *IEEE Trans. on Control Systems Technology,* vol. 4, no. 2, pp. 105-124, 1996.

23. W. Sweet. The glass cockpit. *IEEE Spectrum,* pages 30–38, September 1995.

24. Y. Weiss. Motion segmentation using EM – a short tutorial. http://www-bcs.mit.edu/people/yweiss/tutorials.html, 1997.

25. B. Williams and P.P. Nayak. A model-based approach to reactive self-configuring systems. In *Proceedings of the Thirteenth National Conference on Artificial Intelligence (AAAI-96),* pages 971–978, 1996.

Decidability and Complexity
Results for Timed Automata
and Semi-linear Hybrid Automata

Joseph S. Miller *

Department of Mathematics
Cornell University
Ithaca, NY 14853
jmiller@math.cornell.edu

Abstract. We define a new class of hybrid automata for which reachability is decidable—a proper superclass of the *initialized rectangular hybrid automata*—by taking *parallel compositions* of simple components. Attempting to generalize, we encounter timed automata with algebraic constants. We show that reachability is undecidable for these *algebraic timed automata* by simulating two-counter Minsky machines. Modifying the construction to apply to *parametric timed automata*, we reprove the undecidability of the emptiness problem, and then distinguish the dense and discrete-time cases with a new result. The algorithmic complexity—both classical and parametric—of one-clock parametric timed automata is also examined. We finish with a table of computability-theoretic complexity results, including that the existence of a Zeno run is Σ_1^1-complete for semi-linear hybrid automata; it is too complex to be expressed in first-order arithmetic.

1   Introduction

Though the bulk of this paper will be given over to undecidability results, our initial motivation is the extension, even by a small amount, of the class of hybrid automata for which reachability is known to be decidable. It has been suggested that it is the coupling of continuous variables which leads to undecidability [7]. Parallel composition couples only the discrete dynamics of its components. Thus, arguing informally, if we consider parallel compositions of hybrid automata which obey a sufficient decoupling between discrete and continuous dynamics, then we should be able to circumvent undecidability. We will bring this simple idea to a simple fruition in Sect. 2, but first we must dispose of the preliminaries.

1.1   Hybrid Automata

A hybrid system is a physical system which combines discrete and continuous dynamics. *Hybrid automata* are intended as formal mathematical models of such

---

* Research supported by the ARO under the MURI program "Integrated Approach to Intelligent Systems", grant no. DAA H04-96-1-0341.

systems. The following definition is provided to fix notation for the duration of this paper. Though no standard definition exists, this one is not unusual. Note that the continuous dynamical behavior is expressed by a (non-deterministic) semi-flow, not by vector fields as is more common.

Definition 1. A *hybrid automaton* $\mathcal{A}$ is a tuple $(\mathcal{Q}, \mathcal{E}, \mathcal{X}, \mathcal{I}, \mathcal{S}, \mathfrak{s}, \mathfrak{d}, \mathcal{R}, \Phi)$ such that:

- *[discrete states]* $\mathcal{Q}$ is a finite set
- *[edges]* $\mathcal{E}$ is a finite set
- *[plant states]* $\mathcal{X}$ is any set (usually taken to be a manifold)
- *[invariant set]* $\mathcal{I} \subseteq \mathcal{Q} \times \mathcal{X}$
- *[initial set]* $\mathcal{S} \subseteq \mathcal{Q} \times \mathcal{X}$
- *[source map]* $\mathfrak{s} : \mathcal{E} \to \mathcal{Q}$
- *[destination map]* $\mathfrak{d} : \mathcal{E} \to \mathcal{Q}$
- *[reset relation]* $\mathcal{R} \subseteq \mathcal{X} \times \mathcal{E} \times \mathcal{X}$
- *[semi-flow]* $\Phi : \mathcal{Q} \times \mathcal{X} \times \mathbb{R}_{\geq 0} \to \mathcal{P}(\mathcal{X})$ such that for all $(q, x) \in \mathcal{Q} \times \mathcal{X}$:
  1. $\Phi(q, x, 0) = \{x\}$
  2. $\forall t_1, t_2 \in \mathbb{R}_{\geq 0}\ \Phi(q, x, t_1 + t_2) = \bigcup_{y \in \Phi(q, x, t_1)} \Phi(q, y, t_2)$.

The components of a hybrid automaton $\mathcal{A}$ are written with $\mathcal{A}$ as a superscript, as in $\mathcal{Q}^{\mathcal{A}}$, $\mathfrak{s}^{\mathcal{A}}$ and $\Phi^{\mathcal{A}}$. The superscript may be omitted when the automaton is clear from context. $\mathcal{I}_q$ denotes the invariant set in discrete state q and is taken to be a subset of $\mathcal{X}$. Similarly, $\mathcal{S}_q$, $\mathcal{R}_e$ and Φ_q are given their expected interpretations as subsets of $\mathcal{X}$, $\mathcal{X}^2$ and $\mathcal{X} \times \mathbb{R}_{\geq 0} \times \mathcal{X}$, respectively. Finally, by the *guard* of and edge $e \in \mathcal{E}$ we refer to the support of the reset relation $\mathcal{R}_e$.

Definition 2. A *run* of a hybrid automaton $\mathcal{A}$ is a sequence $(q_0, x_0, f_0, t_0, y_0, e_0, q_1, x_1, f_1, t_1, y_1, e_1, \ldots, e_{n-1}, q_n, x_n, f_n, t_n, y_n)$ such that for all $0 \leq i \leq n$:

- $q_i \in \mathcal{Q}$ • $f_i : [0, t_i] \to \mathcal{X}$
- $x_i, y_i \in \mathcal{X}$ • $f_i(0) = x_i$ and $f_i(t_i) = y_i$
- $(q_0, x_0) \in \mathcal{S}$ • $\forall t \in [0, t_i)\ f_i(t) \in \mathcal{I}$
- $t_i \in \mathbb{R}_{\geq 0}$ • $\forall s, t \in [0, t_i]\ s < t \longrightarrow f_i(t) \in \Phi(q_i, f_i(s), t - s)$

and for all $0 \leq i < n$:

- $e_i \in \mathcal{E}$
- $\mathfrak{s}(e_i) = q_i$ and $\mathfrak{d}(e_i) = q_{i+1}$
- $(y_i, e_i, x_{i+1}) \in \mathcal{R}$.

In Sect. 5 we will generalize the notion of run both by allowing the final time interval to be infinite and by allowing infinite sequences of transitions. Until then, finite runs will be more convenient.

Definition 3. The *semi-linear* (resp. *semi-algebraic*) subsets of $\mathbb{R}^n$ are formed by taking boolean combinations of sets defined by linear (resp. algebraic) equalities and inequalities with rational coefficients.

Definition 4. By *semi-linear hybrid automata* (SLHA) we mean that elusive class of automata which has been variously known as polyhedral and—to the consternation of control theorists—as linear. $\mathcal{A}$ is an *n-dimensional* SLHA if:
- $\mathcal{X}^{\mathcal{A}} = \mathbb{R}^n$ for some n
- for every $q \in \mathcal{Q}^{\mathcal{A}}$ and $e \in \mathcal{E}^{\mathcal{A}}$, the projected components $\mathcal{I}_q^{\mathcal{A}}$, $\mathcal{S}_q^{\mathcal{A}}$, $\mathcal{R}_e^{\mathcal{A}}$ and $\Phi_q^{\mathcal{A}}$ are semi-linear subsets of $\mathbb{R}^n$, $\mathbb{R}^n$, $\mathbb{R}^{2n}$ and $\mathbb{R}^{2n+1}$, respectively.

Semi-algebraic hybrid automata are defined analogously.

1.2   Annotated Hybrid Automata

It will be convenient to add a layer of abstraction to our hybrid automata. An *annotation* associates to each edge an *event* and to each discrete state a nonempty set of possible *conditions*. These annotations do not affect the behavior of the automaton but will be used when we define the *timed language* of an automaton and when we define the operation of *parallel composition*.

Definition 5. An *annotated hybrid automaton* $\mathcal{A}$ is a hybrid automaton with four additional components $(\Sigma, \Gamma, \mathfrak{e}, \mathfrak{c})$:
- [*events*] Σ is a finite set
- [*conditions*] Γ is a finite set
- [*event assignment*] $\mathfrak{e} : \mathcal{E} \to \Sigma$
- [*condition assignment*] $\mathfrak{c} : \mathcal{Q} \to \mathcal{P}(\Gamma)$ such that $\forall q \in \mathcal{Q}\; \mathfrak{c}(q) \neq \emptyset$.

Definition 6. To each run $(q_0,\; x_0,\; f_0,\; t_0,\; y_0,\; e_0,\; q_1,\; x_1,\; f_1,\; t_1,\; y_1,\; e_1,\; \ldots,\; e_{n-1},\; q_n,\; x_n,\; f_n,\; t_n,\; y_n)$ of an annotated hybrid automaton $\mathcal{A}$, we associate an *annotated run* $(c_0,\; t_0,\; v_0,\; c_1,\; t_1,\; v_1,\; \ldots,\; c_n,\; t_n)$ such that:
- for all $0 \leq i \leq n$, $c_i \in \mathfrak{c}(q_i)$
- for all $0 \leq i < n$, $v_i = \mathfrak{e}(e_i)$.

The *timed language* $\mathcal{L}(\mathcal{A})$ of an annotated hybrid automaton $\mathcal{A}$ is set of all annotated runs of $\mathcal{A}$.

The following equivalence relation will be important.

Definition 7. We say that the annotated hybrid automata $\mathcal{A}$ and $\mathcal{B}$ are *language equivalent* iff:
- $\Sigma^{\mathcal{A}} = \Sigma^{\mathcal{B}}$
- $\mathcal{L}(\mathcal{A}) = \mathcal{L}(\mathcal{B})$.

We denote language equivalence by $\mathcal{A} \sim_{\mathfrak{le}} \mathcal{B}$.

Remark 1. Invoking symmetry, one might expect the requirement that $\Gamma^{\mathcal{A}} = \Gamma^{\mathcal{B}}$ in the definition of language equivalence. We disclude this requirement because it is unnecessary, though it would not falsify the results that follow. The interested reader should note in Sect. 1.3 that the set Γ does not play a very important role in parallel composition, while Σ is crucial.

By the *reachability problem* for an annotated hybrid automaton $\mathcal{A}$, we mean the problem of determining which conditions $c \in \Gamma^{\mathcal{A}}$ occur on *some* annotated run. This ensures that language equivalent hybrid automata have equivalent reachability problems. Of course, the reachability of a discrete state can be detected with a suitable annotation and we may suppress explicit mention of annotations when discussing reachability. We say that the reachability problem is decidable for a class $\mathcal{K}$ if there is an algorithm which uniformly solves the reachability problem for every member of $\mathcal{K}$.

1.3 Parallel Composition

Given two annotated hybrid automata we define a product automaton called the *parallel composition*. Conceptually, a run of the parallel composition is comprised of simultaneous runs of the component automata which are independent except that:

- They must synchronize on shared events.
- The only product states that are permitted are those for which the restrictions on conditions are jointly satisfiable.

Definition 8. We define the *parallel composition* $\mathcal{A} \parallel \mathcal{B}$ of the annotated hybrid automata $\mathcal{A}$ and $\mathcal{B}$ in two stages. First, we define a synchronized product automaton $\mathcal{A} \otimes \mathcal{B}$ such that:

- $\mathcal{Q} = \mathcal{Q}^{\mathcal{A}} \times \mathcal{Q}^{\mathcal{B}}$
- $\mathcal{E} = \mathcal{E}_1 \cup \mathcal{E}_2 \cup \mathcal{E}_3$ where:
$$\mathcal{E}_1 = \{(e_1, q_2) \in \mathcal{E}^{\mathcal{A}} \times \mathcal{Q}^{\mathcal{B}} \mid \mathfrak{e}^{\mathcal{A}}(e_1) \notin \Sigma^{\mathcal{B}}\}$$
$$\mathcal{E}_2 = \{(q_1, e_2) \in \mathcal{Q}^{\mathcal{A}} \times \mathcal{E}^{\mathcal{B}} \mid \mathfrak{e}^{\mathcal{B}}(e_2) \notin \Sigma^{\mathcal{A}}\}$$
$$\mathcal{E}_3 = \{(e_1, e_2) \in \mathcal{E}^{\mathcal{A}} \times \mathcal{E}^{\mathcal{B}} \mid \mathfrak{e}^{\mathcal{A}}(e_1) = \mathfrak{e}^{\mathcal{B}}(e_2)\}$$
- $\mathcal{X} = \mathcal{X}^{\mathcal{A}} \times \mathcal{X}^{\mathcal{B}}$
- $\mathcal{I} = \{((q_1, q_2), (x_1, x_2)) \in \mathcal{Q} \times \mathcal{X} \mid (q_1, x_1) \in \mathcal{I}^{\mathcal{A}} \wedge (q_2, x_2) \in \mathcal{I}^{\mathcal{B}}\}$
- $\mathcal{S} = \{((q_1, q_2), (x_1, x_2)) \in \mathcal{Q} \times \mathcal{X} \mid (q_1, x_1) \in \mathcal{S}^{\mathcal{A}} \wedge (q_2, x_2) \in \mathcal{S}^{\mathcal{B}}\}$
- $\mathfrak{s}(c_1, c_2) = \begin{cases} (\mathfrak{s}^{\mathcal{A}}(c_1), c_2) & \text{if } c_2 \in \mathcal{Q}^{\mathcal{B}} \\ (c_1, \mathfrak{s}^{\mathcal{B}}(c_2)) & \text{if } c_1 \in \mathcal{Q}^{\mathcal{A}} \\ (\mathfrak{s}^{\mathcal{A}}(c_1), \mathfrak{s}^{\mathcal{B}}(c_2)) & \text{otherwise} \end{cases}$
- $\mathfrak{d}(c_1, c_2) = \begin{cases} (\mathfrak{d}^{\mathcal{A}}(c_1), c_2) & \text{if } c_2 \in \mathcal{Q}^{\mathcal{B}} \\ (c_1, \mathfrak{d}^{\mathcal{B}}(c_2)) & \text{if } c_1 \in \mathcal{Q}^{\mathcal{A}} \\ (\mathfrak{d}^{\mathcal{A}}(c_1), \mathfrak{d}^{\mathcal{B}}(c_2)) & \text{otherwise} \end{cases}$
- $\mathcal{R} = \{((x_1, x_2), (c_1, c_2), (y_1, y_2)) \in \mathcal{X} \times \mathcal{E} \times \mathcal{X} \mid$
  $((c_1 \in \mathcal{Q}^{\mathcal{A}} \wedge x_1 = y_1) \text{ or } (c_1 \in \mathcal{E}^{\mathcal{A}} \wedge (x_1, c_1, y_1) \in \mathcal{R}^{\mathcal{A}})) \text{ and}$
  $((c_2 \in \mathcal{Q}^{\mathcal{B}} \wedge x_2 = y_2) \text{ or } (c_2 \in \mathcal{E}^{\mathcal{B}} \wedge (x_2, c_2, y_2) \in \mathcal{R}^{\mathcal{B}}))\}$
- $\Phi((q_1, q_2), (x_1, x_2), r) = \Phi^{\mathcal{A}}(q_1, x_1, r) \times \Phi^{\mathcal{B}}(q_2, x_2, r)$.

$\mathcal{A} \otimes \mathcal{B}$ is annotated as follows:

- $\Sigma = \Sigma^{\mathcal{A}} \cup \Sigma^{\mathcal{B}}$
- $\Gamma = \Gamma^{\mathcal{A}} \cap \Gamma^{\mathcal{B}}$
- $\mathfrak{e}(c_1, c_2) = \begin{cases} \mathfrak{e}^{\mathcal{A}}(c_1) & \text{if } c_1 \in \mathcal{E}^{\mathcal{A}} \\ \mathfrak{e}^{\mathcal{B}}(c_2) & \text{otherwise} \end{cases}$
- $\mathfrak{c}(q_1, q_2) = \mathfrak{c}^{\mathcal{A}}(q_1) \cap \mathfrak{c}^{\mathcal{B}}(q_2)$.

The second stage in the formation of $\mathcal{A} \parallel \mathcal{B}$ is to discard all discrete states $q \in \mathcal{Q}^{\mathcal{A} \otimes \mathcal{B}}$ such that $c^{\mathcal{A} \otimes \mathcal{B}}(q) = \emptyset$. This ensures that $\mathcal{A} \parallel \mathcal{B}$ is an annotated hybrid automaton and completes the construction.

Remark 2. Parallel composition is commutative and associative (up to isomorphism). Therefore we can, and will, refer to the parallel composition of several annotated hybrid automata without fear of ambiguity.

The concept of parallel composition defined here is nowise new. Conditions are just an alternative to the *propositional constraints* that commonly arise in the temporal logic literature. The novelty is not in our definition, but in the use we will make of parallel composition—to define a new class of hybrid automata for which the reachability problem is decidable. The following simple relationship between language equivalence and parallel composition will be a key ingredient; it will allow us to do reductions component-wise.

Lemma 1. *If $\mathcal{A} \sim_{\text{le}} \mathcal{A}'$ and $\mathcal{B} \sim_{\text{le}} \mathcal{B}'$ then $\mathcal{A} \parallel \mathcal{B} \sim_{\text{le}} \mathcal{A}' \parallel \mathcal{B}'$.*

2   A New Decidable Class

Definition 9. If $\mathcal{K}$ is a class of hybrid automata, then the *parallel closure $\mathcal{K}^{\parallel}$* is the class of all parallel compositions of all annotations of the elements from $\mathcal{K}$.

Definition 10.
Clock Components:
Let $\mathcal{C}$ be the class of 1-dimensional SLHA such that $\Phi(q, t, x) = x + t$, the plant state is zero in all initial states, and each edge satisfies either:
    (a)   zero reset
or (b)   identity reset

Rectangular Components:
Let $\mathcal{R}$ be the class of 1-dimensional SLHA such that $\Phi(q, t, x) = x + tI_q$, where I_q is an interval for each q, and such that each edge satisfies either:
    (a)   constant set-valued reset map
or (b)   identity reset and
        source and destination have the same flow

Deterministic Components:
Let $\mathcal{D}$ be the class of SLHA with deterministic flows and finite initial set such that each edge satisfies either:
    (a)   constant (single-valued) reset map
or (b)   identity reset and
        source and destination have the same flow

Nondeterministic Components:
Let $\mathcal{N}$ be the class of SLHA such that each edge satisfies either:
  (a) constant set-valued reset map
or (b) identity reset and
        trivial guard and
        source and destination have the same flow and invariant set

The reader is probably already familiar with $\mathcal{C}^{\|}$ and $\mathcal{R}^{\|}$, though our presentation is somewhat unusual. They are, respectively, *timed automata* [1] and *initialized rectangular hybrid automata* [12,7]. Both of these classes are known to have decidable reachability problems.

Lemma 2.

1. *If $\mathcal{A} \in \mathcal{R}$ then every annotation of $\mathcal{A}$ is language equivalent to a two clock timed automaton.*
2. *If $\mathcal{A} \in \mathcal{D} \cup \mathcal{N}$ then every annotation of $\mathcal{A}$ is language equivalent to an annotation of a clock component.*

Part (1) is contained in [12] while Part (2) offers no real difficulty. Combining Lemma 1 with Lemma 2 and the decidability of reachability for timed automata, the following theorem is immediate.

Theorem 1. *Reachability is decidable for $(\mathcal{R} \cup \mathcal{D} \cup \mathcal{N})^{\|}$.*

Note that $(\mathcal{R} \cup \mathcal{D} \cup \mathcal{N})^{\|}$ is a proper superclass of the initialized rectangular hybrid automata, and that the possibility of further extension remains open. New building blocks may be added easily; they will slip right into place, as long as they are language equivalent to timed automata. Admittedly, this is a severe restriction.

3   Irrational Timed Automata

The semi-algebraic sets share many of the nice properties of the semi-linear sets [14]; in particular, they are closed under projection [13] and the boolean operations. So it is natural to ask if the results of the preceding section remain true in this more general context.

Definition 11. *We use $\mathcal{C}_{SA}$, $\mathcal{R}_{SA}$, $\mathcal{D}_{SA}$ and $\mathcal{N}_{SA}$ for the generalizations of $\mathcal{C}$, $\mathcal{R}$, $\mathcal{D}$ and $\mathcal{N}$ to semi-algebraic hybrid automata.*

As before, we can prove that every automaton $\mathcal{A} \in (\mathcal{R}_{SA} \cup \mathcal{D}_{SA} \cup \mathcal{N}_{SA})^{\|}$ is language equivalent to an automaton $\mathcal{A}' \in \mathcal{C}_{SA}^{\|}$. But note that $\mathcal{A}'$ is *not* necessarily a timed automaton; its constants are arbitrary algebraic numbers and may be irrational. So we are led to ask if reachability remains decidable for *algebraic timed automata*. Unfortunately, it does not.

Theorem 2. *Reachability is undecidable for $\mathcal{C}_{SA}^{\|}$.*

Before preceding with a proof of this theorem, there is further motivation. Reachability is decidable for several classes of hybrid systems, for example [8] and [9]; we focus on two. We have already mentioned the initialized rectangular hybrid automata, and even offered a modest generalization. The second class contains the semi-algebraically defined hybrid automata with constant (set-valued) reset maps, which are proven to have computable finite bisimulations in [10]. To what extent can these classes be combined while preserving the decidability of reachability? Algebraic timed automata represent, in our opinion, a simple midpoint between these two classes, and in this light, the undecidability of the reachability problem presents an obstacle to a natural unification.

3.1   Minsky Machines and Undecidability

We prove our main theorem in more generality to illustrate that undecidability does not arise from some subtle property of the algebraics. Rather, it is a consequence of irrationality. This generality will also be useful in Sect. 4.

Definition 12. Given $\mathcal{S} \subseteq \mathbb{R}$, the class $\mathcal{T}_{\mathcal{S}}$ of *irrational timed automata over $\mathcal{S}$* is the generalization of timed automata in which the guards and state invariants are allowed to have constants from $\mathbb{Q} \cup \mathcal{S}$.

In particular, $\mathcal{C}_{SA}^{\|} = \mathcal{T}_{\mathbb{A}}$ is the class of *algebraic timed automata*, where $\mathbb{A}$ is the set of all algebraic numbers, i.e. real roots of polynomial equations with rational coefficients.

Theorem 3. *Let $\tau \in (1, 2)$ be irrational. Let $\mathcal{S} = \{0, 1, \tau, 3 - \tau\}$. Then the reachability problem for the class $\mathcal{T}_{\mathcal{S}}$ is undecidable.*

Our proof of undecidability closely follows the technique in [7], where the undecidability of several slight generalizations of timed automata is proved. In particular, we proceed by reducing the halting problem for two-counter Minsky machines to the reachability problem for the class $\mathcal{T}_{\mathcal{S}}$. Before presenting this reduction, we give a definition of two-counter machines. It is well known that the halting problem for two-counter machines is undecidable [11].

Definition 13. A *two-counter Minsky machine* is finite state machine with two natural number counters c_1 and c_2. Each machine state has an associated command which is executed when the machine is in that state. Possible commands are:
  - increment c_i and go to n
  - decrement c_i and go to n; if $c_i = 0$ then it is unchanged
  - if c_i is zero go to n, otherwise go to m
  - halt

where $i \in \{1, 2\}$ and n, m are machine states. There is a distinguished start state and the machine begins its execution with both counters set to zero.

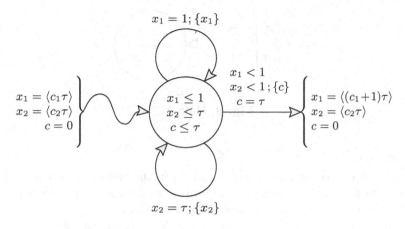

Fig. 1. Increment c_1

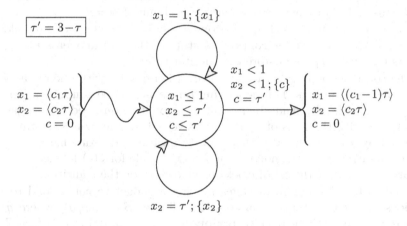

Fig. 2. Decrement c_1

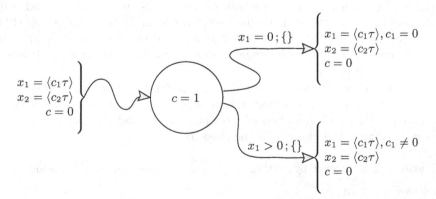

Fig. 3. Test if $c_1 = 0$

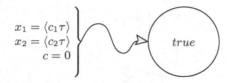

$$x_1 = \langle c_1 \tau \rangle$$
$$x_2 = \langle c_2 \tau \rangle$$
$$c = 0$$

true

Fig. 4. Halt

Proof (Theorem 3).

Let $\mathcal{A}$ be a two-counter machine. To simplify the encoding we can assume that it never decrements a counter containing zero. Of course, any two-counter machine can easily be modified to meet this restriction.

Let $\langle x \rangle$ denote the non-integer part of $x \in \mathbb{R}$. In particular, for any x, $0 \leq \langle x \rangle < 1$. We will encode the values of the counters c_1 and c_2 in continuous variables x_1 and x_2 by representing the natural number n by the real number $\langle n\tau \rangle$. Because τ is irrational, $\langle n\tau \rangle = \langle m\tau \rangle$ if and only if $n = m$.

We now construct a timed automaton $\mathcal{A}^* \in \mathcal{T}_{\mathcal{S}}$. It will have three clocks components. We represent the continuous state of these components by x_1, x_2 and c. As indicated, x_1 an x_2 store the counter values.

In the construction of $\mathcal{A}^*$, each state of $\mathcal{A}$ is replaced with one of the four gadgets illustrated in Figs. 1–4, depending on its associated command. For example, a state with command "Increment c_2" would be replaced by the gadget in Fig. 1, but with the roles of x_1 and x_2 reversed. In the figures, a state q is represented by a node labeled with the state invariant $\mathcal{I}_q$. An edge e is represented by an arrow from the node for $\mathfrak{s}(e)$ to the node for $\mathfrak{d}(e)$ labeled by both the guard for e and by the set of clocks reset to zero by the transition.

We define the destination of edges leaving a gadget to correspond to the transitions in the two-state machine $\mathcal{A}$. Finally, let $\mathcal{S} = (q_0, \mathbf{0})$, where q_0 is the discrete state in the gadget corresponding to the initial state of $\mathcal{A}$. This completely specifies a timed automaton $\mathcal{A}^*$.

The reader is encouraged to carefully examine Figs. 1–4 to understand why the gadgets that they depict have the asserted effects. It is worth noting that Fig. 2 is the same as Fig. 1 except that τ is replaced by $\tau' = 3 - \tau \in (1, 2)$. It is also worth noting that the each gadget is defined to guarantee that $c = 0$ when the next gadget is entered.

By construction, the two-counter machine $\mathcal{A}$ halts if and only if there is a reachable state of $\mathcal{A}^*$ which corresponds to a halting state of $\mathcal{A}$. As mentioned above, the halting problem for two-counter machines is undecidable. This proves that reachability is undecidable for the class $\mathcal{T}_{\mathcal{S}}$. $\qquad\square$

Theorem 2 is proved by letting $\tau = \sqrt{2}$ in Theorem 3 and noting that $\mathcal{T}_{\{0,1,\sqrt{2},3-\sqrt{2}\}} \subseteq \mathcal{C}_{SA}^{\parallel}$.

3.2   Further Results

Both $\mathcal{R}_{SA}$ and $\mathcal{D}_{SA}$ are extensions of $\mathcal{C}_{SA}$. Therefore, the undecidability of reachability for $\mathcal{R}_{SA}^{\parallel}$ and $\mathcal{D}_{SA}^{\parallel}$ follows from Theorem 2. On the other hand, $\mathcal{N}_{SA}^{\parallel}$ requires a different proof. The following results are each proved by refining of the construction for Theorem 3. The gadgets become rather complicated to circumvent the additional restrictions, but no other change is necessary.

Theorem 4.

1. *Reachability is undecidable for $\mathcal{N}_{SA}^{\parallel}$*
2. *Let $\tau > 0$ be irrational. Then reachability is undecidable for $\mathcal{T}_{\{\tau,1\}}$ (with as few as three clocks).*

Definition 14.   Given $\mathcal{S}_1, \mathcal{S}_2 \subseteq \mathbb{R}$, let $\mathcal{T}_{\mathcal{S}_1, \mathcal{S}_2}$ be the class of irrational timed automata with the first clock constrained by constants from $\mathcal{S}_1$ and the remaining clocks constrained by constants from $\mathcal{S}_2$.

Theorem 5. *Let $\tau > 0$ be irrational. Reachability is undecidable for $\mathcal{T}_{\{\tau\},\{1\}}$ (with as few as four clocks).*

Before moving on, one simple decidability result should be mentioned.

Theorem 6. *Reachability for one-clock timed automata over $\mathbb{R}$ depends only on the order of the constants (including zero), and is decidable given that order.*

4   Parametric Timed Automata

Without significant modification, the undecidability results of the previous section carry over to the context of *parametric timed automata*. These automata—introduced in "Parametric Real-time Reasoning" [2]—allow us to express a more sophisticated range of synthesis and verification questions, but their most basic properties turn out to be undecidable [2]. After discussing the connection between parametric timed automata and timed automata with irrational constraints, we state a new undecidability result and then examine the complexity of the one-clock case, for which reachability *is* decidable.

Definition 15.

(a) *Parametric timed automata* are a generalization of timed automata in which the guards and state invariants are allowed to have constants from $\mathbb{Q} \cup \Psi$, where Ψ is a set of parameter variables.
(b) Let $\mathcal{A}$ be a parametric timed automaton with parameters from Ψ and let $\lambda : \Psi \to \mathbb{Q}$. Then $\mathcal{A}_\lambda$ is the timed automaton that results from using λ to substitute for the parameters in $\mathcal{A}$.

(c) If q is a state of $\mathcal{A}$, then $\Gamma_q(\mathcal{A})$ is the subset of parameter space for which $\mathcal{A}$ has a run reaching q. In other words:

$$\Gamma_q(\mathcal{A}) = \{\lambda : \Psi \to \mathbb{Q} \mid q \text{ is a reachable state of } \mathcal{A}_\lambda\}.$$

Now consider what would happen were τ a rational number in the proof of Theorem 3. In particular, let $\tau = a/b \in (1, 2)$, where a and b are relatively prime. As long as our virtual Minsky machine keeps its counter values below b, nothing can go wrong. But a counter value of b is indistinguishable from zero; we have an overflow error. Such an error is easy to detect if we always test for zero after incrementing a counter. Thus, we can correctly simulate the Minsky machine as long as the counter values remain small and suspend the simulation when an overflow error is detected.

At the risk of stating the obvious, note that if a Minsky machine halts then its counters remain bounded. Also note that the rational numbers in the interval $(1, 2)$ have arbitrarily large denominators (in reduced form). Therefore, a Minsky machine halts if and only if that fact is detected by the simulation for *some* rational $\tau \in (1, 2)$. With a few simple details swept under the rug, this is all it takes to translate the theorems of the last section into theorems about parametric timed automata. Under this translation, Theorem 4.2 becomes:

Theorem 7. *The emptiness of $\Gamma_q(\mathcal{A})$ is undecidable for the class of parametric timed automata with three clocks and one parameter.*

This is not an essentially new result. That the emptiness of $\Gamma_q(\mathcal{A})$ for parametric timed automata is undecidable was proved in [2]. The proof given there uses three clocks and six parameters but has the advantage of working for both the dense-time and discrete-time cases. The translation of Theorem 5 is more interesting.

Theorem 8. *The emptiness of $\Gamma_q(\mathcal{A})$ is undecidable for the class of parametric timed automata with only one clock constrained by parameters.*

The corresponding problem is decidable for discrete-time [2], so we have exposed a divergence in the expressive power of dense-time and discrete-time parametric timed automata.

Before leaving the subject of parametric timed automata, let us turn our attention to the one-clock case. Let $\mathcal{A}$ be a one-clock parametric timed automaton. As was noted before, the time-abstract runs of a one-clock timed automaton depend only on the order of the constants. So, to calculate $\Gamma_q(\mathcal{A})$, we simply determine the reachability of q in $\mathcal{A}_\lambda$ for sample assignments λ corresponding to every possible ordering of the constants (including zero) and parameters. Unfortunately, the number of such orderings[1] grows exponentially in the number of parameters.

---

[1] Let π_m^n be the number of (non-strict) orderings that can be formed from n parameters with respect to m distinct constants. The following formulae generalizations of

We conclude with a number of observations about the complexity of the problem of determining emptiness for one-clock parametric timed automata. Both standard algorithmic complexity and parametric complexity [5] results are considered.

Theorem 9. *Consider the problem of determining the non-emptiness of* $\Gamma_q(\mathcal{A})$ *for the class of one-clock parametric timed automata.*

1. *The problem is NP-complete.*
2. *For any fixed k bounding the number of parameters, the problem can be solved in polynomial time.*
3. *Parameterized by the number of parameters, the problem is W[SAT]-hard. Note that it is strongly suspected that W[SAT] is a proper superset of the fixed parameter tractable (FPT) problems [5].*
4. *Parameterized by the number of both constants and parameters, the problem is FPT.*

5   The Complexity of Questions About SLHA

This last section deviates from the course of the paper thus far. The only connection it bears to the earlier sections is the reliance on Minsky machine simulations; they play a central role in proving the hardness directions of all of the completeness results that follow.

Definition 16. A *maximal* run is any run that can not be extended. It either has an infinite number of transitions, ends with an infinite time interval spent in the same discrete state, or reaches a state from which it can neither flow nor jump. A maximal run is said to be *jump-infinite* if it makes an infinite number of discrete transitions; otherwise *jump-finite*. It is *time-infinite* if its time intervals sum to infinity; otherwise *time-finite*. A maximal run is *blocking* if it is both

---

those in [6], where the sequence $\{\pi_0^n\}_{n\in\mathbb{N}}$ of *preferential arrangements* is studied.

$$\pi_m^n = \frac{1}{2}\sum_{k=0}^{\infty}\binom{k}{m}k^n 2^{-k}$$

$$\pi_m^n = 2m\pi_m^{n-1} + \sum_{k=0}^{n-1}\binom{n}{k}\pi_m^k$$

Finally, writing $f \sim_n g$ to denote $\lim_{n\to\infty} f/g = 1$,

$$(\forall m) \quad \pi_m^n \sim_n \frac{(n+m)!}{2m!\ln^{n+m+1}2}$$

$$(\forall n) \quad \pi_m^n \sim_m (2m)^n.$$

Table 1. The complexity of detecting various types of runs

type of run	Semi-linear Hybrid Automata (SLHA)	Compact SLHA	Deterministic SLHA	Deterministic and Non-blocking SLHA	Non-blocking SLHA
Zeno	Σ^1_1-complete		Σ_2-complete		Σ^1_1-complete
time-finite					
non-Zeno			Π_2-complete		
time-infinite, jump-infinite					
time-infinite					
jump-infinite					
infinite	Π_1-complete			always	
arbitrarily long					
arbitrarily long blocking	Π_2-complete			never	
blocking					
time-infinite, jump-finite	Σ_1-complete				
jump-finite					

Table 2. Supplement to Table 1

	Semi-linear Hybrid Automata (SLHA)	Weakly Deterministic SLHA	Deterministic SLHA
Zeno run	Σ^1_1-complete		Σ_2-complete
time-infinite run for every initial state	Π^1_2-complete	Π^1_1-complete	Π_2-complete

time-finite and jump-finite; otherwise we call it *infinite*. Finally, a maximal run is *Zeno* if it is time-finite but jump-infinite.

A hybrid automaton is said to have *arbitrarily long runs* if for each $n \in \mathbb{N}$ there is either a run that makes at least n transitions or a run with a duration of at least n.

Definition 17. An SLHA is *compact* if all of its defining regions are compact. An SLHA with at most one initial state and at most one possible evolution from each state is called *deterministic*, and an SLHA with at least one initial state and at least one evolution from each state is called *non-blocking*.

Table 1 gives, for different classes of SLHA, the complexity of determining whether certain types of runs exist. It is only a sampling of complexity results. Further questions might prove interesting; for example, "Is there an time-infinite run for every initial state?"

Also, our definition of determinism is very restrictive. A more reasonable property is that there is at most one evolution from each state, with no restriction made on the initial set. We call this property *weak determinism*. Table 2 shows that questions may be much harder for weakly deterministic SLHA than for deterministic SLHA. The complexity of many questions matches that of general SLHA, but this is not always the case.

As a closing note, the Zeno phenomenon is exploited in [3] and [4] to show that the reachability problem for dynamical systems with piecewise constant derivatives is arithmetic and hyper-arithmetic, respectively.

References

1. Rajeev Alur and David L. Dill. A theory of timed automata. *Theoret. Comput. Sci.*, 126(2):183–235, 1994.
2. Rajeev Alur, Thomas A. Henzinger, and Moshe Y. Vardi. Parametric real-time reasoning. In *Proceedings of the Twenty-Fifth Annual ACM Symposium on the Theory of Computing*, pages 592–601, San Diego, California, 16–18 May 1993.
3. Eugene Asarin and Oded Maler. Achilles and the tortoise climbing up the arithmetical hierarchy. *Journal of Computer and System Sciences*, 57(3):389–398, December 1998.
4. Olivier Bournez. Achilles and the Tortoise climbing up the hyper-arithmetical hierarchy. *Theoretical Computer Science*, 210(1):21–71, January 1999.
5. R. G. Downey and M. R. Fellows. *Parameterized complexity*. Springer-Verlag, New York, 1999.
6. O. A. Gross. Preferential arrangements. *Amer. Math. Monthly*, 69:4–8, 1962.
7. Thomas A. Henzinger, Peter W. Kopke, Anuj Puri, and Pravin Varaiya. What's decidable about hybrid automata? *J. Comput. System Sci.*, 57(1):94–124, 1998. 27th Annual ACM Symposium on the Theory of Computing (STOC'95) (Las Vegas, NV).
8. Y. Kesten, A. Pnueli, J. Sifakis, and S. Yovine. Integration graphs: a class of decidable hybrid systems. In *Hybrid systems*, pages 179–208. Springer, Berlin, 1993.

9. M. Kourjanski and P. Varaiya. A class of rectangular hybrid systems with computable reach set. *Lecture Notes in Computer Science*, 1273:228–234, 1997.
10. G. Lafferriere, G. J. Pappas, and S. Sastry. O-minimal hybrid systems. Technical Report UCB/ERL M98/29, Department of Electrical Engineering and Computer Science, University of California at Berkeley, May 1998.
11. Marvin L. Minsky. Recursive unsolvability of Post's problem of "tag" and other topics in theory of Turing machines. *Ann. of Math. (2)*, 74:437–455, 1961.
12. Anuj Puri and Pravin Varaiya. Decidability of hybrid systems with rectangular differential inclusions. In *Computer aided verification (Stanford, CA, 1994)*, pages 95–104. Springer, Berlin, 1994.
13. Alfred Tarski. *A Decision Method for Elementary Algebra and Geometry*. RAND Corporation, Santa Monica, Calif., 1948.
14. Lou van den Dries. *Tame topology and o-minimal structures*. Cambridge University Press, Cambridge, 1998.

Level Set Methods for Computation in Hybrid Systems*

Ian Mitchell[1] and Claire J. Tomlin[2]

[1] Scientific Computing and Computational Mathematics Program,
Gates 2B, Stanford University, Stanford CA, 94305
mitchell@sccm.stanford.edu
[2] Department of Aeronautics and Astronautics,
250 Durand, Stanford University, Stanford CA, 94305
tomlin@leland.stanford.edu

Abstract. Reachability analysis is frequently used to study the safety of control systems. We present an implementation of an exact reachability operator for nonlinear hybrid systems. After a brief review of a previously presented algorithm for determining reachable sets and synthesizing control laws—upon whose theory the new implementation rests—an equivalent formulation is developed of the key equations governing the continuous state reachability. The new formulation is implemented using level set methods, and its effectiveness is shown by the numerical solution of three examples.

1 Introduction

The reachability operator, a function or algorithm that can determine the evolution of sets of trajectories, is key in the synthesis and verification of controllers for continuous, discrete or hybrid systems. Regardless of whether reachability appears implicitly, such as in the generation of invariant sets, or explicitly, no technique for determining safe control systems can avoid its use. It is natural that methods for its accurate, automatic computation are attracting considerable attention.

Reachability analysis of hybrid systems has been investigated by both the computer science and control communities. Methods have been developed by computer scientists for computing reachable sets for timed automata [1] and linear hybrid automata [2], for which computation is based on the propagation of polygonal sets under constant rate dynamics. Tools have been developed to perform such calculations automatically [3,4], and to synthesize controllers in such a framework [5,6]. Control theorists have extended reachability tools from continuous state and time dynamical systems theory to incorporate discrete switches [7,8,9,10,11]. However, the efficient computation of reachable sets for hybrid systems with nonlinear dynamics remains a difficult problem to solve.

---

* Research supported by DARPA under the Software Enabled Control Program (AFRL contract F33615-99-C-3014), and by a Frederick E. Terman Faculty Award.

N. Lynch and B. Krogh (Eds.): HSCC 2000, LNCS 1790, pp. 310–323, 2000.

Numerical techniques which over-approximate the nonlinear dynamics with linear dynamics [12], or which over-approximate the reachable sets [13,14,15,16], have recently been developed.

In this paper, we present an implementation of an exact reachability operator for nonlinear hybrid systems. An algorithm which synthesizes control laws for such systems based on the Hamilton-Jacobi equation [9,10,11] is reviewed, and then a new Hamilton-Jacobi formulation with superior numerical properties is developed and proved to be equivalent. While level set techniques were previously investigated for the solution of such equations in [17], we have added several improvements to the basic level set algorithm. Examples from [11] demonstrate the results of applying the new algorithm to the new equations—examples which have never previously been solved computationally.

2 Deriving Reachable Sets in Hybrid Automata

In [11], an algorithm is presented which characterizes the reachable set of a nonlinear hybrid automaton (with desired safety properties) as that whose boundary is the zero level set of a particular Hamilton-Jacobi equation. The algorithm also computes the continuous and discrete control laws to maximize the safe operating region. In this section, we briefly review this hybrid system model and reachability algorithm, and then present a second characterization using a similar Hamilton-Jacobi algorithm with better numerical properties.

2.1 Hybrid Automata and Hamilton-Jacobi Equations

A **hybrid automaton** is defined as

$$H = ((Q \times X), (U \times D), (\Sigma_u \times \Sigma_d), f, \delta, Inv, \Omega) \qquad (1)$$

where Q is a finite set of discrete states, $X = \mathbb{R}^n$, $U \subseteq \mathbb{R}^{n_u}$ is the set of continuous control inputs, $D \subseteq \mathbb{R}^{n_d}$ is the set of continuous disturbances, $\Sigma = \Sigma_u \times \Sigma_d$ is a finite set of actions, where Σ_u denotes the set of discrete control inputs, and Σ_d the set of discrete disturbance inputs, $f : Q \times X \times U \times D \to \mathbb{R}^n$ defines the flow of continuous trajectories, $\delta : Q \times X \times \Sigma_u \times \Sigma_d \to 2^{Q \times X}$ is the discrete transition function, $Inv \subseteq Q \times X$ is the invariant associated to each discrete state, and Ω is an acceptance condition—here $\Omega = (\Box F)$, meaning that the state of the system must remain within a set $F \subseteq Q \times X$. We denote $\mathcal{U}$ as the set of piecewise continuous functions from $\mathbb{R}$ to U, and $\mathcal{D}$ the set of piecewise continuous functions from $\mathbb{R}$ to D.

Three operators are defined:

$$Pre_u(K) = \{(q,x) \in Q \times X \mid \exists \sigma_u \in \Sigma_u \; \forall \sigma_d \in \Sigma_d \; \delta(q,x,\sigma_u,\sigma_d) \subseteq K\} \cap K$$
$$Pre_d(K) = \{(q,x) \in Q \times X \mid \forall \sigma_u \in \Sigma_u \; \exists \sigma_d \in \Sigma_d \; \delta(q,x,\sigma_u,\sigma_d) \cap K^c \neq \emptyset\} \cup K^c$$
$$Reach(G,E) = \{(q,x) \in Q \times X \mid \forall u \in \mathcal{U} \; \exists d \in \mathcal{D} \text{ and } t \geq 0 \text{ such that}$$
$$(q(t),x(t)) \in G \text{ and } (q(s),x(s)) \in Inv \setminus E \text{ for } s \in [0,t]\}$$

where $K \subseteq Q \times X$; $G, E \subseteq X$; and $(q(s), x(s))$ is the continuous state trajectory of $\dot{x} = f(q(s), x(s), u(s), d(s))$ starting at (q, x). The set $Reach(G, E)$ describes those states from which, for all $u(\cdot) \in \mathcal{U}$, there exists a $d(\cdot) \in \mathcal{D}$, such that the state trajectory $(q(s), x(s))$ can be driven to a "bad" set G while avoiding an "escape" set E. With these definitions in place, the algorithm for reachability analysis for hybrid systems proceeds as follows [10,11]:

> Let $W^0 = F, W^{-1} = \emptyset, i = 0$.
> While $W^i \neq W^{i-1}$ do
> $\qquad W^{i-1} = W^i \setminus Reach(Pred_d(W^i), Pre_u(W^i)))$
> $\qquad i = i - 1$
> end

If the algorithm terminates after a finite number of steps, then the fixed point W^* is the largest set of states for which the control $(u(\cdot), \upsilon_u[\cdot])$ can guarantee that the state of the hybrid system remains inside F despite the action of the disturbance $(d(\cdot), \sigma_d[\cdot])$. In order to implement this algorithm, Pre_u, Pre_d, and $Reach$ need to be computed. The calculation of Pre_u and Pre_d requires inversion of the transition relation δ subject to the quantifiers $\exists$ and $\forall$. The computation of $Reach$ requires an algorithm for determining the set of initial conditions from which trajectories can reach one set, avoiding a second set along the way. Our focus in this paper is on numeric computation of the latter operator.

Let $l_G : X \to \mathbb{R}$ and $l_E : X \to \mathbb{R}$ be differentiable functions such that $G \triangleq \{x \in X | l_G(x) \leq 0\}$ and $E \triangleq \{x \in X | l_E(x) \leq 0\}$. Consider the following system of interconnected Hamilton-Jacobi equations [11,17]:

$$-\frac{\partial J_G^*(x,t)}{\partial t} = \begin{cases} H_G^*(x, \frac{\partial J_G^*(x,t)}{\partial x}), & \text{for } \{x \in X \mid J_G^*(x,t) > 0\}, \\ \min\{0, H_G^*(x, \frac{\partial J_G^*(x,t)}{\partial x})\}, & \text{for } \{x \in X \mid J_G^*(x,t) \leq 0\} \end{cases} \quad (2)$$

$$-\frac{\partial J_E^*(x,t)}{\partial t} = \begin{cases} H_E^*(x, \frac{\partial J_E^*(x,t)}{\partial x}), & \text{for } \{x \in X \mid J_E^*(x,t) > 0\}, \\ \min\{0, H_E^*(x, \frac{\partial J_E^*(x,t)}{\partial x})\}, & \text{for } \{x \in X \mid J_E^*(x,t) \leq 0\} \end{cases} \quad (3)$$

where $J_G^*(x, u(\cdot), d(\cdot), 0) = l_G(x)$ and $J_E^*(x, u(\cdot), d(\cdot), 0) = l_E(x)$, and

$$H_G^*(x, \frac{\partial J_G^*}{\partial x}) = \begin{cases} 0, & \text{for } \{x \in X \mid J_E^*(x,t) \leq 0\} \\ \max_{u \in U} \min_{d \in D} \frac{\partial J_G^*}{\partial x} f(x, u, d), & \text{otherwise} \end{cases} \quad (4)$$

$$H_E^*(x, \frac{\partial J_E^*}{\partial x}) = \begin{cases} 0, & \text{for } \{x \in X \mid J_G^*(x,t) \leq 0\} \\ \min_{u \in U} \max_{d \in D} \frac{\partial J_E^*}{\partial x} f(x, u, d), & \text{otherwise} \end{cases} \quad (5)$$

Theorem 1 (Characterization of Reach-Avoid [11]) *Assume that $J_G^*(x,t)$ ($J_E^*(x,t)$ respectively) satisfies the Hamilton-Jacobi equation (2) ((3) respectively), and that it converges uniformly in x as $t \to -\infty$ to a function $J_G^*(x)$ ($J_E^*(x)$ respectively). Then,*

$$Reach(G, E) = \{x \in X \mid J_G^*(x) < 0\} \quad (6)$$

Proof. Please see [11].   □

By our convention, we assume that the unsafe sets, defined as G° and its backwards reachable set under (2)-(5), are open; and safe sets, defined as E and its backwards reachable set, are closed.

2.2   An Equivalent Hamilton-Jacobi Formulation

Although the *Reach* operator can be computed by solving the equations (2)–(5), in practice the discontinuous right hand sides of the equations introduce serious numerical instabilities into the computation. Consider instead the standard form of the Hamilton-Jacobi equation:

$$-\frac{\partial J_G(x,t)}{\partial t} = H_G(x, \frac{\partial J_G(x,t)}{\partial x}) = \max_{u \in U} \min_{d \in D} \frac{\partial J_G}{\partial x} f(x,u,d), \tag{7}$$

$$-\frac{\partial J_E(x,t)}{\partial t} = H_E(x, \frac{\partial J_E(x,t)}{\partial x}) = \min_{u \in U} \max_{d \in D} \frac{\partial J_E}{\partial x} f(x,u,d), \tag{8}$$

with the same initial conditions as those used for J_G^* and J_E^*: $J_G(x,0) = l_G(x)$ and $J_E(x,0) = l_E(x)$. Now let:

$$J_G^{\min}(x,t) = \min_{\tau \in [t,0]} J_G(x,\tau), \tag{9}$$

$$J_E^{\min}(x,t) = \min_{\tau \in [t,0]} J_E(x,\tau), \tag{10}$$

$$J_G(x,t) \geq -J_E^{\min}(x,t), \tag{11}$$

$$J_E(x,t) \geq -J_G^{\min}(x,t). \tag{12}$$

Constraints (9) and (10) replace the "min" on the right hand side of equations (2) and (3), thus ensuring that sublevel sets of $J_G^{\min}(x,t)$ and $J_E^{\min}(x,t)$ do not shrink as time flows backwards; constraints (11) and (12) replace the "freezing" of the Hamiltonian on the right hand sides of equations (4) and (5) and ensure that the interiors of the two sets do not overlap, since for a given $x \in X$, if $J_E^{\min}(x,t) < 0$, then (11) will force $J_G(x,t) \geq 0$; conversely, if $J_G^{\min}(x,t) < 0$ then $J_E(x,t) \geq 0$.

Lemma 1 (Equivalence of Solutions) *The solution $J_G^*(x,t)$ to (2)–(5), and the solution $J_G^{\min}(x,t)$ to (7)–(12), are equivalent in that, for any $x \in X$, they satisfy one of*

$$J_G^*(x,t) \leq 0 \text{ if and only if } J_G^{\min}(x,t) \leq 0 \tag{13}$$

$$J_G^*(x,t) < 0 \text{ if and only if } J_G^{\min}(x,t) < 0 \tag{14}$$

for all $t \leq 0$.

Proof. We choose a particular $x \in X$ and assume that the computation starts at final time $t = 0$ and works backwards into negative time. Also, assume that the interiors of the initial sets do not intersect: $G^\circ \cap E^\circ = \emptyset$.

Case 1 (x **is in** G **at** $t = 0$). Thus $l_G(x) \leq 0$, which implies from (2) $\forall t < 0$ that $J_G^*(x, t) \leq 0$ and from (9) that $J_G^{\min}(x, t) \leq 0$. Thus, for such x, (13) holds.

Case 2 (x **is in** E **at** $t = 0$). Thus $l_E(x) \leq 0$, meaning that $J_E^*(x, 0) \leq 0$ and $J_E(x, 0) \leq 0$, and in addition, due to our assumption that the interiors of the initial sets are disjoint, $J_G^*(x, 0) \geq 0$ and $J_G(x, 0) \geq 0$. By (3), $\forall t < 0$ $J_E^*(x, t) \leq 0$, and so by (4) $J_G^*(x, t) = J_G^*(x, 0) \geq 0$. In our new formulation, $J_E(x, 0) \leq 0$ implies $J_E^{\min}(x, t) \leq 0$ $\forall t \leq 0$; by (11) $J_G(x, t) \geq 0$, which in turn implies $J_G^{\min}(x, t) \geq 0, \forall t < 0$. Thus, $\forall t \leq 0$, $J_G^*(x, t) \geq 0$ and $J_G^{\min}(x, t) \geq 0$. By the contrapositive, for such x, (14) is true.

Case 3 (x **is outside both** G **and** E **at** $t = 0$). Thus, $l_G(x) > 0$ and $l_E(x) > 0$. Now, for all $t \leq 0$, x will remain outside both the reach and avoid sets as long as the following constraints are satisfied:

$$\begin{array}{ll} J_G^*(x, t) > 0 \\ J_G^{\min}(x, t) > 0 \end{array} \text{ and } \begin{array}{ll} J_E^*(x, t) > 0 \\ J_E^{\min}(x, t) > 0 \end{array} \tag{15}$$

For an x under these conditions, (13) is trivially true. Furthermore, while this situation holds, the constrained PDEs (2)–(5) are equivalent to the PDEs and constraints (7)–(12), and so $J_G^*(x, t) = J_G(x, t)$ and $J_E^*(x, t) = J_E(x, t)$. Now consider what will happen if the boundary of one or both of the reach or avoid sets reaches x. Choose $\tau < 0$ to be the first time t that either boundary reaches x.

If $J_G^*(x, \tau) = J_G(x, \tau) = 0$, then (2) guarantees $J_G^*(x, t) \leq 0$ for $t \leq \tau$ and (9) guarantees $J_G^{\min}(x, t) \leq 0$ for $t \leq \tau$. Consequently, for such x, (13) holds $\forall t \leq 0$.

By choice of τ, we know that if $J_E^*(x, \tau) = J_E(x, \tau) = 0$, then $J_G^*(x, \tau) \geq 0$ and $J_G^{\min}(x, \tau) \geq 0$. By (3), $\forall t \leq \tau$, $J_E^*(x, t) \leq 0$, which implies by (4) that $J_G^*(x, t) \geq 0$. Since $J_E(x, \tau) = 0$ implies $\forall t \leq \tau$ that $J_E^{\min}(x, \tau) \leq 0$, (11) requires $\forall t \leq \tau$ that $J_G(x, t) \geq 0$, and so $J_G^{\min}(x, t) \geq 0$. Therefore, for such x, (13) holds for $\tau < t \leq 0$ and (14) holds for $t \leq \tau$.   □

We wish to use Lemma 1 and Theorem 1 to claim

$$Reach(G, E) = \{x \in X \mid J_G^{\min}(x, t) < 0\}.$$

However, the two cases (13) and (14) allowed by Lemma 1 must be reconciled before such a claim is true. We do so by making the assumption that the sets defined by (13) are the closures of the sets defined by (14)[1].

Given this assumption, the formulation (7)–(12) provides a characterization of the reach-avoid operator which is numerically more stable than (2)–(5). While the new formulation does smooth out the solution of the Hamilton-Jacobi equations, it is worth noting that discontinuities in u, d, or f will still lead to non-smooth solutions of (7)–(12), and that even if these system parameters are all smooth, it is possible for discontinuous "shocks" to develop as the solution evolves.

---

[1] This assumption will hold true as long as the functions J_G^* and $J_G^{\min}$ do not develop plateaus. It turns out to be prudent to avoid plateaus for numerical reasons as well, and we describe a method to avoid their formation in the next section.

3   Computing Reachable Sets

The continuous Hamilton-Jacobi partial differential equation appears frequently in applied mathematics, and so numerical methods for its solution have been well studied [18]. In particular, a set of algorithms called level set methods [19,20] have been developed to study the propagation of moving interfaces and boundaries using these equations.

A numerical algorithm to solve the Hamilton-Jacobi equations (7)–(12) was developed in [17]; however, the emergence of numerical instabilities meant that the reach set could be computed for only a few dozen timesteps, and even over that short period, sharp edges tended to become rounded by diffusion. Armed with the better behaved (7)–(12) and a new level set implementation, we are able to tackle more complex examples below, tracking the reach set over any finite time interval without significant loss to diffusion.

3.1   Level Set Method Design

The basic method for solving (7) and (8) is the same as that described in [17]: a first-order, upwinding, finite difference scheme that produces an approximation of the viscosity solution to the Hamilton-Jacobi equation [20,21,22]. We outline several details of our implementation.

Initial conditions: A characteristic of level set methods is that the "level set function" (we use J in the following to represent J_G in (7) or J_E in (8)) is defined as the distance to the boundary being tracked, where distance is negative on the inside of the boundary. Such a definition is compatible with the analysis in the previous section, and so we adopt it for our level set functions.

Boundary conditions: The spatial derivatives in the Hamilton-Jacobi equation are approximated at a grid point by taking differences between the function values at neighboring grid points. For points at the edge of the finite grid, this procedure breaks down. Typical level set methods use Neumann boundary conditions ($\frac{\partial J(x,t)}{\partial n} = 0$, where n is an outward pointing normal) to determine the value of grid points on the boundary. This procedure tends to introduce plateaus to the level set function J close to the boundary, so that it no longer properly measures the distance to the boundary.

Enforcing the constraints: To enforce the constraints (11) and (12), a "max" operator is applied: at each timestep t, for all x,

$$J_G(x,t) = \max(J_G(x,t), -J_E^{\min}(x,t))$$

and similarly for $J_E(x,t)$. This procedure, called *masking J_G with $J_E^{\min}$*, is used in level set methods to ensure that the moving boundary represented by J_G does not enter the forbidden region defined by $J_E^{\min}$ (since $J_E^{\min}(x,t) \leq 0 \implies J_G(x,t) \geq 0$).

An additional complication arises from the discrete timesteps taken by the numeric solver: it is possible for the constraints (11) and (12) to become violated since the various J functions are changing over time and the masking procedure is only applied at the end of each timestep. A conservative solution is to compute (7)–(12) in the order:

$$\text{compute } J_G(x, t - \Delta t) \text{ from Hamilton-Jacobi equation,}$$
$$\text{compute } J_E(x, t - \Delta t) \text{ from Hamilton-Jacobi equation,}$$
$$J_G(x, t - \Delta t) = \max(J_G(x, t - \Delta t), -J_E^{\min}(x, t))$$
$$J_G^{\min}(x, t - \Delta t) = \min(J_G(x, t - \Delta t), J_G^{\min}(x, t))$$
$$J_E(x, t - \Delta t) = \max(J_E(x, t - \Delta t), -J_G^{\min}(x, t - \Delta t))$$
$$J_E^{\min}(x, t - \Delta t) = \min(J_E(x, t - \Delta t), J_E^{\min}(x, t))$$

Masking J_G with $J_E^{\min}$ from the previous timestep, but masking J_E with $J_G^{\min}$ from the current timestep ensures that if the reach and avoid sets grow together and overlap, the reach (unsafe) set is over-approximated, and the avoid (safe) set is under-approximated.

Reinitialization: Level set methods attempt to maintain the level set function as a distance measure to the boundary as it evolves. Numeric solutions tend to distort the distance function considerably: the level set function becomes distorted by limited precision computations, discretization and the Neumann boundary conditions. Because the zero level set is the only information of importance to us, a procedure which resets the level set function so that it correctly measures the distance to the current zero level set—without changing the shape of that level set—would smooth out numerical errors in the level set function and yet leave its important data unharmed. This process, called reinitialization, is accomplished in the examples below by running a few discrete timesteps of a solver for the partial differential equation

$$\frac{\partial J_G(x, t)}{\partial t} = \text{sign}(J_G(x, t))(1 - |\text{grad}(J_G(x, t))|)$$

(and similarly for J_E). This process restores the property $|\text{grad}(J_G(x, t))| \approx 1$ near the zero level set, so that J_G is smoothed to approximate a distance measure.

3.2   A Single State, Straight Flight Example

Consider an example representing two aircraft flying at a fixed altitude and constant heading. Each aircraft is allowed to choose its own speed from a given range of values; we control one aircraft and the other is considered the disturbance. Using relative coordinates, in which the controlled aircraft is at the origin with a heading angle of zero, the dynamics of the system are described by

$$\dot{x}_r = -u + d\cos\psi_r, \qquad \dot{y}_r = d\sin\psi_r, \qquad \dot{\psi}_r = 0, \qquad (16)$$

where x_r and y_r are the relative spatial coordinates, and ψ_r is the relative heading. The controller fails if the disturbance aircraft manages to enter a circle of radius five units centered at the controlled aircraft at the origin, so $l_G(x) = x_r^2 + y_r^2 - 5^2$.

If the control (speed of the controlled aircraft) is restricted to $u \in U = [\underline{u}, \overline{u}] \subset \mathbb{R}^+$ and the disturbance (speed of the disturbance aircraft) is restricted to $d \in D = [\underline{d}, \overline{d}] \subset \mathbb{R}^+$, then it was shown in [11,23] that the optimal control and worst disturbance are

$$u^* = \begin{cases} \underline{u}, & \text{if } x_r > 0, \\ \overline{u}, & \text{if } x_r < 0, \end{cases} \qquad d^* = \begin{cases} \underline{d}, & \text{if } (x_r \cos \psi_r + y_r \sin \psi_r) > 0, \\ \overline{d}, & \text{if } (x_r \cos \psi_r + y_r \sin \psi_r) < 0. \end{cases} \qquad (17)$$

Because there is only a single discrete state, the controlled aircraft has no discrete action to force an unsafe continuous state to become safe, and so the avoid set is empty. Given the definition of the unsafe set $G = \{x \in X | l_G(x) \leq 0\}$, the set of unsafe states $Reach(G, \emptyset)$ is shown shaded in Figure 1. The parameters for the example were chosen to be the normalized values:

$$\psi_r = \frac{7\pi}{12}, \qquad U = [\underline{u}, \overline{u}] = [2, 4], \qquad D = [\underline{d}, \overline{d}] = [1, 5].$$

The dashed circle shows the initial unsafe set G, and the grey arrows show the flow field (16) induced by the optimal control choices (17). Notice that the level set algorithm resolves the sharp corners of $Reach(G, \emptyset)$ at the points where u^* or d^* switch.

This example and those below were coded in Matlab 5.3 on an unloaded Sun UltraSparc 10 (a 300 MHz UltraSparc processor with 512 KB cache and 128 MB main memory). Figure 1 was produced from a run with grid spacing $\Delta x = 0.1$ (requiring about 63000 grid points). The 360 timesteps took just under four minutes to complete.

3.3   A Three State Example

This example again features the collision avoidance maneuvers of two aircraft at fixed altitude; however, the control is now allowed to initiate a discrete change of state for the system. As shown in Figure 2, the aircraft begin in straight flight at a fixed relative heading (mode 1). At some time, the control may switch both aircraft into mode 2; at which point each makes an instantaneous heading change of 90°, and begins a circular flight path. After completing a semicircular arc in π time units, both aircraft switch to mode 3, make another instantaneous 90° turn, and resume their original headings from mode 1.

The dynamics for the system are shown in Figure 3. In this example, the controller has only a single action: the switch from mode 1 to mode 2. The speed of both aircraft is constant, and the only disturbance action is the uncontrolled switch from mode 2 to mode 3, which occurs a fixed time after mode 2 is entered; the variable z in mode 2 is simply a clock to enforce this switch. The parameters used in the run below are

$$\psi_r = \frac{2\pi}{3} = 120°, \qquad u^* = 3, \qquad d^* = 4.$$

More details on this example can be found in [9,11].

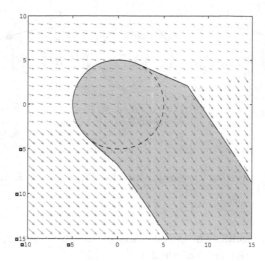

Fig. 1. Shaded Region represents $Reach(G, \emptyset)$ for the Straight Flight Single State Example

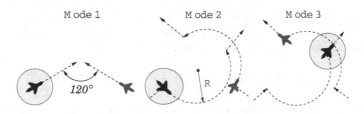

Fig. 2. Aircraft Behavior in the Three Modes

Running the reachability analysis algorithm to compute W^* requires computing the Pre_d and Pre_u operators for each mode. Let R_i^k be the set of unsafe states computed for mode i in iteration k; in other words, the projection of $Reach(Pre_d(W^{k+1}), Pre_u(W^{k+1}))$ onto the continuous state space of mode i for iteration $k < 0$ (let $R_i^0 = G$ to handle the $k = 0$ case). Then the set of safe states at iteration $k < 0$ can be written as $W^k = (\cup_{j=k}^0 \cup_{i=1}^3 R_i^j)^c$. Define the collision set as before: $G = \{x \in X | l_G(x) \le 0\}$, where $l_G(x) = x_r^2 + y_r^2 - 5^2$. We can then deduce the precursor operators.

- For mode 3, there are no discrete actions. This mode may be inhabited for any length of time. The projections of the precursor operators onto the continuous state space of mode 3 are:

$$Pre_u(W^k) = \emptyset, \qquad Pre_d(W^k) = R_3^k.$$

- For mode 2, an uncontrolled discrete action switches the system to mode 3, and there are no controlled discrete actions. This mode is inhabited for

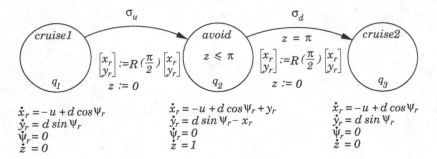

Fig. 3. System Dynamics for the Three Mode Example

exactly π time units. The projections of the precursor operators onto the continuous state space of mode 2 are:

$$Pre_u(W^k) = \emptyset, \qquad Pre_d(W^k) = (R_3^k \text{ rotated } \tfrac{\pi}{2}) \cup R_2^k.$$

- For mode 1, a controlled discrete action switches the system to mode 2, and there are no uncontrolled discrete actions. This mode may be inhabited for any length of time. The projections of the precursor operators onto the continuous state space of mode 1 are:

$$Pre_u(W^k) = (R_2^k \text{ rotated } \tfrac{\pi}{2})^c, \qquad Pre_d(W^k) = R_1^k.$$

Figure 4 shows the results of the reach-avoid computation at each iteration for each mode; unsafe states (complement of W^k) are shaded. The set R_i^k appears in column i and row k. A fixed point W^* of safe states is computed after three iterations, and the corresponding bad states of the fixed point $(W^*)^c$ are shaded in the final row of plots.

The unsafe region for mode 1 is the most interesting—as long as the disturbance aircraft is not in this region, the control may initiate the switch to mode 2 and have confidence that the remainder of the maneuver will be carried out safely. The width of the unbounded portion of the unsafe set is controlled by the radius of the turn in mode 2, and can be removed entirely by making the radius large enough.

The four iterations of this simulation, with a grid spacing of $\Delta x = 0.1$ (or about 90000 grid points) each required about 1400 timesteps; for stability reasons, mode 2 was slightly more than half of the work. Wall clock time was about 75 minutes.

3.4   A Three Dimensional Example

To show that this technique extends easily to higher dimensions, we look at a final aircraft collision avoidance scenario. The model is very similar to that examined

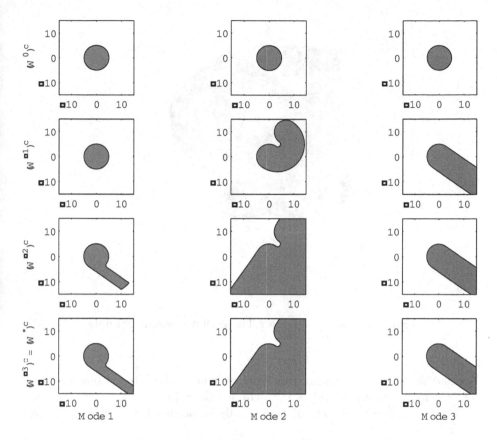

Fig. 4. Unsafe Sets for Three Mode Example

in the first example, except that this time we allow the relative heading of the aircraft to change. Relative angle $\psi_r \in [0, 2\pi)$ is thus our third dimension.

We fix the airspeed of the control aircraft at v_1 and that of the disturbance aircraft at v_2. The control and disturbance inputs are now the angular velocity of the aircraft: $u \in U = [\underline{\omega_1}, \overline{\omega_1}] \subset \mathbb{R}$ and $d \in D = [\underline{\omega_2}, \overline{\omega_2}] \subset \mathbb{R}$. The model is

$$\dot{x}_r = -v_1 + v_2 \cos \psi_r + u y_r, \qquad \dot{y}_r = v_2 \sin \psi_r - u x_r, \qquad \dot{\psi}_r = d - u,$$

For the case where

$$\underline{\omega_1} = \underline{\omega_2} = -1, \qquad \overline{\omega_1} = \overline{\omega_2} = +1,$$

it was shown in [11, pp. 60-62] that the optimal control and disturbance are given by

$$u^* = \text{sign}\left(y_r \frac{\partial J_G}{\partial x_r} - x_r \frac{\partial J_G}{\partial y_r} - \frac{\partial J_G}{\partial \psi_r} \right) \qquad d^* = -\text{sign}\left(\frac{\partial J_G}{\partial \psi_r} \right)$$

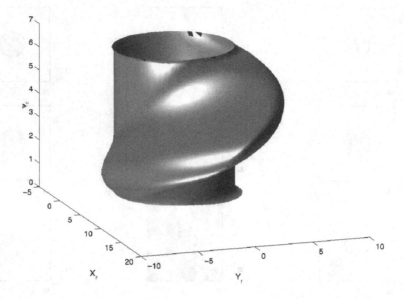

Fig. 5. Unsafe Region for Three Dimensional Example

Because there is only a single discrete state and no discrete actions, the avoid set is empty; the unsafe set is the cylinder $G = \{x \in X | l_G(x) \leq 0\}$ where $l_G(x) = x_r^2 + y_r^2 - 5^2$. A view of $Reach(G, \emptyset)$ for airspeed $v_1 = v_2 = 5$ is shown in Figure 5.

Extending the level set code to three dimensions was painless—a new index for all matrices and a set of boundary conditions (periodic in ψ_r) had to be added. Visualization of the zero sublevel set becomes considerably trickier, but it can be done with Matlab's new isosurface tools. With a grid spacing of $\Delta x = 0.2$ (approximately 400000 grid points), the 400 timesteps required to generate Figure 5 took about 80 minutes to complete.

4   Research Directions

We have presented a numerical algorithm for computing reachable sets of hybrid automata. The algorithm handles nonlinear dynamics with discontinuities, as illustrated by example calculations of both continuous and multi-mode aircraft conflict resolution maneuvers. We are currently investigating further in several directions.

For the examples above, the discrete predecessor maps Pre were determined by hand and hard-coded into the scripts which computed the continuous reach-avoid operator. It is necessary to automatically compute those maps; this will require elimination of existential and universal quantifiers over the set of discrete actions.

As with all finite difference methods, this implementation finds an approximation to the actual solution of the Hamilton-Jacobi equation. In fact, the final example provides proof of the dangers of such approximations: the helical bulge of the unsafe set shown in Figure 5 is computed to protrude farther out if grid spacing is reduced. Methods to quantify the error between exact and approximate reachable sets have not been developed, yet are crucial for proving safety properties. In the reach-avoid calculation, we could use information about error to provide an over-approximation of the unsafe set and an under-approximation of the safe set.

In [9,10,11], control laws are synthesized assuming that the reach and avoid sets are computed exactly. The implications of set approximation on this process must be evaluated.

As can be seen from the final example, these techniques extend easily to higher dimensions—beyond three dimensions visualization becomes impossible, but the basic level set algorithm remains the same. Of major concern, though, is the exponential growth in the number of grid points as dimension increases. Because the timestep depends on the grid size, using rectilinear gridding with a grid spacing of h in d dimensions requires $\mathcal{O}(h^{d+1})$ work. However, we are currently investigating techniques which will lead to considerable time savings: using compiled code instead of Matlab, computing only on grid points near the zero level set (effectively reducing the dimension of the problem by one), taking advantage of the abundant opportunities for parallelism in the algorithm, and projecting higher dimensional sets onto lower dimensional subspaces.

References

1. R. Alur and D. Dill, "A theory of timed automata," *Theoretical Computer Science*, vol. 126, pp. 183–235, 1994.
2. R. Alur, C. Courcoubetis, T. A. Henzinger, and P.-H. Ho, "Hybrid automata: An algorithmic approach to the specification and verification of hybrid systems," in *Hybrid Systems* (R. L. Grossman, A. Nerode, A. P. Ravn, and H. Rischel, eds.), LNCS, pp. 366–392, New York: Springer Verlag, 1993.
3. T. A. Henzinger, P. H. Ho, and H. Wong-Toi, "A user guide to HYTECH," in *TACAS 95: Tools and Algorithms for the Construction and Analysis of Systems* (E. Brinksma, W. Cleaveland, K. Larsen, T. Margaria, and B. Steffen, eds.), no. 1019 in LNCS, pp. 41–71, Springer Verlag, 1995.
4. C. Daws, A. Olivero, S. Tripakis, and S. Yovine, "The tool KRONOS," in *Hybrid Systems III, Verification and Control*, no. 1066 in LNCS, pp. 208–219, Springer Verlag, 1996.
5. O. Maler, A. Pnueli, and J. Sifakis, "On the synthesis of discrete controllers for timed systems," in *STACS 95: Theoretical Aspects of Computer Science* (E. W. Mayr and C. Puech, eds.), no. 900 in LNCS, pp. 229–242, Munich: Springer Verlag, 1995.
6. H. Wong-Toi, "The synthesis of controllers for linear hybrid automata," in *Proceedings of the IEEE Conference on Decision and Control*, (San Diego, CA), 1997.
7. A. Deshpande, *Control of Hybrid Systems*. PhD thesis, Department of Electrical Engineering and Computer Sciences, University of California at Berkeley, 1994.

8. J. Lygeros, *Hierarchical, Hybrid Control of Large Scale Systems*. PhD thesis, Department of Electrical Engineering and Computer Sciences, University of California at Berkeley, 1996.

9. C. Tomlin, J. Lygeros, and S. Sastry, "Synthesizing controllers for nonlinear hybrid systems," in *Hybrid Systems: Computation and Control* (T. Henzinger and S. Sastry, eds.), no. 1386 in LNCS, pp. 360–373, New York: Springer Verlag, 1998.

10. J. Lygeros, C. Tomlin, and S. Sastry, "On controller synthesis for nonlinear hybrid systems," in *Proceedings of the IEEE Conference on Decision and Control*, (Tampa, FL), pp. 2101–2106, 1998.

11. C. J. Tomlin, *Hybrid Control of Air Traffic Management Systems*. PhD thesis, Department of Electrical Engineering, University of California, Berkeley, 1998.

12. M. Greenstreet and I. Mitchell, "Integrating projections," in *Hybrid Systems: Computation and Control* (S. Sastry and T. Henzinger, eds.), no. 1386 in LNCS, pp. 159–174, Springer Verlag, 1998.

13. T. Dang and O. Maler, "Reachability analysis via face lifting," in *Hybrid Systems: Computation and Control* (S. Sastry and T. Henzinger, eds.), no. 1386 in LNCS, pp. 96–109, Springer Verlag, 1998.

14. A. Chutinan and B. H. Krogh, "Verification of polyhedral-invariant hybrid automata using polygonal flow pipe approximations," in *Hybrid Systems: Computation and Control* (F. Vaandrager and J. H. van Schuppen, eds.), no. 1569 in LNCS, pp. 76–90, New York: Springer Verlag, 1999.

15. A. B. Kurzhanski and P. Varaiya, "Ellipsoidal techniques for reachability analysis," in *Hybrid Systems: Computation and Control* (B. Krogh and N. Lynch, eds.), LNCS (these proceedings), Springer Verlag, 2000.

16. O. Botchkarev and S. Tripakis, "Verification of hybrid systems with linear differential inclusions using ellipsoidal approximations," in *Hybrid Systems: Computation and Control* (B. Krogh and N. Lynch, eds.), LNCS (these proceedings), Springer Verlag, 2000.

17. C. Tomlin, J. Lygeros, and S. Sastry, "Computing controllers for nonlinear hybrid systems," in *Hybrid Systems: Computation and Control* (F. Vaandrager and J. H. van Schuppen, eds.), no. 1569 in LNCS, pp. 238–255, New York: Springer Verlag, 1999.

18. M. Bardi and I. Capuzzo-Dolcetta, *Optimal Control and Viscosity Solutions of Hamilton-Jacobi-Bellman equations*. Boston: Birkh auser, 1997.

19. S. Osher and J. A. Sethian, "Fronts propagating with curvature-dependent speed: Algorithms based on Hamilton-Jacobi formulations," *Journal of Computational Physics*, vol. 79, pp. 12–49, 1988.

20. J. A. Sethian, *Level Set Methods: Evolving Interfaces in Geometry, Fluid Mechanics, Computer Vision, and Materials Science*. New York: Cambridge University Press, 1996.

21. M. G. Crandall and P.-L. Lions, "Viscosity solutions of Hamilton-Jacobi equations," *Transactions of the American Mathematical Society*, vol. 277, no. 1, pp. 1–42, 1983.

22. M. G. Crandall, L. C. Evans, and P.-L. Lions, "Some properties of viscosity solutions of Hamilton-Jacobi equations," *Transactions of the American Mathematical Society*, vol. 282, no. 2, pp. 487–502, 1984.

23. C. Tomlin, G. J. Pappas, and S. Sastry, "Conflict resolution for air traffic management: A case study in multi-agent hybrid systems," *IEEE Transactions on Automatic Control*, vol. 43, pp. 509–521, April 1998.

Towards Procedures for Systematically Deriving Hybrid Models of Complex Systems

Pieter J. Mosterman[1] and Gautam Biswas[2]

[1] Institute of Robotics and Mechatronics
DLR Oberpfaffenhofen, P.O. Box 1116, D-82230 Wessling, Germany.
Pieter.J.Mosterman@dlr.de
[2] Department of Electrical Engineering and Computer Science
Box 1679 Sta B, Vanderbilt University, Nashville, TN 37235, U.S.A.
biswas@vuse.vanderbilt.edu

Abstract. In many cases, complex system behaviors are naturally modeled as nonlinear differential equations. However, these equations are often hard to analyze because of "stiffness" in their numerical behavior and the difficulty in generating and interpreting higher order phenomena. Engineers often reduce model complexity by transforming the nonlinear systems to piecewise linear models about operating points. Each operating point corresponds to a mode of operation, and a discrete event switching structure is added to implement the mode transitions during behavior generation. This paper presents a methodology for systematically deriving mixed continuous and discrete, i.e., *hybrid* models from a nonlinear ODE system model. A complete switching specification and state vector update function is derived by combining piecewise linearization with singular perturbation approaches and transient analysis. The model derivation procedure is then cast into the phase space transition ontology that we developed in earlier work. This provides a systematic mechanism for characterizing discrete transition models that result from model simplification techniques. Overall, this is a first step towards automated model reduction and simplification of complex high order nonlinear systems.

1 Introduction

Systems and control engineers often apply simplification techniques when modeling and analyzing complex physical systems that include components like valves, pumps, and diodes, and phenomena such as friction effects [3]. To avoid complex nonlinearities and stiffness caused by steep slopes in the behavior, these components are modeled to exhibit switching behavior. This results in the overall system model generating piecewise continuous behaviors and discrete transitions, i.e., *hybrid* behaviors. *Hybrid automata* [1] have been employed as a computational mechanism for implementing these models, with a discrete control structure defining the switching between *modes* or states of the automata. Each mode has an associated set of ordinary differential equations (ODEs) that governs continuous behavior evolution in that mode. Events associated with the

N. Lynch and B. Krogh (Eds.): HSCC 2000, LNCS 1790, pp. 324–337, 2000.

mode switching generate *actions* that may produce discontinuous changes in state variables.

Consider the hydraulic actuator illustrated in Fig. 1. The valve at the top of the cylinder controls oil flow into and out of the cylinder, and the flow rate is a function of the control pressure p_{in}. The flow of oil determines the position of the piston in the cylinder, and this in turn determines the position of the load, e.g., the elevator control surface of an airplane. To prevent damage to the actuator system, a relief valve on the left side of the cylinder opens when the pressure in the cylinder exceeds a certain value.

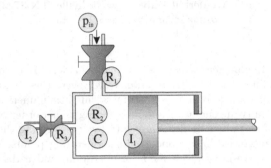

Fig. 1. Model parameters of a hydraulic actuator.

If the valve behaviors are approximated and simplified to be discrete, the actuator can be modeled as a hybrid automata with four states: α_{00}, both valves closed, α_{01}, relief valve open and control valve closed, α_{10}, control valve open and relief valve closed, and α_{11}, both valves open. The dynamic behavior in each of these modes can be derived from the actuator parameters, that include R_1, the resistance of the open control valve, R_2, the internal dissipation parameter for the oil, R_3, the resistance of the open relief valve, C, the oil elasticity, I_1, the piston inertia, and I_2, the relief valve fluid inertia.

System modelers often employ simplification techniques that involve dropping very small and very large parameters that do not play a significant role in gross system behavior. Applying this approach to the actuator system, parameters associated with the oil, R_2 and C, may be removed to reduce the order of the system model. For the simplified model, the dynamic behavior models for the different modes are given in Table 1, where f_1 is the piston velocity and f_2 is the fluid flow rate through the relief valve. The control valve and the relief valve are the two components in the actuator that are modeled to have discrete transitions from open to closed, and vice versa. An external control variable, u_v, determines the opening and closing of the control valve (e.g., the valve is closed when $u_v < 0$). The relief valve opens when $p > p_{th}$.

For mode α_{00}, there is no oil flow into the cylinder, therefore, the entry action, i.e., the initial conditions that have to be satisfied on entry into this mode, includes the constraint, $f_1 = 0$. The entry action for mode α_{01} is more compli-

Table 1. Mode specification table.

mode	$\dot{x} = f(x, u)$	entry action
α_{00}	$\dot{f}_1 = 0$ $\dot{f}_2 = 0$	$f_1 = 0$ $f_2 = 0$
α_{01}	$\dot{f}_1 = -\frac{R_3}{I_1+I_2} f_1$ $\dot{f}_2 = -\frac{R_3}{I_1+I_2} f_2$	$f_1 = \frac{1}{I_1+I_2}(I_1 f_1 - I_2 f_2)$ $f_2 = \frac{1}{I_1+I_2}(I_1 f_1 - I_2 f_2)$
α_{10}	$\dot{f}_1 = p_{in} - \frac{R_1}{I_1} f_1$ $\dot{f}_2 = 0$	$f_2 = 0$
α_{11}	$\dot{f}_1 = \frac{1}{I_1}(p_{in} - R_1(f_1 + f_2))$ $\dot{f}_2 = \frac{1}{I_2}(p_{in} - R_1(f_1 + f_2) - R_3 f_2)$	

cated. In this mode, f_1 and f_2 are algebraically related ($f_1 = -f_2$). The initial values for f_1 and f_2 have to be initialized using this constraint, but one equation is not sufficient to solve for their values. Additional constraints presented in Section 5 are used to define the entry action listed in Table 1.

In the past, engineers have used *ad hoc* approaches to handle transitions between piecewise models, however, even for the simple example above this may lead to incorrect model definitions. In Section 5 systematic analysis shows that the entry actions as specified in Table 1 are incomplete, and demonstrates how the correct state mapping as derived by a structured approach is much more complex. This shows that deriving the correct event structures and corresponding actions at mode transitions is more involved for systems with complex interactions among their subsystems.

This paper develops a structured approach to analyzing complex nonlinear models, applying systematic abstraction and simplification mechanisms to create simpler multiple piecewise continuous models. The price we pay in achieving this reduction is the introduction of complex discrete components in the hybrid model of the system. The two main steps in this procedure are illustrated in Fig. 2. We start with the complex continuous nonlinear model of the system. Step 1 applies simplification techniques to convert the nonlinear models to simpler piecewise continuous (possibly linear) behavior models. The result is a hybrid model whose state variable values are continuous, but the time derivatives may be discontinuous. This is equivalent to a C^0 hybrid model with sets of differential equations defining the behaviors in individual modes, and a function, γ, that defines transitions between the modes. Step 2 applies techniques like *singular perturbation* [3] and *eigenvalue analysis* [8] that remove large and small parameters from the models, and thus eliminate steep transitions in the behaviors within modes. The resultant models combine three components: (i) a reduced order ODE model, f, (ii) the discrete event mode transition function, γ, and (iii) the state transition function, g, that captures the discontinuous state variable value changes between modes.

The derivation process for g can be described by two basic actions in hybrid models of physical systems:

Fig. 2. Abstraction levels.

1. a *manifold projection* that results from the generated algebraic constraints, and
2. an *aborted projection* because detailed continuous projection behavior causes further discrete changes.

We use this framework to derive a computational model of the resulting hybrid system as a hybrid automata extended with branch points (junctions) to model the immediate consecutive discrete events and actions. A phase space analysis illustrates these concepts, and allows us to relate the results back to an ontology of phase space transition behavior presented in previous work [6].

2   The Approach

Consider a nonlinear system with state equations of the form

$$\dot{x} = A(x)x + B(x)u. \tag{1}$$

System designers and analysts often simplify the above model by identifying operating regions of interest within the behavior space, called *modes*. Such modes may be the result of design decisions, e.g., the take-off, cruise, and landing modes of aircraft fly-by-wire systems, or determined from component models that make up the system, e.g., by taking into account the open and closed states of the valves in the actuator system. Modes can also be identified by the discrete control actions of supervisory controllers. Along with mode identification, transitions between modes, α, are also defined (cf. Table 1). Most often, the purpose for breaking up complex behaviors into modes of operating regions is so that the system model can be linearized within each mode, i.e.,

$$f_{\alpha_i} : \dot{x} = A_{\alpha_i}x + B_{\alpha_i}u. \tag{2}$$

The result is a set of piecewise models that together define the behavior space of interest, with transition conditions between pairs of modes, α_i and α_{i+1} given by the function

$$\gamma_{\alpha_i}^{\alpha_{i+1}} : C_{\alpha_i}x + D_{\alpha_i}u > 0. \tag{3}$$

Model reduction techniques, such as singular perturbation and eigenvalue based techniques, are readily applicable to the linearized systems. They provide systematic methodologies to reduce the order of each piecewise model. Applying singular perturbation, a small parameter, ϵ, is removed from the model by letting its value tend to 0. This requires the formulation

$$
\begin{aligned}
f_{\alpha_i} &: \dot{x} = A_{\alpha_i}(\epsilon)x + B_{\alpha_i}(\epsilon)u \\
\gamma_{\alpha_i}^{\alpha_{i+1}} &: C_{\alpha_i}(\epsilon)x + D_{\alpha_i}(\epsilon)u > 0.
\end{aligned}
\tag{4}
$$

In this formulation, slow and fast variables can be separated according to

$$
f_{\alpha_i} : \begin{cases} \dot{y} = A_{\alpha_i}^y(\epsilon)x + B_{\alpha_i}^y(\epsilon)u \\ \epsilon\dot{z} = A_{\alpha_i}^z(\epsilon)x + B_{\alpha_i}^z(\epsilon)u. \end{cases}
\tag{5}
$$

Making $\epsilon \to 0$ leads to equations of the form $z = f(y, u)$. Assuming that the system of algebraic equations is non singular, z can be substituted in the equation for $\dot{y}$ to derive an explicit reduced order ODE system. However, if $\epsilon \to 0$ leads to a singular solution, $0 = f(y, u)$, system behavior is now defined by an implicit system of differential and algebraic equations (DAE), and the variable vector y may also include a fast component. In the limit, this fast behavior is replaced by an instantaneous projection $y^+ = g_{\alpha_{i+1}}(y, u)$, where y^+ is the initial value in mode α_{i+1}, $y_{\alpha_{i+1}}^0 = y^+$, and y is the value of the reduced order system in mode α_i when $\gamma_{\alpha_i}^{\alpha_{i+1}} > 0$ was first satisfied.

Similarly, when the system of equations becomes singular for $\epsilon \to 0$, a state vector transformation may be required to achieve the desired separation and this may require a projection, $g_\alpha(x, u)$. We discuss this in greater detail in Section 5. In general, it may be difficult to derive the transformation by analytic methods. Information about the physical system can be invoked to assist in deriving the solution. The projection can be found by boundary behavior analysis of the detailed model, i.e., with the ϵ parameter. As an alternative, or if this detailed model is not available, the projection can be computed by integrating the instantaneous field dynamics [4] and by subspace iteration [7]. These are implementations based on the use of the reducing subspaces of the Kronecker Canonical Form [2] to capture the state projection.

The resulting model contains the reduced order specification of continuous behavior, f, the transition conditions, γ, and the projection equations, g,

$$
\begin{aligned}
f_{\alpha_i} &: \dot{y} = A'_{\alpha_i}y + B'_{\alpha_i}u \\
\gamma_{\alpha_i}^{\alpha_{i+1}} &: C'_{\alpha_i}y + D'_{\alpha_i}u > 0 \\
g_{\alpha_{i+1}} &: y_{\alpha_{i+1}}^0 = E_{\alpha_{i+1}}y + F_{\alpha_{i+1}}u
\end{aligned}
\tag{6}
$$

We study the effects of the order reduction technique on the state vector transfer function and transition conditions in this paper. Detailed analysis may be required when variables that constitute the γ function exhibit impulsive behavior. To identify such behavior, γ can be expressed in terms of $\dot{y}$. If any of the variables in the y vector are part of the algebraic constraints that develop when $\epsilon \to 0$, they produce impulses. Detailed study may reveal the need for an additional transition modes to be introduced in the mode transition behavior. This transitional mode exists only at a point in time, and has no specification for continuous behavior. Furthermore, some transitional modes may have no effect on the state vector. In previous work, we termed these transitional modes *pinnacles* and *mythical modes*, respectively. A phase space analysis conducted in this paper establishes the relation between this approach and our established ontology for phase space transition behavior [6].

3   A Piecewise Model

In a nonlinear continuous ODE model of the hydraulic actuator, the nonlinear characteristics of the externally controlled valve and the relief valve can be modeled as shown in Fig. 3. Including the oil parameters (R_2 and C) results in the fifth order nonlinear ODE

$$
\begin{bmatrix} \dot{f}_1 \\ \dot{f}_2 \\ \dot{p}_1 \\ \dot{s}_1 \\ \dot{s}_2 \end{bmatrix} = \begin{bmatrix} \frac{-R_1(s_1)R_2}{D_1} & \frac{-R_1(s_1)R_2}{D_1} & \frac{R_1(s_1)}{D_1} & 0 & 0 \\ \frac{-R_1(s_1)R_2}{D_2} & \frac{-R_1(s_1)R_2-R_3(p,s_2)(R_1(s_1)+R_2)}{D_2} & \frac{R_1}{D_2} & 0 & 0 \\ \frac{-R_1}{D_3} & \frac{-R_1(s_1)}{D_3} & \frac{-1}{D_3} & 0 & 0 \\ 0 & 0 & 0 & 0 & 0 \\ 0 & 0 & 0 & 0 & 0 \end{bmatrix} \begin{bmatrix} f_1 \\ f_2 \\ p_1 \\ s_1 \\ s_2 \end{bmatrix} + \begin{bmatrix} \frac{R_2 p_{in}}{D_1} \\ \frac{R_2 p_{in}}{D_2} \\ \frac{p_{in}}{D_3} \\ u_v \\ u_r \end{bmatrix}
$$

$$(7)$$

with $D_1 = I_1(R_1(s_1) + R_2)$, $D_2 = I_2(R_1(s_1) + R_2)$, and $D_3 = C(R_1(s_1) + R_2)$. The variable u_v is externally controlled, and $u_r = \frac{1}{1+e^{-a(|p|-p_{th})}}$ represents a function that approaches a step when $a \to \infty$. The two state variables, s_1 and s_2, provide a parametric representation model for the detailed continuous switching behavior of the two valves. The cylinder oil pressure, p, expressed in terms of the state variables, is:

$$
p = \frac{R_1}{R_1 + R_2}(p_1 + \frac{R_2}{R_1}p_{in} - R_2(f_1 + f_2)). \tag{8}
$$

When p approaches $\pm p_{th}$, u_r becomes positive and the valve opens by switching to another behavior dimension ($s_2 > 0$). Since u_r is always positive, the valve does not close, once it is opened. Therefore, transitions from α_{01} to α_{00} and α_{10} are not defined in Table 2. For the same reason, there are no transitions from α_{11} to α_{00} and α_{10}.

Piecewise linearization of $R_1(s_1)$ and $R_3(p, s_2)$ into regions of high resistance, $R_{i,h}$, and low resistance, $R_{i,l}$, is defined as:

$$
\begin{aligned}
R_1 &= \text{ if } (u_v < 0) \text{ then } R_{1,h} \text{ else } R_{1,l} \\
R_3 &= \text{ if } (R_3 = R_{3,l} \text{ or } p > p_{th}) \text{ then } R_{3,l} \text{ else } R_{3,h}
\end{aligned} \tag{9}
$$

This allows for removal of the states s_1 and s_2 from the system model, resulting in a linear ODE model with four global modes: $\alpha_{00} \to \{R_{1,h}, R_{3,h}\}$, $\alpha_{01} \to \{R_{1,h}, R_{3,l}\}$, $\alpha_{10} \to \{R_{1,l}, R_{3,h}\}$, and $\alpha_{11} \to \{R_{1,l}, R_{3,l}\}$. The transitions between the modes are specified in Table 2.

4   From Complex to Simpler ODEs

The parameters $R_{1,h}$ and $R_{3,h}$ in the piecewise models are large compared to the other system parameters. The singular perturbation approach can be applied to

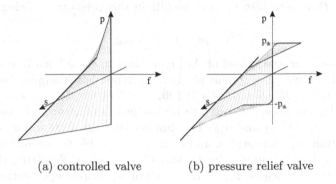

(a) controlled valve   (b) pressure relief valve

Fig. 3. Nonlinear valve resistance characteristics.

Table 2. Mode transition table.

present mode	next mode			
	α_{00}	α_{01}	α_{10}	α_{11}
α_{00}		$p > p_{th}$	$u_v > 0$	$p > p_{th} \wedge u_v > 0$
α_{01}				$u_v > 0$
α_{10}	$u_v < 0$	$u_v < 0 \wedge p > p_{th}$		$p > p_{th}$
α_{11}		$u_v < 0$		

remove these large parameters ($R_{1,h} \to \infty$ and $R_{3,h} \to \infty$) and arrive at simpler reduced order ODEs for each mode. To simplify notation, we set $R_1 = R_{1,l}$ and $R_3 = R_{3,l}$. The dynamic behavior in α_{01} is derived by $R_{1,h} \to \infty$ and can be expressed as:

$$f_{\alpha_{01}} : \begin{bmatrix} \dot{f_1} \\ \dot{f_2} \\ \dot{p_1} \end{bmatrix} = \begin{bmatrix} -\frac{R_2}{I_1} & -\frac{R_2}{I_1} & \frac{1}{I_1} \\ -\frac{R_2}{I_2} & -\frac{R_2+R_3}{I_2} & \frac{1}{I_2} \\ -\frac{1}{C} & -\frac{1}{C} & 0 \end{bmatrix} \begin{bmatrix} f_1 \\ f_2 \\ p_1 \end{bmatrix} \tag{10}$$

From Table 2 it is clear that this abstraction does not affect the switching constraints that define further discrete transitions out of this mode. The only transition out of this mode, $\alpha_{01} \to \alpha_{11}$, is governed by the external variable u_v ($u_v > 0$).

For α_{10}, $R_{3,h} \to \infty$ implies $f_2 = 0$ and behavior reduces to a second order system

$$f_{\alpha_{10}} : \begin{bmatrix} \dot{f_1} \\ \dot{p_1} \end{bmatrix} = \begin{bmatrix} -\frac{R_1 R_2}{I_1(R_1+R_2)} & \frac{R_1}{I_1(R_1+R_2)} \\ -\frac{R_1}{C(R_1+R_2)} & -\frac{1}{C(R_1+R_2)} \end{bmatrix} \begin{bmatrix} f_1 \\ p_1 \end{bmatrix} + \begin{bmatrix} \frac{R_2}{I_1(R_1+R_2)} \\ \frac{1}{C(R_1+R_2)} \end{bmatrix} \begin{bmatrix} p_{in} \end{bmatrix}. \tag{11}$$

In both modes, the pressure, p, in the switching condition is given by Eq. (8). The reduced behavior in α_{00} is given by an autonomous second order system

$$f_{\alpha_{00}} : \begin{bmatrix} \dot{f_1} \\ \dot{p_1} \end{bmatrix} = \begin{bmatrix} -\frac{R_2}{I_1} & \frac{1}{I_1} \\ -\frac{1}{C} & 0 \end{bmatrix} \begin{bmatrix} f_1 \\ p_1 \end{bmatrix}. \tag{12}$$

Introducing $R_{1,h} \to \infty$ into Eq. (8), results in this pressure, p, being expressed as

$$p = p_1 - R_2 f_1 > p_{th}. \tag{13}$$

It turns out that the spread of the eigenvalues in these linearized, simplified, and reduced systems of equations is still quite large. For example, given parameter values, $R_{1,l} = R_{3,l} = 0.01$, $R_2 = 100$, $C = 5 \cdot 10^{-6}$, $I_1 = 1$, and $I_2 = 0.01$, one of the eigenvalues is computed to be five orders of magnitude less than the others in the modes α_{01} and α_{10}. This implies that the system still operates at two widely differing time scales, and it may be possible to simplify the system model further by abstracting the R_2 and C parameters. Applying this change will affect the state variable p_1, which is part of the switching condition, $p > p_{th}$. This requires a detailed study of the switching characteristics.

5   The State Mapping

The application of singular perturbation methods to the model in the last section with $\frac{1}{R_2}$ and C as the small parameters, replaces some differential equations by algebraic constraints. For example, the α_{01} mode is reduced from a 3rd order to a 1st order system, whereas mode α_{00} is reduced from a 2nd to a 0th order (purely algebraic) system. This may cause state variable values to change discontinuously during mode transitions.

5.1   Jump into Mode α_{01}

When $R_2 \to \infty$, Eq. (10) becomes a singular system of equations with $-f_1 - f_2 = 0$. In phase space, this algebraic relation constitutes a manifold to which behavior is confined. The dynamic system behavior on this manifold is derived by applying a transformation, $x = I_1 f_1 - I_2 f_2$, which gives $\dot{x} = R_3 f_2$. Substituting for $f_1 (= -f_2)$ in the expression for x and eliminating f_2 yields

$$\dot{x} = -\frac{R_3}{I_1 + I_2} x. \tag{14}$$

If mode α_{01} is entered at a point not on this manifold, an instantaneous projection in the impulse space has to be executed to satisfy the manifold constraint. The impulse space can be derived by integrating the dynamic behavior in Eq. (14) over an infinitesimal interval from t to t^+, which gives $I_1(f_1^+ - f_1) - I_2(f_2^+ - f_2) = 0$ [4]. Combined with the manifold constraint at t^+, $f_1^+ = -f_2^+$, this computes the projection to be

$$g_{\alpha_{01}} : f_1^+ = \frac{1}{I_1 + I_2}(I_1 f_1 - I_2 f_2). \tag{15}$$

Table 2 shows that the transition conditions for α_{01} are not affected by the variables f_1 and f_2, therefore, no further analysis is required.

5.2   The Jump into Mode α_{00}

In mode α_{00}, $R_2 \rightarrow \infty$ produces $f_1 = 0$. Again, this constitutes a manifold in phase space and transition into α_{00} requires a projection,

$$g_{\alpha_{00}} : f_1^+ = 0. \tag{16}$$

However, analysis of the detailed model indicates that the switching condition from mode α_{00} to mode α_{01} in Eq. (13) may be activated before f_1 becomes 0. Therefore, this transition condition needs to be analyzed more precisely.

When $R_2 \rightarrow \infty$, the switching condition in Eq. (13) becomes singular, and the value for p cannot be determined from the state variables. Therefore, p has to be expressed in terms of the time derivatives of the states. For this system, Eqs. (12) and (13) yield, $p = \frac{1}{I_1}\dot{f_1}$. The $f_1 = 0$ constraint corresponds to a discontinuous change in f_1, therefore, $\dot{f_1}$ may produce an impulse.

Impulse behavior is too coarse an approximation of the underlying detailed continuous transient. A more refined analysis solves the detailed differential equation in the time domain. The characteristic polynomial of $f_{\alpha_{00}}$ has two roots

$$\lambda_{1,2} = \frac{1}{2I_1}\left(-R_2 \pm \sqrt{R_2^2 - \frac{4I_1}{C}}\right). \tag{17}$$

Assuming complex eigenvalues[1] ($\lambda_{1,2} = \lambda_r \pm j\lambda_i$), the pressure variable in mode α_{00} is ($t_0 = 0$ for notational convenience)

$$p_1(t) = e^{\lambda_r t}\left(p_1(0)cos(\lambda_i t) + \frac{1}{\lambda_i}(\dot{p}_1(0) - \lambda_r p_1(0))sin(\lambda_i t)\right). \tag{18}$$

Applying a third order Taylor series approximation yields ($\dot{p}_1 = -\frac{1}{C}f_1$),

$$p_1(t) = \left(1 + \lambda_r t + \frac{\lambda_r^2 t^2}{2}\right)\left(p_1(0)\left(1 - \frac{(\lambda_i t)^2}{2}\right) + \left(-\frac{f_1(0)}{C} - \lambda_r p_1(0)\right)t\right). \tag{19}$$

The switching condition is based on $p = p_1 - R_2 f_1(t)$, where $f_1(t)$ is used instead of $f_1(0)$ because the value of f_1 changes during the time interval in which $p_1(t)$ rises and it may be different from $f_1(0)$ when $p(t)$ reaches p_{th}.

This condition can be used to check if $p > p_{th}$, and if so, the time, $t_s = f_t(p_1, f_1, p_{th})$, at which this constraint becomes true. This value can then be used in the expression for f_1 to derive the discontinuous change upon switching. Abbreviating $f_1(0)$ and $p_1(0)$ as f_1 and p_1, respectively, and using $a = (-\frac{f_1}{C} - \lambda_r \frac{p_1 + f_1 R_2 R_2}{I_1} - \frac{p_1 R_2}{I_1} + \lambda_r R_2 f_1)/\lambda_i$, $b = p_1 - R_2 f_1$, and $c = -p_{th}$, the solution is given by

$$t_s = \frac{-(b-c)}{a\lambda_i + c\lambda_r} + \frac{\frac{1}{2}c(\lambda_r^2 + \lambda_i^2)(b-c)^2}{(a\lambda_i + c\lambda_r)^3}. \tag{20}$$

Substituting t_s in the expression for $f_1(t)$ in α_{00} results in the state mapping

$$g_{p,\alpha_{00}} : f_1^+ = e^{\lambda_r t_s}\left(f_1 cos(\lambda_i t_s) + \left(-\frac{R_2}{I_1}f_1 + \frac{p_1}{I_1} - \lambda_r f_1\right)\frac{sin(\lambda_i t_s)}{\lambda_i}\right), \tag{21}$$

---

[1] Analysis of real eigenvalues is similar.

where $f_1 = f_1(t_0)$ and f_1^+ the value of f_1 when α_{00} is exited because $p > p_{th}$.

This is graphically depicted in Fig. 4 for a third and fourth order Taylor approximation of f_1, $f_1^{p,3}$ and $f_1^{p,4}$, respectively.[2] Here an initial positioning maneuver of the piston is aborted at t_0, which causes the relief valve to open at t_s. The error between the third and fourth order approximations is shown by ϵ_3 and ϵ_4, respectively.

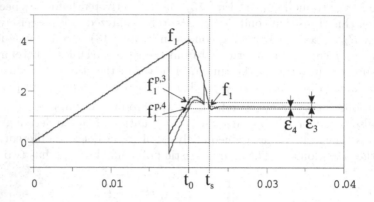

Fig. 4. Value of f_1 at t_s for a detailed model and its predictions at t_0.

5.3 A Computational Model

The discrete transition model that results from the abstractions of the detailed continuous behavior can be modeled by the extended hybrid automata structure in Fig. 5. The traditional hybrid automata is extended by junctions (indicated by small circles). When an event triggers a transition to a junction, the events on each of the exiting transitions from the junction are evaluated, resulting in an immediate second transition.

In this model, when the external control valve closes ($u_v < 0$), the time t_s at which the relief valve opens is computed by $f_t(p_1, f_1, p_{th})$ using the detailed continuous transient. If this computation returns a value $t_s \geq 0$, control is switched to the lower branch, else control switches to the branch at the right. This last branch indicates that the system moves to the field description for α_{00}, and, therefore, requires a consistent projection of the state variables (i.e., $f_1 = 0$). If the lower branch is taken, first the effect of the quick pressure build-up and corresponding flow decrease has to be accounted for by executing $f_1 = g_{p,\alpha_{00}}(f_1, p_1, t_s)$. This results in a new value for f_1 when the continuous behavior in α_{01} is activated. Again, behavior in this mode is subject to manifold constraints, and the corresponding projection $f_1 = g_{\alpha_{01}}(f_1, f_2)$ takes place

---

[2] The predictions are computed during a short time interval around t_0 to avoid singularities that exist over the entire range. Note that the values only need to be computed at t_0.

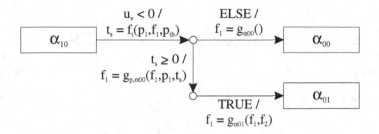

Fig. 5. Complex discrete switching structure.

before α_{01} is activated to ensure values are consistent in this mode. Note that in α_{01}, p_1 is not a state variable but derived from $p_1 = p_{in} - f_1 R_1$. Further, the systematically derived control structure is more complex as compared to the transitions in Table 2.

6   Phase Space Transition Behavior

The mode and discontinuous state changes can now be characterized in terms of a phase space transition ontology. In other work [6], three principal transition functions were analyzed in phase space: (i) transition to a mythical mode, (ii) transition to a pinnacle, and (iii) transition to a continuous mode.

When switching to α_{00}, the two possible scenarios are

1. $p < p_{th}$ in which case a projection of f_1 onto $f_1 = 0$ occurs, and the system remains in α_{00}. This represents a transition to a *continuous* mode.
2. $p \geq p_{th}$ in which case
   - there may be a distinct drop in f_1 before switching to α_{01}. This is a transition to a *pinnacle*, or
   - the switch to α_{01} has occurred before any significant change in f_1 occurs. This represents a transition to a *mythical mode*.

The switch to α_{01} may also include a discontinuous state change because of the manifold projection that immediately follows the pinnacle or mythical mode.

Figure 6 shows the phase space transition behavior for two values of C in a C^0 hybrid model with parameter values $R_{1,l} = R_{3,l} = 0.01$, $R_{1,h} = R_{3,h} = 1 \cdot 10^7$, $R_2 = 100$, $I_1 = 1$, $I_2 = 0.01$, and $p_{th} = 1000$. Velocity f_1 is plotted on the x-axis and pressure p is plotted on the y-axis. The discontinuous approximations are superimposed by dotted lines.[3] When the control valve closes, f_1 has value 4, and the pressure in the cylinder starts to rise quickly (Fig. 4 depicts the time domain behavior). This behavior consists of an immediate change in p caused by the term $f_1 R_2$, and a quick continuous change because of the pressure build up.

---

[3] These approximations are not simulation results.

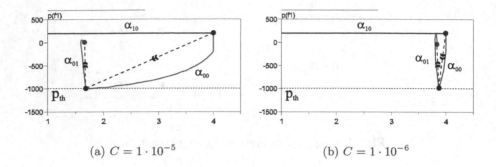

(a) $C = 1 \cdot 10^{-5}$ (b) $C = 1 \cdot 10^{-6}$

Fig. 6. Dominant C phase space switching behavior.

If the absolute pressure exceeds 1000, the mode switch to α_{01} occurs. In this mode, there is another quick change in f_1, this time governed by the dependency between I_1 and I_2. Because I_1 is several orders of magnitude larger than I_2, only a small change in f_1 occurs. The interaction between the three state variables in the C^0 hybrid model, f_1, f_2, and p_1, causes oscillatory behavior in α_{01}. This is clearly seen in Fig. 6(b). The discontinuous jump does not include this behavior, but immediately reaches the final value. The phase space behaviors present examples of two consecutive discontinuous state variable value changes that are of different types. The intermediate value of f_1 is achieved in a pinnacle mode, and the final value is governed by a manifold projection. Note that the pinnacle is crucial in computing the correct final value of the variable in α_{01}, when continuous behavior resumes.

If $R_2 > 250$, $f_1 R_2$ becomes the dominant factor in the phase space transition behavior, as shown in Fig. 7 for $C = 1 \cdot 10^{-5}$ and $R_2 = 500$. The consecutive switch to α_{01} follows immediately after the switch to α_{00}. As a consequence, α_{00} does not affect the value of f_1, therefore, this is a mythical mode. Mode α_{00} is not intrinsically a mythical mode because the state variable values when the mode is entered determine whether it is exited immediately. Only in such situations mythical behavior occurs. The projection in α_{01} that follows is shown more clearly in Fig. 7(b) for a larger value of I_2. For these parameters, Eq. (15) verifies the value $f_1^+ = 2$ ($f_1 = 4$, $f_2 = 0$) and confirms that larger values of I_2 have a greater effect on the magnitude change of f_1. Again, the fast oscillatory behavior of the manifold projection is abstracted away in the discontinuous approximation.

7   Conclusions

This paper shows how nonlinear and high order system models can be systematically reduced to piecewise linear systems with more uniform time scales of behavior. The resultant hybrid model is obtained in two steps: (i) C^0 continuous with piecewise simpler behavior and switching conditions, and (ii) piecewise

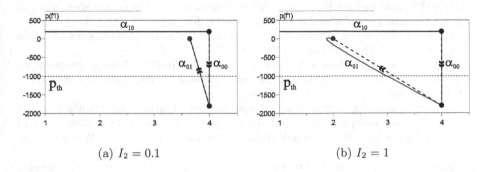

(a) $I_2 = 0.1$ (b) $I_2 = 1$

Fig. 7. Dominant R_2 phase space switching behavior.

continuous reduced order behavior with switching conditions and discontinuous changes in state variable values. The reduction in continuous domain complexity is gained at the cost of increasingly complex discrete event control structures. Because of the intricacies in defining the switching conditions and the corresponding jumps in the variable values, *ad hoc* modeling schemes can often produce erroneous results. This is most likely to happen when jumps occur in state variable values, caused by the introduction of algebraic constraints. The manifold projections that result may be aborted because intermediate variable values derived from the detailed dynamic models indicate that further immediate transitions occur. These concepts are illustrated by analysis of phase space behavior of a hydraulic actuator.

The approach fits into our ontology for describing transition behavior in phase spaces that we have established in previous work [6]. We hope to extend this approach to systematic procedures for automated model reduction of complex nonlinear systems into simpler hybrid representations. A longer term goal of this work is to develop real time models of complex systems so that they may be employed in hybrid observers for Fault Detection and Isolation (FDI) studies of complex nonlinear systems [5].

Acknowledgements

Pieter J. Mosterman is supported by a grant from the DFG Schwerpunktprogramm KONDISK. Gautam Biswas is supported by grants from HP Labs, and the DARPA Software Enabled Control program.

References

1. Rajeev Alur, Costas Courcoubetis, Thomas A. Henzinger, and Pei-Hsin Ho. Hybrid automata: An algorithmic approach to the specification and verification of hybrid systems. In R.L. Grossman, A. Nerode, A.P. Ravn, and H. Rischel, editors, *Lecture Notes in Computer Science*, volume 736, pages 209–229. Springer-Verlag, 1993.

2. James Demmel and Bo Kågström. Stably computing the Kronecker structure and reducing subspaces of singular pencils $A - \lambda B$ for uncertain data. In J. Cullum and R. A. Willoughby, editors, *Large Scale Eigenvalue Problems*. Elsevier Science Publishers B.V. (North-Holland), 1986.
3. Petar V. Kokotović, Hassan K. Khalil, and John O'Reilly. *Singular Perturbation Methods in Control: Analysis and Design*. Academic Press, London, 1986. ISBN 0-12-417635-6.
4. Pieter J. Mosterman. State Space Projection onto Linear DAE Manifolds Using Conservation Principles. Technical Report #R262-98, Institute of Robotics and System Dynamics, DLR Oberpfaffenhofen, P.O. Box 1116, D-82230 Wessling, Germany, 1998.
5. Pieter J. Mosterman and Gautam Biswas. Building Hybrid Observers for Complex Dynamic Systems using Model Abstractions. In Frits W. Vaandrager and Jan H. van Schuppen, editors, *Hybrid Systems: Computation and Control*, pages 178–192, 1999. Lecture Notes in Computer Science; Vol. 1569.
6. Pieter J. Mosterman, Feng Zhao, and Gautam Biswas. An Ontology for Transitions in Physical Dynamic Systems. In *AAAI98*, July 1998.
7. A. J. van der Schaft and J. M. Schumacher. The complementary-slackness of hybrid systems. *Math. Contr. Signals Syst.*, (9):266–301, 1996.
8. Andreas Varga. On modal techniques for model reduction. Technical Report TR R136-93, Institute of Robotics and System Dynamics, DLR Oberpfaffenhofen, P.O. Box 1116, D-82230 Wessling, Germany, 1993.

Computing Optimal Operation Schemes for Chemical Plants in Multi-batch Mode*

Peter Niebert and Sergio Yovine

VERIMAG
2 Av. de Vignate, 38610 Gières, France
{Peter.Niebert,Sergio.Yovine}@imag.fr

Abstract. We propose a computer-aided methodology to automatically generate time optimal production schemes for chemical batch plants operating in multi-batch mode. Our approach is based on the following principles: (1) the plant is modeled at the level of process operations whose behavior is specified by timed automata, (2) the optimal production schemes are generated using algorithms for reachability analysis of timed automata implemented in OPENKRONOS, (3) the output of the verification tool is post-processed to derive high-level control code. We apply our methodology to the batch plant at the University of Dortmund. The automatically computed operation schemes turned out to be more efficient than the previously used handwritten ones.

1 Introduction

A chemical batch plant consists of a collection of containers, reactors, pipes, valves, pumps, etc., for storing, transporting, processing and transforming raw materials to obtain a final chemical product. A plant is also equipped with an integrated hardware and software architecture for controlling and supervising its operation. Generally, batch plants are operated in multi-batch mode where several products are manufactured concurrently. The structure and operation of batch plants are standardized in the norm ISA S88.01 [7]. A central notion of the ISA S88.01 standard is that of a *plant-independent recipe*, a description of "abstract" processing steps (e.g., mixing, heating, cooling) leading to a production goal. For a specific batch plant, it is the task of the control engineer to construct an *operation scheme*, that is, a "concrete" arrangement of *process operations* of the plant (e.g., mixing materials A and B in container C, emptying out container C into container D, cooling the content of container C), that realize a given plant-independent recipe. Process operations are actually carried out by sequences of low-level *process actions* (e.g., opening and closing valves, starting and stopping pumps), which are commanded by hardware or software that implements each process operation as a procedure that is invoked by a high-level control program.

---

* This work has been supported by EC Esprit-LTR Project 26270 VHS.

N. Lynch and B. Krogh (Eds.): HSCC 2000, LNCS 1790, pp. 338–351, 2000.

An operation scheme should make an efficient use of the resources and satisfy all the constraints required for a safe and correct functioning of the plant. The usual approach followed by the control engineer to find an optimal operation scheme out of a plant-independent recipe is somewhat analogous to the one followed by the cook to make a cake out of a recipe found in a cookbook. Roughly, it consists in first providing a *plant-dependent recipe* made up of a (partially) ordered set of process operations of the plant that can be done to realize the plant-independent one. The plant-dependent recipe specifies a set of possible, though eventually conflicting and not necessarily optimal, sequences of operations, together with constraints on the usage of shared resources. In the terms of our analogy with the cook, this step corresponds to determining which particular kitchen-ware to use and how to use it (e.g., for how long, how many times, under which conditions, when, to do what, etc.) in order to realize the different actions specified in the recipe. The second step consists in finding an optimal sequence that meets the constraints. Typically, the plant-dependent recipe is given as an acyclic directed graph. Standard (combinatorial) optimization techniques are used to find the best path in the graph. Notice that such sequence may not exist, which indeed means that the plant-dependent recipe cannot be realized by that operation scheme. Again, this is similar to what the cook does. She figures out that certain steps of the recipe can be done concurrently, e.g., cooking the cake in the oven and preparing the chocolate sauce to cover the cake, that others are conflicting due to the existence of limited resources, e.g., preparing the paste and the chantilly, and eventually that the cake cannot be made at all, e.g., because the chosen oven does not heat enough. In this approach, much of the intrinsic complexity of the problem needs to be taken care of by the engineer (or the cook) during the specification of the plant-dependent recipe. However, even for a single batch, constructing that recipe is not a trivial task and it becomes significantly more difficult for multi-batch processing (i.e., making several cakes concurrently on the same kitchen), specially for complex batch plants exhibiting a high degree of parallelism.

The first aim of this work is to alleviate the task of the engineer, at the cost of eventually using computationally more expensive techniques, by providing computer-aided support to *automatically generate an operation scheme for multi-batch processing*, without having to specify a plant-dependent recipe, but only a mapping between "abstract" processing steps and production goals into one or more "concrete" process operations that can effectively realize them. The second objective is to *automatically derive the control program that carries out, in real time, the operation scheme on the control architecture of the plant*.

The basic underlying idea to do so is to require the engineer to provide an "operational" model of the plant, together with (1) the production goal, which may consists of an arbitrary number of batches, and (2) the optimization criterion, typically shortest overall production time. The major difficulty that arises here concerns the modeling of the operation of the plant at a level of abstraction suitable for both the recipe and the control architecture (hardware and software) of the plant. In general, such a modeling framework would need to take into account both the discrete events and the continuous chemical and physical phenomena, leading to the need of using a *hybrid* model, e.g., hybrid automata [1]. However, if we consider the problem at the level of process operations, it is possible to use simpler models, e.g., timed automata [2], that abstract away most of the details of the complex continuous behaviors, while preserving all timing and concurrency constraints relevant to the operation of the plant. This is mainly due to the fact that: (1) many recipes are indeed described in terms of quantities of raw materials and timing constraints on abstract processing steps such as mixing, heating, cooling, etc., and (2) the execution time of process operations can, generally, be estimated quite accurately.

In order to achieve our goal, we propose a methodology based on the following principles, models and tools:

1. The plant is modeled at the level of process operations whose behaviors are specified by timed automata extended with shared variables.
2. The optimal production schemes are generated using the algorithms for reachability analysis of timed automata implemented in the VERIMAG timing verification toolsuite OPENKRONOS [5].
3. The output of the verification tool is post-processed (1) to visualize the operation schemes in different ways (e.g., Gantt and Hasse diagrams), and (2) to derive high-level control code.

In order to illustrate the feasibility of the approach in practice, we apply it to a case study: The chemical batch plant of the Dortmund Process Control Laboratory [11]. We derive time- and resource optimal schedules for several number of batches. Moreover, the operation schemes computed by the tool turned out to be more efficient than the ones obtained using the "classical" cook-like method described before.

The rest of the paper is structured as follows. In Section 2 we sketch a framework for modeling the operation of batch plants at the level of abstraction of process operations. In Section 3 we describe the case study. In Section 4 we present our approach for searching for optimal operation schemes using reachability analysis. In Section 5 we report on the experimental results obtained for the case study. In Section 6 we discuss current work concerning the integration of the approach into the control architecture of the plant.

2   Modeling Chemical Batch Plants

We model a plant by specifying:

- A collection of *resources*. In principle every single valve, tube and container may be involved in the execution of process operations, but there are many devices which are reasonably used together[1], e.g., a valve is only used in combination with its containing pipe. As a rule, the most important devices are the containers. In almost all cases, the surrounding pipes of a container

---

[1] These groups are called units in the standard.

are only used in combination with transfers from or to this container. Thus, a typical resource would be a *container*. Also, we can abstract away those resources which are only used in combination with other (modeled) resources and do not contribute to the state of the system on a macroscopic level.

− A collection of *possible discrete contents* of each resource (typically concerning mass or volume, temperature, chemical phases ...). The process operations are assumed to perform discrete transitions on the state space of container contents. Hence, it is usually possible to give a discrete and finite representation of the contents of a container as it occurs before or after the execution of process operations. In particular the possible values must encode intermediate products in recipes to allow to map recipes onto the plant.

− A collection of *process operations*. Each process operation is furthermore associated with:

  • A name (the name of a PLC control routine).
  • A collection of resources used by the operation.
  • A condition for the enabledness of the operation, which depends on the states of the involved resources.
  • A function representing the transition on the states of the involved resources.
  • A function to estimate the time consumption of the operation on the bases of the container states before the operation.

Based on this semi-formal description, a formal specification is derived as a network of communicating timed automata [2] appropriately extended with shared variables. To obtain the formal model we proceed as follows:

− The availability of each resource is modeled by a boolean variable.
− The contents of resources are correspondingly modeled by shared variables (volume, temperature, ...) over finite domains.
− Each operation is realized by a *timed automaton* which has (in addition to the shared variables) two control locations *non-active* and *active*, as well as a *start* transition and a *finish* transition.

  • The *start* transition which depends on the guard for the operation and the availability of resources, reserves the resources and starts the clock.
  • The *finish* transition which depends on the duration constraints on the clock, changes the values of the variables modeling the content of the containers, and releases the resources.
  • Invariant constraints associated with control locations guarantee that transitions must occur within the predefined time bounds.

The reader is referred to [2] for a detailed description of the formal semantics of the timed automaton model. In the next section we informally present the semantics through an illustrative example.

3   Case Study: Modeling

We model here the chemical batch plant from the Dortmund Process Control Laboratory [11]. An overview of the plant and the architecture of the integrated control system is given in Figure 1.

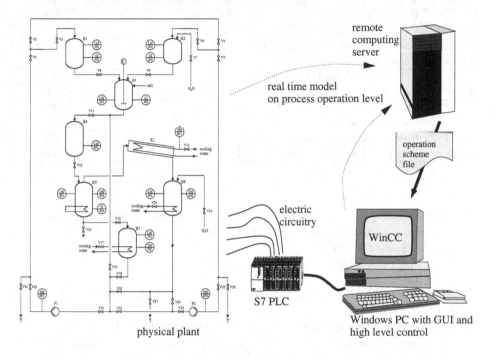

Fig. 1. Control architecture of the plant.

The plant consists of seven containers, namely B1 to B7. Containers B1, B2 and B4 are ordinary ones. Container B3 has a device for mixing. Container B5 is the evaporator connected to a condenser. The condensed steam flows into container B6. Both B6 and B7 are attached to a cooling system.

There are essentially three levels in the control. The lowest level concerns the physical control elements, such as sensors, valves and electric devices for pumping, heating and stirring. On top of this, there exist a number of basic control routines implemented on a Siemens S7 PLC (and described by SFCs, sequential function charts, in [11]), which realise process operations. These control routines are invoked from the higher layer by an operator or by a control program running on a PC.

The plant-independent recipe is as follows:

1. Produce highly concentrated brine by manually adding salt to tap water.
2. Mix it with demineralised or tap water to produce a medium concentration.

Table 1. Process operations of the plant.

operation	description	duration
B2	fill B2 with water from tap	10s/1
B3KA	Fill 4l of pure water into B3	320s
B3KB	Manually add NaCL into B3 and mix until concentration is 5g/l	600s
B3U	Pump concentrated solution from B3 to B1	420s
B3A	Fill 4l of concentrated brine from B1 into B3	320s
B3B	Thin down concentration in B3 to 3g/l	240s
B3B4	Fill solution from B3 into B4	600s
B4B5	Fill solution from B4 into B5	330s
EVAP	Evaporate and condensate from B5 to B6 until high conc. reached	1500s
B5B7	Fill hot concentrate from B5 to B7	260s
B7	Cool solution in B7	600s
B7B1	Pump up solution from B7 to B1	220s
B6B2	Pump up pure water from B6 to B2	240s

3. Heat and evaporate water out of this medium solution such as to return to the high concentration and condensate the vapor and capture the condensate (demineralised water).
4. Cool down the resulting solution.

A batch is finished when the highly concentrated remainder after evaporation is cooled down.

The actual modeling follows precisely the scheme indicated in Section 2. Our model allows for the maximal exploitation of parallelism in the plant, which is of great importance for efficient multi-batch execution. The process operations are listed in Table 1. The duration estimations are derived from experimental values listed in [11]. [2] Due to lack of space, we do not present here the full model, but focus on the description of the variables that have been used and the specification of some of the more illustrative process operations.

For each container Bi there is (1) a boolean variable Bi that models the availability of the container, and (2) a discrete variable Vi ranging over a finite domain, modeling the relevant values of volumes of liquid in the container, e.g., 0 (empty) and 4 (4l) for B1, and the interval of values from 0 to 6 for container B2. For container B3, there is a discrete variable C3, modeling the significant values of the concentration, namely 0g/l (demineralized), 3g/l (medium) and 5g/l (high). For container B7, the variable H7 is used to model the two possible estimations of the temperature of its content, namely hot and cold. In order to determine the number of finished batches, we use the additional variable count, ranging over the natural numbers.

---

[2] For the sake of simplicity, operations for occasional rinsing, which do not contribute to the production, have been omitted. Furthermore, we have split some operations involving B3 into two parts to introduce more potential parallelism into the system.

Table 2. Specification of the containers.

container	volume	temperature	concentration
B1	$V1 : \{0,4\}$		
B2	$V2 : [0,6]$		
B3	$V3 : \{0,4,7\}$		$C3 : \{demineralized, medium, high\}$
B4	$V4 : [0,7]$		
B5	$V5 : \{0,4,7\}$		
B6	$V6 : \{0,3\}$		
B7	$V7 : \{0,4\}$	$H7 : \{hot, cold\}$	

```
des(0,2,2)
(0, [B7 ∧ H7=hot] B7start B7:=false CL7:=0, 1)
(1, [CL7=60] B7finish B7:=true H7:=cold count:=count+1, 0)
```

Fig. 2. Timed automaton modeling process operation B7.

The timed automaton modeling the operation of cooling the content of B7 is depicted[3] in Figure 2. The automaton has two control locations, namely 0, the initial one, and 1, and two transitions, labeled B7start and B7finish. The guard of B7start checks if the container B7 is available and hot. In such case, the transition is said to be *enabled* and can be executed. When doing so, container B7 is reserved for exclusive use, by setting B7 to false, the clock CL7 is reset to 0 to start measuring the duration of the cooling process, and the automaton moves to location 1. In this location, the automaton waits until the corresponding clock has reached 60, modeling the 600s of cooling, releases B7 by setting B7 to true, changes H7 to cold, and moves back to location 0. Transition B7finish indeed models the completion of a batch, which consists in obtaining a cold, highly concentrated brine in container B7. Therefore, the value of count, representing the number of already produced batches, is updated when B7finish is executed.

```
des(0,2,2)
(0, [B5 ∧ B7 ∧ V5=4 ∧ V7=0] B5B7start B5:=false B7:=false CL5:=0, 1)
(1, [CL5=26] B5B7finish B5:=true B7:=true V5:=0 V7:=4 H7:=true, 0)
```

Fig. 3. Timed automaton modeling process operation B5B7.

The automaton modeling the operation of emptying out the content of B5 into B7 is depicted in Figure 3. It has two control locations and two transitions.

---

[3] The complete syntax of the input language of OPENKRONOS can be found in http://www-verimag.imag.fr/DIST_SYS/SMI.

In location 0, it waits for B5 and B7 to be free, the volume of B5 to be 4l, and B7 to be empty. In such case, it can move to location 1, while blocking the use of both containers, and resetting the clock CL5 to start counting the time spent in the operation. When the value of CL5 reaches 26 (modeling the 260s required for emptying out B5 into B7), the automaton moves back to location 0, releases the containers, and updates the values of the variables modeling the contents (V5 and V7 become 0 and 7, respectively).

4   Synthesis of Optimal Operation Schemes

The problem we are interested in solving is the following: Given a specification of a batch-processing plant as a set O of process operations and a number N of (identical) batches to be produced, find an operation scheme, i.e., a partial order of process operations $\pi = (O', \leq)$ where O' is a multiset of elements of O (allowing for multiple occurrances of the same operation), that executes the required N batches. Notice that we search for a partial order of operation instances that allows for parallel execution of independent operations. Besides, the operation scheme π is also required to satisfy some "optimality" criteria related to the time spent to finish the N batches, the number of resources used, etc.

At first glance, this problem can be viewed as a particular instance of the more general problem of "controller synthesis" stated as follows: Given a plant P and a property S, construct a controller C that "forces" P to meet S by disabling some (controllable) behaviors of P. A technique for solving this problem in the context of discrete-event systems has been first proposed in [14]. This technique has recently been extended to timed systems in [10], but the currently available prototype [3] is not able to deal with large systems like the one we are considering here.

Fortunately, model-checking provides us with means to look at the problem from another angle. Indeed, since the execution times (more precisely, the upper bounds of the execution times) of the process operations are known and the operation scheme to be calculated is *finite*, a solution to the problem can be obtained by using a reachability algorithm capable of providing a (timed) sequence of start and stop transitions that reaches the desired goal.

Certainly, any "standard" reachability algorithm will allow us to find some operation scheme (if at least one exists) but not necessarily an optimal one. An algorithm for solving the "optimal-controller synthesis" problem has been recently proposed in [4], but not yet implemented. However, a much simpler "ad-hoc" solution can be devised by making use of the knowledge we have about (1) the particular search method used by the reachability algorithm and (2) the structure of the plant.

Using a breadth-first exploration of the reachability graph ensures that the operation scheme found makes optimal use of the resources, in the sense that no other operation scheme can achieve the same goal executing fewer process operations. However, the operation scheme might not be the fastest, that is, there might be another one that performs the same number of process operations in less time. One possibility to overcome this problem is to guess an upper bound T for the completion time and to iterate the reachability exploration by appropriately increasing or decreasing the time horizon according to the result obtained in the previous iteration. By applying this strategy, the optimal operation scheme can be obtained in $log(T)$ number of iterations.

Still, the size of the state-space to be explored might be a serious obstacle and it is advisable to exploit the knowledge of the plant to try to overcome it. For example, in the case study, it is easy to see that optimal operation schemes must intensively use resource B5. Appropriate use of this information has indeed revealed to be vital for solving the problem.

This approach can be automated using state-of-the-art verification tools for timed automata, such as OPENKRONOS [5]. In particular, for the case study we have used a discrete-time BDD-based reachability algorithm [4]. The result of the timing analysis is a *timed trace*, which is a sequence consisting of *ticks*, representing the elapsed time units, and of transitions of the automaton, representing the beginning (e.g., B7start) and termination (e.g., B7finish) of process operations.

Such a trace can be visualized as a Gantt-like diagram, where the operation instances are visualized as blocks in a two dimensional diagram, one dimension for the resources (blocked by the operation), the other for time. Notice that in this context some operations may use several resources.

However, we are looking for a partial order of operations, which we can reconstruct out of the timed trace: By specification, two instances of process operations a and b using a common resource R will not be active in parallel, i.e. if we find the start event of a in the trace before the start event of b, then also the finish event of a comes before the start event of b. In this case we say that operation instance b *depends* on operation instance a ($a <' b$). The *transitive closure* of these dependencies between arbitrary occurrences of operations in the scheme gives us a partial order, the operation scheme.

The actual output of this algorithm is the Hasse diagram of the partial order, without any timing information. This partial order serves as input to a high-level control software, as will be explained in Section 6. We consider the temporal information here only as justification, why this partial order of operations is realisable under the timing constraints imposed by the specification. No timing is needed for control here.

5   Case Study: Synthesis

The plant-dependent recipe for the production of cold, highly concentrated brine described in [11] (and designed for manual operation by students) consists of two phases:

---

[4] A BDD is a compact data structure for encoding boolean predicates (e.g., sets of states). The reader is referred to [9] for more details.

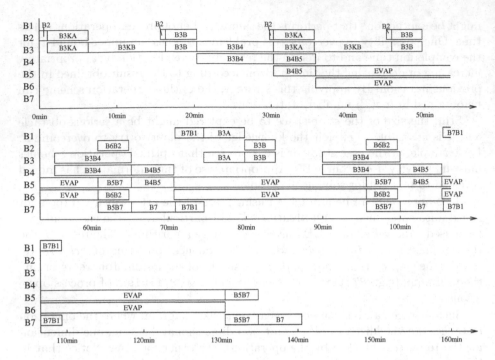

Fig. 4. The timed trace calculated for three batches as Gantt diagram.

- A preparatory phase which aims to place highly concentrated brine in container B1 and water in B2: B2, B3KA, B3KB, B3U, B2.
- A cyclic scheme to produce a single batch with the invariant that before and after this scheme there is concentrated brine in B1 and water in B2: B3A, B3B, B3B4, B4B5, EVAP, B5B7, B7 to obtain the final product, and finally execute B6B2, B7B1 to recycle material.

Notice, that the recipe allows for very little parallelism for a single batch (only process operations B7 and B7B1 can be executed in parallel with B6B2 and, if the next batch is taken into account, B3A can be executed in parallel), underexploiting the capacity of the plant. Obtaining a better performance requires using another recipe which cannot be obtained by re-scheduling the given one.

Instead of trying to *schedule* several instances of *this* recepy, we had the computer search for operation schemes for the production of several batches (i.e. for observing operation B7 several times).

Figure 4 and Figure 5 respectively show the Gantt-like and Hasse[5] diagrams of the computed operation scheme for three batches.

It is interesting to note that the calculated operation scheme requires less time and less operations than one described in [11]. First, the second B2 operation can

---

[5] The picture of the Hasse diagram is obtained using the *graphviz* tool from Bell Labs.

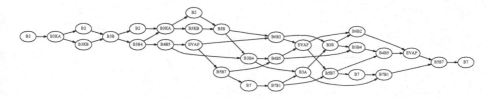

Fig. 5. The operation scheme for three batches as Hasse diagram.

be done concurrently with B3KB since they do not share any resource. Second, the sequence B3U, B3A, is actually never used. Indeed, this sequence is redundant because the concentrated brine is mixed in B3 and there is no need to pump it to B1 (operation B3U) to re-pump it later back into B3 (operation B3A), a cycle both taking a significant amount of time and making a non-efficient use of the resources.

Another interesting observation is that, to achieve optimal performances with respect to time, it is sufficient to manually produce only two portions of concentrated brine for the first two batches. The following ones will re-use the brine produced by the previous ones and stored in B1. This is because the operations involving container B5, in particular the heating but also the re-fillings into and from B5, take long enough for the preparation of the next batch in B4 out of the already produced brine. This is a robust feature of the plant. Actually, we have experimented with an imaginary stronger heating system performing the evaporation in half the time with the same result. Evidently, the period in the optimal continuous operation of the plant is the sum of the times of the operations involving B5. Hence, changes to the hardware affecting other parts of the system, such as a hose recently incorporated into B4 to speed up the refilling from B3 to B4 only speed up the production of the first batch. In contrast, any improvement on the three operations involving B5 will directly speed up the multi-batch production.

Concerning the performances of the tool, Figure 6 shows the measured computation times and memory requirements of OpenKronos for a growing number of batches. Intermediate curves between the discrete values give an impression of the growth of requirements relatively to the search depth (i.e., number of batches). These results were obtained with a 200MHz PentiumPro linux system with 500MB of main memory. With a growing number of batches (and thus a growing search depth) the memory and time consumption rises, but the size of the BDD representing the set of reached states stops growing after about 6 batches and then constantly requires about 200MB of storage. The BDD of the whole reachable state space has about 400K nodes representing about 16 billion of states. For more than 6 batches the execution time nevertheless rises linearly with the number of batches required. For an experiment with 32 batches the calculation took about 9 hours.

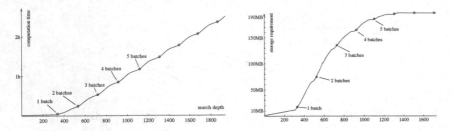

Fig. 6. Time and space performances of OPENKRONOS.

6   Integration into the Control Architecture of the Plant

The last step towards an end-to-end deployment of the proposed methodology consists in being able to fed the computed optimal operation scheme into the control software.

The high level control is realised by a *dispatcher*, which is in charge of triggering process operations. Ideally, the dispatcher would start the process operations at the starting times figuring in the Gantt-diagram and the operations would finish at the given times. However, the times in the model are estimations and not neccessarily exact. In short, the dispatcher can control the beginning of an operation but not its end!

Therefore, the dispatcher uses the operation scheme – the causal order – without any timing information and starts operations following the strategy *as soon as possible*:

- Initially, the dispatcher invokes the process operations without predecessor in the partial order.
- Afterwards it is called by events that indicate the termination of a process operation. Then the dispatcher removes that process operation from the partial order and triggers all those operations, which have no predecessor now.
- When the partial order is empty, the dispatcher signals the completed processing of the operation scheme.

The important properties of this strategy are:

- Under the assumption that reality follows exactly the model in operation times, the operation scheme is actually optimal.
- Under the assumption that the operation times given were safe upper bounds, also the whole process as controlled by the dispatcher satisfies the global time bound.
- Even if an operation should take longer than expected, the process can continue. It will just wait and on the whole possibly take longer than expected.

Of course, there are systems where this simple strategy does not always work. For instance, upper bounds on some operations may require to start a certain process operation with delay.

The control architecture of the plant is depicted in Figure 1. The plant is equipped with a Siemens Step 7 control system consisting of an S7 PLC which is connected to a Windows PC. The PC runs Step 7 for development of PLC programs and WinCC, a PC based component in the control that serves for visualization, data acquisition, as well as *high-level control*. WinCC communicates with the PLC over a communication line and is programmed to communicate with the PLC software via events concerning variables on the PLC. Process operations are implemented as PLC control programs that can be invoked from WinCC, either by manual user interaction (clicking on a button) or by WinCC routines written in C. The PLC routines in turn communicate to WinCC when they are finished.

The operation scheme is fed to the dispatcher, which is realized as a C-callback function of the WinCC runtime system. On the whole, the operation is performed as follows.

- From *manual mode*, the operator requests automatic operation from the GUI by pushing a corresponding button and chosing the desired operation scheme via a file dialogue. The system then starts the dispatcher and goes to *automatic mode*.
- When the operation scheme is finished, the dispatcher returns control to *manual mode*.

Safety requirements are met by allowing the operator to interrupt the dispatcher and return to manual mode. Obviously, this action requires manually returning the system to a consistent state.

7   Concluding Remarks

We have proposed an approach to automatically generate high-level control code for multi-batch processing using timed automata and their associated reachability analysis tools. Certainly, timed automata and model-checking algorithms can also be used to verify the correctness of low-level control programs (PLC routines) as it is done for the Dortmund plant in [8,6].

We believe that the experimental results obtained with the case study are encouraging. They illustrate the applicability of the approach in practice. The "off the shelf" tool (OPENKRONOS) we used was chosen for mainly two reasons: (1) immediate availability and (2) natural support of the modeling framework. Of course, the methodology does not depend on this choice. Indeed, there may be specialized algorithms, which eventually perform better on the particular class of models used. For instance, since it is usually possible to predict a bound on the size of the operation scheme (in total time as well as number of operations), it is possible to state the problem as an NP problem and then to apply a SAT solver (e.g., [12]).

We are currently working on the integration of our methodology into the Dortmund plant in the context of the European Project ESPRIT-LTR VHS [13].

Acknowledgements

Thanks go to Stefan Kowalewski, Nanette Bauer and André Deparade for their patient explanations on the Dortmund plant as well as the problem we finally approached; to Angelika Mader for discussions on the structure of the control program for the plant; to Thao Dang for general discussions about verification tasks for this plant; to Marius Bozga for helping with the use of the BDD-based verification tool; and to Oded Maler for many discussions about numerous related topics.

References

1. R. Alur, C. Courcoubetis, N. Halbwachs, T. Henzinger, P. Ho, X. Nicollin, A. Olivero, J. Sifakis, and S. Yovine. The algorithmic analysis of hybrid systems. *Theoretical Computer Science*, 138:3–34, 1995.
2. R. Alur and D. L. Dill. A theory of timed automata. *Theoretical Computer Science*, 126(2):183–235, April 1994.
3. K. Altisen, G. Goessler, A. Pnueli, J. Sifakis, S. Tripakis, and S .Yovine. A framework for scheduler synthesis. In Proc. RTSS'99, Phoenix, AZ, USA, December 1999. IEEE Computer Society Press.
4. E. Asarin and O. Maler. As soon as possible: Time optimal control for timed automata. In F. Vaandrager and J. van Schuppen, editors, *Hybrid Systems: Computation and Control*, volume 1569 of *LNCS*, pages 19–30. Springer, Mars 1999.
5. M. Bozga, C. Daws, O. Maler, A. Olivero, S. Tripakis, and S. Yovine. KRONOS: a model-checking tool for real-time systems. In *CAV*, LNCS, 1998.
6. R. Huuck. Verifying Timing Aspects of VHS Case Study 1. Tech. Report. University of Kiel, May 1999.
7. ISA S88.01 Batch Control Part I, Models and Terminology. ANSI/ISA, 1995.
8. K. Kristoffersen, K. Larsen, P. Pettersson, and C. Weise. Experimental Batch Plant - VHS Case Study 1 using Timed Automata and UPPAAL. Tech. Report, BRICS, Denmark, May 1999.
9. K.L. McMillan. *Symbolic Model-Checking: an Approach to the State-Explosion problem*, Kluwer, 1993.
10. A. Pnueli E. Asarin, O. Maler and J. Sifakis. Controller synthesis for timed automata. In *Proc. System Structure and Control*, pages 469–474. IFAC, Elsevier, July 1998.
11. S. Kowalewski. Description of VHS case study 1 "Experimental Batch Plant". Draft. University of Dortmund, Germany, July 1998.
12. M. Säflund and G. Stalmarck. Modeling and verifying systems and software in propositional logic. In *SAFECOMP'90*, pages 31–36, 1990.
13. European Project ESPRIT-LTR VHS. http://www-verimag.imag.fr/VHS.
14. W. M. Wonham and P. J. Ramadge. On the supremal controllable sublanguage of a given language. *SIAM J. Control and Optimization*, 25(3):637–659, May 1987.

Hybrid Systems Verification
by Location Elimination

Andreas Nonnengart

Deutsches Forschungszentrum für Künstliche Intelligenz (DFKI GmbH)
Im Stadtwald, Geb. 36, 66123 Saarbrücken, Germany
http://www.dfki.de/~nonnenga/
Andreas.Nonnengart@dfki.de

Abstract. In this paper we propose a verification method for hybrid systems that is based on a successive elimination of the various system locations involved. Briefly, with each such elimination we compute a weakest precondition (strongest postcondition) on the predecessor (successor) locations such that the property to be proved cannot be violated. Experiments show that this approach is particularly interesting in cases where a standard reachability analysis would require to travel often through some of the given system locations.

1 Introduction

Hybrid Systems are real-time systems that are embedded in analog environments. They contain discrete and continuous components and interact with the physical world through sensors and actuators. A common model for hybrid systems can be found in hybrid automata. These are finite graphs whose nodes correspond to global states as illustrated in the famous "Leaking Gas Burner" example [ACH+95]:

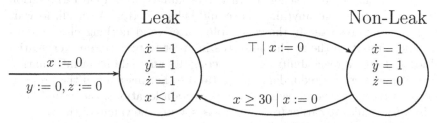

Nodes "Leak" and "Non-Leak" represent discrete locations, whereas x, y, and z are data variables. Each location may contain a location invariant ($x \leq 1$ in the example) and the continuous activity which describes how the values of the data variables change in time. In the above example the value of x and y increase by 1 per time unit (say, second), i.e., the first derivative of the function describing the behavior of x and y over time is the constant 1. z also increases by one per second in location "Leak", however, it remains unchanged ($\dot{z} = 0$) in location "Non-Leak". Edges are annotated with guards and discrete actions. Guards form

N. Lynch and B. Krogh (Eds.): HSCC 2000, LNCS 1790, pp. 352–365, 2000.

a constraint on the data variables to hold if a transition via the corresponding edge is to be performed. The discrete action specifies how the data variables are to be changed after taking the transition. In the above example the guard of the edge from "Leak" to "Non-Leak" is logical truth ($\top$), i.e., no special condition has to be fulfilled, whereas the guard of the edge from "Non-Leak" to "Leak" is $x \geq 30$, i.e., this edge may only be taken in case the value of the data variable x is at least 30. The discrete action for both edges is to reset x to 0.

A *computation* of such an automaton is a sequence of state changes (steps). Within each step the system state evolves continuously according to a dynamical law until a transition from one node to another one occurs.

Since hybrid systems typically operate in safety-critical situations, the development of rigorous analysis techniques is of high importance. In the last ten years several proposals for a verification methodology for hybrid systems arose [ACD90, ACH+95, ACHH93, AD94, AH92, AHH96, AHS96, ANKS95, GNRR93, Hen96, HNSY92]. Most of them are based on a so-called (forward or backward) reachability analysis. Intuitively, a forward reachability analysis (for safety properties) performs the following operation: starting from the initial situation (state), all possible (immediate) time and edge successor states are computed. Then the resulting set is reentered as an input to compute further time and edge successors, and so on. This will go on until no further new states can be derived (reached). Provided this procedure at all terminates, it ultimately comes up with the set of all reachable states (reachable from the initial state) which may be used to check the property to be proved.

Backward reachability, on the other hand, starts from a description of the states that do *not* fulfill the property to be proved and tries to compute all possible predecessor states, i.e., all the states from which one of the unsafe states could possibly be reached. Again, upon termination, it ends up with a set of states all of whose elements may lead to an unsafe situation, and what remains to be done is to check whether the initial state is contained in this set or not.

At the first glance, at least for forward reachability, and upon termination, it can hardly be seen that anything else could behave better. After all, forward reachability computes exactly the reachable states (and nothing else) and we need to know about all the reachable states in order to perform our verification task. Indeed, forward reachability does not compute any redundant information. However, it may perform redundant computations. For instance, if the property to be proven requires several passes through certain locations, the actual effect is usually very similar to earlier (or later) passes. Only the values of the variables involved might vary a bit, although in some more or less regular way.

The purpose of the approach proposed in this paper is to show how to gain such a knowledge and how to take advantage of it, i.e., to compute the behavior of locations once and for all and to forget about these locations later on. This can result in certain extra properties to be proved for the other locations (weakest preconditions or strongest postconditions) that take over the responsibility of the location just eliminated. The following example illustrates this:

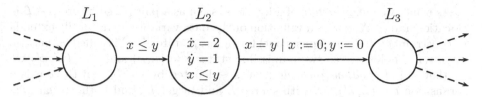

Let us assume that we want to prove that $x+y \leq 10$ is an invariant of the system (from which we can only see a small portion in this illustration). It might very well be that any reachability analysis will have to go through location L_2 several times and therefore we want to eliminate this very location by computing a weakest precondition on location L_1 that guarantees that L_2 could impossibly violate the desired property. The approach to be presented in this paper will come up with the following result: the invariant indeed holds for the whole system if and only if the invariant holds for the "simplified" system

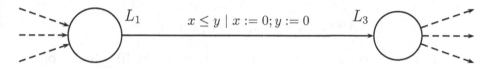

provided we can guarantee that $x \leq y \rightarrow 2y \leq x + 5$ whenever we are within location L_1. In what follows we describe formally how such a condition can be computed and how this method can be used as a verification tool for hybrid systems.

2   Preliminaries

Given a fixed variable set X we define the set CT of *Constraint Terms* (over the variable set X) as the smallest set containing the reals and the set X and that is closed under addition, subtraction and multiplication with reals. The set CF of *Constraint Formulas* (over the variable set X) is defined as the smallest set containing $\top$, $\bot$ (logical truth and falsity respectively), $t_1 > t_2$, $t_1 \geq t_2$, $t_1 < t_2$, $t_1 \leq t_2$, and $t_1 = t_2$ (where t_1 and t_2 are constraint terms), and that is closed under logical conjunction.

A hybrid system is a tuple of the form $\mathcal{H} = (X, \mathcal{L}, \mathcal{E}, dif, inv, guard, act)$ where X is a finite set of real-valued *data variables*, $\mathcal{L}$ is a finite set of *locations*, i.e., nodes of a graph, $\mathcal{E} \subseteq \mathcal{L} \times \mathcal{L}$ is a finite (multi)set of *transitions*, i.e., edges of the graph with nodes from $\mathcal{L}$, *dif*: $\mathcal{L} \times X \mapsto$ CT is a mapping that associates with each location and each data variable a constraint term (over X), *inv*: $\mathcal{L} \mapsto$ CF is a mapping that associates with each location a constraint formula, representing the location invariant, *guard*: $\mathcal{E} \mapsto$ CF is a mapping that associates with each edge a constraint formula, representing the enabling condition for this transition, and, *act*: $\mathcal{E} \times X \mapsto$ CT is a mapping that associates with each edge and each data variable a constraint term, representing the value of the variable after traveling along the edge.

As usual, we define a *state* of a hybrid system as a pair (L, ϕ) where $L \in \mathcal{L}$ is a location and $\phi\colon X \mapsto \mathbb{R}$ is a valuation of the data variables. ϕ naturally extends to (constraint) terms and (constraint) formulas. A state (L, ϕ) is called *admissible* if $\phi(inv(L))$ holds. Given two admissible states $s = (L, \phi)$ and $s' = (L', \phi')$ we say that s' is *transition-reachable* from s – denoted by $s \overset{tr}{\mapsto} s'$ – if there exists a transition $t = (L, L') \in \mathcal{E}$ with source L and target L', and both $\phi(guard(t))$ and $\phi'(x) = \phi(act(t, x))$ for each $x \in X$. We call s' *timely-reachable* from s with delay δ – denoted by $s \overset{\delta}{\mapsto} s'$ – where δ is a non-negative real number, if $L = L'$ and for each $x \in X$ there exists a differentiable function $f_x\colon [0, \delta] \mapsto \mathbb{R}$, with the first derivative $\dot{f}_x\colon (0, \delta) \mapsto \mathbb{R}$, such that (1) $f_x(0) = \phi(x)$ and $f_x(\delta) = \phi'(x)$ and (2) for all $\epsilon \in \mathbb{R}$ with $0 < \epsilon < \delta$: both $inv(L)[x_1/f_{x_1}(\epsilon), \ldots, x_n/f_{x_n}(\epsilon)]$ and $\dot{f}_x(\epsilon) = dif(L, x)[x_1/f_{x_1}(\epsilon), \ldots, x_n/f_{x_n}(\epsilon)]$ are true. s' is *timely-reachable* from s – denoted by $s \overset{*}{\mapsto} s'$ – if there exists a non-negative $\delta \in \mathbb{R}$ such that $s \overset{\delta}{\mapsto} s'$. s' is said to be *reachable* from s if $(s, s') \in (\overset{*}{\mapsto} \cup \overset{tr}{\mapsto})^*$. A *run* ρ of $\mathcal{H}$ with initial state $\sigma_0 = (L_0, \phi_0)$ is a maximal sequence of states represented as

$$\rho = \sigma_0 \overset{t_0}{\underset{f_0}{\mapsto}} \sigma_1 \overset{t_1}{\underset{f_1}{\mapsto}} \sigma_2 \overset{t_2}{\underset{f_2}{\mapsto}} \sigma_3 \overset{t_3}{\underset{f_3}{\mapsto}} \cdots$$

where $t_i \in \mathbb{R}^{\geq 0}$ and $f_i\colon [0, t_i] \mapsto (X \mapsto \mathbb{R})$, such that (i) $f_i(0) = \phi_i$, (ii) $inv(L_i)[X/f_i(t)(X)]$ holds for all $0 \leq t \leq t_i$, (iii) $(L_i, f_i(t_i)) \overset{tr}{\mapsto} \sigma_{i+1}$ and (iv) for all $0 \leq t' \leq t' + \delta \leq t_i : (L_i, f_i(t')) \overset{\delta}{\mapsto} (L_i, f_i(t' + \delta))$. The set of states contained in such a run ρ is given as $\{(L_i, f_i(t)) \mid t \in \mathbb{R}, 0 \leq t \leq t_i\}$. The set of all runs of a hybrid system $\mathcal{H}$ with initial state σ is denoted by $runs(\mathcal{H}, \sigma)$. A *position* π of a run $\rho = \sigma_0 \overset{t_0}{\underset{f_0}{\mapsto}} \sigma_1 \overset{t_1}{\underset{f_1}{\mapsto}} \sigma_2 \overset{t_2}{\underset{f_2}{\mapsto}} \sigma_3 \overset{t_3}{\underset{f_3}{\mapsto}} \cdots$ is a pair $\pi = (i, r) \in \mathbb{N} \times \mathbb{R}$ such that $0 \leq r \leq t_i$. Positions are ordered lexicographically, i.e., $(i, r) < (j, s)$ if and only if $i < j$ or $(i = j$ and $r < s)$. Also, $(i, r) \leq (j, s)$ if and only if $(i, r) < (j, s)$ or $(i = j$ and $r = s)$. By $\rho(\pi)$ with $\pi = (i, r)$ we denote the state $(L_i, f_i(r))$.

A run is said to be *non-zeno* if $\sum t_i$ diverges. In the sequel we assume that the runs of the hybrid system under consideration are all non-zeno.[1]

In order to formulate properties of hybrid systems we consider (a fragment of) ICTL, the Integrator Computation Tree Logic [AHH96]. For simplicity we omit the Until-operators in this paper. Their introduction does not cause much more effort, though (see [Non99]).

Given some hybrid system with locations $\mathcal{L}$ and data variables X, the set of ICTL formulas is defined as the smallest set containing all constraint formulas from CF over X, all location names from $\mathcal{L}$, and that is closed under the usual boolean connectives together with temporal (ICTL) operators AG, EF, EG, and AF. Moreover, if Φ is an ICTL formula, z is a new data variable, and $\{L_1, \ldots, L_n\} \in \mathcal{L}$ then $z^{\{L_1, \ldots, L_n\}}.\Phi$ is an ICTL formula as well. Intuitively, the

---

[1] The assumption of non-zenoness implies that hybrid systems are deadlock-free, i.e., there is no reachable state that has no successor. So-called livelocks, however, are not excluded. This means that we absolutely allow states which have only themselves as future states. The latter case just states that the situation does not change in time, whereas the former case (deadlock) would claim that time itself has come to an end.

temporal operators $AG\,\Phi$, $AF\,\Phi$, $EG\,\Phi$, $EF\,\Phi$ mean "always", "inevitably", "possibly always", and "possibly" respectively. Their formal semantics with respect to hybrid systems is defined below.

Definition 1. *Given a hybrid system* $\mathcal{H} = (X, \mathcal{L}, \mathcal{E}, \mathrm{dif}, \mathrm{inv}, \mathrm{guard}, \mathrm{act})$ *and a state* $\sigma = (L, \phi)$, *the semantics of ICTL with respect to* $\mathcal{H}$ *and* σ *is defined as:*

$\mathcal{H}, \sigma \models c$ *iff* $\models \phi(c)$ *for constraint formula* c

$\mathcal{H}, \sigma \models N$ *iff locations* N *and* L *are identical*

$\mathcal{H}, \sigma \models \neg\Phi$ *iff* $\mathcal{H}, \sigma \not\models \Phi$

$\mathcal{H}, \sigma \models \Phi \wedge \Psi$ *iff* $\mathcal{H}, \sigma \models \Phi$ & $\mathcal{H}, \sigma \models \Psi$
                 and similarly for the other boolean connectives

$\mathcal{H}, \sigma \models AG\,\Phi$ *iff* $\forall\rho\,(\rho \in \mathrm{runs}(\mathcal{H}, \sigma) \Rightarrow \forall\pi\,(\pi \in \mathrm{positions}(\rho) \Rightarrow \mathcal{H}, \rho(\pi) \models \Phi))$

$\mathcal{H}, \sigma \models EG\,\Phi$ *iff* $\exists\rho\,(\rho \in \mathrm{runs}(\mathcal{H}, \sigma)$ & $\forall\pi\,(\pi \in \mathrm{positions}(\rho) \Rightarrow \mathcal{H}, \rho(\pi) \models \Phi))$

$\mathcal{H}, \sigma \models z^{\mathcal{N}}.\Phi$ *iff* $\mathcal{H}^{z^{\mathcal{N}}}, (L, \phi[z/0]) \models \Phi$,   *where* $\mathcal{N} \subseteq \mathcal{L}$,

where $EF\,\Phi \equiv \neg AG\,\neg\Phi$ *and* $AF\,\Phi \equiv \neg EG\,\neg\Phi$. *By* $\mathcal{H}^{z^{\{L_1,\ldots,L_n\}}}$ *we mean the extended system we obtain by adding the new clock* z *(initialized with 0) which is supposed to run with slope 1 within locations* $L_1, \ldots, L_n$ *and with slope 0, i.e., it is stopped, for all other locations.*

Given a variable valuation ϕ *we define the new valuation* $\phi[z/0]$ *as*

$$\phi[z/0](x) = \begin{cases} \phi(x) & \text{if } x \neq z \\ 0 & \text{otherwise.} \end{cases}$$

3 The Verification Approach

Here we restrict our view to *linear* hybrid systems, where $\mathrm{dif}(L, x)$ is a constant, say k_L^x, for each location L and data variable x. This restriction can easily be weakened to *rectangular* hybrid systems (where $\mathrm{dif}(L, x)$ is given as an interval of reals) without any real effort. For a better readability we denote sequences of the form $x_1 + k_L^{x_1}\delta, \ldots, x_n + k_L^{x_n}\delta$ by $X + k_L^X \delta$ where $X = \{x_1, \ldots, x_n\}$, and, similarly we mean $L(\mathrm{act}(T, x_1), \ldots, \mathrm{act}(T, x_n))$ whenever we write $L(\mathrm{act}(T, X))$.

3.1 First-Order Theories for Reachability and Inevitability

As usual, an *interpretation* $\Im = (\mathcal{D}, \Im_{\mathcal{L}}, \phi)$ for a first-order theory associated with a hybrid system $\mathcal{H}$ with locations $\mathcal{L}$ has a fixed domain $\mathcal{D}$ (the reals or the rationals, say), a valuation ϕ for the data variables in X, and a meaning function $\Im_{\mathcal{L}}$ for the locations in $\mathcal{L}$ such that $\Im_{\mathcal{L}}(L) \in \mathcal{D}^n$, where n is the number of data variables in X. A *model* of a formula Φ is an interpretation satisfying this formula.

We often also speak of a model as a set of ground atoms of the form

$$\{L(\Im(t_1), \ldots, \Im(t_n)) \mid \Im \models L(t_1, \ldots, t_n)\}, \text{where } t_i \text{ are constraint terms}$$

where $\Im$ is a model in the above sense. Interpretations (models) are partially ordered by set-inclusion. A *minimal model* of Φ is a model of Φ such that there exists no proper subset of it that also satisfies Φ.

We now define two different kind of first-order theories for a given hybrid system: one that is responsible for the possible states, and one that is responsible for the unavoidable states.

Definition 2 (Reachability Theory). *Let* $\mathcal{H} = (X, \mathcal{L}, \mathcal{E}, \text{dif}, \text{inv}, \text{guard}, \text{act})$ *be a hybrid system. For each* $L \in \mathcal{L}$ *we define the first-order theory*

$$\forall X \; L(X) \rightarrow \begin{cases} \text{inv}(L) \; \wedge \\ \forall \delta \; (\delta \geq 0 \wedge \text{inv}(L)[X/X + k_L^X \delta] \rightarrow L(X + k_L^X \delta)) \; \wedge \\ \bigwedge_{T=(L,N)\in\mathcal{E}} \text{guard}(T) \rightarrow N(\text{act}(T, X)) \end{cases}$$

as the local reachability theory *of* L *in* $\mathcal{H}$, $\mathcal{R}_\mathcal{H}^L$ *for short. By the* reachability theory *of* $\mathcal{H}$ – *which we call* $\mathcal{R}_\mathcal{H}$, *or simply* $\mathcal{R}$ *if* $\mathcal{H}$ *is clear from the context* – *we understand the conjunction of all local reachability theories, i.e.,* $\mathcal{R}_\mathcal{H} = \bigwedge_{L \in \mathcal{L}} \mathcal{R}_\mathcal{H}^L$.

What the reachability theory expresses is that (for each location L) (i) the location invariant must hold, (ii) that there is a possible time transition, and (iii) for each outgoing edge: if the enabling guard is true then the target location can be reached provided the corresponding discrete action has been peformed.

Definition 3 (Inevitability Theory). *Let* $\mathcal{H} = (X, \mathcal{L}, \mathcal{E}, \text{dif}, \text{inv}, \text{guard}, \text{act})$ *be a hybrid system. For each* $L \in \mathcal{L}$ *we define the first-order theory*

$$\forall X \; L(X) \rightarrow \text{inv}(L)$$

$$\forall X \; L(X) \rightarrow \begin{cases} \forall \delta \; \delta \geq 0 \rightarrow L(X + k_L^X \delta) \; \vee \\ \exists \delta \begin{cases} \delta \geq 0 \; \wedge \\ \forall \delta' \; 0 \leq \delta' \leq \delta \rightarrow L(X + k_L^X \delta') \; \wedge \\ \bigvee_{T=(L,N)\in\mathcal{E}} \begin{array}{l} \text{guard}(T)[X/X + k_L^X \delta] \wedge \\ N(\text{act}(T, X)[X/X + k_L^X \delta]) \end{array} \end{cases} \end{cases}$$

as the local inevitability theory *of* L *in* $\mathcal{H}$, $\mathcal{I}_\mathcal{H}^L$ *for short. By the* inevitability theory *of* $\mathcal{H}$ – *which we call* $\mathcal{I}_\mathcal{H}$, *or simply* $\mathcal{I}$ *if* $\mathcal{H}$ *is clear from the context* – *we understand the conjunction of all local inevitability theories, i.e.,* $\mathcal{I}_\mathcal{H} = \bigwedge_{L \in \mathcal{L}} \mathcal{I}_\mathcal{H}^L$.

The inevitability theory might require some more explanation. In a sense it expresses (for each given state) between which possibilities the system can choose. The first part of any local inevitability theory is trivial. It just guarantees the mere fact that for each location predicate the corresponding location invariant is supposed to hold. The second part is more complicated and more interesting.

Note that, given an arbitrary state represented by the location predicate $L(X)$, either the system remains forever in this location, i.e., $\forall \delta\ \delta \geq 0 \rightarrow L(X + k_L^X \delta)$, or it will sooner or later leave this very location. In the latter case we know that there is a time delay δ after which one of the guards of the outgoing edges is true and until then the system remains within location L. This is exactly what is expressed by the complicated second part of the local inevitability theories.

Intuitively, the reachability theory tells us what *can* be done in certain situations (states), whereas the inevitability theory describes what *must* be done, it collects all the immediate future possibilities.

The importance of these two theories will become apparent from the following Lemma.

Lemma 1. *Given a hybrid system $\mathcal{H}$ and an initial state (L, ϕ).*

- *The (unique) minimal model of $L(\phi(X)) \wedge \mathcal{R}_{\mathcal{H}}$ corresponds to the set of states that are reachable from (L, ϕ) in the hybrid system $\mathcal{H}$.*
- *Each minimal model of $L(\phi(X)) \wedge \mathcal{I}_{\mathcal{H}}$ corresponds to the set of states of one of the runs of $\mathcal{H}$.*
- *The set of states of each run of $\mathcal{H}$ forms a model of $L(\phi(X)) \wedge \mathcal{I}_{\mathcal{H}}$.*

Proof. Can be found in [Non99].

The above lemma provides us with a formal connection between the reachability theory (inevitability theory) and the reachable (inevitable) states. Briefly, Φ holds always $(AG\ \Phi)$ iff Φ holds in all reachable states iff Φ holds for every element in the unique minimal model of the reachability theory (together with the initial state) iff Φ holds for every element of some model of the reachability theory (together with the initial state). This observation leads to the following definition and main theorem.

Definition 4 (Characteristic Constraint Formulas). *The (second-order) formula associated with an ICTL formula Φ, the hybrid system $\mathcal{H}$, and location L, $\lceil \Phi \rceil_{\mathcal{H}}^{L(X)}$, representing a characteristic constraint formula for Φ given $\mathcal{H}$ in L, is recursively defined by*

$$\lceil c \rceil_{\mathcal{H}}^{L(X)} = c$$

$$\lceil L' \rceil_{\mathcal{H}}^{L(X)} = \begin{cases} \top & \text{if } L \text{ and } L' \text{ are identical} \\ \bot & \text{otherwise} \end{cases}$$

$$\lceil \neg \Phi \rceil_{\mathcal{H}}^{L(X)} = \neg \lceil \Phi \rceil_{\mathcal{H}}^{L(X)}$$

$$\lceil \Phi \wedge \Psi \rceil_{\mathcal{H}}^{L(X)} = \lceil \Phi \rceil_{\mathcal{H}}^{L(X)} \wedge \lceil \Psi \rceil_{\mathcal{H}}^{L(X)}$$

and similarly for the other boolean connectives

$$\lceil z^{\mathcal{N}}.\Phi \rceil_{\mathcal{H}}^{L(X)} = \lceil \Phi \rceil_{\mathcal{H}:\mathcal{N}}^{L(X,0)} \text{ where } \mathcal{N} \subseteq \mathcal{L}$$

$$\lceil AG\ \Phi \rceil_{\mathcal{H}}^{L(X)} = \exists L_1, \dots, L_n\ L(X) \wedge \mathcal{R}_{\mathcal{H}} \wedge \bigwedge_{N \in \mathcal{L}} \forall X\ N(X) \rightarrow \lceil \Phi \rceil_{\mathcal{H}}^{N(X)}$$

$$\lceil EG\ \Phi \rceil_{\mathcal{H}}^{L(X)} = \exists L_1, \dots, L_n\ L(X) \wedge \mathcal{I}_{\mathcal{H}} \wedge \bigwedge_{N \in \mathcal{L}} \forall X\ N(X) \rightarrow \lceil \Phi \rceil_{\mathcal{H}}^{N(X)}$$

Theorem 1. *Given a hybrid system $\mathcal{H}$ with data variables X, an initial state (L, ϕ) and an ICTL formula Φ. Then*

$$\mathcal{H}, (L, \phi) \models \Phi \qquad \text{iff} \qquad \models \phi \left(\lceil \Phi \rceil_{\mathcal{H}}^{L(X)} \right)$$

Proof. By induction on the structure of Φ.
The base cases are trivial. Also in case of a boolean connective there are no problems at all. The induction steps are exemplified for the case $\Phi = AG\,\Psi$. For the other cases see [Non99].

$\mathcal{H}, (L, \phi) \models AG\,\Psi$

   iff $\mathcal{H}, \sigma \models \Psi$ for every σ reachable from (L, ϕ)

   iff $\forall \sigma \; ((L, \phi), \sigma) \in (\overset{*}{\mapsto} \cup \overset{\text{tr}}{\mapsto})^* \Rightarrow \mathcal{H}, \sigma \models \Psi$

   iff $\forall N, \phi' \; N(\phi'(X)) \in minMod(L(\phi(X)) \wedge \mathcal{R}_{\mathcal{H}}) \Rightarrow \mathcal{H}, (N, \phi') \models \Psi$
        (Lemma 1)

   iff $\exists \mathfrak{S} \; \mathfrak{S} \models L(\phi(X)) \wedge \mathcal{R}_{\mathcal{H}} \; \& \; \forall N, \phi' \; (N, \phi') \in \mathfrak{S} \Rightarrow \mathcal{H}, (N, \phi') \models \Psi$

   iff $\exists \mathfrak{S} \; \mathfrak{S} \models L(\phi(X)) \wedge \mathcal{R}_{\mathcal{H}} \; \& \; \forall N, \phi' \; (N, \phi') \in \mathfrak{S} \Rightarrow \models \phi' \left(\lceil \Psi \rceil_{\mathcal{H}}^{N(X)} \right)$
        (induction hypothesis)

   iff $\exists \mathfrak{S} \; \mathfrak{S} \models L(\phi(X)) \wedge \mathcal{R}_{\mathcal{H}} \; \& \; \mathfrak{S} \models \bigwedge_{N \in \mathcal{L}} \forall X \; N(X) \rightarrow \lceil \Psi \rceil_{\mathcal{H}}^{N(X)}$

   iff $\exists \mathfrak{S} \; \mathfrak{S} \models L(\phi(X)) \wedge \mathcal{R}_{\mathcal{H}} \wedge \bigwedge_{N \in \mathcal{L}} \forall X \; N(X) \rightarrow \lceil \Psi \rceil_{\mathcal{H}}^{N(X)}$

   iff $\models \exists L_1, \dots, L_n \; L(\phi(X)) \wedge \mathcal{R}_{\mathcal{H}} \wedge \bigwedge_{N \in \mathcal{L}} \forall X \; N(X) \rightarrow \lceil \Psi \rceil_{\mathcal{H}}^{N(X)}$

   iff $\models \phi \left(\lceil AG\,\Psi \rceil_{\mathcal{H}}^{L(X)} \right)$

3.2 Eliminating Locations

Theorem 1 tells us that we can solve a hybrid system verification problem by proving the satisfiability of some suitable first-order theory, or equivalently, by showing the validity of some corresponding second-order formula. The *Elimination Theorem* below helps us in this respect, for it allows us to transform a given second-order formula into an equivalent first-order formula (if this is at all possible).[2]

---

[2] In general, second-order formulas do not necessarily have a (finite or infinite) first-order equivalent. However, in case the second-order formula is of the form $\exists P \; \Phi$, where Φ is a first-order formula that is Horn in P, then we know that there exists a first-order equivalent (which may be infinite, though). Note that proving Safety properties, i.e. proving the validity of second-order formulas involving reachability theories, are just of this form.

Theorem 2 (Elimination Theorem). *Let Φ and Ψ be two first-order formulas that are positive with respect to the predicate symbol L. Then*

$$\exists L \; \big(\forall \overline{x} \, (L(\overline{x}) \to \Phi) \wedge \Psi\big) \quad \equiv \quad \Psi \left[L(\overline{a}) / \, (\nu L(\overline{x}).\Phi)\frac{\overline{x}}{\overline{a}} \right]$$

where $\;\nu L(\overline{x}).\Phi(L) = \bigwedge_{i \leq \omega} \Phi^i(\top) \quad$ *with* $\quad \Phi^0(\top) = \top, \Phi^{n+1}(\top) = \Phi(\Phi^n(\top))$

The proof of this Theorem can be found in [NS95] (but also see [NS99] and [NOS99]). There also some generalizations and dual forms are examined. For the purpose of this paper, however, the above form suffices.

The Elimination Theorem tells us that any second-order formula of the form $\exists L \; \big(\forall \overline{x} \, (L(\overline{x}) \to \Phi) \wedge \Psi\big)$ is equivalent to Ψ with every occurrence of L (with actual argument list $\overline{a}$) within Ψ replaced by the greatest fixpoint of Φ (after instantiating the abstract parameters with the actual arguments). Notice that the second-order formulas we are dealing with are indeed of the form required. Therefore, with each application of the Elimination Theorem we get rid of one of the existentially quantified predicate symbols. Now, since these predicate symbols are just the location names of the hybrid system under consideration, each application of the Elimination Theorem also eliminates one of the locations.

Evidently, it cannot be guaranteed that the fixpoint computation will terminate in general. However, it can easily be shown (see [Non99]) that in case we are about to eliminate a location which has no outgoing edge leading to itself, the fixpoint computation will definitely terminate after two iterations.

Coming back to the example on page 353 where we wanted to examine the effect of eliminating location L_2, we now know that we have to compute – in fact, find a first-order equivalent for – the second-order formula

$$\exists L_2 \left[\begin{array}{l} \forall x, y \; L_1(x,y) \to x \leq y \to L_2(x,y) \; \wedge \\[4pt] \forall x, y \; L_2(x,y) \to \left\{ \begin{array}{l} x \leq y \; \wedge \\ x + y \leq 10 \; \wedge \\ \forall \delta \; (\delta \geq 0 \wedge x + 2\delta \leq y + \delta \to L_2(x+2\delta, y+\delta)) \; \wedge \\ x = y \to L_3(0,0) \end{array} \right. \end{array} \right\}$$

The five conjuncts of the above second-order formula describe the transition from L_1 to L_2, the location invariant for L_2, the property to be proved, the time transition for location L_2, and the edge transition from L_2 to L_3 respectively. In order to apply the Elimination Theorem to this second-order formula, let $\Psi = \forall x, y \; L_1(x,y) \to x \leq y \to L_2(x,y)$ and $\Phi = x \leq y \wedge x + y \leq 10 \wedge \forall \delta \; (\delta \geq 0 \wedge x + 2\delta \leq y + \delta \to L_2(x+2\delta, y+\delta)) \wedge x = y \to L_3(0,0)$. Then $\Phi^0(\top) = \top$, $\Phi^1(\top) = x \leq y \wedge x + y \leq 10 \wedge x = y \to L_3(0,0)$, $\Phi^2(\top) = \Phi^1(\top) \wedge x \leq y \to 2y \leq 5 + x$, and $\Phi^3(\top) = \Phi^2(\top)$ as can easily be checked by the reader. We thus have found the fixpoint and substitute it for L_2 in Ψ resulting in

$$\forall x, y \; L_1(x,y) \to x \leq y \to L_3(0,0)$$
$$\forall x, y \; L_1(x,y) \to x \leq y \to 2y \leq x + 5$$

The first formula describes just the new edge to be introduced. The second formula, on the other hand, tells us about the necessary and sufficient condition on the data variables for location L_1 such that it would be impossible to violate $x + y \leq 10$ in location L_2.

3.3   Examples and Experimental Results

There exists a prototype implementation of the Elimination Approach (for proving safety-properties) written in Sicstus-Prolog with the CLP(Q,R)-library for constraint handling. It has been tested on a lot of examples known from the literature (or taken from hybrid system verifier distributions). The experimental results can briefly be summarized as follows: as already expected in the introduction, standard forward reachability (if it at all terminates) can hardly be beaten in case we are about to prove safety properties for non-trivial systems that only require a single pass through the reachable locations. This is the case for instance for the famous "audio-protocol"-example.[3] For other, unfortunately still trivial systems like the "Leaking Gas Burner" or the "Billiards"-example, the Elimination Approach showed a slightly better behavior than standard reachability analysis.[4] However, in such cases, where safety properties can be proved in milliseconds anyway, this can hardly be called evidence. The lack of non-trivial hybrid system in the literature that require several passes through some of their locations made us compose our own examples. They are designed as simple as possible such that they may serve to illustrate the effect of the Elimination Approach compared to reachability analysis methods. Two such examples are given below.

A Silly Multiplier. This is an example within which three numbers a, b, and c are to be multiplied and the final product is stored in the data variable p. The multiplication is performed by successively adding 1 to p, similar to the nested for-loop

$$\text{for } (w = 0; w < c; w{+}{+})$$
$$\text{for } (v = 0; v < b; v{+}{+})$$
$$\text{for } (u = 0; u < a; u{+}{+})$$
$$p := p + 1$$

---

[3] In case of more trivial such examples like the "Water-Level-Monitor" or the "Railroad-Gate-Controller" there is not so much of a difference.

[4] It should be noted here that it is in fact very easy to compare the Elimination Approach with standard reachability analysis methods for, in a sense, reachability analysis can be viewed as a special case of the Elimination Approach: we just have to move the location names to the argument list as a further additional argument (leaving a single unique dummy predicate as the only remaining predicate symbol). This then leads to backward reachability, whereas, by using a dual form of the Elimination Theorem, we can also get forward reachability (see also [Non99]).

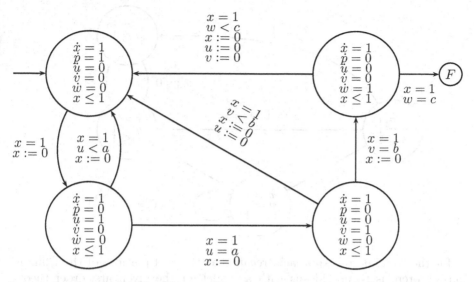

It is to be shown that the location F can be reached – after all, as soon as F is reached, the data variable p contains the multiplication result we are interested in. (Backward or forward) reachability analysis in a sense simulates the behavior of the multiplier. I.e., since this system is fully deterministic, it takes a walk through the whole computation. Evidently, this is very time consuming even if we only attempted to compute $10 \times 10 \times 10$. Now, compare this with the Elimination Approach. It takes two steps to eliminate the top left location, another $2a$ iterations for the bottom left location, some $2b$ iterations for the bottom right location, and finally, $2c$ iterations for the top right location. We then end up with a single location, namely F, for which the data variable p is initialized with the product $a \times b \times c$. Thus, even for big numbers a, b, and c the elimination approach is able to fulfill its task in a very short amount of time, whereas any kind of reachability analysis requires much more effort (essentially $a + b + c$ versus $a \times b \times c$).

An Example Where Reachability Fails. The particularity about the following example is that it contains an "impossible" transition, i.e., one of the locations – the one at the bottom – is unreachable because the guard $(y = 2)$ of the transition that may lead to this very location will never be enabled. In a sense, forward reachability analysis detects this, though rather indirectly, for it never tries to compute states involving this location. However, forward reachability nevertheless does not terminate, since the data variable y may become arbitrarily big and therefore the fixpoint computation finds new possible states in each iteration. At the first glance, backward reachability might have a better chance. Suppose we were about to prove that $x \leq y$ is an overall invariant of the system. If there were not the bottom location, backward reachability would have no problem to detect that the invariant indeed holds. But it this very location (or actually the transition that leads to itself) that prevents backward reachability from termination.

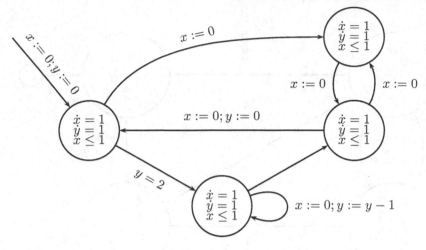

As for the Elimination Approach, recall that the real purpose of the Elimination Theorem is to provide us with a model for the given first-order theory. The Elimination Theorem is just one of the possible means we can imagine that may help us finding such a model. There are, however, cases for which we can (even by mere syntactic examinations) find out that a given first-order formula indeed has a (trivial) model. This is the case, for instance, if we reach a situation (maybe after having performed some eliminations) where every conjunct of the intermediate result contains a negative location predicate literal. In this case we can conclude that all locations that have not yet been eliminated are unreachable. This is in fact very common. The famous "Railroad-Gate-Controller" for example, has 22 (compound) locations. But only 7 of them are in fact reachable. Our implementation of the Elimination Approach first eliminates these seven reachable locations and then detects that all others are unreachable and terminates with success. The same happens for the example above: first the top three locations get eliminated and then the unreachability of the final (bottom) location is detected. Thus the system terminates with success.

4 Summary and Conclusion

We presented a hybrid systems verification methodology that is based on a successive location elimination procedure that allows us to simplify the system until it becomes trivial. This approach turns out to be particularly useful in case standard forward or backward reachability analysis methods would require to pass some of the locations several times (long lasting loops). There are even examples where a reachability analysis fails, whereas the Elimination Approach terminates successfully. However, there also are some non-trivial examples where forward reachability seems unbeatable, namely systems which contain quite a huge amount of locations, yet require merely a single pass in order to compute the set of reachable states. This observation leads to the conjecture that it might make sense to suitably combine the two methodologies: in case a non-trivial system (with many locations) contains long-lasting loops that prevent a reachability analysis from termination within some reasonable amount of time, it certainly makes sense to try and eliminate such loops with the help of the Elimination Approach and maybe to proceed with the reachability analysis after that.

References

[ACD90]   R. Alur, C. Courcoubetis, and D. L. Dill. Model checking for real-time systems. In *Proceedings of the 5th Annual Symposium on Logic in Computer Science*, pages 414–425, 1990.

[ACH+95] R. Alur, C. Courcoubetis, N. Halbwachs, T. A. Henzinger, P.-H. Ho, X. Nicollin, A. Olivero, J. Sifaksi, and S. Yovine. The algorithmic analysis of hybrid systems. *Theoretical Computer Science*, 138:3–34, 1995.

[ACHH93] R. Alur, C. Courcoubetis, T. A. Henzinger, and P.-H. Ho. Hybrid automata: An algorithmic approach to the specification and verification of hybrid systems. In R. L. Grossman, A. Nerode, A. P. Ravn, and H. Rischel, editors, *Hybrid Systems*, pages 209–229. Springer Verlag, Lecture Notes in Computer Science, vol. 736, 1993.

[AD94]   R. Alur and D. L. Dill. A theory of timed automata. *Theoretical Computer Science*, 126:183–235, 1994.

[AH92]   R. Alur and T. A. Henzinger. Logics and models of real-time: A survey. In J.W. de Bakker, K. Huizing, W.-P. de Roever, and G. Rozenberg, editors, *Real Time: Theory in Practice*, pages 74–106. Springer Verlag, New York, LNCS 600, 1992.

[AHH96]   Rajeev Alur, Thomas A. Henzinger, and Pei-Hsin Ho. Automatic symbolic verification of embedded systems. *IEEE Transactions on Software Engineering*, 22(3):181–201, 1996.

[AHS96]   R. Alur, T. A. Henzinger, and E. Sontag, editors. *Hybrid Systems III*. Lecture Notes in Computer Science, Springer Verlag, 1996.

[ANKS95] P. Antsaklis, A. Nerode, W. Kohn, and S. Sastry, editors. *Hybrid Systems II*. Lecture Notes in Computer Science, vol. 999, Springer Verlag, 1995.

[GNRR93] R. L. Grossman, A. Nerode, A. P. Ravn, and H. Rischel, editors. *Hybrid Systems*. Springer Verlag, Lecture Notes in Computer Science, vol. 736, 1993.

[Hen96]   T. A. Henzinger. The theory of hybrid automata. In *Proceedings of the 11th LICS*, pages 278–292. IEEE Comp. Soc. Press, 1996.

[HNSY92] T. A. Henzinger, X. Nicollin, J. Sifakis, and S. Yovine. Symbolic model checking for real-time systems. In *Proceedings of the 7th Annual Symposium on Logic in Computer Science*, pages 394–406. IEEE Computer Society Press, New York, 1992.

[Non99]   Andreas Nonnengart. A deductive model checking approach for hybrid systems. Technical Report MPI-I-1999-2-006, Max-Planck-Institute for Computer Science, Saarbrücken, Germany, November 1999. Available via http://www.mpi-sb.mpg.de/.

[NOS99]   Andreas Nonnengart, Hans Jürgen Ohlbach, and Andrzej Szałas. Elimination of predicate quantifiers. In Hans Jürgen Ohlbach and Uwe Reyle, editors, *Logic, Language and Reasoning – Essays in Honour of Dov Gabbay*. Kluwer, Dordrecht, Netherlands, 1999. ISBN: 0-7923-5687-X.

[NS95]   Andreas Nonnengart and Andrzej Szałas. A fixpoint approach to second-order quantifier elimination with applications to correspondence theory. Technical Report MPI-I-95-2-007, Max-Planck-Institute for Computer Science, Saarbrücken, Germany, March 1995. Available via: http://www.mpi-sb.mpg.de/.

[NS99]   Andreas Nonnengart and Andrzej Szalas. A fixpoint approach to second-order quantifier elimination with applications to correspondence theory. in: [Orł99], 1999.

[Orł99]   Ewa Orłowska, editor. *Logic at Work: Essays Dedicated to the Memory of Helena Rasiowa*, volume 24 of *Studies in Fuzziness and Soft Computing*. Physica-Verlag, c/o Springer Verlag, 1999. ISBN: 3-7908-1164-5.

A Dynamic Bayesian Network Approach to Tracking Using Learned Switching Dynamic Models

Vladimir Pavlović*, James M. Rehg, and Tat-Jen Cham

Compaq Computer Corporation
Cambridge Research Lab
Cambridge, MA 02139
{vladimir,rehg,tjc}@crl.dec.com

Abstract. Switching linear dynamic systems (SLDS) attempt to describe a complex nonlinear dynamic system with a succession of linear models indexed by a switching variable. Unfortunately, despite SLDS's simplicity exact state and parameter estimation are still intractable. Recently, a broad class of learning and inference algorithms for time-series models have been successfully cast in the framework of dynamic Bayesian networks (DBNs). This paper describes a novel DBN-based SLDS model. A key feature of our approach are two approximate inference techniques for overcoming the intractability of exact inference in SLDS. As an example, we apply our model to the human figure motion analysis. We present experimental results for learning figure dynamics from video data and show promising results for tracking, interpolation, synthesis, and classification using learned models.

1 Introduction

Many natural processes have complex, highly nonlinear and time-varying dynamics. For instance, economic trends, maneuvering targets, and the human figure all exhibit complex and rich dynamic behavior. Dynamics are essential to the analysis of these processes as well as to their realistic prediction (forecasting) and synthesis (simulation). Dynamic models can provide a powerful cue in the presence of missing/multiple measurements and measurement noise. A dynamic model imposes additional structure on the state space by specifying which state trajectories are possible (or probable) and by specifying the speed at which a trajectory evolves.

Unfortunately, state and parameter estimation problems in complex dynamic models can be a daunting task. State estimation in non-linear models is usually cast in frameworks whose origins lay in the theory of extended Kalman filters (c.f. [1]). Parameter estimation of such highly nonlinear models is often a result

---

* Contact author: Vladimir Pavlović, Compaq Computer Corp., Cambridge Research Lab, 1 Kendall Sq., Cambridge, MA 02139, phone (617) 551-7699, fax (617) 551-7650, e-mail vladimir@crl.dec.com

N. Lynch and B. Krogh (Eds.): HSCC 2000, LNCS 1790, pp. 366–380, 2000.

of tedious measurements and expert knowledge about the problem. For instance, consider the human figure modeling in the field of biomechanics. The dynamics of the figure are the result of its mass distribution, joint torques produced by the motor control system, and reaction forces resulting from contact with the environment (e.g. the floor). Research efforts in biomechanics, rehabilitation, and sports medicine have resulted in complex, specialized models of human motion (c.f. [11].) Such complex models have been used successfully to simulate [10] and to track human body motion [27].

This paper explores the alternative method of learning dynamic models from a training corpus of observed state space trajectories. In cases where sufficient training data is available, the learning approach promises flexibility and generality. A wide range of learning algorithms can be cast in the framework of dynamic Bayesian networks (DBNs) [7], a subclass of now famous Bayesian network models (c.f. [23, 13]). DBNs generalize two well-known signal modeling tools: Kalman filters [1] for continuous state linear dynamic systems (LDS) and Hidden Markov Models (HMMs) [24] for classification of discrete state sequences.

The DBN framework provides two distinct benefits: First, a broad variety of modeling schemes can be conceptualized in a single framework with an intuitively-appealing graphical notation (see Figure 1 for an example). Second, a broad corpus of exact and approximate statistical inference and learning techniques from the Bayesian network literature can be applied to dynamical systems. In particular, it has been shown that estimation in LDSs and inference in HMMs are special cases of inference in DBNs.

The focus of this paper is on a subclass of DBN models called Switching Linear Dynamic Systems [2, 26, 17, 9, 22]. Intuitively, these models attempt to describe a complex nonlinear dynamic system with a succession of linear models that are indexed by a switching variable. While other approaches such as learning weighted combinations of linear models are possible, the switching approach has an appealing simplicity and is naturally suited to the case where the dynamics are time-varying.

This paper makes two contributions. First, we derive two efficient algorithms for approximate state estimation in SLDSs. An approximate Viterbi inference algorithm and a structured variational inference algorithm are cast as in the framework of DBN inference. Second, we demonstrate the application of the SLDS framework to modeling the human figure dynamics. In particular, we demonstrate the learning of switching models of fronto-parallel walking and jogging motion from video data. We demonstrate the application of these learned models to segmentation, synthesis, and tracking tasks.

2   Switching Linear Dynamic System Model

Consider an SLDS described using the following set of continuous and discrete state-space equations:

$$x_{t+1} = A(s_{t+1})x_t + v_{t+1}(s_{t+1}), \quad y_t = Cx_t + w_t$$

for the continuous-valued linear dynamic system (LDS), and

$$Pr(s_{t+1}|s_t) = s'_{t+1}\Pi s_t$$

for the discrete switching model. We assumed that the LDS models a Gauss-Markov process with state noise $v_t(s_t) \sim \mathcal{N}(0, Q(s_t))$, measurement noise $w_t \sim \mathcal{N}(0, R)$, and initial state $x_0 \sim \mathcal{N}(x_0(s_0), Q_0(s_0))$. The switching model is assumed to be a discrete first order Markov process. State variables of this model are written as s_t. They belong to the set of S discrete symbols $\{e_0, \ldots, e_{S-1}\}$, where e_i is the unit vector of dimension S with a non-zero element in the i-th position. The switching model is defined with the state transition matrix Π whose elements are $\Pi(i,j) = Pr(s_{t+1} = e_i|s_t = e_j)$, and an initial state distribution π_0.

Coupling between the LDS and the switching process stems from the dependency of the LDS parameters A and Q on the switching process state s_t. Namely,

$$A(s_t = e_i) = A_i, \quad Q(s_t = e_i) = Q_i$$

In other words, switching state s_t determines which of S possible plant models is used at time t.

The complex state space representation is equivalently depicted by the DBN dependency graph in Figure 1. The dependency graph implies that the *joint*

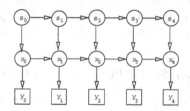

Fig. 1. Bayesian network representation (dependency graph) of an SLDS of duration five. s denote instances of the discrete valued switching states. x and y are continuous valued LDS states and measurements. Arcs in the graph show dependencies among variables.

distribution P over the variables of the SLDS can be written as

$$P(\mathcal{Y}_T, \mathcal{X}_T, \mathcal{S}_T) =$$
$$Pr(s_0) \prod_{t=1}^{T-1} Pr(s_t|s_{t-1}) Pr(x_0|s_0) \prod_{t=1}^{T-1} Pr(x_t|x_{t-1}, s_t) \prod_{t=0}^{T-1} Pr(y_t|x_t),$$

where $\mathcal{Y}_T, \mathcal{X}_T$, and $\mathcal{S}_T$ denote the sequences (of length T) of observations and hidden state variables. For instance, $\mathcal{Y}_T = \{y_0, \ldots, y_{T-1}\}$. From the Gauss-Markov assumption on the LDS (e.g. $x_{t+1}|x_t, s_{t+1}=e_i \sim \mathcal{N}(A_i x_t, Q_i)$) and recalling the Markov switching model assumption, the joint pdf of the SLDS of duration T can be easily defined.

2.1   Hidden State Inference and Estimation

The goal of inference in complex DBNs is to estimate the posterior probability of the hidden states of the system (s_t and x_t) given some known sequence of observations $\mathcal{Y}_T$ and the known model parameters. Namely, we need to find the posterior

$$P(\mathcal{X}_T, \mathcal{S}_T | \mathcal{Y}_T) = Pr(\mathcal{X}_T, \mathcal{S}_T | \mathcal{Y}_T),$$

or , equivalently, its *sufficient statistics*. Given the form of P it is easy to show that these are the first and the second order statistics: mean and covariance among hidden states $x_t, x_{t-1}, s_t, s_{t-1}$.

If there were no switching dynamics, the inference would be straightforward – we could infer $\mathcal{X}_T$ from $\mathcal{Y}_T$ using LDS inference (RTS smoothing [25]). However, the presence of switching dynamics embedded in matrix Π makes exact inference more complicated. To see that, assume that the initial distribution of x_0 at $t = 0$ is Gaussian, at $t = 1$ the pdf of the physical system state x_1 becomes a mixture of S Gaussian pdfs since we need to marginalize over S possible but unknown plant models. At time t we will have a mixture of St Gaussians, which is clearly intractable for even moderate sequence lengths. It is therefore necessary to explore approximate inference techniques that will result in a tractable learning method.

2.2   Approximate Inference Using Viterbi Approximation

The task of Viterbi approximation approach is to find the most likely sequence of switching states s_t for a given observation sequence $\mathcal{Y}_T$. If the best sequence of switching states is denoted $\mathcal{S}_T^*$ we can then approximate the desired posterior $P(\mathcal{X}_T, \mathcal{S}_T | \mathcal{Y}_T)$ as[1]

$$P(\mathcal{X}_T, \mathcal{S}_T | \mathcal{Y}_T) = P(\mathcal{X}_T | \mathcal{S}_T, \mathcal{Y}_T) P(\mathcal{S}_T | \mathcal{Y}_T) \approx P(\mathcal{X}_T | \mathcal{S}_T, \mathcal{Y}_T)\, \delta(\mathcal{S}_T - \mathcal{S}_T^*),$$

i.e. the switching sequence posterior $P(\mathcal{S}_T | \mathcal{Y}_T)$ was approximated by its mode. It is well known how to apply Viterbi inference to discrete state hidden Markov models [24] and continuous state Gauss-Markov models [16]. Here we develop an algorithm for approximate Viterbi inference in SLDSs.

More formally, we are looking for the switching sequence $\mathcal{S}_T^*$ such that

$$\mathcal{S}_T^* = \arg\max_{\mathcal{S}_T} P(\mathcal{S}_T | \mathcal{Y}_T).$$

It is easily to shown that a (suboptimal) solution to this problem can be obtain by recursive optimization of the probability of the best sequence at time t

$$
\begin{aligned}
J_{t,i} &= \max_{\mathcal{S}_{t-1}} P\left(\mathcal{S}_{t-1}, s_t = e_i, \mathcal{Y}_t\right) \\
&\approx \max_j \Big\{ P\left(y_t | s_t = e_i, s_{t-1} = e_j, S_{t-2}^*(j), \mathcal{Y}_{t-1}\right) P\left(s_t = e_i | s_{t-1} = e_j\right) \\
&\qquad \max_{\mathcal{S}_{t-2}} P\left(\mathcal{S}_{t-2}, s_{t-1} = e_j, \mathcal{Y}_{t-1}\right) \Big\}
\end{aligned}
\tag{1}
$$

[1] $\delta(x) = 1$ for $x = \emptyset$ and zero otherwise.

Here $S_{t-2}^*(i)$ is the "best" switching sequence up to time $t-1$ when SLDS is in state i at time $t-1$, $S_{t-2}^*(i) = \arg\max_{S_{t-2}} J_{t-1,i}$.

To find the likelihood term $P(y_t)$ in Equation 1 note that concurrently with the recursion for each pair of consecutive switching state i, j at times $t, t-1$ one can update the statistics [2] of continuous states of LDS based on current "best" switching sequences using the Kalman filter inference (c.f. [1]). For instance,

$$\hat{x}_{t|t,i} \triangleq \langle x_t | \mathcal{Y}_t, s_t = e_i, S_{t-1}^*(i) \rangle$$

$$\hat{x}_{t|t-1,i,j} \triangleq \langle x_t | \mathcal{Y}_{t-1}, s_t = e_i, s_{t-1} = e_j, S_{t-2}^*(j) \rangle$$

$$\hat{x}_{t|t,i,j} \triangleq \langle x_t | \mathcal{Y}_t, s_t = e_i, s_{t-1} = e_j, S_{t-2}^*(j) \rangle.$$

$\Sigma_{t|t,i} \Sigma_{t|t-1,i,j} \Sigma_{t|t,i,j}$ denote corresponding second order statistics. The likelihood term can then be easily computed as the probability of innovation $y_t - C\hat{x}_{t|t-1,i,j}$ of $j \to i$ transition, which has normal distribution with mean $C\hat{x}_{t|t-1,i,j}$ and variance $C\Sigma_{t|t-1,i,j}C' + R$.

The index of the most likely state j^* at $t-1$ corresponding to the maximum in Equation 1 is kept for every state i at time t in the state transition record $\psi_{t-1,i} = j*$. LDS statistics corresponding to $j*$ are updated accordingly, $\hat{x}_{t|t,i} = \hat{x}_{t|t,i,\psi_{t-1,i}} \Sigma_{t|t,i} = \Sigma_{t|t,i,\psi_{t-1,i}}$. Once all T observations have been fused to the "best" switching state sequence is the one that ends in $i_{T-1}^* = \arg\min_i J_{T-1,i}$. States of this sequence can be traced back through the state transition record $\psi_{t-1,i}$, $i_t^* = \psi_{t,i_{t+1}^*}$.

Once the "best" switching sequence is known, the switching model's sufficient statistics are simply $\langle s_t \rangle = e_{i_t^*}$ and $\langle s_t s_{t-1}' \rangle = e_{i_t^*} e_{i_{t-1}^*}'$. Sufficient LDS statistics for this switching sequence can be easily obtained using Rauch-Tung-Streiber (RTS) fixed interval smoothing [1].

The Viterbi inference algorithm for SLDSs can now be summarized as

---

Find most likely state sequence $\mathcal{S}_T^*$ using recursion in 1;
Find DBN's sufficient statistics for $P(\mathcal{S}_T | \mathcal{Y}_T) = \delta(\mathcal{S}_T - \mathcal{S}_T^*)$;

Approximate Inference Using Structured Variational Inference   General structured variational inference technique for Bayesian networks is described in [14]. The idea behind this method is to find distribution Q which is in some sense close to the desired conditional distribution P, but is easier to compute. One can then employ Q as an approximation of P, $P(\mathcal{X}_T, \mathcal{S}_T | \mathcal{Y}_T) \approx Q(\mathcal{X}_T, \mathcal{S}_T | \mathcal{Y}_T)$.

Namely, for a given set of observations $\mathcal{Y}_T$, a distribution $Q(\mathcal{X}_T, \mathcal{S}_T | \eta, \mathcal{Y}_T)$ with an additional set of *variational parameters* η is defined such that Kullback–Leibler divergence between $Q(\mathcal{X}_T, \mathcal{S}_T | \eta, \mathcal{Y}_T)$ and $P(\mathcal{X}_T, \mathcal{S}_T | \mathcal{Y}_T)$ is minimized

---

[2] LDS statistics are state means and covariances. $\langle . \rangle$ denotes the expectation operator.

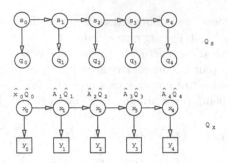

Fig. 2. Factorization of the original SLDS. Factorization reduces the fully coupled model into a seemingly decoupled pair of a HMM (Q_s) and a LDS (Q_x).

with respect to η:

$$\eta^* = \arg\min_{\eta} \sum_{\mathcal{S}_T} \int_{\mathcal{X}_T} Q(\mathcal{X}_T, \mathcal{S}_T | \eta, \mathcal{Y}_T) \log \frac{P(\mathcal{X}_T, \mathcal{S}_T | \mathcal{Y}_T)}{Q(\mathcal{X}_T, \mathcal{S}_T | \eta, \mathcal{Y}_T)}.$$

The dependency structure of Q is chosen such that it closely resembles the dependency structure of the original distribution P. However, unlike P the dependency structure of Q *must* allow a computationally efficient inference. In our case we define Q by decoupling the switching and LDS portions of SLDS as shown in Figure 2. The two subgraphs of the original network are a hidden Markov model (HMM) Q_S with variational parameters $\{q_0, \ldots, q_{T-1}\}$ and a time-varying LDS Q_X with variational parameters $\{\hat{x}_0, \hat{A}_0, \ldots, \hat{A}_{T-1}, \hat{Q}_0, \ldots, \hat{Q}_{T-1}\}$. Because the subgraphs are *decoupled*, inference can be performed for each submodel separately, $Q(\mathcal{X}_T, \mathcal{S}_T | \eta, \mathcal{Y}_T) = Q_X(\mathcal{X}_T | \eta, \mathcal{Y}_T) \, Q_S(\mathcal{S}_T | \eta)$. This is also reflected in the sufficient statistics of the posterior defined by the approximating network, e.g. $\langle x_t x_t' s_t \rangle = \langle x_t x_t' \rangle \langle s_t \rangle$.

The optimal values of the variational parameters η are obtained by minimizing the KL-divergence w.r.t. η. This leads to, for instance, the following recursive expression for the LDS Q_X's state transition matrix

$$\hat{A}_t = \hat{Q}_t \sum_{i=0}^{S-1} Q_i^{-1} A_i \langle s_t(i) \rangle. \tag{2}$$

Similarly, an expression can be found for optimal HMM variational parameters

$$\log q_t(i) = -\frac{1}{2} \left\langle (x_t - A_i x_{t-1})' \hat{Q}_i^{-1} (x_t - A_i x_{t-1}) \right\rangle - \frac{1}{2} \log |\hat{Q}_{t,i}|. \tag{3}$$

To obtain the expectation terms $\langle s_t \rangle = Pr(s_t | q_0, \ldots, q_{T-1})$ we use the inference in the HMM with output "probabilities" q_t [24]. Similarly, to obtain $\langle x_t \rangle = E[x_t | \mathcal{Y}_T]$ we perform LDS inference in the decoupled time-varying LDS via RTS smoothing. Since $\hat{A}_t, \hat{Q}_t$ in the decoupled LDS Q_X depends on $\langle s_t \rangle$ from the decoupled HMM Q_S and q_t depends on $\langle x_t \rangle, \langle x_t x_t' \rangle, \langle x_t x_{t-1}' \rangle$ from the decoupled LDS, the optimal parameter equations (e.g. 2 and 3) together with the inference solutions in the decoupled models form a set of fixed-point equations. Solution of this fixed-point set yields a tractable approximation to the intractable inference of the original fully coupled SLDS.

The variational inference algorithm for fully coupled SLDSs can now be summarized as:

```
error = ∞;
Initialize ⟨s_t⟩;
while (error > maxError) {
        Find Q̂_t, Â_t, x̂_0 from ⟨s_t⟩;
        Estimate ⟨x_t⟩, ⟨x_t x_t'⟩ and ⟨x_t x_{t-1}'⟩ from y_t
            using time-varying LDS inference;
        Find q_t from ⟨x_t⟩, ⟨x_t x_t'⟩ and ⟨x_t x_{t-1}'⟩;
        Estimate ⟨s_t⟩ from q_t using HMM inference.
        Update approximation error (KL divergence);
}
```

2.3   Maximum Likelihood Learning of Complex DBNs

Learning in complex DBNs can be formulated as the problem of ML learning in general Bayesian networks. Hence, a generalized EM algorithm [21] can be used to find optimal values of DBN parameters $\{A, C, Q, R, \Pi, \pi_0\}$. The expectation (E) step of EM is the inference itself. We outlined two approximate inference algorithms in the previous section. Note that the variational inference algorithm does not necessarily lead to non-decreasing likelihood of data in the EM (even though it usually does so in practice.) On the other hand, (a bound on) likelihood is guaranteed not to decrease when structural variational inference is used. See [14] for more details.

Given the sufficient statistics from the inference phase, *parameter update equations* in maximization (M) step are obtained by maximizing $\langle \log P(\mathcal{X}_T, \mathcal{S}_T, \mathcal{Y}_T) \rangle$ with respect to the parameter of interest. For instance, updated values of the state transition parameters are easily shown to be

$$\hat{A}_i = \left(\sum_{t=1}^{T-1} \langle x_t x_{t-1}' s_t(i) \rangle \right) \left(\sum_{t=1}^{T-1} \langle x_{t-1} x_{t-1}' s_t(i) \rangle \right)^{-1}$$

$$\hat{\Pi} = \left(\sum_{t=1}^{T-1} \langle s_t s_{t-1}' \rangle \right) \operatorname{diag} \left(\sum_{t=1}^{T-1} \langle s_t \rangle \right)^{-1}.$$

All the variable statistics are evaluated before updating any parameters. Notice that the above equations represent a generalization of the parameter update equations of classical (non-switching) LDS models [8].

3 Previous Work

SLDS models and their equivalents have been studied in statistics, time-series modeling, and target tracking since early 1970's. Bar-Shalom [2] and Kim [17] have developed a number of approximate pseudo-Bayesian inference techniques based on mixture component truncation or collapsing is SLDSs. They did not address the issue of learning system parameters. Shumway and Stoffer [26] presented a systematic view of inference and learning in SLDS while assuming known prior switching state distributions at each time instance, $Pr(s_t) = \pi_t(i)$ and no temporal dependency between switching states. Krishnamurthy and Evans [18] imposed Markov dynamics on the switching model. However, they assumed that noisy measurements of the switching states are available.

Ghahramani [9] introduced a DBN-framework for learning and approximate inference in one class of SLDS models. His underlying model differs from ours in assuming the presence of S independent, white noise-driven LDSs whose measurements are selected by the Markov switching process. An alternative input-switching LDS model was proposed by Pavlovic et al. [22] and utilized for mouse motion classification. A switching model framework for particle filters is described in [12] and applied to dynamics learning in [3]. Manifold learning [5] is another approach to constraining the set of allowable trajectories within a high dimensional state space. An HMM-based approach is described in [4].

4 Experiments

We applied our DBN-based SLDS framework to the modeling of motion of the human figure. Most current models of the human figure dynamics belong to one of two model groups. One assumes highly complex, hand-crafted biomechanical models. This approach has been used successfully to produce computer graphics animations of human motion [10] and to track upper body motion in a user-interface setting [27]. On the other end of the spectrum are simple LDS models. Most previous figure trackers which have used a dynamic model employed a simple smoothness prior such as a constant velocity Kalman filter [15].

Two categories of fronto-parallel motion were present in our data: walking and jogging. Fronto-parallel motions exhibit interesting dynamics and are free from the difficulties of 3-D reconstruction. Experiments can be conducted easily using a single video source, while self-occlusions and cluttered backgrounds make the tracking problem non-trivial.

We adopted the 2-D Scaled Prismatic Model proposed by Morris and Rehg [19] to describe the kinematics of the figure. The kinematic model lies in the image plane, with each link having one degree of freedom (DOF) in rotation and another DOF in length. A chain of SPM transforms can model the image displacement and foreshortening effects produced by 3-D rigid links. The appearance of each link in the image is described by a template of pixels which is manually initialized and deformed by the link's DOF's.

In our figure tracking experiments we analyzed the motion of the legs, torso, and head, and ignoring the arms. Our kinematic model had eight DOF's, corresponding to rotations at the knees, hip, and neck. A sample configuration of our figure model is shown in Figure 4.2.

4.1   Classification

The first task we addressed was learning an SLDS model for walking and running. Our training set consisted of 18 sequences of six individuals jogging (two examples of three people) and walking at a moderate pace (two examples of six people.) Each sequence was approximately 50 frames duration. The training data consisted of the joint angle states of the SPM in each image frame, which was obtained manually.

Each of the two motion types were each modeled as multi-state[3] SLDSs and then combined into a single complex SLDS. Measurement matrix in all cases was assumed to be identity, $C = I$. Initial state segmentation within each motion type was obtained using unsupervised clustering in a state space of some simple dynamics model (e.g. constant velocity model.) Parameters of the model $(A, Q, R, x_0, \Pi, \pi_0)$ were then reestimated using the EM-learning framework with approximate Viterbi inference. This yielded refined segmentation of switching states within each of the models. An example of the learned switching state sequence within a single "jog" training example is shown in Figure 3(a).

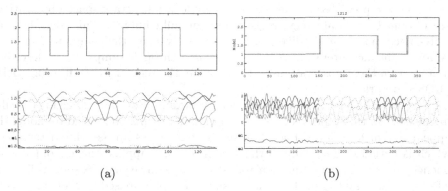

(a)   (b)

Fig. 3. (a) Segmentation of two-state SLDS model states within single "jog" motion sequence. (b) Segmentation of mixed walking/running sequence. Top graph shows correct segmentation (dotted red line) and estimated segmentation (solid blue line). Bottom graph depicts the segmentation of the estimated LDS states.

To test the classification ability of our learned model we next considered segmentation of sequences of *complex* motion, i.e., motion consisting of alter-

---

[3] We explored SLDS models with two to six states.

nations of "jog" and "walk."[4] Identification of different motion "regimes" was conducted using the approximate Viterbi inference. Estimates of "best" switching states $\langle s_t \rangle$ indicated which of the two models can be considered to be driving the corresponding motion segment. One example of this segmentation is depicted in Figure 3(b).

Classification experiments on a set of 20 test sequences gave an error rate[5] of 2.9% over a total of 8312 classified data points.

Additional classification experiments were performed using the structured variational inference technique. Figure 4 depicts state estimates and variational parameters in iterations 1 through 4 of variational inference. Initial uncertain switching state distribution $\langle s \rangle$ leads to low variational state noise variance $\hat{Q}$ (whose determinant is indicated by $|Q_v|$ in Figure 4) and low variational state transition matrix (whose determinant is indicated by $|A_v|$ in Figure 4). Through further iterations the variational inference algorithm converges to the true switching state sequence.

4.2   Tracking

A second experiment explored the utility of the SLDS model in improving tracking of the human figure from video. The difficulty in this case is that feature (joint angle) measurements are not readily available from a sequence of image intensities. Hence, we use the SLDS as a multi-hypothesis predictor that initializes multiple local template searches in the image space. Instead of choosing S^2 multiple hypotheses $\hat{x}_{t|t-1,i,j}$ at each time step we pick the best S hypothesis with the highest switching probability, i.e., $\hat{x}_{t|t-1,i,i_t^*}$ where $i_t^* = \arg\max_j \{ \Pi(i,j) J_{t-1,j} \}$.

Given the predicted means for the figure locations, state-space observations are obtained by local image registration, or hill-climbing. This identifies the state-space modes in the likelihood function given by the template model. A larger set of measurements could be explored through sampling, as described in [6]. Given these observations of figure state, the regular SLDS filtering yields SLDS state priors.

Figure 5 shows stills from a representative example of SLDS tracking of walking motion. In this experiment, simple template features were used to model the appearance of the figure. Each link in the model has an associated template, which is initialized manually in the first frame and applied throughout the sequence. Template features are not robust to appearance changes such as lighting effects or the wrinkling of cloth. As a result, a template-based tracker can benefit substantially from an accurate dynamical model.

A constant velocity predictor does poorly in this case, leading to tracking failure by frame seven (shown in Figure 4.b). The learned SLDS model gives improved predictions leading to more robust tracking.

---

[4] Test sequences were constructed by concatenating in random order randomly selected and noise corrupted training sequences. Transitions between sequences were smoothed using B-spline smoothing.

[5] Classification error was defined as the difference between inferred segmentation and true segmentation accumulated over all sequences, $e = \sum_{t=0}^{T-1} | \langle s_t \rangle - s_{\text{true},t} |$.

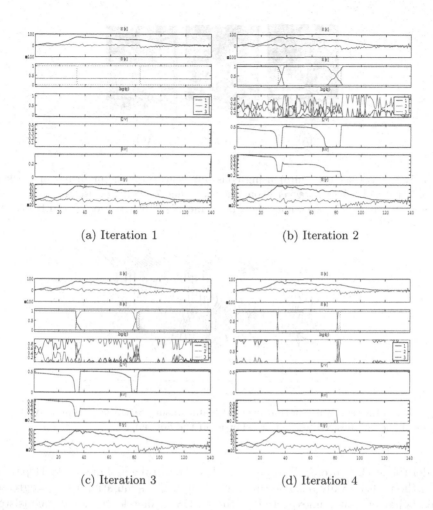

(a) Iteration 1 (b) Iteration 2

(c) Iteration 3 (d) Iteration 4

Fig. 4. Iterations of variational inference. The graphs depict: continuous state estimates $\langle x \rangle$, switching state estimates $\langle s \rangle$, HMM variational parameter $\log q$, determinants of LDS variational parameters $|Q_v|$ and $|A_v|$, and the true (dotted) and estimated measurements $E[y]$.

4.3 Synthesis and Interpolation

In Section 2 we introduced SLDS as a *generative* model. Nonetheless, SLDS is most commonly employed as a classifier (e.g. Section 4.1.) To test the power of the learned SLDS framework we examined its use in synthesizing realistic–looking motion sequences and interpolating motion between missing frames.

In the first set of experiments the learned walk/jog SLDS was used to generate a "synthetic walk." Two stick figure motion sequences of the noise driven model are shown in Figure 6. Depending on the amount of noise used to drive the

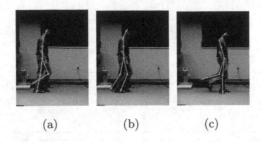

(a)   (b)   (c)

Fig. 5. (a) Tracker (in white) using constant velocity predictor drifts off track by frame 7. (b) SLDS-based tracker is on track at frame 7. Model (switching state) 3 has the highest likelihood. Black lines show prior mean and observation. (c) SLDS tracker at frame 20.

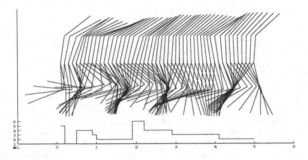

Fig. 6. Synthesized walk motion over 50 frames using SLDS as a generative model. States of the synthesized motion are shown on the bottom.

model the stick figure exhibits more or less "natural"–looking walk. Departure from the realistic walk becomes more evident as the simulation time progresses. This behavior is not unexpected as the SLDS in fact learns locally consistent motion patterns.

Another realistic situation may call for filling-in a small number of missing frames from a large motion sequence. SLDS can then be utilized as an interpolation function. In a set of experiments we employed the learned walk/jog model to interpolate a walk motion over two sequences with missing frames (see Figure 7.) The visual quality of the interpolation and the motion synthesized from it was high (left column in Figure 7.) As expected, the sparseness of the measurement set had definite bearing on this quality.

5   Conclusions

We have introduced a new approach to dynamics learning based on switching linear models. We have proposed two approximation techniques, Viterbi and structural variational inference, which overcomes the exponential complexity of

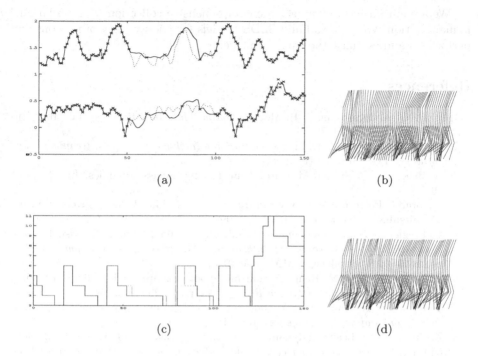

(a)   (b)

(c)   (d)

Fig. 7. SLDS as an interpolation function. A motion sequences with missing measurements between frames 50 and 100 was interpolated using an SLDS model. Symbols 'x' in figure (a) indicate known measurement points. Solid lines show interpolated joint angle values. Dotted lines indicate ground truth (smoothing with no missing measurements.) Figure (c) depicts corresponding SLDS states. Stick figure motion generated from interpolated data is shown in figure (d). Figure (b) shows true stick figure motion.

exact inference. Simplicity of approximate Viterbi inference is contrasted by the lack of an exact bound on the approximation error. This is a problem in general with greedy Viterbi-style approximations, as well as with Markov chain Monte Carlo methods [20]. On the other hand, more complex structured variational inference guarantees minimization of this error by considering global approximation of intractable SLDS distribution.

Our preliminary experiments have demonstrated promising results in classification of human motion, improved visual tracking performance, and motion synthesis and interpolation using our SLDS framework. We demonstrated accurate discrimination between walking and jogging motions. We showed that SLDS models provide more robust tracking performance than simple constant velocity predictors. The fact that these models can be learned from data may be an important advantage in figure tracking, where accurate physics-based dynamical models may be prohibitively complex.

We are currently building a more comprehensive collection of frontoparallel human motion. We plan to build SLDS models for wide variety of motions and performers and evaluate their performance.

References

[1] B. D. O. Anderson and J. B. Moore. *Optimal filtering.* Prentice-Hall, Inc., Englewood Cliffs, NJ, 1979.

[2] Y. Bar-Shalom and X.-R. Li. *Estimation and tracking: principles, techniques, and software.* YBS, Storrs, CT, 1998.

[3] A. Blake, B. North, and M. Isard. Learning multi-class dynamics. In *NIPS '98*, 1998.

[4] M. Brand. Pattern discovery via entropy minimization. Technical Report TR98-21, Mitsubishi Electric Research Lab, 1998. Available at *http://www.merl.com*.

[5] C. Bregler and S. M. Omohundro. Nonlinear manifold learning for visual speech recognition. In *Proceedings of International Conference on Computer Vision*, pages 494–499, Cambridge, MA, June 1995.

[6] T.-J. Cham and J. M. Rehg. A multiple hypothesis approach to figure tracking. In *Computer Vision and Pattern Recognition*, pages 239–245, 1999.

[7] T. Dean and K. Kanazawa. A model for reasoning about persistance and causation. *Computational Intelligence*, 5(3), 1989.

[8] Z. Ghahramani. Learning dynamic Bayesian networks. In C. L. Giles and M. Gori, editors, *Adaptive processing of temporal information*, Lecture notes in artificial intelligence. Springer-Verlag, 1997.

[9] Z. Ghahramani and G. E. Hinton. Switching state-space models. submitted for publication, 1998.

[10] J. K. Hodgins, W. L. Wooten, D. C. Brogan, and J. F. O'Brien. Animating human athletics. In *Computer Graphics (Proc. SIGGRAPH '95)*, pages 71–78, 1995.

[11] V. T. Inman, H. J. Ralston, and F. Todd. *Human Walking.* Williams and Wilkins, 1981.

[12] M. Isard and A. Blake. A mixed-state CONDENSATION tracker with automatic model-switching. In *Proceedings of International Conference on Computer Vision*, pages 107–112, Bombay, India, 1998.

[13] F. V. Jensen. *An introduction to Bayesian Networks.* Springer-Verlag, 1995.

[14] M. I. Jordan, Z. Ghahramani, T. S. Jaakkola, and L. K. Saul. An introduction to variational methods for graphical models. In M. I. Jordan, editor, *Learning in graphical models*. Kluwer Academic Publishers, 1998.

[15] I. A. Kakadiaris and D. Metaxas. Model-based estimation of 3D human motion with occlusion based on active multi-viewpoint selection. In *Computer Vision and Pattern Recognition*, pages 81–87, San Fransisco, CA, June 18-20 1996.

[16] R. E. Kalman and R. S. Bucy. New results in linear filtering and prediction. *Journal of Basic Engineering (ASME)*, D(83):95–108, 1961.

[17] C.-J. Kim. Dynamic linear models with markov-switching. *Journal of Econometrics*, 60:1–22, 1994.

[18] V. Krishnamurthy and J. Evans. Finite-dimensional filters for passive tracking of markov jump linear systems. *Automatica*, 34(6):765–770, 1998.

[19] D. D. Morris and J. M. Rehg. Singularity analysis for articulated object tracking. In *Computer Vision and Pattern Recognition*, pages 289–296, Santa Barbara, CA, June 23-25 1998.

[20] R. M. Neal. Connectionist learning of belief networks. *Artificial Intelligence*, pages 71–113, 1992.

[21] R. M. Neal and G. E. Hinton. A new view of the EM algorithm that justifies incremental and other variants. In M. Jordan, editor, *Learning in graphical models*, pages 355–368. Kluwer Academic Publishers, 1998.

[22] V. Pavlovic, B. Frey, and T. S. Huang. Time-series classification using mixed-state dynamic Bayesian networks. In *Computer Vision and Pattern Recognition*, pages 609–615, June 1999.

[23] J. Pearl. *Probabilistic reasoning in intelligent systems*. Morgan Kaufmann, San Mateo, CA, 1998.

[24] L. R. Rabiner and B. Juang. *Fundamentals of Speech Recognition*. Prentice Hall, Englewood Cliffs, New Jersey, USA, 1993.

[25] H. E. Rauch. Solutions to the linear smoothing problem. *IEEE Trans. Automatic Control*, AC-8(4):371–372, October 1963.

[26] R. H. Shumway and D. S. Stoffer. Dynamic linear models with switching. *Journal of the American Statistical Association*, 86(415):763–769, September 1991.

[27] C. R. Wren and A. P. Pentland. Dynamic models of human motion. In *Proceeding of the Third International Conference on Automatic Face and Gesture Recognition*, pages 22–27, Nara, Japan, 1998.

Stability of Hybrid Systems Using LMIs
– A Gear-Box Application

Stefan Pettersson and Bengt Lennartson

Control Engineering Lab, Chalmers University of Technology
S-412 96 Gothenburg, Sweden
sp@s2.chalmers.se, bl@s2.chalmers.se

Abstract. This paper includes an application consisting of an automatic gear-box and cruise controller which naturally is modelled as a hybrid system including state jumps in the continuous state of the controller. Motivated by this application, we extend existing stability results to include state jumps as well. The proposed stability results are based on Lyapunov techniques. The search for the (piecewise quadratic) Lyapunov functions is formulated as a linear matrix inequality (LMI) problem. It is shown how the proposed stability analysis is applied to the automatic gear-box and cruise controller.

1 Introduction

Many physical systems today are modeled by interacting continuous and discrete event systems. Such *hybrid systems* contain both continuous and discrete states that influence the dynamic behavior. There is a lot of interest in these kind of systems today, since a large number of systems are neither pure continuous nor discrete but a combination. This is mostly due to the growing use of computers in the control of physical plants but also as a result of the hybrid nature of many physical processes. Physical systems suitably modeled as hybrid systems are for instance the management of a fishery resource [13], computer disk system [9], motion control systems [5], robotics [6], power systems [10], systems in classical mechanics [4], air traffic management [18] and automated vehicles [12].

This paper includes an application consisting of an automatic gear-box and cruise controller. Both the automatic gear-box (plant) and the cruise controller (controller) are naturally modeled as hybrid systems which interact to control the velocity at a desired value. The automatic gear-box is modeled as simple as possible, with the velocity and the gear position as continuous and discrete state respectively. The discrete state is changed when the velocity reaches different values, which affects the continuous dynamics. The cruise controller consists of a PI-controller, where the continuous state is the integrator state, implying that the velocity converges to the desired state despite the influence of disturbances. To obtain a comfortable ride, there are restrictions imposed on the derivative of the acceleration. This implies that the gain in the controller must have different values for different gear positions, implying that it acts as a discrete state

N. Lynch and B. Krogh (Eds.): HSCC 2000, LNCS 1790, pp. 381–395, 2000.

with different values. Furthermore, the restrictions also imply state jumps in the integrator state at the times when the gear position is changed (bumpless transfer).

Many stability results for hybrid systems using a Lyapunov approach have been proposed in the literature; see for instance [14, 7, 3, 21]. However, none of the approaches considers the additional complexity to include also state jumps in the analysis. Therefore, stability results applicable to hybrid systems with state jumps are proposed in this paper. The problem to find the different local (piecewise quadratic) Lyapunov functions is formulated as a linear matrix inequality (LMI) problem [2], for which there exists numerical software [8].

The proposed stability results in this paper can be generalized to consider even larger classes of hybrid systems than the application model. Such results have in fact been carried out in the Ph.D. thesis [15]. However, to keep the paper short and reduce the complexity, possible generalizations will not be discussed herein but we refer to the thesis for interested readers.

The outline of this paper is as follows: The application is given in detail in the next section motivating the use of stability results for hybrid systems including state jumps. Section 3 proposes conditions for exponential stability. It is shown how the stability result can be formulated as a linear matrix inequality (LMI) problem in Section 4. The paper is concluded by showing how the proposed stability result is applied to the gear-box application.

2   Application and Hybrid Model

The proposed theory in this paper is motivated by the following application:

2.1   Gear-Box Application

A motor together with transmission through a gear-box is naturally modeled by continuous and discrete states. Nonlinear models describing the dynamics of a vehicle with throttle angle as input are given in [19]. Continuous state variables are the manifold pressure and velocity of the vehicle, and a discrete state variable is the gear position. In this example, a satisfactory model illustrating the hybrid behavior is obtained by assuming that the input signal is the torque T out from the motor (hence, all dynamics in the motor are neglected). The gear-box transforms the torque T and angular velocity ω according to

$$T_1 = pT \text{ and } \omega_1 = \frac{1}{p}\omega, \tag{1}$$

where T_1 is the torque and ω_1 is the angular velocity of the wheels, and p is the gear position. If the radius of the wheel is r, the force F accelerating the vehicle and the velocity v of the vehicle becomes

$$F = T_1/r \text{ and } v = r\omega_1. \tag{2}$$

The vehicle acceleration is according to Newton's law of motion

$$M\dot{v} = F - F_l, \tag{3}$$

where M is the weight of the car and F_l is the load force induced from the road. If it is assumed that F_l is proportional to the square of the vehicle velocity v and the road angle is α, this force can be modeled as

$$F_l = kv^2 \text{sign}\, v + Mg \sin \alpha. \tag{4}$$

By combining (1), (2) and (4) into (3), the vehicle dynamics is given by

$$\dot{v} = \frac{p_r T}{M} - \frac{k}{M} v^2 \text{sign}\, v - g \sin \alpha,$$
$$\omega = p_r v, \tag{5}$$

where $p_r = p/r$ is assumed to take values in the discrete set $\{p_{r_1}, p_{r_2}, p_{r_3}, p_{r_4}\}$, $p_{r_1} > p_{r_2} > p_{r_3} > p_{r_4}$. Hence, there are four possible discrete gear positions, where p_{r_1} corresponds to gear 1, p_{r_2} to gear 2 and so on.

In this illustrative example, the automatic gear-box is designed in such a way that the change of gear occurs if the engine rotational speed exceeds ω_{high}, implying a higher gear (if not already gear 4), or goes below ω_{low}, implying a lower gear (if not already gear 1). Depending on the gear, the values ω_{high} and ω_{low} corresponds to different velocities of the vehicle; see (5). The desired behavior is obtained by changing gear position at velocities given by the switch sets

$$S_{i,i+1} = \{v \in \Re \mid v = \frac{1}{p_{r_i}} \omega_{high}\} \text{ and } S_{i+1,i} = \{v \in \Re \mid v = \frac{1}{p_{r_{i+1}}} \omega_{low}\}, i = 1, 2, 3,$$

where $S_{i,i+1}$ denotes gear position changes from i to $i+1$ and vice versa for $S_{i+1,i}$.

The cruise controller is designed in the following way. The torque T consists of the terms

$$T = T_P + T_I + \frac{k}{p_r} v^2 \text{sign}\, v \tag{6}$$

where

$$T_P = K_r(v_{ref} - v),$$
$$\dot{T}_I = \frac{K_r}{T_r}(v_{ref} - v), \tag{7}$$

and v_{ref} is the desired velocity. Hence, the cruise controller (6) is essentially a PI-controller which compensates for the nonlinearity due to the load force. If the closed-loop system is (asymptotically) stable, the integrator part of the controller implies that the vehicle velocity v converges to the desired velocity for stationary input values v_{ref} despite the influence of a constant road angle α. Every time a new value of the desired velocity v_{ref} is given by the driver, the integrator state T_I is put to zero.

Besides stabilizing the closed-loop system, the parameters K_r and T_r should be selected in such a way that a desirable performance is obtained. A comfortable ride is maintained if the acceleration is limited to $|\dot{v}| \leq 2m/s^2$; cf. [1]. This condition restricts the gain K_r. In the design of the integration time T_r, there is a tradeoff between fast convergence and small overshoot.

Besides conditions on the acceleration, it is desirable also to have restrictions on the derivative of the acceleration since abrupt changes of this variable can be quite uncomfortable. One reason for possible abrupt changes of $\ddot{v}$ occurs when the gear position is changed. If t_k denotes the time when the change of gear occurs and t_k^- and t_k^+ denote the times just before and after that time, and K_r takes values in the set $\{K_{r_1}, K_{r_2}, K_{r_3}, K_{r_4}\}$, where K_{r_1} corresponds to gear 1, and so on, then (5) and (7) imply that there are no abrupt changes of $\ddot{v}$ due to a change of gear if

$$
\begin{aligned}
p_{r_i} K_{r_i} &= p_{r_{i+1}} K_{r_{i+1}} \\
p_{r_i} T_I(t_k^-) &= p_{r_{i+1}} T_I(t_k^+) \quad \text{gear } i \text{ to } i+1 \qquad i = 1, 2, 3. \\
p_{r_{i+1}} T_I(t_k^-) &= p_{r_i} T_I(t_k^+) \quad \text{gear } i+1 \text{ to } i
\end{aligned}
\tag{8}
$$

Hence, by designing the gain parameters $K_{r_1}, \ldots, K_{r_4}$ and abruptly changing the value of the T_I-variable such that (8) is satisfied, discontinuities in $\ddot{v}$ (and hence $\dot{v}$) due to change of gear position are avoided. The jump in the state variable T_I avoiding jumps in the control signal T is commonly called bumpless transfer [11].

Let the numerical values be equal to: $p_{r_1} = 50$, $p_{r_2} = 32$, $p_{r_3} = 20$, $p_{r_4} = 14$, $k = 0.7$, $M = 1500$, $g = 10$, $\omega_{low} = 230$, $\omega_{high} = 500$, $K_{r_1} = 3.75$, $K_{r_2} = 5.86$, $K_{r_3} = 9.37$, $K_{r_4} = 13.39$, $T_r = 40$, $v_{ref} = 30$, $m(0) = m_3$, $v(0) = 14$ and $T_I(0) = 0$. For a specified desired velocity v_{ref} the system converges exponentially to v_{ref}, which will be verified by LMIs after the stability theory.

2.2   Hybrid Model

The hybrid model in the application has the form:

$$
\begin{aligned}
\dot{x} &= A(m)x, \\
x^+ &= \psi(x, m), \\
m^+ &= \phi(x, m),
\end{aligned}
\tag{9}
$$

where $x \in \Re^n$ is the continuous state and $m \in \mathcal{M} = \{m_1, \ldots, m_N\}$ is the discrete state. The hybrid state space H is the Cartesian product $\Re^n \times \mathcal{M}$. The continuous dynamics is given by a linear differential equation, including the possibility of expressing state jumps by the function $\psi : \Re^n \times \mathcal{M} \to \Re^n$, and $\phi : \Re^n \times \mathcal{M} \to \mathcal{M}$ is a function describing the dynamics of the discrete state. The notation x^+ and m^+ means the next state of x and m respectively. The hybrid system described in (9) is autonomous, i.e. there are no external inputs affecting the dynamics. This may be the result when external inputs are feed back functions of the continuous and discrete state, which is the case in the application.

The discrete state changes when x and m take certain values can, instead of being described by a function ϕ, be expressed by a number of *switch sets* $S_{i,j}$ (as in the application), which are related to ϕ according to

$$S_{i,j} = \{x \in \Re^n \mid m_j = \phi(x, m_i)\}, \quad i \in I_N, \ j \in I_N,$$

where N is the number of elements in $\mathcal{M}$, and $I_N = \{1, 2, \ldots, N\}$; cf. [20, 17]. Hence, the switch sets indicate where in the continuous state space $\Re^n$ the discrete state m_i changes to m_j. It is usual to specify only those switch sets that cause a change of discrete state variable from m_i to m_j where $m_i \neq m_j$. The switch sets are often given as geometrical hypersurfaces or hyperplanes (as in the application). Similarly, the set of states where state jumps occur for the different discrete states can equivalently be described by sets

$$J_i = \{x \in \Re^n \mid x^+ = \psi(x, m_i)\}, \quad i \in I_N.$$

It is assumed that $S_{i,j}$ and J_i coincide in this paper and the next continuous state is related to the previous one according to

$$x^+ = G(m_i)x \tag{10}$$

at these states.

In the stability analysis given next, it is assumed that there only is a finite number of switches in finite time. Hence, the continuous and discrete dynamics is well behaved.

3 Exponential Stability

We are now prepared to show exponential stability of hybrid systems including state jumps.

3.1 Region Partitioning

To show stability, $\Omega \subseteq H$ of the hybrid state-space is partitioned into ℓ disjoint regions. If for a given initial point in Ω, t_k, $k = 1, 2 \ldots$ are the consecutive times when the trajectory passes from one region to another, it is assumed that the *partitioning* is made in such a way that t_k is strictly less than t_{k+1}, i.e. $t_k < t_{k+1}$.

Let Ω be a hybrid set. Then, the following projection sets are defined:

$$\begin{aligned}
\Omega^x &= \{x \in \Re^n \mid (x, m) \in \Omega\}, \\
\Omega^{x,m_i} &= \{x \in \Re^n \mid (x, m_i) \in \Omega\}, \\
\Omega^m &= \{m \in \mathcal{M} \mid (x, m) \in \Omega\}.
\end{aligned} \tag{11}$$

Hence, Ω^x and Ω^{x,m_i} are sets consisting of continuous states while Ω^m consists of discrete states.

Assume that a trajectory satisfies (9) for any initial value in H_0. Let $\varepsilon > 0$ andefine the *neighboring regions* $\Lambda_{q,r}$, $q \in I_\ell$, $r \in I_\ell$, $(q \neq r)$ by

$$\Lambda_{q,r} = \{(x,m) \in \Omega \mid \exists t > 0 \text{ such that } (x(t - \varepsilon), m(t - \varepsilon)) \in \Omega_q \text{ and}$$
$$(x(t + \varepsilon), m(t + \varepsilon)) \in \Omega_r, \text{ when } \varepsilon \to 0\},$$

which are sets where trajectories pass from Ω_q to Ω_r. Let

$$I_\Lambda = \{(q,r) \mid \Lambda_{q,r} \neq \emptyset\},$$

which is the set of tuples indicating that there is at least one point for which the trajectory passes from Ω_q to Ω_r.

Let $V_q = x^T P_q x$, $q \in I_\ell$, be a quadratic function which is used as a measure of the system's (abstract) energy in region Ω_q. Let the overall energy be defined as

$$V(x) = V_q(x) \quad \text{when} \quad (x,m) \in \Omega_q, \tag{12}$$

which, in general, is a discontinuous function at the neighboring regions $\Lambda_{q,r}$, $(q,r) \in I_\Lambda$. Since it is assumed that the partitioning is made in such a way that $t_k < t_{k+1}$ for every trajectory with initial point in Ω, it is ensured that the overall energy defined in (12) is *piecewise continuous* as a function of time. The time derivative of $V_q(x)$ in region Ω_q can be written as:

$$\dot{V}_q(x) = A(m_i)^T P_q + P_q A(m_i), \quad x \in \Omega_q^{x,m_i}, \ m_i \in \Omega_q^m,$$

using the projection sets in (11).

3.2   Exponential Stability Conditions

Definition 1. *The region of exponential attraction $R(k_1, k_2)$ of a hybrid system (9) is the set of initial hybrid states for which the continuous trajectory exponentially converges to the origin according to*

$$R(k_1, k_2) = \{(x_0, m_0) \in H_0 \mid \|x(t)\| \leq k_1 e^{-k_2 t} \|x_0\|, \ t \geq 0, \ k_1 > 0, k_2 > 0\}.$$

Exponential stability in a region can be verified by the following theorem. The proof of this theorem can be found in [16, 15].

Theorem 1. *If there exist P_q, $q \in I_\ell$, and constants $\alpha > 0$ and $\beta > 0$, such that*

1. $x \in \Omega_q^x$, $\alpha x^T x \leq x^T P_q x \leq \beta x^T x$, $q \in I_\ell$
2. $x \in \Omega_q^{x,m_i}$, $x^T (A(m_i)^T P_q + P_q A(m_i))x \leq -x^T x$, $m_i \in \Omega_q^m, q \in I_\ell$
3. $x \in \Lambda_{q,r}^x$, $x^T G(m_i)^T P_r G(m_i)x \leq x^T P_q x$, $m_i \in \Omega_q^m, (q,r) \in I_\Lambda$

then the equilibrium point 0 is exponentially stable in the sense of Lyapunov. If the assumptions hold globally, then the equilibrium point 0 is globally exponentially stable.

If V is defined as in (12), then

$$R_c = \{(x, m) \in H \mid V(x) \le c\} \subseteq R(k_1, k_2),$$

where $R_c \subseteq \Omega$ and

$$k_1 = \left(\frac{\beta}{\alpha}\right)^{\frac{1}{2}}, \quad k_2 = \frac{1}{2\beta}. \tag{13}$$

The left-hand side of the first condition guarantees that each *local* quadratic Lyapunov function is positive. The righ-hand side is introduced to calculate the upper bound of the exponential convergence rate. The second and third condition guarantee that the overall energy V (12) decreases, both in regions (second condition) and when another region is entered (third condition).

To show exponential stability, the existence of P_q's satisfying the stability conditions has to be verified. This can be done by solving an LMI problem, described next.

4   Linear Matrix Inequalities

All conditions of the stability theorem are constrained to be satisfied, not in the entire state space but in a part of the continuous state space. The first condition is restricted to the region Ω_q^x, the second condition is restricted to Ω_q^{x,m_i} and the third condition is restricted to $\Lambda_{q,r}^x$. It is now described how the constrained conditions can be replaced by unconstrained conditions, by first expressing the regions by positive (quadratic) functions and then using a general technique called the S-procedure to obtain an unconstrained condition. This procedure is first explained in general terms and then applied to the constrained conditions in the stability theorem.

4.1   From Constrained Conditions to Unconstrained Conditions

Replacement of Constraint to Regions with Constraint by Functions.
Assume that $F^0(x) : \Re^n \to \Re$ is a function having unknown variables which are to be decided, satisfying the condition

$$F^0(x) \ge 0 \text{ for all } x \text{ in the region } \mathcal{R}. \tag{14}$$

Assume that $F^k(x) : \Re^n \to \Re$, $k \in I_\kappa$, are known functions satisfying

$$F^k(x) \ge 0, \ k \in I_\kappa, \text{ for all } x \text{ in the region } \mathcal{R}.$$

Condition (14) can then be replaced with the possibly stronger condition

$$F^0(x) \ge 0 \text{ for all } x \text{ satisfying } F^k(x) \ge 0, \ k \in I_\kappa. \tag{15}$$

Hence, the condition $F^0(x) \ge 0$ constrained to the region $\mathcal{R}$ has been replaced by constraints by the functions $F^k(x) \ge 0$, $k \in I_\kappa$. The replacement of $\mathcal{R}$ by functions $F^k(x) \ge 0$ is illustrated in Figure 1.

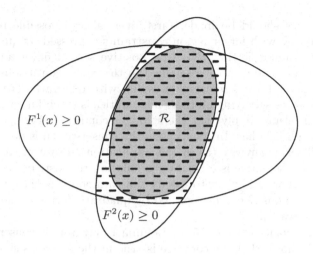

Fig. 1. The shaded region $\mathcal{R}$ is replaced with a region described by all x satisfying $F^1(x) \geq 0$ and $F^2(x) \geq 0$, which is the dashed region.

$\mathcal{S}$-procedure. It is possible to replace the constrained condition (15) by a condition without constraints by introducing additional variables $\lambda^k \geq 0$, $k \in I_\kappa$ in the following way:

Lemma 1. [2] *If there exist $\lambda^k \geq 0$, $k \in I_\kappa$, such that*

$$\forall x \in \Re^n, \; F^0(x) \geq \sum_{k=1}^{\kappa} \lambda^k F^k(x), \tag{16}$$

then (15) holds.

The proof follows directly by noting that the right-hand side of (16) is greater or equal to zero for all x satisfying $F^k(x) \geq 0$, $k \in I_\kappa$, since $\lambda^k \geq 0$, $k \in I_\kappa$.

The constrained condition (14) has been replaced by the unconstrained condition in Lemma 1. In the case of quadratic functions

$$F^k(x) = x^T Q^k x, \quad k = 0, \dots, \kappa, \tag{17}$$

where $Q^k = (Q^k)^T \in \Re^n \times \Re^n$, the condition (16) can be written as an LMI:

$$Q^0 \geq \sum_{k=1}^{\kappa} \lambda^k Q^k, \quad \lambda^k \geq 0, \; k \in I_\kappa, \tag{18}$$

The finesse of formulating the LMI condition as (18) instead of only $Q^0 \geq 0$ is that the condition $x^T Q^0 x \geq 0$ does not have to be fulfilled in the entire state space, implying that Q^0 has to be positive semi-definite, but only in a part of the state space where at least all $x^T Q^k x \geq 0$ are satisfied. The unknown matrix variable Q^0 and the different λ^k can be found by solving the LMI problem (18).

Some remarks should be made. First, it is always possible to replace an arbitrary region $\mathcal{R}$ with larger region constraints expressed by quadratic forms $F^k(x) \geq 0$, by simply letting $F^k(x)$ be positive semi-definite functions. This implies that $F^0(x) \geq 0$ has to be satisfied in the entire continuous state space. However, one should avoid replacing a region with quadratic forms $F^k(x) \geq 0$ such that the states satisfying all these inequalities is much larger than $\mathcal{R}$, since this conservatism may imply that the stronger condition (15) does not have a solution although (14) has. In some cases, the conservatism is no problem since a solution will exist anyway (as in the application). However, in other cases, not being too conservative is crucial for a solution to exist. In Section 4.3 it is explained more thoroughly how to specify the parameters in the quadratic forms (17) such that regions $\mathcal{R}$ can be replaced by quadratic forms $F^k(x) \geq 0$ without too much conservatism.

Second, the replacement of (15) by Lemma 1 may also be conservative. However, it can be shown that the converse is true in the case of a single quadratic form, $\kappa = 1$ [2], provided that there is some x such that $F^1(x) > 0$.

Third, in the case of hypersurfaces defined by $F^k(x) = 0$, $k \in I_\kappa$, it is not necessary to require that the different λ^k, $k \in I_\kappa$, have to be greater or equal to zero in Lemma 1, since this lemma holds despite the sign of these constants.

The above procedure is now applied to the constrained conditions in the stability theorem.

Stability Conditions. All conditions of the stability theorem are described by $F^0(x) \geq 0$, where $F^0(x)$ is a quadratic function defined as in (17). The first and second conditions in the stability theorems are restricted to regions Ω_q^x and Ω_q^{x,m_i} respectively. These conditions can be replaced by unconstrained conditions of the form (18). Matrices Q^k corresponding to regions Ω_q^x are denoted Q_q^k and regions Ω_q^{x,m_i} are denoted Q_{q,m_i}^k.

The third condition is restricted to hypersurfaces $\Lambda_{q,r}^x$. When these are given by $F^k(x) = 0$, $k \in I_\kappa$, where each $F^k(x)$ has the form (17), there will be no restrictions on the additional variables λ^k in (18). However, if some switch surface cannot exactly be described by $F^k(x) = 0$, $k \in I_\kappa$, then it is possible to include such a region with quadratic functions satisfying $F^k(x) \geq 0$, in which case the additional variables λ^k in (18) have to be greater or equal to zero. Matrices Q^k corresponding to hypersurfaces $\Lambda_{q,r}^x$ are denoted $Q_{q,r}^k$.

4.2   LMIs for Hybrid Systems with Linear Vector Fields

In the case of verifying exponential stability, it may be desirable not only to find a solution but to search for a solution that gives a better estimate of the convergence rate k_2 in (13). This can be achieved by searching for a solution where β is minimized. The LMI problem then becomes as follows:

LMI Problem If there is a solution to

$\min \beta$ subject to

0. $\alpha > 0$, $\mu_q^k \geq 0$, $\nu_q^k \geq 0$, $\vartheta_{q,m_i}^k \geq 0$, $q \in I_\ell$
1. $\alpha I + \sum_{k=1}^{\kappa_q} \mu_q^k Q_q^k \leq P_q \leq \beta I - \sum_{k=1}^{\kappa_q} \nu_q^k Q_q^k$, $q \in I_\ell$
2. $A(m_i)^T P_q + P_q A(m_i) + \sum_{k=1}^{\kappa_{q,m_i}} \vartheta_{q,m_i}^k Q_{q,m_i}^k \leq -I$, $m_i \in \Omega_q^m$, $q \in I_\ell$
3. $G(m_i)^T P_r G(m_i) + \sum_{k=1}^{\kappa_{q,r}} \eta_{q,r}^k Q_{q,r}^k \leq P_q$, $(q,r) \in I_\Lambda$

then the equilibrium point 0 is exponentially stable in the sense of Lyapunov.

The variables α, μ_q, ν_q, ϑ_{q,m_i} and matrices P_q are unknowns, while the different Q:s are known matrices corresponding to the different local regions where the conditions have to be valid. The convergencerate rate is estimated as in Theorem 1.

The LMI formulation can be extended to consider affine vector fields by extending the quadratic local Lyapunov functions from $V_q(x) = x^T P_q x$ to $V_q(x) = \pi_q + 2p_q^T x + x^T P_q x$. Nonlinear vector fields may also be handled by slightly modifying the second condition in the LMI problem [15].

4.3   Describing Regions by Quadratic Forms

It is now explained how to specify the parameters in the quadratic forms (17) such that the set of states satisfying all quadratic forms $F^k(x) \geq 0$ includes the set $\mathcal{R}$. We are focusing on regions partitioned by hyperplanes. More general regions are discussed in [15].

Quadratic Forms Describing Half-Planes. If a region $\mathcal{R}$ containing the origin is given by the set of states restricted by two half-planes

$$(c^a)^T x \geq 0 \text{ and } (c^b)^T x \geq 0,$$

then $\mathcal{R}$ will be described by a quadratic form

$$x^T Q^1 x \geq 0, \text{ where } Q^1 = c^a (c^b)^T + c^b (c^a)^T. \tag{19}$$

The set of states satisfying a quadratic form (19) has the property that if x_1 satisfies the inequality so does $-x_1$; see Figure 2.

If the dimension n is equal to two, there is no reason for replacing a region $\mathcal{R}$ by a quadratic form (19) described by more than two hyperplanes, since the set of states satisfying several half-planes can equivalently be described by only two half-planes. However, this is reasonable in higher dimensions. In this case, the quadratic forms $x^T Q^k x \geq 0$ are obtained by taking all possible combinations of two different half-planes. This results in $\frac{\varrho(\varrho-1)}{2}$ different quadratic forms as in (19) and hence also variables λ^k in (16), where ϱ is the number of half-planes. There is no reason to add the combinations of the same half-planes since these quadratic forms are greater or equal to zero for all states.

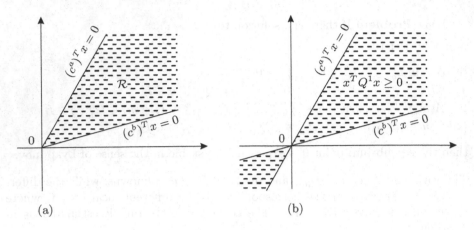

Fig. 2. (a) Region (dashed) $\mathcal{R}$ restricted by hyperplanes $(c^a)^T x \geq 0$ and $(c^b)^T x \geq 0$. (b) Region (dashed) of states satisfying $x^T Q^1 x \geq 0$, where $Q^1 = c^a(c^b)^T + c^b(c^a)^T$.

The reason for specifying a number of quadratic inequalities instead of only one, in case $\mathcal{R}$ is restricted by several half-planes, is that $\mathcal{R}$ cannot exactly be described by the set of states given by a quadratic form greater or equal to zero, even if $\mathcal{R}$ is symmetric around the origin (meaning that if $x \in \mathcal{R}$ then $-x \in \mathcal{R}$; cf. Figure 2b). The set of states satisfying $x^T Q^1 x \geq 0$ for a quadratic form given by any combination of two half-planes describing $\mathcal{R}$ will be strictly larger than $\mathcal{R}$. Since it cannot be said that the set of states given by one quadratic form $x^T Q^1 x \geq 0$ is better than another, all reasonable combinations are specified. The variables λ^k obtained by solving the resulting LMI problem (18) then decide the quadratic form such that $\mathcal{R}$ most suitably is replaced by the set of states satisfying the quadratic inequality.

Quadratic Forms Describing Hyperplanes. The quadratic forms equal to zero at a hyperplane can be obtained as follows. Assume that $\mathcal{R}$ is given by the set of states satisfying a hyperplane

$$c^T x = 0, \tag{20}$$

where $c = [c^1 \ldots c^n]^T \in \Re^n$. The states satisfying (20) also satisfy

$$2(\lambda^T x)^T (c^T x) = 0 \tag{21}$$

where $\lambda = [\lambda^1, \ldots, \lambda^n]^T \in \Re^n$ are arbitrary additional variables. The equality in (21) can be written as

$$x^T \lambda c^T x + x^T c \lambda^T x = \sum_{k=1}^{n} \lambda^k x^T Q^k x = 0$$

where

$$Q^k = e^k c^T + c(e^k)^T,$$

and e^k is a column vector with n elements such that

$$e^k(i) = \begin{cases} 1 \ i = k, \\ 0 \ i \neq k, \end{cases}$$

where i means the i:th element of e^k.

5 Stability of the Gear-Box Application

We are now prepared to show stability of the gear-box application. By denoting $\Delta v = v_{ref} - v$ and $\Delta T_I = T_I$, the closed-loop dynamics becomes

$$\begin{bmatrix} \Delta \dot{v} \\ \Delta \dot{T}_I \end{bmatrix} = \begin{bmatrix} -p_r K_r / M & -p_r / M \\ K_r / T_r & 0 \end{bmatrix} \begin{bmatrix} \Delta v \\ \Delta T_I \end{bmatrix},$$

where $M = 1500$, $T_r = 40$, $p_r \in \{50, 32, 20, 14\}$, $K_r \in \{3.75, 5.86, 9.37, 13.39\}$ and $p_r K_r = 187.5$ for all discrete states. For a specified desired velocity v_{ref} ($= 30\,\text{m/s}$) the system converges exponentially to v_{ref}, illustrated in Figure 3, and formally proven next.

If it is first assumed that there are no state jumps in T_I, stability can be shown by a single partitioning, implying a single Lyapunov function common for all discrete states. This results in a solution

$$P = \begin{bmatrix} 255.589 & 72.262 \\ 72.262 & 40.822 \end{bmatrix}$$

satisfying the conditions in the LMI problem. Hence, the hybrid system is globally exponentially stable without state jumps. The optimal value of $\beta = 277.6388$.

If the state jumps are included in the dynamics, they occur when the discrete state is changed. Trajectories satisfying the condition $T_I > K_r(v - v_{ref})$ cross the switch set $S_{i,i+1}$ and $S_{i+1,i}$ from left to right (see Figur 3) and oppositely for $T_I < K_r(v - v_{ref})$. In the operating region of this cruise controller ($T_I(0)$ is always put to zero when a new desired velocity is given), the gear shiftings will always occur from lower to higher gear when the first condition is satisfied and conversely in the second case. Hence, the third condition of the LMI problem is formulated such that the energy decreases passing from gears i to $i+1$ satisfying $T_I > K_r(v - v_{ref})$ and gears $i+1$ to i satisfying $T_I < K_r(v - v_{ref})$.

Consider the case when the trajectories start in the first region. The jump condition (8) on the form (10) gives

$$G(m_i) = \begin{bmatrix} 1 & 0 \\ 0 & \frac{p_{r_{i+1}}}{p_{r_i}} \end{bmatrix} \qquad i = 1, 2, 3.$$

According to the LMI problem, the third LMI condition then becomes

$$G(m_i)^T P G(m_i) \leq P,$$

which is equivalent to

$$p^{2,2} \left(\frac{p_{r_{i+1}}}{p_{r_i}} \right)^2 \leq p^{2,2},$$

where $p^{2,2}$ is the $(2,2)$ element of P. Since $\left(\frac{p_{r_{i+1}}}{p_{r_i}} \right)^2 < 1$, the energy will decrease due to the state jumps for any quadratic function $x^T P x$. Therefore, the same solution as above verifies stability in this case. However, when the trajectories start in the second region, the jump condition (8) is the same as above except that p_{r_i} and $p_{r_{i+1}}$ change position. In this case, there will not exist any solution since $\left(\frac{p_{r_i}}{p_{r_{i+1}}} \right)^2 > 1$.

To overcome this problem, the state space is further partitioned to verify exponential stability. One quadratic candidate Lyapunov-like function is associated with each of the discrete states. The switch surfaces $\Lambda_{i,i+1}$ then coincide with $S_{i,i+1}$ for $i = 1, 2, 3$. Solving the LMI problem leads to a solution

$$P_1 = \begin{bmatrix} 304.082 & 87.089 \\ 87.089 & 376.934 \end{bmatrix}, P_2 = \begin{bmatrix} 248.013 & 79.625 \\ 79.625 & 144.215 \end{bmatrix},$$
$$P_3 = \begin{bmatrix} 212.101 & 59.328 \\ 59.328 & 53.788 \end{bmatrix}, P_4 = \begin{bmatrix} 147.495 & 53.571 \\ 53.571 & 24.112 \end{bmatrix}.$$

Hence, the hybrid system is exponentially stable also in the case of state jumps. The optimal value of $\beta = 439.9$. The level curves for the local quadratic Lyapunov functions are shown in Figure 3.

6   Conclusions

A gear-box application has served as a motivation for investigating stability of hybrid systems including state jumps. None of the results reported in the literature can deal with this additional complexity. Stability results for hybrid systems including state jumps are proposed in this paper using Lyapunov techniques. It has been shown how to formulate the search for the (piecewise quadratic) Lyapunov functions as a linear matrix inequality (LMI) problem. The theory has been applied to the gear-box application to formally show stability.

References

[1] A. Björnberg. Design of control algorithms for intelligent cruise control. Technical Report 184L, Control Engineering Lab, Chalmers University of Technology, 1994.

[2] S. Boyd, L. El Ghaoui, E. Feron, and V. Balakrishnan. *Linear Matrix Inequalities in System and Control Theory*. SIAM, 1994.

[3] M. S. Branicky. Stability of switched and hybrid systems. In *Proc. of the 33rd IEEE Conference on Decision and Control*, pages 3498–3503, Lake Buena Vista, Florida, 1994.

[4] R. Brockett. Hybrid systems in classical mechanics. In *Proc. of 13th IFAC*, pages c:473–476, 1996.

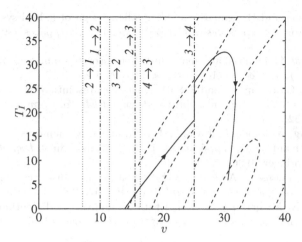

Fig. 3. The phase plot of v and T_I during a simulation of 80 seconds. The dash-dotted lines (-·) are the hyperplanes when the gear shifts from lower to higher gears $(i \rightarrow i+1)$ and oppositely $(i+1 \rightarrow i)$ at the dotted lines (··). The dashed lines $(--)$ are the level curves of the local quadratic Lyapunov functions. The continuous trajectories cross these curves such that the energy decreases all the time, verifying that the system is exponentially stable.

[5] R. W. Brockett. Hybrid models for motion control systems. In H. L. Trentelman and J. C. Willems, editors, *Essays on Control: Perspectives in the Theory and its Applications*, chapter 2, pages 29–53. Birkhäuser, 1993.

[6] M. Doğruel, S. Drakunov, and Ü. Özguner. Sliding mode control in discrete state systems. In *Proceedings of the 32nd CDC*, pages 1194–99, 1993.

[7] M. Doğruel and Ü. Özgüner. Stability of hybrid systems. In *IEEE International Symposium on Intelligent Control*, pages 129–134, 1994.

[8] P. Gahinet, A. Nemirovski, A. J. Laub, and M. Chilali. *LMI Control Toolbox, For use with MATLAB*. The Math Works Inc., 1995.

[9] A. Göllü and P. Varaiya. Hybrid dynamical systems. In *Proc. of the 28th IEEE Conference on Decision and Control*, pages 2708–2712, Tampa, Florida, Dec. 1989.

[10] I. A. Hiskens. Analysis tools for power systems – contending with nonlinearities. *Proceedings of the IEEE*, 83(11):1573–1587, 1995.

[11] W. S. Levine, editor. *The Control Handbook*. CRC Press, 1996.

[12] J. Lygeros, D. N. Godbole, and S. Sastry. Verified hybrid controllers for automated vehicles. *IEEE Trans. on Automatic Control*, 43(4):522–539, 1998.

[13] P. Peleties and R. DeCarlo. Modeling of interacting continuous time and discrete event systems: An example. In *Twenty-Sixth Annual Allerton Conference on Communication, Control and Computing*, pages 1150–9, 1988.

[14] P. Peleties and R. DeCarlo. Asymptotic stability of m-switched systems using Lyapunov-like functions. In *Proc. of the American Control Conference*, pages 1679–1684, Boston, 1991.

[15] S. Pettersson. *Analysis and Design of Hybrid Systems*. PhD thesis, Control Engineering Laboratory, Chalmers University of Technology, 1999.

[16] S. Pettersson and B. Lennartson. Controller design of hybrid systems. In Oded Maler, editor, *Lecture Notes in Computer Science 1201*, pages 240–254. Springer, 1997.

[17] L. Tavernini. Differential automata and their discrete simulators. *Nonlinear Analysis, Theory, Methods & Applications*, 11(6):665–83, 1987.

[18] C. Tomlin, G. J. Pappas, and S. Sastry. Conflict resolution for air traffic management: A study in multiagent hybrid systems. *IEEE Trans. on Automatic Control*, 43(4):509–521, 1998.

[19] L. Y. Wang, A. Beudoun, J. Cook, J. Sun, and I. Kolmanovsky. Optimal hybrid control with automotive applications. In A. S. Morse, editor, *Logic-based Switching Control*. Springer, 1996.

[20] H. S. Witsenhausen. A class of hybrid-state continuous-time dynamic systems. *IEEE Trans. on Automatic Control*, 11(2):161–67, 1966.

[21] H. Ye, A. N. Michel, and L. Hou. Stability analysis of discontinuous dynamical systems with applications. In *Proc. of 13th IFAC*, pages E:461–466, 1996.

Invariance of Approximating Automata for Piecewise Linear Systems with Uncertainties

Jacob Roll

Division of Automatic Control, Dept. of Electrical Engineering, Linköping University,
SE-581 83 Linköping, Sweden, email: roll@isy.liu.se

Abstract. A special class of hybrid systems, that occurs in many applications, are the piecewise linear systems. Due to their nonlinearity, they may often be difficult to analyse. Therefore, different approximating methods have been developed for analysis, verification and control design. This paper considers one such method, and gives a method for investigating how sensitive it is to changes in the dynamics of the underlying linear subsystems. This method can be used either for robustness analysis or for control design.

1 Introduction

Piecewise linear systems constitute a special class of hybrid systems. These systems consist of several linear (or rather affine) subsystems, between which switchings occur at different occasions. In this paper, the piecewise linear systems will be on the form

$$\dot{x} = A^i x + b^i , \quad x \in X^i , \quad i = 1, \ldots, N . \tag{1}$$

This implies that the dynamics of a trajectory $x(t)$ just depends on x, not on t or on any external input. The different regions X^i are assumed to be *polyhedra*, i.e., regions defined by linear inequalities. Systems of this kind occur in many applications. A very simple example of a piecewise linear system could be a linear system, controlled by linear feedback, but where the control signal is bounded.

Since piecewise linear systems, like other hybrid systems, are highly nonlinear, they might be difficult to analyse. Several approximating methods for analysis, verification, and control design have therefore been developed for different classes of hybrid systems, e.g., [2, 3, 4], [5], [7], [8], [10].

The problems considered in this paper arise for example when considering robustness aspects of the method proposed in [5], which is a method for verification of piecewise linear switched systems. In this method, the behaviour of the vector field $\dot{x}(t)$ at the borders of the regions X^i is analysed. Specifically, questions such as "At a given face of the polyhedron X^i, is there a point, x_0, such that $\dot{x}_0$ is pointing out of X^i, or are all trajectories at this face going into X^i?" are answered (this kind of computations has also been used by others, e.g., by [6]). The information obtained is used to determine which transitions between different regions are possible, which transitions are guaranteed to occur

N. Lynch and B. Krogh (Eds.): HSCC 2000, LNCS 1790, pp. 396–406, 2000.

non-deterministically (i.e., one transition out of a set of transitions from a given polyhedron is guaranteed to occur) and which are not. Then finite automata are constructed, showing the guaranteed or possible transitions. The finite automata give an approximation of the system, and can be used for different kinds of verification. For example, we can guarantee that certain states in the original system are not reachable from some other initial states, by proving that there is no sequence of possible transitions in the finite automata, taking the system state from the region of the initial states to the region of the final states.

Like all other methods mentioned above, the method in [5] assumes that a model of the system is given. It would be desirable to be able to determine how sensitive the approximating automata are to changes in the underlying linear subsystems. Such information could be used to get a measure of how robust the verification process is to model errors, or as an aid in a control design process, if we would like to adjust the system dynamics without losing the verified property. Sometimes we would only be interested in that some crucial transitions should not change, whereas in other cases we might want the entire approximating automata to remain invariant.

Since the approximating method considers the behaviour of $\dot{x}(t)$ at the borders of the regions X^i, we must determine how this behaviour changes with varying A^i and b^i. That is the topic of this paper.

2   Notation and Problem Formulation

The systems considered in this paper are on the form

$$\dot{x} = (A^i - \Delta^i)x + b^i - \delta^i, \quad x \in X^i, \quad i = 1, \ldots, N, \tag{2}$$

where $A^i \in \mathbb{R}^{n \times n}$, $b^i \in \mathbb{R}^n$ and $X^i \subset \mathbb{R}^n$ are given, while $\Delta^i \in \mathbb{R}^{n \times n}$, $\delta^i \in \mathbb{R}^n$ can be viewed either as uncertainties in the model, or as matrices of our choice. To begin with, the regions X^i will be *polytopes*, i.e., they are bounded regions defined by

$$X^i = \{x \in \mathbb{R}^n \mid C^i x \preccurlyeq d^i\}, \tag{3}$$

where $C^i \in \mathbb{R}^{m^i \times n}$, $m^i > n$, and $\preccurlyeq$ denotes componentwise inequality. The case of unbounded regions is considered in Sect. 3.2.

We will use the notation C_l^i for the lth row of C^i, and for example d_l^i for the lth element of a vector d^i. We will also introduce the notation

$$X_l^i = \{x \in \mathbb{R}^n \mid C^i x \preccurlyeq d^i, C_l^i x = d_l^i\}, \tag{4}$$

i.e., X_l^i is the face of the polytope corresponding to equality in the lth constraint.

Now consider one of the polytopes, say X^0, and how the trajectories $x(t)$ behave inside it. At a given face of the polytope, say X_m^0, and for given Δ^0 and δ^0, there are three different options for the qualitative behaviour of the trajectories (see Fig. 1):

1. They are all exiting the polytope.
2. They are all entering the polytope.
3. There exists (at least) one point in X_m^0, where the trajectories are parallel to the polytope face. In this latter case we may have some trajectories exiting the polytope and others entering it through the same face.

Since the system is linear inside the polytope, the trajectories are smooth, and therefore these three cases are the only possible options. The three different cases will lead to different approximating automata. An interesting question is: How much could Δ^0 and δ^0 change, without changing the qualitative behaviour at each face of the polytope, i.e., without affecting the approximating automata? In other words, for what values of Δ^0 and δ^0 do the different cases occur?

In the following section these problems are solved, and an example is given in Sect. 6.

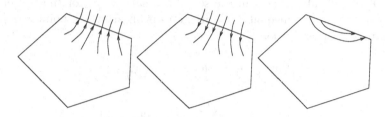

Fig. 1. Three options for the behaviour of the trajectories in the vicinity of a polytope face.

3   Solutions to the Problems

Since C_m^0 is a normal vector of the polytope face X_m^0, we can easily see that the three cases from the previous section correspond to the three problems

1. $C_m^0 \dot{x} > 0$ for all $x \in X_m^0$,
2. $C_m^0 \dot{x} < 0$ for all $x \in X_m^0$,
3. $C_m^0 \dot{x} = 0$ for some $x \in X_m^0$,

so our task is to find the sets of solutions (in Δ^0 and δ^0) to all these problems.

Let us begin with the first problem. By using (2) we can rewrite it as

$$C_m^0[(A^0 - \Delta^0)x + (b^0 - \delta^0)] > 0 \quad \text{for all } x \in X_m^0 , \tag{5}$$

or

$$C_m^0(A^0 x + b^0) > C_m^0(\Delta^0 x + \delta^0) \quad \text{for all } x \in X_m^0 , \tag{6}$$

This last form has a natural interpretation: On the left hand side we have the nominal flow through the face of the polytope, and the right hand side is the part of the flow that is affected by the variable matrices Δ^0 and δ^0. What (6) tells us is that the variable flow must be made small enough; otherwise we will get a total flow in the other direction from what was specified.

To get a solution to (6) we need to find a *direct representation* of X_m^0:

$$X_m^0 = \{\sum_{j=1}^{r} \lambda_j v_j \mid \lambda_j \in \mathbb{R}, \ \lambda_j \geq 0, \ \sum_{j=1}^{r} \lambda_j = 1\} \ . \tag{7}$$

Here $v_j \in \mathbb{R}^n$, $j = 1, \ldots r$ are the corners of X_m^0. In words, the direct representation means that we write each point of X_m^0 as a "weighted mean", or *convex combination*, of the corners.

Now, the set of solutions is given by

$$S = \{(\Delta^0, \delta^0) \mid C_m^0(A^0 v_j + b^0) > C_m^0(\Delta^0 v_j + \delta^0), \ j = 1, \ldots, r\} \ . \tag{8}$$

Proof. Obviously, all (Δ^0, δ^0) in the set above satisfy (6) for the corners v_j of X_m^0. To show that the inequality is satisfied for an arbitrary point $x \in X_m^0$, we use the direct representation (7):

$$C_m^0(A^0 x + b^0) = C_m^0(A^0 \sum_{j=1}^{r} \lambda_j v_j + b^0) =$$

$$= \sum_{j=1}^{r} \lambda_j C_m^0(A^0 v_j + b^0) >$$

$$> \sum_{j=1}^{r} \lambda_j C_m^0(\Delta^0 v_j + \delta^0) =$$

$$= C_m^0(\Delta^0 \sum_{j=1}^{r} \lambda_j v_j + \delta^0) =$$

$$= C_m^0(\Delta^0 x + \delta^0)$$

□

Note that the solution set is a polyhedron in the space $\mathbb{R}^{n \times n} \times \mathbb{R}^n$, and therefore convex.

The second problem ($C_m^0 \dot{x} < 0$ for all $x \in X_m^0$) is treated in the same way as the first. The solution set of the third problem can then be obtained as the complement of the first two solution sets.

3.1 Multiple Requirements

So far, we have only been looking at one single polytope face. In most cases, the requirements may stipulate that several transitions of an approximating automaton should remain invariant. This case is easily handled by partitioning the problem into subproblems of the form treated above, and then taking the intersection of the solution sets as the solution set for the entire problem. How many transitions we need to consider will depend on the system and what we want to verify. For example, if all we are interested in is keeping the state on one side of a hyperplane, we only need to consider transitions through this hyperplane. It should be noted that considering fewer transitions will lead to a larger – and therefore less conservative – solution set, and will also require less computations.

3.2   Unbounded Polyhedra

In some cases we might need to consider not only regions that are polytopes, but also include the case that X^0 is an unbounded polyhedron. This means that also X_m^0 could be unbounded. A direct representation of X_m^0 would then be:

$$X_m^0 = \{\sum_{j=1}^{r+h} \lambda_j v_j \mid \lambda_j \in \mathbb{R}, \ \lambda_j \geq 0, \ \sum_{j=1}^{r} \lambda_j = 1\} \ . \tag{9}$$

As before, $v_j \in \mathbb{R}^n$, $j = 1, \ldots r$ are the corners of X_m^0, but in addition we have h vectors, $v_{r+1}, \ldots, v_{r+h}$, which are parallel to the unbounded edges of X_m^0. Note that $\lambda_{r+1}, \ldots, \lambda_{r+h}$ are not included in the set of λ_j that should sum up to one; they can be arbitrarily large.

In this case, the solution set S' for the first problem is given by

$$\begin{aligned} S' = \{(\Delta^0, \delta^0) \mid & C_m^0(A^0 v_j + b^0) > C_m^0(\Delta^0 v_j + \delta^0), \ \ j = 1, \ldots, r; \\ & C_m^0 A^0 v_{r+j} \geq C_m^0 \Delta^0 v_{r+j}, \ \ j = 1, \ldots, h\} \ . \end{aligned} \tag{10}$$

The proof is analogous to the bounded case.

4   Interpretations

Perhaps the most obvious interpretation is to view Δ^i and δ^i as uncertainties due to model errors and/or noise. The algorithm then provides bounds for the uncertainties for the requirements of the approximating automata to hold. For natural reasons, the bounds may be very asymmetric, indicating that the system is more sensitive to certain types of model errors than to others.

The problem formulation is quite general in that no structure of Δ^i and δ^i is assumed. If the uncertainty has some structure, we can parametrise Δ^i and δ^i accordingly, thereby reducing the dimensionality and simplifying the problem. For example, if $\Delta^i = 0$ we get a model with additive noise:

$$\dot{x} = A^i x + b^i - \delta^i, \quad x \in X^i, \quad i = 1, \ldots, N \ . \tag{11}$$

It should be noted that this case is much easier to solve than the general one: It turns out that it can be reduced to solving two LP problems. For further details, see [9].

Another parametrisation is used in the example in Sect. 6. In this parametrisation some of the elements of Δ^i and δ^i are common to several polyhedra.

An alternative interpretation is to consider Δ^i and δ^i as parameters of our choice, to be used for control design. A natural parametrisation would then be $\delta^i = 0$, $\Delta^i = B^i L^i$, where B^i is a fixed vector that depends on the system, while we can choose L^i freely. In this way we get (piecewise) linear state feedback control, and the problem becomes that of finding the linear state feedback vectors L^i that make our system fulfil the requirements on the approximating automata.

5   Computational Complexity

From Sect. 3 we know that once we know a direct representation of X_m^i, it is trivial to divide $\mathbb{R}^{n \times n} \times \mathbb{R}^n$ into the three solution sets for the three different problems. Conversely, if we want the solutions to be written as intersections of halfspaces (as in (8)), we need to know the direct representation of X_m^i. Therefore the computational complexity for this problem is essentially identical to that of finding the direct representation. Unfortunately, the number of vectors needed in such a representation grows very quickly with the size of the problem. An upper bound for the number of corners in a polytope can be calculated in the following way: In a corner, n linearly independent faces meet (where n is still the dimension of the state space). Since the polytope has m faces, the maximal number of corners cannot be larger than $\binom{m}{n}$.

However, if we restrict ourselves to the case where the polyhedra are formed by the state space being divided by hyperplanes, it is fairly easy (but still quite time-consuming) to calculate the direct representation of all the polyhedra once and for all. The total number of corners is then bounded above by $\binom{M}{n}$, where M is the number of separating hyperplanes.

6   Example: A Chemical Reactor

To demonstrate the properties of this kind of problems, we can look at a simple example. In [5], a (fictional) chemical reactor is modeled, and a control strategy is proposed, after which some properties are verified. Here we assume that some of the parameter values are uncertain, and try to determine how large errors can be tolerated before the verification is not valid any more.

6.1   System Model

A figure of the chemical reactor is shown in Fig. 2. It consists of a tank containing a mixture of two fluids. When a certain temperature is reached, an exothermal reaction between the two fluids starts, giving the desired product. The temperature can be controlled by a heater and a cooler. There is also a blender helping to mix the fluid. The mixture is provided through an inflow valve. There is also a draining valve. The valves can be either open or closed.

The system model derived in [5] has two continuous state variables: the fluid level x_1 and the temperature x_2. Furthermore, there are six control signals, each

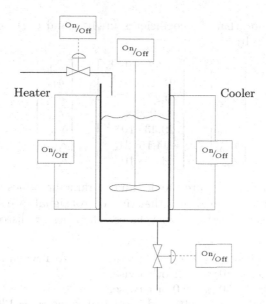

Fig. 2. A schematic figure of the chemical reactor

one taking a value in $\{0,1\}$. They are described in Table 1. It could be worth mentioning that u_r is an artificial, uncontrollable signal that indicates whether or not the reaction is in progress.

Table 1. Inputs to the chemical reactor

Signal	Interpretation
u_b	blender signal
u_i	inflow valve signal
u_d	draining valve signal
u_h	heater signal
u_c	cooler signal
u_r	reaction signal

The plant dynamics is described by

$$\dot{x} = A(u)x + b(u) \ , \tag{12}$$

where

$$A(u) = \begin{bmatrix} -a_h u_d & 0 \\ 0 & -(a_{T_1}(1 - u_b) + a_{T_2} u_b) \end{bmatrix} \tag{13}$$

$$b(u) = \begin{bmatrix} b_h u_i \\ b_{heat} u_h + b_{cool} u_c + b_{reac} u_r \end{bmatrix} \ . \tag{14}$$

Here we will assume that the coefficients in $A(u)$ and $b(u)$ are uncertain, and that they are given by

$$
\begin{bmatrix} a_h \\ a_{T_1} \\ a_{T_2} \\ b_h \\ b_{heat} \\ b_{cool} \\ b_{reac} \end{bmatrix} = \begin{bmatrix} 1.23 \cdot 10^{-3} \\ 0.15 \cdot 10^{-3} \\ 0.22 \cdot 10^{-3} \\ 9.838 \\ 29.43 \cdot 10^{-3} \\ -44.15 \cdot 10^{-3} \\ 44.15 \cdot 10^{-3} \end{bmatrix} - \begin{bmatrix} \delta_{ah} \\ \delta_{T_1} \\ \delta_{T_2} \\ \delta_{bh} \\ \delta_{heat} \\ \delta_{cool} \\ \delta_{reac} \end{bmatrix} , \tag{15}
$$

where the numerical values are the nominal parameter values used in [5].

The controller is designed such that the control signals are switched on or off when the state reaches certain hyperplanes. The rules are listed below:

1. $u_b = 0$ when $x_1 < 3$, $u_b = 1$ otherwise.
2. u_i is set to 0 when $25x_1 + x_2 = 300$, and is set to 1 when $25x_1 + x_2 = 250$.
3. $u_d = 0$ when $x_2 < 50$, $u_d = 1$ otherwise.
4. $u_h = 1$ when $x_2 < 50$, $u_h = 0$ otherwise.
5. u_c is set to 0 when $x_2 = 110$, and is set to 1 when $x_2 = 130$:
6. $u_r = 0$ when $x_2 < 50$, $u_r = 1$ otherwise.

Note that the system contains hysteresis in u_i and u_c. This is handled by considering each polytope where the hysteresis occurs as two polytopes with two different subsystems.

The switching hyperplanes and an example trajectory are shown in Fig. 3. For further details concerning the system model, see [5].

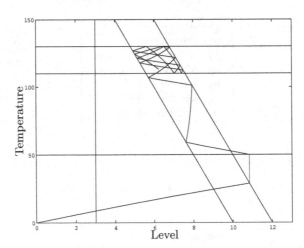

Fig. 3. The switching hyperplanes and an example trajectory.

6.2   What to Verify

There are certain requirements on the controller, which are verified in [5]. These are:

1. The temperature should stay between 0 and 150.
2. The tank must not be empty, and it must not overflow. The maximum level is 13.
3. There should be an operating region with moderate temperature and fluid level which is invariant. In [5], this region is chosen to be

$$\{x \mid 250 \le 25x_1 + x_2 \le 300,\ 110 \le x_2 \le 130\}\ . \qquad (16)$$

4. The operating region should always be reached from the initial states in finite time. For simplicity, and to avoid introducing additional conservatism, we will not consider this requirement in this paper.

The requirements can be translated to mathematical formulas:
1. (a) $\dot{x}_2 > 0$ when $0 \le x_1 \le 13$, $x_2 = 0$.
   (b) $\dot{x}_2 < 0$ when $0 \le x_1 \le 13$, $x_2 = 150$.
2. (a) $\dot{x}_1 > 0$ when $x_1 = 0$, $0 \le x_2 \le 150$.
   (b) $\dot{x}_1 < 0$ when $x_1 = 13$, $0 \le x_2 \le 150$.
3. (a) $\dot{x}_2 > 0$ when $250 \le 25x_1 + x_2 \le 300$, $x_2 = 110$.
   (b) $\dot{x}_2 < 0$ when $250 \le 25x_1 + x_2 \le 300$, $x_2 = 130$.
   (c) $\begin{bmatrix} 25 & 1 \end{bmatrix} \dot{x} > 0$ when $25x_1 + x_2 = 250$, $110 \le x_2 \le 130$.
   (d) $\begin{bmatrix} 25 & 1 \end{bmatrix} \dot{x} < 0$ when $25x_1 + x_2 = 300$, $110 \le x_2 \le 130$.

We also have to know what linear subsystems $\dot{x}$ will satisfy in the different cases. We get those by considering the control rules.

6.3   Deriving Bounds for Parameter Uncertainties

Since we assume that the parameter values are uncertain, the question is how large the errors can get before the requirements are violated. By using the algorithm in Sect. 3, we can get an exact answer to this question: The errors have to lie in a polyhedron described by

$$
\begin{bmatrix}
0 & 0 & 0 & 1 & 0 & 0 & 0 \\
0 & 0 & 0 & 0 & 1 & 0 & 0 \\
0 & 150 & 0 & 0 & 0 & -1 & -1 \\
0 & 0 & 150 & 0 & 0 & -1 & -1 \\
13 & 0 & 0 & 0 & 0 & 0 & 0 \\
-140 & 0 & -110 & 25 & 0 & 0 & 1 \\
-120 & 0 & -130 & 25 & 0 & 0 & 1 \\
0 & 0 & -110 & 0 & 0 & 0 & 1 \\
-140 & 0 & -110 & 25 & 0 & 1 & 1 \\
-120 & 0 & -130 & 25 & 0 & 1 & 1 \\
0 & 0 & 130 & 0 & 0 & -1 & -1 \\
190 & 0 & 110 & 0 & 0 & 0 & -1 \\
170 & 0 & 130 & 0 & 0 & 0 & -1 \\
190 & 0 & 110 & 0 & 0 & -1 & -1 \\
170 & 0 & 130 & 0 & 0 & -1 & -1
\end{bmatrix}
\begin{bmatrix}
\delta_{\mathrm{ah}} \\
\delta_{T_1} \\
\delta_{T_2} \\
\delta_{\mathrm{bh}} \\
\delta_{\mathrm{heat}} \\
\delta_{\mathrm{cool}} \\
\delta_{\mathrm{reac}}
\end{bmatrix}
<
\begin{bmatrix}
9.8380 \\
0.0294 \\
0.0225 \\
0.0330 \\
0.0160 \\
245.7977 \\
245.8179 \\
0.0200 \\
245.7536 \\
245.7738 \\
0.0286 \\
0.2137 \\
0.1935 \\
0.2579 \\
0.2377
\end{bmatrix}
\ . \qquad (17)
$$

We notice immediately that the polyhedron contains the origin, which means that the nominal system satisfies the requirements. To get a more intuitive feeling for the bounds, one can also consider a subset of the errors and set the other errors to zero. For example, suppose that only a_h and a_{T_2} are uncertain. In order not to violate the requirements, their deviations from the nominal values have to be contained in the polytope shown in Fig. 4. As we can see, δ_{T_2} basically has to lie in the interval $[-0.18 \cdot 10^{-3}, 0.22 \cdot 10^{-3}]$, while δ_h approximately can vary between -1.76 and $1 \cdot 10^{-3}$.

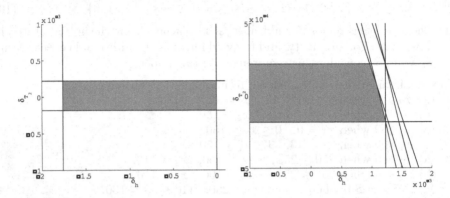

Fig. 4. The allowed region for δ_h and δ_{T_2}, assuming that the other errors are equal to zero. The right image shows a close-up of the region near the origin.

7 Conclusions

We have suggested an approach to investigate how sensitive approximating automata for piecewise linear systems, as described in [5], might be to changes in the underlying subsystems. Section 3 provided the sets of system matrices that satisfy certain demands on the behaviour of the system. As pointed out, these can either be seen as giving a measure of how robust the approximating automata are to uncertainties in the system, or as giving limits for how much the system can be changed, e.g., in a control design process, without altering the overall behaviour described by the approximating automata.

It would be natural to combine these demands with other constraints. For example, when using the state feedback parametrisation described in Sect. 4, one would probably want to find L^i that are optimal in a certain respect. Since the solution sets of the two first problems in Sect. 3 are convex, we can form all sorts of convex optimisation problems, which can be solved very efficiently once we know the direct representations of the polyhedra (see for example [1]).

The theory in this paper is immediately extendable to switched systems as described in [5]. The important thing is that the switch sets are fixed hyperplanes in the state space.

References

[1] Stephen Boyd and Lieven Vandenberghe. Convex optimization. Course Reader for EE364, Introduction to Convex Optimization with Engineering Applications, Stanford University, May 3, 1999.

[2] Alongkrit Chutinan and Bruce H. Krogh. Computing approximating automata for a class of hybrid systems. *Mathematical and Computer Modeling of Dynamical Systems: Special Issue on Discrete Event Models of Continuous Systems*, 1998.

[3] Alongkrit Chutinan and Bruce H. Krogh. Computing approximating automata for a class of linear hybrid systems. In *Hybrid Systems V, Lecture Notes in Computer Science*. Springer-Verlag, 1998.

[4] Alongkrit Chutinan and Bruce H. Krogh. Computing polyhedral approximations to flow pipes for dynamic systems. In *The 37th IEEE Conference on Decision and Control: Session on Synthesis and Verification of Hybrid Control Laws (TM-01)*, 1998.

[5] Valur Einarsson. On verification of switched systems using abstractions. Licentiate Thesis, Department of Electrical Engineering, Linköping University, SE-581 83 Linköping, Sweden. Thesis No. 705, 1998.

[6] Mikael Johansson. *Piecewise Linear Control Systems*. PhD thesis, Department of Automatic Control, Lund Institute of Technology, Lund University, Box 118, SE-221 00 Lund, Sweden, 1999.

[7] T. Moor and J. Raisch. Discrete control of switched linear systems. In *European Control Conference, ECC'99*, 1999.

[8] Patrick Philips, Martin Weiss, and Heinz A. Preisig. Control based on discrete-event models of continuous systems. In *European Control Conference, ECC'99*, 1999.

[9] Jacob Roll. Invariance of approximating automata for piecewise linear systems. Technical Report LiTH-ISY-R-2178, Department of Electrical Engineering, Linköping University, SE-581 83 Linköping, Sweden, 1999.

[10] O. Stursberg and S. Kowalewski. Approximating switched continuous systems by rectangular automata. In *European Control Conference, ECC'99*, 1999.

Decidable Controller Synthesis
for Classes of Linear Systems

Omid Shakernia[1], George J. Pappas[1,2], and Shankar Sastry[1]

[1] Department of EECS, University of California at Berkeley, Berkeley, CA 94704,
{omids,gpappas,sastry}@eecs.berkeley.edu
[2] Department of CIS, University of Pennsylvania, Philadelphia, PA 19104
pappasg@grip.cis.upenn.edu

Abstract. A problem of great interest in the control of hybrid systems is the design of least restrictive controllers for reachability specifications. Controller design typically uses game theoretic methods which compute the region of the state space for which there exists a control such that for all disturbances, an unsafe set is not reached. In general, the computation of the controllers requires the steady state solution of a Hamilton-Jacobi-Isaacs partial differential equation which is very difficult to compute, if it exists. In this paper, we show that for classes of linear systems, the controller synthesis problem is *decidable*: There exists a computational algorithm which, after a finite number of steps, will exactly compute the least restrictive controller. This result is achieved by a very interesting interaction of results from mathematical logic and optimal control.

1 Introduction

Reachability specifications for hybrid systems require the trajectories of a hybrid system to avoid an undesirable region of the state space. One of the most important problems in the control of hybrid systems is the design of least restrictive controllers which satisfy the reachability specifications. This problem has been considered in the context of classical discrete automata [3,15], timed automata [1], linear hybrid automata [18], and general hybrid systems [12]. The framework presented in [12] has been applied to automated vehicles [11], and air traffic management systems [16].

Designing least restrictive controllers for reachability specifications requires computing the set of all initial states for which there exists a control such that for all disturbances, the system will avoid the undesirable region. The least restrictive controller is then a static feedback controller which allows any control value outside this set of initial conditions while allowing all safe control values on the boundary of this set.

The computation of the safe set of initial states for general hybrid systems leads to game theoretic methods, and in particular to the steady state solution to Hamilton-Jacobi-Isaacs equations [12]. In general, these partial differential equations are very difficult to solve. In addition, steady state solutions, if they

N. Lynch and B. Krogh (Eds.): HSCC 2000, LNCS 1790, pp. 407–420, 2000.

exist, may be discontinuous even if the initial problem data is continuous. This is due to the appearance of shocks, and switchings in the optimal control policy.

The above difficulties in the computation of least restrictive controllers naturally raise the following question : *Can we find classes of systems where the game theoretic approach does not require the solution of the Hamilton-Jacobi-Isaacs equation?* In this paper, we give a positive answer to the above question for *normal* linear control systems where the system matrix is either nilpotent or diagonalizable with purely real rational eigenvalues, and with reachability specifications defined by polynomial inequalities. The normality condition requires controllability of the linear system with *each* input and disturbance. This condition ensures that the optimal control and disturbance are well defined, and unique. For the case of real eigenvalues, normality also ensures that the optimal control and disturbance have a finite number of switchings [13].

Our framework first applies Pontryagin's maximum principle to synthesize the optimal control and worst disturbance. The switching behavior of the control and the disturbance is then abstracted by a hybrid system, on which we perform reachability computations. By combining the recent decidability results of [8,9], with the normality condition which guarantees finite number of switchings [13], we show that the least restrictive controller can be *decidably* computed. This interesting interplay of results from mathematical logic and optimal control presents us with the first decidable controller synthesis problem for classes of linear systems.

2   Controller Synthesis Methodology

In this section, we briefly review the least restrictive controller synthesis methodology for dynamical systems as presented in [12]. Consider the dynamical system

$$\dot{x} = f(x, u, d) \tag{1}$$

with state $x \in \mathbb{R}^n$, controls $u \in U \subset \mathbb{R}^{n_u}$, disturbances $d \in D \subset \mathbb{R}^{n_d}$. Suppose there is a *target set* $G \subset \mathbb{R}^n$ which specifies an undesirable region of the state space. In the context of dynamic pursuit-evasion games [2,10], the goal of the disturbance is to *capture* the state by driving it into the target set, while the goal of the controller is to remain in the *safe set* G^c, the complement of G. The target set is described by $G = \{x \in \mathbb{R}^n \mid h(x) < 0\}$, for a smooth function $h : \mathbb{R}^n \to \mathbb{R}$.

Let $\mathcal{U}, \mathcal{D}$ be the set of piecewise continuous functions from $\mathbb{R}$ into U and D respectively. Given an initial condition $x_0 \in \mathbb{R}^n$, input $u(\cdot) \in \mathcal{U}$, and disturbance $d(\cdot) \in \mathcal{D}$, the *flow* of the differential equation (1) is a map $\Phi : \mathbb{R}^n \times \mathcal{U} \times \mathcal{D} \times \mathbb{R} \to \mathbb{R}^n$ given by

$$\Phi(x_0, u(\cdot), d(\cdot), t) = x_0 + \int_0^t f(x(\tau), u(\tau), d(\tau)) d\tau. \tag{2}$$

Clearly, the largest set of *safe initial states* for which the controller can avoid being captured regardless of the disturbance is given by

$$W = \{x_0 \in \mathbb{R}^n \mid \exists u(\cdot) \in \mathcal{U} \; \forall d(\cdot) \in \mathcal{D} \; \forall t \geq 0 \; : \; \Phi(x_0, u(\cdot), d(\cdot), t) \in G^c\}. \quad (3)$$

The set W is called the *maximal controlled invariant* subset of the safe set G^c. In the differential games literature, W is called the *escape set*, since there exists a control policy such that the controller can avoid the target set, and W^c is called the *capture set*. While equation (3) conceptually describes the escape set, it hardly affords a method of computing it. However, the capturability requirement can be encoded by a value function $J : \mathbb{R}^n \times \mathcal{U} \times \mathcal{D} \times \mathbb{R}_- \to \mathbb{R}$, which, given an initial state $x_0 \in \mathbb{R}^n$, $u(\cdot) \in \mathcal{U}$, $d(\cdot) \in \mathcal{D}$ and $t \leq 0$, returns

$$J(x_0, u(\cdot), d(\cdot), t) = h(x(0)).$$

Therefore, the value function is the cost of a trajectory that starts at initial state x_0 at time $t \leq 0$ and evolves according to system equation (1) with input $u(\cdot)$, disturbance $d(\cdot)$, and ends at final state $x(0)$ at time $t = 0$. Since the control tries to avoid G while the disturbance tries to steer the system to G, we naturally arrive at the dynamic game

$$J^*(x_0, t) = \max_{u \in \mathcal{U}} \min_{d \in \mathcal{D}} J(x_0, u(\cdot), d(\cdot), t).$$

J^* is called the optimal value function, since it is the value function corresponding to the optimal controls and disturbances of the dynamic game. The maximal controlled invariant subset of the safe set is described in terms of the optimal value function by

$$W = \{x \in \mathbb{R}^n \mid \min_{t \leq 0} J^*(x, t) \geq 0\}. \quad (4)$$

In order to compute $J^*(x, t)$, we first introduce the *Hamiltonian*

$$H(x, p, u, d) = p^T f(x, u, d), \quad (5)$$

where $p \in \mathbb{R}^n$ is called the *co-state*. The *optimal Hamiltonian* is given by

$$H^*(x, p) = \max_{u \in U} \min_{d \in D} H(x, p, u, d). \quad (6)$$

The computation of $J^*(x, t)$ requires the solution of a modified Hamilton-Jacobi-Isaacs partial differential equation [12]

$$\begin{aligned} J^*(x, 0) &= h(x) \\ -\frac{\partial J^*(x,t)}{\partial t} &= \min\{0, H^*(x, \frac{\partial J^*(x,t)}{\partial x})\}. \end{aligned} \quad (7)$$

Assuming that (7) has a differentiable solution that converges to a function $J_1^*(x)$ as $t \to -\infty$, then the set

$$W = \{x \in \mathbb{R}^n \mid J_1^*(x) \geq 0\} \quad (8)$$

is the maximal controlled invariant subset of the safe set G^c, and the controller $g : \mathbb{R}^n \to 2^U$ defined by

$$
g(x) = \begin{cases} \left\{ u \in U \mid \min_{d \in D} \left(\frac{\partial J_1^*(x)}{\partial x} \right)^T f(x, u, d) \geq 0 \right\} & \text{if } x \in \partial W \\ U & \text{if } x \in W^o \cup W^c \end{cases} \tag{9}
$$

is *least restrictive* controller which renders W invariant [12]. The controller (9) is least restrictive in the sense that if $g\prime : \mathbb{R}^n \to 2^U$ is any other controller that renders W invariant, then $\forall x \in \mathbb{R}^n$ we have $g\prime(x) \subseteq g(x)$.

The main difficulty in the above framework is the *computation* of W. In general, solving the Hamilton-Jacobi-Isaacs equation (7) seems necessary for exactly computing W. However, there are very difficult issues that must be resolved in this case:

1. Existence and uniqueness of solutions,
2. Existence and uniqueness of *steady state* solutions,
3. *Shocks*: non-smooth solutions to smooth problems,
4. Convergence of numerical algorithms.

Given the above difficulties, a natural direction of research is to find classes of systems for which some (or all) of these issues are resolved. In this paper, we adopt this point of view and we will prove the following theorem.

Theorem 1 (Decidable Controller Synthesis). *Consider the controller synthesis problem for the dynamical system*

$$
\dot{x} = Ax + Bu + Ed \tag{10}
$$

with controls $u \in U \subset \mathbb{R}^{n_u}$, disturbances $d \in D \subset \mathbb{R}^{n_d}$ and target set $G \subset \mathbb{R}^n$ given by

$$
G = \{ x \in \mathbb{R}^n \mid h(x) < 0 \}. \tag{11}
$$

Suppose the dynamical system and target set satisfy the following properties:

1. *$A \in \mathbb{Q}^{n \times n}$, $B \in \mathbb{Q}^{n \times n_u}$, $E \in \mathbb{Q}^{n \times n_d}$,*
2. *For each column b_i of B, the pair (A, b_i) is completely controllable,*
3. *For each column e_i of E, the pair (A, e_i) is completely controllable,*
4. *The feasible sets of controls U and disturbances D are compact rectangles with rational vertices, that is $U = \prod_{i=1}^{n_u} [\underline{U}_i, \overline{U}_i]$ and $D = \prod_{i=1}^{n_d} [\underline{D}_i, \overline{D}_i]$*
5. *$h \in \mathbb{Q}[x_1, x_2, ..., x_n]$ and $\frac{\partial h}{\partial x}(x) \neq 0$ when $h(x) = 0$.*

If A is nilpotent or diagonalizable with real rational eigenvalues, then the controller synthesis problem is decidable.

Linear systems that are completely controllable by each component of the input are called *normal* in the optimal control literature. It is well known that time-optimal controllers of normal systems have no *singular conditions*: conditions where the optimal input is undetermined for a finite time interval [6]. In fact, according to the Pontryagin's Maximum Principle [13], for a normal linear system, the time-optimal control *exists*, is *unique*, and is piecewise constant that taking values on the vertices of the feasible input set. Moreover, the optimal control has a *finite* number of switchings if the dynamic matrix A has purely real eigenvalues. These results will be crucial in establishing the well-posedness of our models, and the termination of the following controller synthesis procedure.

Controller Synthesis Methodology

1. *Apply Maximum Principle to obtain the saddle solution of optimal u^*, d^*.*
2. *Construct a hybrid system using the switching logic of optimal u^*, d^*.*
3. *Perform reachability computations on the constructed hybrid system.*
4. *Compute the least restrictive controller.*

In the next sections, we describe in detail each step of the above procedure.

3 Differential Games and the Maximum Principle

In this section, we apply results from differential game theory [2,10] to formulate the optimal control problem for our controller synthesis methodology. The Hamiltonian for the system (10), is given by $H(x, p, u, d) = p^T Ax + p^T Bu + p^T Ed$. The Hamiltonian satisfies the state and co-state differential equations

$$\dot{x} = \frac{\partial H}{\partial p}, \qquad \dot{p} = -\frac{\partial H}{\partial x}^T. \tag{12}$$

Consider the target set $G = \{x \in \mathbb{R}^n \mid h(x) < 0\}$. By setting $p(x, 0) = \frac{\partial h}{\partial x}(x)$, then for every $x \in \partial G$, $p(x, 0)$ is the outward pointing normal to ∂G at x. With this initial condition, the co-state is completely specified by

$$p(x, 0) = \frac{\partial h}{\partial x}(x), \qquad \dot{p}(x, t) = -A^T p(x, t). \tag{13}$$

Since the goal of the controller is to avoid G, the controller tries to maximize the Hamiltonian, while the disturbance tries to minimize it. In this case, the *Isaacs condition* [2], namely

$$\max_{u \in U} \min_{d \in D} H(x, p, u, d) = \min_{d \in D} \max_{u \in U} H(x, p, u, d), \tag{14}$$

is satisfied since the Hamiltonian is *separable, i.e.* $H(x, p, u, d) = H_1(x, p, u) + H_2(x, p, d)$. Satisfaction of the Isaacs condition implies that there exists a *saddle solution* of optimal controls and disturbances (u^*, d^*) such that

$$H(x, p, u, d^*) \leq H(x, p, u^*, d^*) \leq H(x, p, u^*, d).$$

The saddle solution of optimal controls and disturbances u^*, d^* satisfies the well-known Maximum Principle [13]

$$\begin{cases} u^*(x_0, t) \in \arg\max_{u \in U} p(x_0, t)^T Bu \\ d^*(x_0, t) \in \arg\min_{d \in D} p(x_0, t)^T Ed. \end{cases} \tag{15}$$

Equation (15) only constrains the optimal control and disturbance to lie in sets. We will soon see that under the normality condition, these sets are singletons, *i.e.* the optimal control and disturbance are unique. Starting from an initial $x_0 \in \partial G$, the input $u^*(x_0, \cdot)$ is the best the controller can do to avoid G regardless of the actions of the disturbance, while $d^*(x_0, \cdot)$ is the best the disturbance can do to drive the state towards G. These controls and disturbances are generally open-loop (as opposed to feedback) policies and are so-called "bang-bang controls" since they switch among the vertices of the set of admissible controls and disturbances. Notice that due to the separability of the Hamiltonian, the problem of computing a saddle solution to the dynamic game reduces to solving two linear optimal control synthesis problems.

Propositions 1 and 2 are fundamental for establishing the well-posedness of our controller synthesis methodology. The proofs are due to Pontryagin [13] and can be found in many optimal control texts, such as [6].

Proposition 1 (Nonsingular Optimal Control and Disturbance). *If the linear system (10) is normal with respect to both the control and disturbance, then for any $x_0 \in \partial G$, the optimal control $u^*(x_0, \cdot)$ and disturbance $d^*(x_0, \cdot)$ are unique and piece-wise constant taking values on the vertices of U, D.*

Proposition 2 (Finite Switchings of Optimal Control). *If the linear system (10) is normal and A has purely real eigenvalues, then there is a uniform upper bound, independent of x_0 on the number of switchings of the optimal control $u^*(x_0, \cdot)$, and disturbance $d^*(x_0, \cdot)$.*

4   Construction of Hybrid System

The switching policy of the optimal control and disturbance can be naturally abstracted as a hybrid system.

Definition 1 (Hybrid Systems). *A hybrid system is a tuple $H = (X, F, Inv, R)$ where*

- *$X = X_D \times \mathbb{R}^m$ is the state space with $X_D = \{q_0, \ldots, q_{k-1}\}$,*
- *$F : X_D \times \mathbb{R}^m \to \mathbb{R}^m$ assigns to each discrete location $q \in X_D$ a differential equation $\dot{x} = F(q, x)$,*
- *$Inv : X_D \to 2^{\mathbb{R}^m}$ assigns to each discrete location an invariant set $Inv(q) \subseteq \mathbb{R}^m$, and*
- *$R \subseteq X \times X$ is a relation capturing the discrete transitions .*

The elements of X_D are the *discrete* states whereas $x \in \mathbb{R}^m$ is the *continuous* state. Hybrid systems are typically represented as graphs with vertices X_D, and edges E defined by

$$E = \{(q, q') \in X_D \times X_D \mid (q, x, q', x') \in R \text{ for some } x, x' \in \mathbb{R}^m\}.$$

With each edge $e = (q, q') \in E$ we associate a guard set defined as

$$Guard(e) = \{x \in Inv(q) \mid (q, x, q', x') \in R \text{ for some } x' \in \mathbb{R}^m\}$$

and the set valued reset map

$$Reset(e, x) = \{x' \in Inv(q') \mid (q, x, q', x') \in R\}.$$

Due to switched nature of the optimal control and disturbance, in this paper, it will suffice to assume that for all $e \in E$, $Reset(e, x) = x$. Therefore, all reset maps will be the identity map. Furthermore, we do not require the explicit specification of any initial states for our hybrid system.

The solution of the dynamic game played between the control and the disturbance d can be *naturally encoded* by a hybrid system. The optimal controls and disturbances always lie on the vertices of the admissible set of controls and disturbances U and D which are n_u and n_d dimensional rectangles. Thus, there are $2^{n_u} \cdot 2^{n_d}$ possible vector fields associated with the optimal controls and disturbances. We can therefore construct a hybrid system with $2^{n_u} \cdot 2^{n_d}$ discrete states, one for each possible control/disturbance pair.

We naturally encode the discrete states as a string of boolean numbers of length $n_u + n_d$. The first n_u elements encode the value that the i-th component of the optimal control. Similarly the last n_d components encode the value of the optimal disturbance. We adopt the convention that 1 stands for the upper bound ($u_i^* = \overline{U}_i$ or $d_i^* = \overline{D}_i$), and 0 stands for the lower bound ($u_i^* = \underline{U}_i$ or $d_i^* = \underline{D}_i$). For example, in a system with two controls and one disturbance, the discrete state $(0, 0, 1)$ stands for the case where $u_1^* = \underline{U}_1$, $u_2^* = \underline{U}_2$, and $d_1^* = \overline{D}_1$. It is therefore clear that the number of discrete states is $2^{n_u + n_d}$, since X_D contains all such boolean strings. According to which is notationally most convenient in the context, we will refer to discrete state k as either q_k or the boolean string that represents k in binary. That is, for the example above we may refer to discrete state 5 as either q_5 or $(1, 0, 1)$.

Since the optimal control depends on the co-state p, the continuous state associated with the hybrid system is actually $(x, p)^T \in \mathbb{R}^{2n}$. The vector field with each discrete state q_j then

$$\begin{pmatrix} \dot{x} \\ \dot{p} \end{pmatrix} = \begin{pmatrix} A & 0 \\ 0 & -A^T \end{pmatrix} \begin{pmatrix} x \\ p \end{pmatrix} + \begin{pmatrix} B \\ 0 \end{pmatrix} u_{q_j} + \begin{pmatrix} E \\ 0 \end{pmatrix} d_{q_j}, \tag{16}$$

where $u_{q_j} \in \mathbb{R}^{n_u}$ and $d_{q_j} \in \mathbb{R}^{n_d}$ are the constant controls and disturbances associated with discrete state q_j.

Let $(s_1, \ldots, s_{n_u}, t_1, \ldots, t_{n_d}) \in X_D$ where all the s_i and t_i are either zero or one. Consider the formulas

$$I_i^u(s) = \begin{cases} p^T(-A)^{\beta_i(p)} b_i > 0 \text{ if } s = 1 \\ p^T(-A)^{\beta_i(p)} b_i < 0 \text{ if } s = 0 \end{cases} \tag{17}$$

$$I_i^d(s) = \begin{cases} p^T(-A)^{\varepsilon_i(p)} e_i < 0 \text{ if } s = 1 \\ p^T(-A)^{\varepsilon_i(p)} e_i > 0 \text{ if } s = 0, \end{cases} \tag{18}$$

where b_i and e_i are the columns of B and E respectively, and $\beta_i(\cdot), \varepsilon_i(\cdot)$ are the *relative degrees* that are now defined.

Definition 2 (Relative Degree). *The relative degrees of the i-th input and disturbance are functions $\beta_i, \varepsilon_i : \mathbb{R}^n \to \mathbb{Z}$ defined by:*

$$\beta_i(p) = \begin{cases} 0 \text{ if } p^T b_i \neq 0 \\ 1 \text{ if } p^T b_i = 0 \wedge p^T(-A)b_i \neq 0 \\ \vdots \\ j \text{ if } \bigwedge_{k=0}^{j-1} p^T(-A)^k b_i = 0 \wedge p^T(-A)^j b_i \neq 0 \end{cases} \tag{19}$$

$$\varepsilon_i(p) = \begin{cases} 0 \text{ if } p^T e_i \neq 0 \\ 1 \text{ if } p^T e_i = 0 \wedge p^T(-A)e_i \neq 0 \\ \vdots \\ j \text{ if } \bigwedge_{k=0}^{j-1} p^T(-A)^k e_i = 0 \wedge p^T(-A)^j e_i \neq 0. \end{cases} \tag{20}$$

The invariant set associated with discrete state $(s_1, \ldots, s_{n_u}, t_1, \ldots, t_{n_d})$ is simply

$$Inv((s_1, \ldots, s_{n_u}, t_1, \ldots, t_{n_d})) = \bigwedge_{i=1}^{n_u} I_i^u(s_i) \wedge \bigwedge_{j=1}^{n_d} I_i^d(t_j). \tag{21}$$

In other words, the optimal control and disturbance remain the same as long as the signs of all components of $p^T B$ and $p^T E$ do not change. Proposition 1 ensures that components of $p^T B$ and $p^T E$ cannot be zero for nontrivial intervals of time, and, furthermore, if some component of $p^T B$ or $p^T E$ is momentarily zero, the optimal control and disturbance can be uniquely determined by looking at the first nonzero Lie derivative.

Since, in general, the optimal policy can jump from any control/disturbance pair to any other control/disturbance pair, the edge relation E is all of $X_D \times X_D$. Consider discrete states $(s_1^1, \ldots, s_{n_u}^1, t_1^1, \ldots, t_{n_d}^1)$ and $(s_1^2, \ldots, s_{n_u}^2, t_1^2, \ldots, t_{n_d}^2)$ and let J_u be the set of indices i in $\{1, \ldots, n_u\}$ such that $s_i^1 \neq s_i^2$. Thus J_u contains the indices of all control components that switch optimal policy. Similarly define J_d. The guard that enables the transition e from $(s_1^1, \ldots, s_{n_u}^1, t_1^1, \ldots, t_{n_d}^1)$ to $(s_1^2, \ldots, s_{n_u}^2, t_1^2, \ldots, t_{n_d}^2)$ is given by

$$Guard(e) = \bigwedge_{i \in J_u} I_i^u(\overline{s}_i) \wedge \bigwedge_{j \in J_d} I_i^d(\overline{t}_j). \tag{22}$$

where $\overline{s}$ denotes the boolean complement of s.

Notice that for each discrete state, the invariant and the guard depend only on the co-state p. Therefore, there are formulas $Inv_j : \mathbb{R}^n \to \{\text{TRUE, FALSE}\}$ for $j \in \{0, \ldots, 2^{n_u+n_d} - 1\}$ such that

$$Inv(q_j) = \{(x, p)^T \in \mathbb{R}^{2n} \mid Inv_j(p)\} \tag{23}$$

The formulas Inv_j will be used for notational convenience in the reach set computation of the next section. This concludes the specification of the optimal control

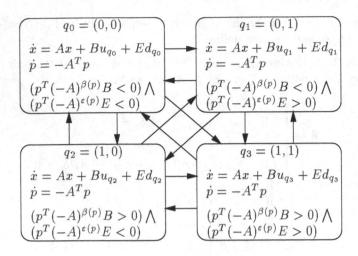

Fig. 1. Natural encoding of game solution as a hybrid system

policy as a hybrid system. Figure 1 shows a block diagram of a hybrid system constructed out of a differential game between one control and one disturbance.

From Propositions 1 and 2 it is straightforward to show that the hybrid system we construct is also well defined in the following sense.

Proposition 3 (Properties of Hybrid System). *The hybrid system constructed above is nonblocking, deterministic, and non-Zeno.*

The problem of computing the maximal controlled invariant set W has thus been transformed to the problem of computing all states of the hybrid system constructed above that the x component of the continuous state can reach G. This reachability computation is the goal of the next section.

5 Reachability Computation

For the vector field associated with each discrete state q_j , we define the *predecessor* operator $\text{Pre}_j : 2^{\mathbb{R}^{2n}} \to 2^{\mathbb{R}^{2n}}$. Suppose a set $K \subset \mathbb{R}^{2n}$ is defined by $K = \{(x, p) \in \mathbb{R}^{2n} \mid P(x, p)\}$. Then $\text{Pre}_j(K)$ is defined by

$$\text{Pre}_j(K) = \{(x, p)^T \in \mathbb{R}^{2n} \mid \exists y \; \exists q \; \exists t \; : \; P(y, q) \wedge t \geq 0 \wedge \; q = e^{-tA^T}p$$
$$\wedge \; y = e^{tA}x + (\textstyle\int_0^t e^{(t-s)A}ds)(Bu_{q_j} + Ed_{q_j}) \quad (24)$$
$$\wedge \; \forall s : 0 \leq s \leq t \Rightarrow \text{Inv}_j(e^{-sA^T}p)\}.$$

An immediate corollary of the main theorem of [9], which is based on the results in [7,8], is the following:

Proposition 4. *Consider a semialgebraic set $K \subset \mathbb{R}^n$ and a dynamic system $\dot{x} = Ax + b$ where $A \in \mathbb{Q}^{n \times n}$, $b \in \mathbb{Q}^n$. If A is nilpotent or diagonalizable with real rational eigenvalues, then computing the states that can reach K is decidable.*

Proof. Suppose the K is defined by $K = \{x \in \mathbb{R}^n \mid P(x)\}$. By defining

$$\phi(x,t) = e^{At}x + \int_0^t e^{A(t-s)}b \, ds,$$

we have that the set of states that can reach K is given by $\{x \in \mathbb{R}^n \mid \exists y \exists t : P(y) \wedge t \geq 0 \wedge y = \phi(x,t)\}$. In order to prove the result, we must show that for each condition on A above, $\phi(x,t)$ can be converted to an equivalent formula in $(\mathbb{R}, <, +, \cdot, 0, 1)$, which is decidable. If A is nilpotent, then each entry of e^{At} is polynomial in t. Therefore each entry of $\phi(x,t)$ is polynomial in t and hence definable in $(\mathbb{R}, <, +, \cdot, 0, 1)$. If A is diagonalizable with real eigenvalues then each entry of e^{At} is a linear combination of the functions $e^{\lambda_i t}$ with λ_i an eigenvalue of A. Since the entries of e^{-As} are linear combinations of $e^{-\lambda_i s}$, after integration the entries of $\phi(x,t)$ are linear combinations of $e^{\lambda_i t}$, $e^{-\lambda_i t}$. If $\lambda \in \mathbb{Q}$, then by the procedure outlined in [9], $\phi(x,t)$ may be converted into an equivalent formula in $(\mathbb{R}, <, +, \cdot, 0, 1)$. $\qquad\square$

An immediate result of Proposition 4 is that the computation of $\text{Pre}_j(K)$ is decidable for each discrete state q_j if K is a semialgebraic set, and A is either nilpotent, or is diagonalizable with real eigenvalues. Notice that if K is semialgebraic, then so is $\text{Pre}_j(K)$.

Now, our goal is to compute all the states of the dynamical game (10) for which the disturbance can drive the state into reach the target set G regardless of the input. In fact, it is only necessary to compute the states that for which the disturbance can drive the state to the "Usable Part" of G:

Definition 3 (Usable Part). *The* Usable Part *(UP) of the target set G is the subset of ∂G for which the disturbance can instantaneously drive the state into G regardless of the control action. Thus UP for the dynamic game (10) and the target set (11) is given by:*

$$UP = \left\{ x \in \partial G \mid \forall u \in U \ \exists d \in D \ \left(\tfrac{\partial h(x)}{\partial x}\right)^T (Ax + Bu + Ed) < 0 \right\}. \quad (25)$$

Since $h(x)$ is a polynomial, then the defining formula (25) for UP is definable in the theory of the reals $(\mathbb{R}, <, +, \cdot, 0, 1)$ which is known to admit quantifier elimination and be decidable [14]. Therefore computing UP is decidable. Since the hybrid system has the identity as its reset map, any trajectory that enters G if and only if is passes through UP. Now, we need to convert our reachability specification of the linear system (10) into a specification for our abstracted hybrid system. To this end, we define the set

$$\widetilde{UP} = \left\{ (x,p) \in \mathbb{R}^{2n} \mid x \in UP, p = \tfrac{\partial h(x)}{\partial x}^T \right\}. \quad (26)$$

Define $\text{Pre} : 2^{\mathbb{R}^{2n}} \to 2^{\mathbb{R}^{2n}}$ such that for a given $K \subset \mathbb{R}^{2n}$, $\text{Pre}(K)$ is the set of all states of the hybrid system that can reach K. It is easily seen that

$$W^c = \{x \in \mathbb{R}^n \mid \exists p : (x,p) \in \text{Pre}(\widetilde{UP}) \vee x \in G\} \quad (27)$$

Therefore, the computation of W is decidable if and only if the computation of $\mathrm{Pre}(\widetilde{\mathrm{UP}})$ is decidable.

We now turn to computing $\mathrm{Pre}(\widetilde{\mathrm{UP}})$. In general, different states on UP may require different optimal control and disturbance values. We therefore partition $\widetilde{\mathrm{UP}}$ into a disjoint union of subsets according to the optimal controls and disturbances. That is, we partition $\widetilde{\mathrm{UP}} = \bigcup_{q \in Q} S_q$, where $Q \subseteq X_D$ and the set S_q contains those state of $\widetilde{\mathrm{UP}}$ for which the optimal control and disturbance of the are represented by discrete state q of the hybrid system. Since there are only $2^{n_u + n_d}$ possible optimal controls and disturbances, the partition is finite. Using similar quantifier elimination arguments, it is straightforward to show that the computation of this partition is decidable. Since we have that

$$\mathrm{Pre}(\widetilde{\mathrm{UP}}) = \mathrm{Pre}\left(\bigcup_{q \in Q} S_q\right) = \bigcup_{q \in Q} \mathrm{Pre}(S_q), \tag{28}$$

we can concentrate of computing $\mathrm{Pre}(S_q)$ for a given $q \in Q$. We know that the initial optimal control and disturbance is equal for all initial conditions in S_q, therefore S_q is contained within the same discrete state.

Theorem 2 (Computation of Maximal Controlled Invariant Set). *Consider a dynamic game $\dot{x} = Ax + Bu + Ed$ with controls $u \in U \subset \mathbb{R}^{n_u}$, disturbances $d \in D \subset \mathbb{R}^{n_d}$ and target set $G \subset \mathbb{R}^n$ given by $G = \{x \in \mathbb{R}^n \mid h(x) < 0\}$. Suppose the system and target set satisfy the following properties:*

1. *$A \in \mathbb{Q}^{n \times n}$, $B \in \mathbb{Q}^{n \times n_u}$, $E \in \mathbb{Q}^{n \times n_d}$,*
2. *For each column b_i of B, the pair (A, b_i) is completely controllable,*
3. *For each column e_i of E, the pair (A, e_i) is completely controllable,*
4. *The feasible sets of controls U and disturbances D are compact rectangles with rational vertices,*
5. *$h \in \mathbb{Q}[x_1, x_2, ..., x_n]$ and $\frac{\partial h}{\partial x}(x) \neq 0$ when $h(x) = 0$.*

If A is nilpotent or diagonalizable with real rational eigenvalues, then the computation of the maximal controlled invariant set W is decidable.

Proof. Due to our partition in equation (28), we have

$$W^c = \{x \in \mathbb{R}^n \mid \exists p : (x, p) \in \mathrm{Pre}(\widetilde{\mathrm{UP}})\} \cup G \tag{29}$$

$$= \bigcup_{q \in Q} \{x \in \mathbb{R}^n \mid \exists p : (x, p) \in \mathrm{Pre}(S_q)\} \cup G \tag{30}$$

where each of the above steps is decidable. Thus it suffices to show that for a given $q \in Q$ computing $\mathrm{Pre}(S_q)$ is decidable.

Due to Proposition 3, we need not worry about any pathologies in the Pre computation. Since, in each discrete state, the optimal input and disturbance are constant we apply Proposition 4 to decidably compute the set of states that can reach S_q for that particular combination of control/disturbance pair. However, the optimal control or disturbance may change and a discrete transition may be taken. The predecessor operator of the discrete jumps is trivial since the reset map of our jumps is the identity map.

Now, if the matrix A has real eigenvalues, then due to Proposition 2 after a finite number of switchings, uniformly in $(x, p)^T \in S_q$, there are no more switches and we can use Proposition 4 one last time. Therefore, the algorithm terminates after a finite number of steps.   □

6   Least Restrictive Controller

Our goal in this section is to compute the least restrictive controller that renders the maximum controlled invariant set W invariant. The result of the previous section is that W is definable in $(\mathbb{R}, <, +, \cdot, 0, 1)$ which is decidable [14]. Since $(\mathbb{R}, <, +, \cdot, 0, 1)$ admits quantifier elimination, we may compute a quantifier-free formula ψ such that $W = \{x \in \mathbb{R}^n \mid \psi(x)\}$. The quantifier elimination that is required in this procedure can be done by the computer logic software systems REDLOG [5] or QEPCAD [4]. The defining formula of the set W may be converted to the so-called *disjunctive normal form* to yield:

$$W = \left\{ x \in \mathbb{R}^n \mid \bigvee_{j=1}^{L} \left(\bigwedge_{k=1}^{M_j} f_{j_k}(x) \; m_{j_k} \; 0 \right) \right\} \tag{31}$$

where $f_{j_k} \in \mathbb{Q}[x_1, \dots, x_n]$ and $m_{j_k} \in \{<, \leq, <, \geq, =, \neq\}$.

Since the least restrictive controller specifies a control action *only on the boundary of W*, our first task is to compute the boundary of W, ∂W. We will need the following lemma from [17].

Lemma 1. *If $W \subset \mathbb{R}^n$ is definable in a decidable theory, then so is the closure $\overline{W}$, the interior W^o, and the boundary ∂W.*

Proof. For a set $W = \{x \in \mathbb{R}^n \mid \psi(x)\}$, the sets $\overline{W}$ and W^o are given by

$$\overline{W} = \{x \in \mathbb{R}^n \mid \forall(y_1, \dots, y_n) \, \forall(z_1, \dots, z_n) : [\wedge_{i=1}^n y_i < x_i < z_i \Rightarrow \tag{32}$$
$$\exists(w_1, \dots, w_n) : \wedge_{i=1}^n y_i < w_i < z_i \wedge \psi(w)]\}$$
$$W^o = \{x \in \mathbb{R}^n \mid \exists(y_1, \dots, y_n) \, \exists(z_1, \dots, z_n) : [\wedge_{i=1}^n y_i < x_i < z_i \wedge \tag{33}$$
$$\forall(w_1, \dots, w_n) : \wedge_{i=1}^n y_i < w_i < z_i \Rightarrow \psi(w)]\}$$

where we use the shorthand notation $(\alpha \Rightarrow \beta) \equiv (\neg\alpha \vee \beta)$. The expressions (32) and (33) are simply the definitions of closure and interior in the usual topology of $\mathbb{R}^n$. Let the defining formulas for $\overline{W}, W^o$ be $\overline{\psi}, \psi^o$ respectively. Then the defining formula for ∂W is simply $\partial\psi \equiv \overline{\psi} \wedge (\neg\psi^o)$. Clearly if ψ is defined in a theory which admits quantifier elimination, then so are $\overline{\psi}, \psi^o$, and $\partial\psi$.   □

From the Lemma 1 we have that ∂W may be defined by a quantifier-free formula

$$\partial W = \{x \in \mathbb{R}^n \mid \partial\psi(x)\}. \tag{34}$$

Since the least restrictive controller only specifies a control action on ∂W, then $g : \mathbb{R}^n \to 2^U$ must be of the form

$$g(x) = \{u \in U \mid \partial\psi(x) \Rightarrow \phi(x, u)\} \tag{35}$$

where $\phi(x, u)$ is a formula to be described below.

Denote $W = \bigcup_{j=1}^{L}(\bigcap_{k=1}^{M_j} W_{j_k})$, and consider the least restrictive controller for a single polynomial constraint $W_{j_k} = \{x \in \mathbb{R}^n \mid f_{j_k}(x) \; m_{j_k} \; 0\}$. For this polynomial constraint, we define the formula

$$\phi_{j_k}(x, u) \equiv \left((f_{j_k}(x) = 0) \Rightarrow \forall d \in D : \frac{\partial f_{j_k}(x)}{\partial x}^T (Ax + Bu + Ed) \; m_{j_k} \; 0 \right). \tag{36}$$

Using equation (36), it is direct to see that the least restrictive controller that renders W_{j_k} invariant is given by $g_{j_k}(x) = \{u \in U \mid \phi_{j_k}(x, u)\}$. This least restrictive controller is simply a re-writing of equation (9) in terms of a decidable formula. Now, the least restrictive controller for $W = \bigcup_{j=1}^{L}(\bigcap_{k=1}^{M_j} W_{j_k})$ must be satisfy each of the of the simpler constraints, and hence is given by the following.

Theorem 3 (Least Restrictive Controller). *For the differential game $\dot{x} = Ax + Bu + Ed$, the least restrictive controller $g : \mathbb{R}^n \to 2^U$ that renders the set $W = \{x \in \mathbb{R}^n \mid \bigvee_{j=1}^{L}(\bigwedge_{k=1}^{M_j} f_{j_k}(x) \; m_{j_k} \; 0)\}$ invariant is given by*

$$g(x) = \left\{ u \in U \mid \partial\psi(x) \Rightarrow \bigvee_{j=1}^{L} \left(\bigwedge_{k=1}^{M_j} \phi_{j_k}(x, u) \right) \right\}, \tag{37}$$

where $\partial\psi$ is the defining formula of ∂W and $\phi_{j_k}(x, u)$ is given by equation (36). If W is definable in a decidable theory, then so is $g(x)$.

Therefore, Theorems 2 and 3 collectively result in Theorem 1.

7   Conclusions

In this paper we have shown that controller synthesis for classes of linear systems with polynomial reachability specifications is decidable. In further research, we will extend the target set G to a semialgebraic set, investigate conditions for semi-decidability in the absence of the normality condition, and extend the continuous decidability results to semidecidability results for classes of linear hybrid systems. In the case of purely imaginary eigenvalues, the problem becomes quickly undecidable unless one remains in a compact region of the state space. The observation along with the results of this paper have a clear and natural connection with o-minimal theories of the reals [7,17], which will explored in future research.

Acknowledgments

This research has been supported by DARPA under grant F33615-98-C-3614, and DARPA/NASA grant NAG2-1214.

References

1. E. Asarin, O. Maler, and A. Pnueli. Symbolic controller synthesis for discrete and timed systems. In P. Antsaklis, W. Kohn, A. Nerode, and S. Sastry, editors, *Hybrid Systems II*, volume 999 of *Lecture Notes in Computer Science*. Springer-Verlag, 1995.
2. T. Başar and G.J. Olsder. *Dynamic Noncooperative Game Theory*. Academic Press, 2nd edition, 1995.
3. A. Church. Logic, arithmetic, and automata. In *Proceedings of the International Congress of Mathematics*, pages 23–35, 1962.
4. G.E. Collins and H. Hong. Partial cylindrical algebraic decomposition for quantifier elimination. *Journal of Symbolic Computation*, 12:299–328, September 1991.
5. A. Dolzman and T. Sturm. REDLOG : Computer algebra meets computer logic. *ACM SIGSAM Bulletin*, 31(2):2–9, June 1997.
6. D.E. Kirk. *Optimal Control Theory, An Introduction*. Prentice Hall, 1970.
7. G. Lafferriere, G. J. Pappas, and S. Sastry. O-minimal hybrid systems. *Mathematics of Control, Signals, and Systems*. To appear.
8. G. Lafferriere, G. J. Pappas, and S. Yovine. A new class of decidable hybrid systems. In *Hybrid Systems : Computation and Control*, volume 1569 of *Lecture Notes in Computer Science*, pages 137–151. Springer Verlag, 1999.
9. G. Lafferriere, G. J. Pappas, and Sergio Yovine. Reachability computation for linear hybrid systems. In *Proceedings of the 14th IFAC World Congress*, volume E, pages 7–12, Beijing, P.R. China, July 1999.
10. J. Lewin. *Differential Games*. Springer Verlag, 1994.
11. J. Lygeros, D.N. Godbole, and S. Sastry. Verified hybrid controllers for automated vehicles. *IEEE Transactions on Automatic Control*, 43(4):522–539, April 1998.
12. J. Lygeros, C. Tomlin, and S.S. Sastry. Controllers for reachability specifications for hybrid systems. *Automatica*, 35(3):349–370, March 1999.
13. L.S. Pontryagin, V. Boltyanskii, R. Gamkrelidze, and E. Mischenko. *The Mathematical Theory of Optimal Processes*. John Wiley & Sons, 1962.
14. A. Tarski. *A decision method for elementary algebra and geometry*. University of California Press, second edition, 1951.
15. W. Thomas. On the synthesis of strategies in infinite games. In Ernst W. Mayr and Claude Puech, editors, *Proceedings of STACS 95, Volume 900 of LNCS*, pages 1–13. Springer Verlag, Munich, 1995.
16. C. Tomlin, G. J. Pappas, and S. Sastry. Conflict resolution for air traffic management : A study in muti-agent hybrid systems. *IEEE Transactions on Automatic Control*, 43(4):509–521, April 1998.
17. L. van den Dries. *Tame Topology and o-minimal structures*. Cambridge University Press, 1998.
18. H. Wong-Toi. The synthesis of controllers for linear hybrid automata. In *Proceedings of the 36th IEEE Conference on Decision and Control*, San Diego, CA, December 1997.

Towards a Geometric Theory of Hybrid Systems[*]

Slobodan N. Simić, Karl Henrik Johansson, Shankar Sastry, and John Lygeros

Department of Electrical Engineering and Computer Sciences
University of California, Berkeley, CA 94720-1770
{simic,johans,sastry,lygeros}@eecs.berkeley.edu
http://www.eecs.berkeley.edu/~{simic,johans,sastry,lygeros}

Abstract. The main purpose of this paper is to introduce a new framework for a global, geometric study of hybrid systems, and demonstrate its usefulness through its application to the analysis of the Zeno phenomenon and stability of hybrid equilibria.

1 Introduction

In this paper we present a unifying approach for treatment of hybrid systems. We define the notions of the *hybrid manifold* (or *hybrifold*) and *hybrid flow*, which enable us to study the hybrid system "in one piece", that is, as a single, generally non-smooth *dynamical system*.

Having established a reasonable framework for the geometric study of hybrid systems as dynamical systems, we focus particularly on the *Zeno phenomenon*, which does not occur in smooth dynamical systems. We study its causes, ways of removing it from the system, and classify it topologically in dimension two.

The last part of the paper deals with stability of isolated hybrid equilibria. We prove a theorem which explains, among others, examples in which a *stable hybrid* equilibrium is composed of *unstable classical* equilibria. Proofs of all statements in the paper can be found in [SJSL].

2 Preliminaries

2.1 Definitions and Examples

Definition 1. *An n-dimensional hybrid system is a 6-tuple* $\mathbf{H} = (Q, E, \mathcal{D}, \mathcal{X},$ $\mathcal{G}, \mathcal{R})$, *where:*

- $Q = \{1, \dots, k\}$ *is the collection of* (discrete) *states of* $\mathbf{H}$, *where* $k \geq 1$ *is an integer;*
- $E \subset Q \times Q$ *is the collection of* edges;

---

[*] This work was supported by the NASA grant NAG-2-1039, the Swedish Foundation for International Cooperation in Research and Higher Education, Telefonaktiebolaget L.M. Ericsson, ONR under N00014-97-1-0946, DARPA under F33615-98-C-3614, and ARO under DAAH04-96-0341.

N. Lynch and B. Krogh (Eds.): HSCC 2000, LNCS 1790, pp. 421–436, 2000.
© Springer-Verlag Berlin Heidelberg 2000

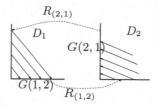

Fig. 1. The water tank example.

– $\mathcal{D} = \{D_i : i \in Q\}$ *is the collection of* domains[1] *of* **H**, *where* $D_i \subset \{i\} \times \mathbb{R}^n$
  for all $i \in Q$;
– $\mathcal{X} = \{X_i : i \in Q\}$ *is the collection of vector fields such that* X_i *is Lipschitz*
  on D_i *for all* $i \in Q$; *we denote the local flow of* X_i *by* $\{\phi_t^i\}$.
– $\mathcal{G} = \{G(e) : e \subset E\}$ *is the collection of* guards, *where for each* $e = (i,j) \in E$,
  $G(e) \subset D_i$;
– $\mathcal{R} = \{R_e : e \in E\}$ *is the collection of* resets, *where for each* $e = (i,j) \in E$,
  R_e *is a relation between elements of* $G(e)$ *and elements of* D_j, *i.e.* $R_e \subset$
  $G(e) \times D_j$.

Remark. If a reset relation R_e is actually a map $G(e) \to D_j$, with $e = (i,j) \in E$,
instead of $(x,y) \in R_e$ we write $y = R_e(x)$. Observe that domains D_i lie in distinct
copies of $\mathbb{R}^n$. However, we will sometimes abuse the notation and consider the
domains as subsets of a single copy of $\mathbb{R}^n$. We also set $D = \bigcup_{i \in Q} D_i$, and
call this set the *total domain* of **H**, and $G = \bigcup_{e \in E} G(e)$, $R = \bigcup_{e \in E} R_e(G(e))$,
$\overline{\mathcal{G}} = \{\overline{G(e)} : e \in E\}$, $\overline{\mathcal{R}} = \{\overline{R_e(G(e))} : e \in E\}$.
Given **H**, the basic idea is that starting from a point in some domain D_i we flow
according to X_i until (and if) we reach some guard $G(i,j)$, then switch via the
reset $R_{(i,j)}$, continue flowing in D_j according to X_j and so on.

Example 1 (Water Tank WT). Here $n = 2$, $k = 2$, $E = \{(1,2),(2,1)\}$, $D_1 =$
$\{1\} \times C$, $D_2 = \{2\} \times C$, where $C = [l_1, \infty) \times [l_2, \infty)$, $X_1 = (w - v_1, -v_2)^T$, $X_2 =$
$(-v_1, w - v_2)^T$, $G(1,2) = \{(1, x_1, x_2) \in D_1 : x_2 = l_2\}$, $G(2,1) = \{(2, x_1, x_2) \in$
$D_2 : x_1 = l_1\}$, and $R_{(1,2)}(1, x_1, l_2) = (2, x_1,, l_2)$, $R_{(2,1)}(2, l_1, x_2) = (1, l_1, x_2)$.

The interpretation is as follows (cf. Fig. 1). For $i \in Q$, x_i denotes the volume
of water in tank i, v_i is the constant rate of flow of water out of tank i, and l_i
is the desired volume of water in tank i. The constant rate of water flow into
the system, dedicated exclusively to one tank at a time, is denoted by w. The
control task is to keep the water volume above l_1 and l_2 (assuming the initial
volumes are above l_1 and l_2 respectively) by a strategy that switches the inflow
to the first tank whenever $x_1 = l_1$ and to the second tank whenever $x_2 = l_2$.

Example 2 (Bouncing Ball BB).
     This is a simplified model of an elastic ball that is bouncing and losing a
fraction of its energy with each bounce. We denote by x_1 its altitude and by

---

[1] In the literature also known as "invariants".

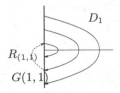

Fig. 2. Bouncing ball.

x_2 its vertical speed. Here $n = 2$, $k = 1$, $E = \{(1,1)\}$, $D_1 = \{(x_1, x_2) : x_1 \geq 0\}$, $X_1(x_1, x_2) = (x_2, -g)^T$, $G(1,1) = \{(0, x_2) : x_2 \leq 0\}$, $R_{(1,1)}(0, x_2) = (0, -cx_2)$, where g is the acceleration due to gravity and $0 < c < 1$ (cf. Fig. 2).

Example 3 (Bouncing m-Ball $BB(m)$).
    The only difference between this and the previous example is that we have m different domains in which the ball can bounce and after each bounce the ball switches to the next domain in a cyclic order. That is, $n = 2$, $k = m > 1$, $E = \{(1,2), (2,3), \ldots, (m-1, m), (m, 1)\}$, and for all $i \in Q$, $D_i = \{i\} \times \{(x_1, x_2) : x_1 \geq 0\}$, $G(i, i+1) = \{i\} \times \{(0, x_2) : x_2 \leq 0\}$, $R_{(i,i+1)}(i, 0, x_2) = (i+1, 0, -cx_2)$, where we conveniently identify $m + 1 := 1$. Note that here the domains are just different copies of the closed right half-plane in $\mathbb{R}^2$.

Example 4 (Ball Bouncing on an N-step Staircase $BBS(N)$).
    Here a ball is bouncing on an N-step staircase. Assume that step $i = 1, \ldots, N$ has width $w_i > 0$ and height $h_i > 0$, and define $\hat{w}_m = \sum_{i=1}^m w_i$ and $\hat{h}_m = \sum_{i=1}^m h_i$. Assume also that the ball loses a proportional amount of its vertical velocity (x_2) with each bounce and that the ball has constant horizontal speed (x_3). Denote by x_1 its vertical position. Then we have: $Q = \{1, \ldots, N+1\}$, $E = \{(i, i) : 1 \leq i \leq N+1\} \cup \{(1,2), \ldots, (N, N+1)\}$, and for $1 \leq i \leq N+1$: $D_i = \{i\} \times [\hat{h}_i, \infty) \times (-\infty, 0] \times (-\infty, \hat{w}_i]$, $G(i, i) = \{(x_1, x_2, x_3) \in D_i : x_1 = \hat{h}_i\}$, $R_{(i,i)}(i, x_1, x_2, x_3) = (i, x_1, -cx_2, x_3)$ and $X_i(x_1, x_2, x_3) = (x_2, -g, v)^T$. Furthermore, for $1 \leq i \leq N$: $G(i, i+1) = \{(x_1, x_2, x_3) \in D_i : x_3 = \hat{w}_i\}$, $R_{(i,i+1)}(i, \mathbf{x}) = (i+1, \mathbf{x})$. For more details see [JLSM].

Example 5 (Two Saddles $S2(\lambda)$).
    Here $n = 2$, $k = 2$, $\lambda > 0$, $E = \{(1,2), (2,1)\}$, the domains are two copies of the square $S = [-1, 1] \times [-1, 1]$, i.e. for $i \in Q$, $D_i = \{i\} \times S$, $X_1(x_1, x_2) = (\lambda x_1, -x_2)^T$, $X_2(x_1, x_2) = (-x_1, \lambda x_2)^T$, $G(1,2) =$ union of the vertical sides of D_1, $G(2,1) =$ union of the horizontal sides of D_2, $R_{(i,j)}(i, x) = (j, x)$, for all $(i, j) \in E$.

Example 6 (Flow on the 2-torus $T^2(\alpha)$).
    We have $\alpha > 0$, $n = 2$, $k = 2$, $E = \{(1,2), (2,1)\}$, $D_i = \{i\} \times K$, where $K = [0,1] \times [0,1]$ is the unit square, $X_1 = X_2 = (1, \alpha)^T$ are constant vector fields, $G(i, i) = \{i\} \times S_{\text{upper}}$, $G(i, j) = \{i\} \times S_{\text{right}}$, $R_{(i,i)}(i, x, 1) =$

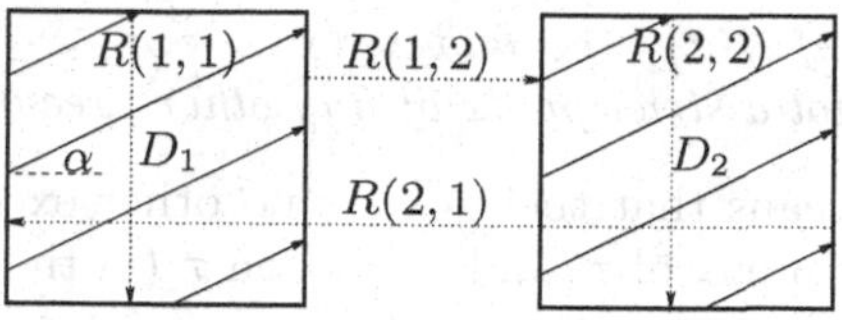

Fig. 3. $T^2(\alpha)$.

$(i, x, 0)$ and $R_{(i,j)}(i, 1, y) = (j, 0, y)$, where $i, j = 1, 2, i \neq j$, $S_{\text{upper}} = [0, 1] \times \{1\}$ and $S_{\text{right}} = \{1\} \times [0, 1)$ denote the (closed) upper and (half-closed) right side of K. Note that $R_{(i,i)}(\{i\} \times S_{\text{upper}}) = \{i\} \times S_{\text{lower}}$ and $R_{(i,j)}(\{i\} \times S_{\text{right}}) = \{j\} \times S_{\text{left}}$, with the obvious meaning of S_{lower} and S_{left}.

If we proceed as is usually done in geometry and identify $\{i\} \times S_{\text{upper}}$ with $\{i\} \times S_{\text{lower}}$ via $R_{(i,i)}$ and $\{i\} \times S_{\text{right}}$ with $\{j\} \times S_{\text{left}}$ via $R_{(i,j)}$ (where $i, j - 1, 2$, $i \neq j$), we obtain the standard 2-torus with a *smooth* flow with slope α on it. This is a baby-version of a construction we will later apply to more general hybrid systems.

Keeping in mind the examples above, we formally define the notion of an execution of a hybrid system.

Definition 2. *A (forward) hybrid time trajectory is a sequence (finite or infinite)* $\tau = \{I_j\}_{j=0}^N$ *of intervals such that* $I_j = [\tau_j, \tau_j']$ *for all* $j \geq 0$ *if the sequence is infinite; if* N *is finite, then* $I_j = [\tau_j, \tau_j']$ *for all* $0 \leq j \leq N - 1$ *and* I_N *is either of the form* $[\tau_N, \tau_N']$ *or* $[\tau_N, \tau_N')$. *The sequences* τ_j *and* τ_j' *satisfy:* $\tau_j \leq \tau_j' = \tau_{j+1}$, *for all* j.

One thinks of τ_j's as time instants when discrete transitions (or switches) from one domain to another take place. If τ is a hybrid time trajectory, we will call N its *size* and denote it by $N(\tau)$. Also, we use $\langle \tau \rangle$ to denote the set $\{0, \ldots, N(\tau)\}$ if $N(\tau)$ is finite, and $\{0, 1, 2, \ldots\}$ if $N(\tau)$ is infinite.

We will say that τ is a *prefix* of an execution $\tau' = \{I_j'\}_{j=0}^{N'}$ if $N \leq N'$ (where the inequality is taken in the extened real number system), and for $0 \leq j < N$, we have $I_j = I_j'$; furthermore, if τ has finite size, then we must also have $I_N \subset I_N'$.

Definition 3. *An execution (or forward execution) of a hybrid system* **H** *is a triple* $\chi = (\tau, q, x)$, *where* τ *is a hybrid time trajectory,* $q : \langle \tau \rangle \to Q$ *is a map, and* $x = \{x_j : j \in \langle \tau \rangle\}$ *is a collection of* C^1 *maps such that* $x_j : I_j \to D_{q(j)}$ *and for all* $t \in I_j$, $\dot{x}_j(t) = X_{q(j)}(x_j(t))$. *Furthermore, for all* $j \in \langle \tau \rangle$, *we have* $(q(j), q(j+1)) \in E$, $x_j(\tau_j') \in G(q(j), q(j+1))$, *and* $(x_j(\tau_j'), x_{j+1}(\tau_{j+1})) \in R_{(q(j), q(j+1))}$.

For an execution $\chi = (\tau, q, x)$, denote by $\tau_\infty(\chi)$ its *execution time*: $\tau_\infty(\chi) = \sum_{j=0}^{N(\tau)} (\tau_j' - \tau_j) = \lim_{j \to N(\tau)} \tau_j' - \tau_0$.

Definition 4. *An execution* χ *is called:*

- *infinite, if* $N(\tau) = \infty$ *or* $\tau_\infty(\chi) = \infty$;

– a Zeno execution if $N(\tau) = \infty$ and $\tau_\infty(\chi) < \infty$;
– maximal if it is not a strict prefix of any other execution of **H**.

The last statement means that there exists no other execution $\chi' = (\tau', q', x')$ such that τ is a strict prefix of τ' and $x = x'$ on τ (in the sense that $x_j = x'_j$ on I_j for all $j \in \langle \tau \rangle$).

Note that in Examples 1 (WT), 2 (BB) and 3 $(BB(m))$ every execution is Zeno. The same can be shown for Examples 4 $(BBS(N))$ if $0 < c < 1$ and 5 $(S2(\lambda))$ if $0 < \lambda < 1$. On the other hand, every execution in Example 6 $(T^2(\alpha))$ is infinite with infinite execution time.

We say that an execution $\chi = (\tau, q, x)$ starts at a point $p \in D$ if $p = x_0(\tau_0)$ and $\tau_0 = 0$. It passes through p if $p = x_j(t)$ for some $j \in \langle \tau \rangle$, $t \in I_j$, $t > \tau_0$.

Given $p \in D$, it is not difficult to see that there are many ways in which a hybrid system can accept several executions starting from or passing through p. For instance, this happens if at least one of the resets is a relation which is not a function.

Definition 5. A hybrid system is called deterministic if for every $p \in D$ there exists at most one maximal execution starting from p. It is called non-blocking if for every $p \in D$ there is at least one infinite execution starting from p.

Necessary and sufficient conditions for a hybrid system to be deterministic and non-blocking can be found in [LJSE]. Roughly speaking, resets have to be functions, guards have to be mutually disjoint and whenever a continuous trajectory of one of the vector fields in $\mathcal{X}$ is about to exit the domain in which it lies, it has to hit a guard.

2.2 Standing Assumptions

From now on we will assume that every hybrid system $\mathbf{H} = (Q, E, \mathcal{D}, \mathcal{X}, \mathcal{G}, \mathcal{R})$ in this paper satisfies the following assumptions.

(A1) **H** is deterministic and non-blocking.[2]
This means that every point in D is the starting point of a unique infinite (and therefore maximal) execution of **H**.

(A2) Each domain D_i is a contractible n-dimensional smooth submanifold of $\mathbb{R}^n$, with piecewise smooth boundary. No two smooth components of the boundary meet at a zero angle.

The non-zero angle requirement eliminates, for instance, cusps in dimension two, but does not eliminate "corners". Thus for domains of a hybrid system we allow disks, half-spaces, rectangles, etc.

(A3) Each guard is a piecewise smooth $(n-1)$-dimensional submanifold of the boundary of the corresponding domain. The boundary of each guard is piecewise smooth (or possibly empty).

---

[2] These assumptions can be relaxed. However, to simplify the exposition and avoid some nonessential technical difficulties in the subsequent construction, we keep them in the present form.

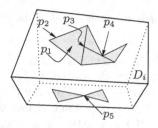

Fig. 4. p_i is of Type (Roman) i $(1 \leq i \leq 4)$.

(A4) *Each reset is a piecewise smooth homeomorphism onto its image. The image of every reset lies in the boundary of the corresponding domain.*

(A5) *Any sets in $\overline{\mathcal{G}} \cup \overline{\mathcal{R}}$ (i.e. closures of guards and images of resets) can intersect only along their boundaries. Furthermore, if $p \in \overline{G} \cup \overline{R}$, then p can be of only one of the following four types (cf. Fig. 4):*

Type I : $p \in \operatorname{int} G \cup \operatorname{int} R$;

Type II : $p \in \partial G \cup \partial R$ *and there exists exactly one set $S \in \overline{\mathcal{G}} \cup \overline{\mathcal{R}}$ which contains* p;

Type III : $p \in \partial G \cup \partial R$ *and there exist sets $S_1, \dots, S_l \in \overline{\mathcal{G}} \cup \overline{\mathcal{R}}$ $(l \geq 2)$ such that $p \in \partial S_1 \cap \dots \cap \partial S_l$ and some neighborhood of p in $S_1 \cup \dots \cup S_l$ is homeomorphic to $\mathbb{R}^{n-1}$;*

Type IV : $p \in \partial G \cup \partial R$ *and there exist sets $S_1, \dots, S_l \in \overline{\mathcal{G}} \cup \overline{\mathcal{R}}$ $(l \geq 2)$ such that $p \in \partial S_1 \cap \dots \cap \partial S_l$ and some neighborhood of p in $S_1 \cup \dots \cup S_l$ is homeomorphic to $\mathbb{R}_+^{n-1}$.*

Assumption (A5) ensures that intersections of guards and images of resets (that is, their closures) are sufficiantly nice. This in particular means that the configuration around p_5 in Fig. 4 is not allowed.

(A6) *For all $e = (i,j) \in E$, X_i points outside D_i along $G(e)$, and X_j is points inside D_j along $\operatorname{im} R_e$.*

This means that if $p \in G(i,j)$, $q = R_{(i,j)}(p)$, then there exists $\epsilon > 0$ such that $\phi_{-t}^i(p) \in \operatorname{int} D_i$ and $\phi_t^j(q) \in \operatorname{int} D_j$, for all $0 < t < \epsilon$, where int denotes the interior of a set. In particular, we have that X_i is transverse to the smooth part of $G(e)$ and X_j is transverse to the smooth part of $\operatorname{im} R_e$, the image of the map R_e.

(A7) *Each reset map R_e extends to a map $\tilde{R}_e$ defined on a neighborhood of $\overline{G(e)}$ (the closure of $G(e)$) in D_i such that $\tilde{R}_e$ is a piecewise smooth homeomorphism onto its image, which, in turn, is a neighborhood of $\operatorname{im} R_e$ in D_j. Each vector field X_i can be smoothly extended to a neighborhood of D_i in $\{i\} \times \mathbb{R}^n$.*

The last one is a fairly technical assumption the need for which will become apparent later. Note that all the examples provided above satisfy this (as well as all other) assumptions. For instance, in Example 2 *(BB)*, we can take $\tilde{R}_{(1,1)}(x_1, x_2) = (x_1, -cx_2)$.

Definition 6. *A hybrid system which satisfies assumptions* (A1) - (A7) *will be called regular.*

Given **H**, define a map $\Phi^{\mathbf{H}} : \Omega_0 \to D$, (where $\Omega_0 \subset \mathbb{R} \times D$ will be specified later) as follows. Let $p \in D$ be arbitrary. Because of (A1), there exists a unique infinite execution $\chi(p) = (\tau, q, x)$ starting at p. For any $0 \le t < \tau_\infty(\chi(p))$ there exist a unique $j \in Q$ such that $t \in [\tau_j, \tau_j')$. Then define $\Phi^{\mathbf{H}}(t, p) = x_j(t)$. To define $\Phi^{\mathbf{H}}(t, p)$ for negative t, set $\Phi^{\mathbf{H}}(t, p) = \Phi^{\mathbf{H}'}(-t, p)$, where **H**$'$ is the *reverse* hybrid system $(Q', E', \mathcal{D}', \mathcal{X}', \mathcal{G}', \mathcal{R}')$ defined by: $Q' = Q$, $\mathcal{D}' = \mathcal{D}$, $X_i' = -X_i$, for all $i \in Q$; $(i, j) \in E'$ if and only if $(j, i) \in E$; and for every $e = (i, j) \in E'$, we have $G'(e) = R_{(j,i)}(G(j, i))$ and $R_e' = R_e^{-1}$.

It can easily be checked that **H**$'$ satisfies (A1) - (A7) if **H** does. Now let Ω_0 be the largest subset of $\mathbb{R} \times D$ on which $\Phi^{\mathbf{H}}$ is defined.

For instance, in Example 2, for any $p \ne \mathbf{0}$, $\Phi^{BB}(t, p) \to \mathbf{0}$, as $t \to \tau_\infty(\chi(p))$, where $\chi(p)$ is the unique infinite execution starting at p. Note, however, that $\chi(\mathbf{0})$ makes no time progress, i.e. $\tau_j = 0$ for all $j \ge 0$, but it involves infinitely many switches at the same, i.e. initial point, which happens to be fixed by the reset map.

Theorem 1. **(a)** Ω_0 *contains a neighborhood of* $\{0\} \times \text{int } D$ *in* $\mathbb{R} \times D$.
(b) *For all* $p \in D$, $\Phi^{\mathbf{H}}(0, p) = p$. *Furthermore,* $\Phi^{\mathbf{H}}(t, \Phi^{\mathbf{H}}(s, p)) = \Phi^{\mathbf{H}}(t + s, p)$, *whenever both sides are defined.*

3   The Hybrid Manifold and Hybrid Flow

The basic idea in construction of the hybrid manifold from a hybrid system is simple: "glue" the closure of each guard to the image of the corresponding extended reset via the extended reset map. Some relatively similar ideas appear in [GJ].

3.1   The Hybrifold

Let **H** be a *regular* hybrid system. On D let $\sim$ be the equivalence relation generated by $p \sim \tilde{R}_e(p)$, for all $e \in E$ and $p \in \overline{G(e)}$. Collapse each equivalence class to a point to obtain the quotient space $M_{\mathbf{H}} = D/\sim$.

Definition 7. *We call* $M_{\mathbf{H}}$ *the hybrid manifold or hybrifold of* **H**.[3]

Denote by π the natural projection $D \to M_{\mathbf{H}}$ which assigns to each p its equivalence class $p/\sim$. Put the *quotient topology* on $M_{\mathbf{H}}$. Recall that this is the smallest topology that makes π continuous, i.e. a set $V \subset M_{\mathbf{H}}$ is open if and only if $\pi^{-1}(V)$ is open in D.

Define the *hybrid flow* of **H**, $\Psi^{\mathbf{H}} : \Omega \to M_{\mathbf{H}}$, by $\Psi^{\mathbf{H}}(t, \pi(p)) = \pi\Phi^{\mathbf{H}}(t, p)$. Here $\Omega = \{(t, \pi(p)) : (t, p) \in \Omega_0\}$. In other words, orbits of $\Psi^{\mathbf{H}}$ are obtained by

---

[3] The authors thank Renaud Dreyer for suggesting the term hybrifold. The term "manifold" will be justified by Theorem 2.

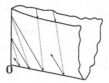

Fig. 5. Hybrifold and an orbit of the hybrid flow for WT.

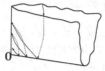

Fig. 6. Hybrifold and an orbit of the hybrid flow for BB.

projecting orbits of $\Phi^{\mathbf{H}}$ by π. By the $\Phi^{\mathbf{H}}$-orbit of p we mean the collection of points $\Phi^{\mathbf{H}}(t, p)$ for all possible t (i.e. all t such that $(t, p) \in \Omega_0$).

Let us run this construction on some of the examples listed above.

Example 7 (WT continued).

Without loss we assume that $l_1 = l_2 = 0$. To obtain M_{WT} we have to identify the x_1-axis from D_1 with the same axis from D_2 via $R_{(1,2)}$ and similarly with the x_2-axis.

It is not difficult to see that M_{WT} is homeomorphic to $\mathbb{R}^2$ (see Fig. 5). However, M_{WT} has a singularity (or "corner") at $\mathbf{0} = \pi(1, 0, 0)$, i.e. π does not define a smooth structure on M_{WT}. Note that every execution starting at $x \neq \mathbf{0}$ converges to $\mathbf{0}$.

Example 8 (BB continued).

Here we have to identify the negative part with the positive part of the x_2-axis. The resulting space M_{BB} is again homeomorphic to $\mathbb{R}^2$ (see Fig. 6), but π again does not define a smooth structure on it. As in the previous example, $\Psi^{BB}(t, x) \to \mathbf{0}$, as $t \to \tau_\infty(\chi(x))$, for all $x \neq \mathbf{0}$.

Example 9 (BB(m) continued).

For simplicity assume $m = 2$. It is not difficult to see that $M_{BB(2)}$ is smooth (in the sense explained above) and *diffeomorphic* to $\mathbb{R}^2$. However, the hybrid flow is not smooth.

Example 10 (S2(λ) continued).

$M_{S2(\lambda)}$ is homeomorphic to the 2-sphere; it is not equipped with a smooth structure by π.

Example 11 (T²(α) continued).

We already observed that $M_{T^2(\alpha)}$ is the standard 2-torus and $\Psi^{T^2(\alpha)}$ is a smooth linear flow on it. If α is rational, then every orbit is closed; if α is irrational, then every orbit is dense in T^2.

Theorem 2. (a) $M_\mathbf{H}$ *defined above is a topological n-manifold with boundary.*
(b) *Both $M_\mathbf{H}$ and its boundary are piecewise smooth.*
(c) *The restriction $\pi|_{\operatorname{int} D} : \operatorname{int} D \to \pi(\operatorname{int} D)$ is a diffeomorphism.*

Recall that M is called a *topological n-manifold with boundary* if it is Hausdorff and every point in M has a neighborhood homeomorphic to either $\mathbb{R}^n$ or the closed upper half-space $\mathbb{R}^n_+ = \{(x_1, \dots, x_n) : x_n \geq 0\}$. Points having the latter property are said to be on the boundary ∂M, which is a topological $(n-1)$-manifold.

3.2   The Hybrid Flow

Let $\Psi := \Psi^\mathbf{H}$ be the hybrid flow of $\mathbf{H}$, as defined above. For each $t \in \mathbb{R}$ and $x \in M_\mathbf{H}$, let $M(t) = \{y \in M_\mathbf{H} : \Psi(t,y) \text{ is defined}\}$, and $J(x) = \{s \in \mathbb{R} : \Psi(s,x) \text{ is defined}\}$. Observe that if $x = \pi(p)$, then $J(x) \cap [0, \infty) = [0, \tau_\infty(\chi(p)))$. Also, for $t > 0$, $M(t)$ contains all points $x = \pi(p)$ such that $\tau_\infty(\chi(p))) > t$. As usual, $\chi(p)$ denotes the unique execution of $\mathbf{H}$ starting at p.

If $M(t)$ is not empty, denote by $\Psi_t : M(t) \to M_\mathbf{H}$ the *time t map of Ψ*, defined by $\Psi_t(x) = \Psi(t,x)$. Recall that a function (in particular, vector field) is said to be smooth on a *closed* set F if it is the restriction of a smooth function defined on a neighborhood of F. Then we have the following theorem.

Theorem 3. *Suppose each vector field X in $\mathcal{X}$ is smooth (in addition to being globally Lipschitz). Then:*

(a) *For each $x \in M_\mathbf{H}$ the map $t \mapsto \Psi_t(x)$ is continuous and, if $J(x)$ is not a single point, piecewise smooth on $J(x)$. More precisely, it is smooth except at (at most) countably many points in $J(x)$. Furthermore, each map Ψ_t is injective.*
(b) *Whenever both sides are defined: $\Psi_t^\mathbf{H}\Psi_s^\mathbf{H}(x) = \Psi_{t+s}^\mathbf{H}(x)$.*
(c) *There is an open and dense subset of Ω on which Ψ is smooth.*

4   ω-Limit Sets and the Zeno Phenomenon

It has to be pointed out that Zeno executions do not arise in physical systems and are a consequence of modeling over-abstraction. Therefore, one wishes to avoid them. However, from a mathematical viewpoint, the Zeno phenomenon poses numerous interesting questions. In this section we show that, in short, the topological cause of Zenoness is a lack of smoothness in the hybrid flow and that the Zeno phenomenon can be removed by smoothing out the hybrifold and the hybrid flow on it.

Definition 8. *A point $y \in M_{\mathbf{H}}$ is called an ω-limit point of $x \in M_{\mathbf{H}}$ if $y = \lim_{m \to \infty} \Psi_{t_m}^{\mathbf{H}}(x)$, for some increasing sequence (t_m) in $J(x)$ such that $t_m \to \tau_\infty(x)$, as $m \to \infty$. The set of all ω-limit points of x is called the ω-limit set of x and is denoted by $\omega(x)$.*

By $\tau_\infty(x)$ we denote the execution time of the unique execution of $\mathbf{H}$ starting from p, where $x = \pi(p)$; that is, $\tau_\infty(x) = \tau_\infty(\chi(p))$. It is easy to check that this is a well defined element of the extended real number system. In other words, ω-limit points for x are accumulation points of the orbit of x.

Suppose $x \in M_{\mathbf{H}}$ and denote by $E_\infty(x)$ the set of discrete transitions which occur infinitely many times in the execution starting from x. If $E_\infty(x)$ is empty, then the orbit of x eventually ends up in a single domain D_i (that is, its image under π in the hybrifold) in which case $\omega(x) \subset \pi(\overline{D_i})$. This means that every point $y \in \omega(x)$ is an accumulation point of the orbit of a single vector field, namely X_i. We will call such a point y, a *pure ω-limit point*.

If $E_\infty(x)$ is nonempty, then every ω-limit point for x is a result of both the continuous *and* discrete (i.e. hybrid) dynamics of $\mathbf{H}$ and will accordingly be called a *hybrid ω-limit point of x.*

Theorem 4. *For every $x \in M_{\mathbf{H}}$, $\omega(x)$ is invariant with respect to the hybrid flow. That is, if $y \in \omega(x)$, then $\Psi_t^{\mathbf{H}}(y) \in \omega(x)$, for all $t \in J(y)$.*

4.1   Properties of Zeno Executions

Definition 9. *A point $z \in M_{\mathbf{H}}$ is called a Zeno state for x if $z \in \omega(x)$ and $\tau_\infty(x) < \infty$.*

We will also refer to points in $\pi^{-1}(z)$ as *Zeno states* in $\mathbf{H}$. For example, the "origin" of M_{WT} (as well as M_{BB} and $M_{BB(2)}$) is a Zeno state for every point. Moreover, for each x, $\omega(x)$ contains only one Zeno state. We now show this is always the case.

Theorem 5. *If the execution starting from $x \in M_{\mathbf{H}}$ is Zeno, then $\omega(x)$ consists of exactly one Zeno state for x and $\omega(x) \subset \bigcap_{e \in E_\infty(x)} \pi(\overline{G(e)})$.*

Note than in all the Zeno examples above none of the flows involved in creating the Zeno state has an equilibrium at the Zeno state. The following lemma shows that this is not a coincidence.

Lemma 1. *A Zeno state is not a standard equilibrium (cf. Def. 12). More specifically, if $z \in M_{\mathbf{H}}$ is a Zeno state, then for every $p \in \pi^{-1}(z)$, if $p \in D_i$, then $X_i(p) \neq 0$.*

Example 12 (equilibrium + cusp = Zeno).
   Consider the following one-domain hybrid system: $D = \{(x, y) \in \mathbb{R}^2 : y \geq 0, \ -f(y) \leq x \leq f(y)\}$, $G = \{(-f(y), y) : y \geq 0\}$, $R(-f(y), y) = (f(cy), cy)$, $X(x, y) = (-x - y, x - y)^T$. Here $0 < c < 1$, $f : [0, \infty) \to [0, \infty)$ is a smooth

function such that $f(0) = 0$ and for all $y \geq 0$, $f(y) \leq y^2$. In particular, $f'(0) = 0$, which means that D has a cusp at $\mathbf{0}$. It is not difficult to check that $\mathbf{0}$ is a Zeno state despite the fact that it is an equlibrium for X. This shows the importance of geometry of domains and assumption (A2).

Theorem 6. *Suppose* $\mathbf{H}$ *is a hybrid system such that its hybrid flow* $\Psi^{\mathbf{H}}$ *is smooth. (This in particular means that its hybrifold* $M_{\mathbf{H}}$ *is smooth.) Then* $\mathbf{H}$ *admits no Zeno executions or equivalently, there are no Zeno states in* $M_{\mathbf{H}}$.

In general it may not be easy to check whether, given $\mathbf{H}$, the hybrifold $M_{\mathbf{H}}$ is smooth. Even if it were, non-smoothness of the hybrid flow may cause Zeno (cf. $BB(2)$). However, the following result provides an easily verifiable criterion for smoothness of $\Psi^{\mathbf{H}}$.

Theorem 7. *Suppose that* $M_{\mathbf{H}}$ *is smooth and for every* $e = (i,j) \in E$, X_i *and* X_j *are* $\tilde{R}_e$-*related on* $\overline{G(e)}$. *That is, for every* $p \in \overline{G(e)}$: $T\tilde{R}_e(X_i(p)) = X_j(\tilde{R}_e(p))$. *Then the hybrid flow is smooth.*

Example 13. Consider $BB(2)$. Here we have: $X_1(x_1, x_2) = (x_2, -g)^T = X_2$, $\tilde{R}_{(i,j)}(i, x_1, x_2) = (j, x_1, -cx_2)$, where $(i,j) = (1,2)$ or $(2,1)$. It is easily seen that $T\tilde{R}_{(1,2)}(X_1) \neq X_2$. Recall that the hybrid flow for $BB(2)$ is not smooth.

Example 14. It is not difficult to check that in case of $T^2(\alpha)$, the condition from Theorem 7 is satisfied for every $\alpha > 0$. Thus $T^2(\alpha)$ does not admit Zeno, as was already shown above.

Corollary 1. *If* $\mathbf{H}$ *is a hybrid system satifying condition from Theorem 7, then* $\mathbf{H}$ *accepts no Zeno executions.*

4.2   Removal of Zeno

Suppose that $\mathbf{H}$ is a regular hybrid system and that $z \in M_{\mathbf{H}}$ is a Zeno state. We have seen that $M_{\mathbf{H}}$ in a certain sense has a singularity at z. Consider the following ways of removing such singularities.

Smoothing. Suppose that $M_{\mathbf{H}}$ can be equipped with a smooth structure which induces the same topology as the original one and denote the smoothed hybrifold by $M_{\mathbf{H}}^{smooth}$ (cf. Fig. 7). Note that $M_{\mathbf{H}}$ and $M_{\mathbf{H}}^{smooth}$ are homeomorphic. It is not guaranteed that the hybrid flow $\Psi^{\mathbf{H}}$ will be smooth on $M_{\mathbf{H}}^{smooth}$. If, however, $\Psi^{\mathbf{H}}$ *is* smooth with respect to the differentiable structure on $M_{\mathbf{H}}^{smooth}$, then Theorem 6 implies that there are no Zeno states in $M_{\mathbf{H}}^{smooth}$. We say that we have removed Zeno by *smoothing*.

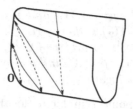

Fig. 7. Smoothed water tank M_{WT}^{smooth}.

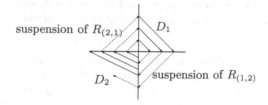

Fig. 8. ϵ-suspended water tank $S^\epsilon M_{WT}$.

Hybrid suspension. [4] The basic idea is to "interpolate" executions between guards and images of corresponding resets, i.e. to make "instantaneous" discrete transitions given by reset maps "last" some time ϵ. The constructions goes as follows. Let $\epsilon > 0$ be arbitrary and assume $e = (i, j) \in E$. Instead of gluing $\overline{G(e)}$ to $\overline{\mathrm{im}\, R_e}$ via $\tilde{R}_e$, first enlarge the domain D_i by $D_i^\epsilon = D_i \cup (\overline{G(e)} \times [0, \epsilon])$, and then identify $(p, \epsilon) \sim \tilde{R}_e(p)$, for every $p \in \overline{G(e)}$. Denote the space obtained by this identification for all $e \in E$ by $S^\epsilon M_{\mathbf{H}}$ and by π^ϵ the quotient (i.e. identification) map. On each $\overline{G(e)} \times [0, \epsilon]$, consider the trivial "vertical" flow: $(p, s, t) \mapsto (p, s + t)$ $(p \in \overline{G(e)}, 0 \le s \le \epsilon, t \in \mathbb{R})$. Denote by $S^\epsilon \Psi^{\mathbf{H}}$ the flow on $S^\epsilon M_{\mathbf{H}}$ obtained by projecting via π^ϵ this flow (for each $e \in E$) as well as $\Phi^{\mathbf{H}}$. We will call $S^\epsilon M_{\mathbf{H}}$ the ϵ-*suspended hybrid manifold* and $S^\epsilon \Psi^{\mathbf{H}}$ the associated ϵ-*suspended hybrid flow* (see Fig. 8). (This construction resembles the standard suspension of a map; cf. e.g. [PdM].) It is immediate by construction that for ever $\epsilon > 0$, $S^\epsilon \Psi^{\mathbf{H}}$ accepts no Zeno-type executions.

5 Conjugacy of Hybrid Systems and Classification of Zeno States in Dimension Two

In this section we discuss the following question: when are two hybrid systems qualitatively the same? For that purpose we borrow the notion of conjugacy from the theory of dynamical systems. Roughly speaking, two dynamical systems are conjugate if their phase portraits look qualitatively (or topologically) the same. Similarly, two hybrid systems are conjugate if their hybrid flows are conjugate. We now make this more precise.

---

[4] We thank Morris W. Hirsch for suggesting this idea in a recent conversation.

Definition 10. *Two hybrid systems $\mathbf{H_1}$ and $\mathbf{H_2}$ are said to be topologically conjugate (denoted by $\mathbf{H_1} \approx \mathbf{H_2}$) if there exists a homeomorphism $h : M_{\mathbf{H_1}} \to M_{\mathbf{H_2}}$ which sends orbits of $\Psi^{\mathbf{H_1}}$ to orbits of $\Psi^{\mathbf{H_2}}$. If $M_{\mathbf{H_1}}$ and $M_{\mathbf{H_2}}$ happen to be smooth manifolds of class C^r ($r \geq 1$) and h is a C^r diffeomorphism, then $\mathbf{H_1}$ and $\mathbf{H_2}$ are said to be C^r-conjugate.*

As usual, by the *orbit* of a point x under a (local) flow $\{\phi_t\}$ we mean the set of points $\phi_t(x)$ for all t for which $\phi_t(x)$ is defined. We usually think of h as a change of coordinates so that two hybrid systems are topologically conjugate if their hybrid flows are the same up to a continuous coordinate change. Note that conjugacy does not necessarily preserve the time parameter t. If it does, it is called *equivalence*.

Example 15. WT is topologically conjugate to BB. This can be seen by suitably projecting M_{WT} and M_{BB} onto $\mathbb{R}^2$ so that both Ψ^{WT} and Ψ^{BB} look like a spiral sink at the origin. For more details, see [SJSL]. We will see later that in dimension two this picture is typical.

Example 16. $T^2(1)$ is not conjugate to $T^2(\sqrt{2})$. Even though the hybrifold for both hybrid systems is the same (the 2-torus), every orbit of $T^2(1)$ is closed, while every orbit of $T^2(\sqrt{2})$ is dense in T^2.

Even though it is not possible to classify all hybrid systems up to conjugacy (this attempt fails even for smooth dynamical systems), the next theorem shows that near a Zeno state, every 2-dimensional hybrid flow looks like Ψ^{WT} near $\mathbf{0}$.

Theorem 8. *Let $\mathbf{H}$ be a 2-dimensional hybrid system and suppose that $z \in M_{\mathbf{H}}$ is a Zeno state. Then there is a neighborhood U of z in $M_{\mathbf{H}}$ and a neighborhood V of $\mathbf{0}$ in M_{WT} such that $\Psi^{\mathbf{H}}|_U$ is topologically conjugate to $\Psi^{WT}|_V$.*

6 Stability of Hybrid Equilibria

Recall that if ϕ_t is a local flow generated by a smooth vector field X on some set U (in $\mathbb{R}^n$ or any manifold), then $p \in U$ is an *equilibrium* for X (equivalently: for ϕ_t) if $X(p) = 0$ (equivalently: if $\phi_t(p) = p$ for all $t \in \mathbb{R}$). In case of a hybrid system there is usually more than one vector field at play, and even in the case when there is only one, resets are involved in generating the hybrid dynamics. Taking this into account we define a hybrid equilibrium as follows.

Definition 11. *Let $\mathbf{H}$ be a hybrid system. A point $x \in M_{\mathbf{H}}$ is called an (hybrid) equilibrium for the hybrid flow $\Psi^{\mathbf{H}}$ if $\Psi^{\mathbf{H}}(t, x) = x$ for all $t \in J(x)$.*

Equivalently, $x \in M_{\mathbf{H}}$ is a hybrid equilibrium if the hybrid dynamics of $\mathbf{H}$, consisting of reset maps and local flows of $\mathbf{H}$, map $\pi^{-1}(x)$ to itself. For example, any Zeno state is a hybrid equilibrium despite Lemma 1; however, hybrid dynamics make no time progress at this kind of equilibrium. The following definition distinguishes those hybrid equilibria which are created from equilibria of vector fields in $\mathbf{H}$ in the standard sense.

Definition 12. *A point $x \in M_H$ is called a* standard equilibrium *for Ψ^H if it is a hybrid equilibrium and for each $p \in \pi^{-1}(x)$, if $p \in D_i$, then p is an equilibrium for X_i (i.e. $X_i(p) = 0$). It is called a* pure equilibrium *if it is standard and belongs to $\pi(int\,D)$.*

Note that the only dynamics involved in creating a pure equilibrium are those of a single vector field. We now define the notions of (Lyapunov) stability and asymptotic stability of hybrid equilibria in analogy with those from dynamical systems.

Definition 13. *An equilibrium x_* of Ψ^H is called* (Lyapunov) stable *if for every neighborhood U of x_* in M_H there exists a neighborhood V of x_* in U such that for every $x \in V$, $\Psi_t^H(x) \in U$ for all $t \in [0, \tau_\infty(x))$. If V can be chosen so that in addition to the properties described above, $\lim_{t \to \tau_\infty(x)} \Psi_t^H(x) = x_*$, then x_* is* asymptotically stable.

Example 17. There are well known 2-dimensional hybrid systems (and they are also not difficult to construct from scratch; cf. [SJSL]) with a standard hybrid equilibrium which can described as follows: stable + stable = unstable, or unstable + stable = stable, or unstable + unstable = stable. This means that (in the case of the first example) the unstable hybrid equilibrium in question is created by stable equilibria for the vector fields at play in the hybrid system. These examples show us that extra caution is needed in analyzing stability of hybrid equilibria.

In the subsequent text, we use the following notation: if X is a vector field on a manifold M with local flow ϕ_t and $f : M \to \mathbb{R}$ a function, Xf will denote the derivative of f in the direction of X: $(Xf)(x) = Tf(X(x))$. For a map $h : (A, d_A) \to (B, d_B)$ between metric spaces, let $\text{Lip}_p(f) = \sup_{q \in A - \{p\}} \frac{d_B(f(q), f(p))}{d_A(q, p)}$. This is the *Lipschitz constant of f at p*.

The following theorem is an analog of the linearization theorem for stability of equilibria of a single dynamical system. In the hybrid case, the linearized data include, besides the derivatives of the vector fields at the equilibrium, the tangent spaces at the equilibrium of guards and images of resets involved in the hybrid dynamics near the equilibrium. Here, for a manifold A with boundary and $p \in \partial A$, we denote by $T_p^+ A$ the set of all vectors $v \in T_p A$ which point inside A (i.e. there exists $\epsilon > 0$ and a smooth curve $c : [0, \epsilon] \to A$ such that $c(0) = p$, $\dot{c}(0) = v$ and $c(t) \in A - \partial A$ for $0 < t \leq \epsilon$).

Theorem 9 (Stability via Linearization).
Let $x_* \in M_H$ be an isolated standard equilibrium for Ψ^H and $\pi^{-1}(x_*) = \{p_1, \ldots, p_l\}$, where $p_j \in D_{i_j}$ and $1 \leq j \leq l$. Suppose that there exists a bounded neighborhood W of x_* and for each $1 \leq j \leq l$ a smooth function $f_j : U_j - \{p_j\} \to \mathbb{R}$, where U_j is a neighborhood of $D_{i_j} \cap \pi^{-1}(W)$ in $\{i_j\} \times \mathbb{R}^n$, such that:

(a) *$p_j \in A_j \cap B_j$, where $A_j = \overline{\text{im}R_{(i_{j-1}, i_j)}} \cap U_j$, $B_j = \overline{G(i_j, i_{j+1})} \cap U_j$, for all $1 \leq j \leq l$. Assume further that A_j and B_j are differentiable at p_j.*

(b) $a_j^- \leq f_j \leq a_j^+$ on A_j, and $B_j = \overline{f_j^{-1}(b_j)}$, for all j, for some numbers $a_j^- \leq a_j^+ < b_j$.

(c) $0 < m_j^- \leq X_{i_j} f_j \leq m_j^+$ on $U_j - \{p_j\}$ $(1 \leq j \leq l)$.

(d) For each j there exists $\tau_j > 0$ such that $e^{\tau_j L_j}(T_{p_j}^+ A_j) \subset T_{p_j}^+ B_j$, where $L_j = T_{p_j} X_{i_j}$.

For $1 \leq j \leq l$, let S_j be an $n \times (n-1)$-matrix whose columns form an orthonormal basis for $T_{p_j} A_j$ and belong to $T_{p_j}^+ A_j$. Let

$$\mu_j = \sqrt{\lambda_{\max}[(e^{\tau_j L_j} S_j)^T e^{\tau_j L_j} S_j]},$$

and $\nu_j = \|T_{p_j} R_{(i_j, i_{j+1})}\|$. Define $\eta_{\mathbf{H}}(x_*) = \prod_{j=1}^{l} \mu_j \nu_j$. If $\eta_{\mathbf{H}}(x_*) < 1$, then x_* is an asymptotically stable hybrid equilibrium. If $\dim \mathbf{H} = 2$ and $\eta_{\mathbf{H}}(x_*) > 1$, then x_* is unstable.

Remarks.

(i) Condition (b) says that B_j is the closure of a level set of f_j while A_j is "almost" a level set of f_j. The function f_j measures the progress trajectories of X_{i_j} make towards B_j, starting from A_j.

(ii) Condition (c) says that the time-τ_j map of the linearization of the flow of X_{i_j} at p_j (i.e. $T\phi_t^{i_j}$) maps $T_{p_j}^+ A_j$ to $T_{p_j}^+ B_j$. This means that at least on the level of linearizations, B_j is reachable from A_j in a bounded amount of time.

(iii) Note that (unlike in [B] and [MH]) it is not necessary to integrate any vector fields and that all the input data of the theorem are computable (even though finding f_j's and τ_j's may be difficult).

Example 18. Define a 3-dimensional hybrid system $\mathbf{H}$ by: $D_1 = \{1\} \times S$, $D_2 = \{2\} \times \mathbb{R}^3 - S$, where $S = \{(x, y, z) : x \geq 0, \ y \geq x^2, \ z \in \mathbb{R}\} \cup \{(x, y, x) : x \leq 0, \ y \geq -x(x - c), \ z \in \mathbb{R}\}$, and $G(1, 2) = \{(x, y, z \in D_1 : y = x^2\}$, $G(2, 1) = \{(x, y, z) \in D_2 : y = -x(x - c)\}$, for some constant c. Let $X_1(x, y, z) = (-x - y, x - y, -\lambda_1 z)$ and $X_2(x, y, z) = (x - y, x + y, \lambda_2 z)$, where $0 < \lambda_2 \leq 1 \leq \lambda_1$. Then it is not difficult to check that $\eta_{\mathbf{H}}(0) = e^{-2\gamma}$, where $\gamma = \arctan c$, so if $c > 0$, then $\mathbf{0}$ is asymptotically stable.

Example 19. Let $\mathbf{H}$ be a 3-dimensional hybrid system with $D_1 = \{1\} \times K \times \mathbb{R}$ and $D_2 = \{2\} \times \mathbb{R}^2 - K \times \mathbb{R}$, where $K = [0, \infty) \times [0, \infty)$. Let $G(1, 2) = \{(x, y, z) \in D_1 : x = 0\}$, $G(2, 1) = \{(x, y, z) \in D_2 : y = 0\}$, and $X_1(x, y, z) = (x - y, x + y, -\lambda_1 z)$, $X_2(x, y, z) = (-x - y, x - y, \lambda_2 z)$, where $\lambda_1, \lambda_2 > 0$. The resets are identity maps.

Then the full trajectories of X_1 are spirals around the z-axis which increase in radius and converge to the xy-plane. The full trajectories of X_2 are also spirals around the z-axis, but they decrease in radius and diverge from the xy-plane. It is not difficult to check that, with notation from Theorem 9, $\mu_1 = e^{\pi/2}$, $\mu_2 = e^{3\pi\lambda_2/2}$, so $\eta_{\mathbf{H}}(0) > 1$ and the theorem is inconclusive.

However, the flows can be decoupled into their xy- and z-parts the analysis of which shows that if $\lambda_1 > 3\lambda_2$, then $\mathbf{0}$ is an asymptotically stable hybrid equilibrium of $\mathbf{H}$. The reason Theorem 9 does not provide the same answer, intuitively speaking, is because it is not able to measure the small amount of contraction around $\mathbf{0}$ in the flows of both X_1 and X_2, which turns out to be sufficient for asymptotic stability. Namely, on $G(2,1)$ the flow of X_1 contracts in only one direction (and expands in the other) and similarly for the flow of X_2 on $G(1,2)$.

References

B.   M.S. Branicky, Multiple Lyapunov functions and other analysis tools for switched and hybrid systems, *IEEE Trans. on Automatic Control*, 43(4), 475-482, 1998

GJ.   J. Guckenheimer and S. Johnson, Planar hybrid systems, in *Hybrid Systems and Autonomous Control Workshop*, Cornell University, Ithaca, NY, 1994

JELS.   K. H. Johansson, M. Egerstedt, J. Lygeros and S. Sastry, On the Regularization of Zeno hybrid automata, *Systems & Control Letters*, 38, 141-150, 1999

JLSM.   K.H. Johansson, J. Lygeros, S. Sastry and M. Egerstedt: Simulation of Zeno hybrid automata, *IEEE Conference on Decision and Control*, Phoenix, AZ, 1999

LJSE.   J. Lygeros, K.H. Johansson, S. Sastry and M. Egerstedt, On the existence of executions of hybrid automata, in *IEEE Conference on Decision and Control*, Phoenix, AZ, 1999

LLN.   J. Lygeros et al., Hybrid Systems: Modeling, Analysis and Control, *Lecture Notes and Class Projects for EE291E*, Spring 1999, Mem. No. UCB/ERL M99/34

LJZS.   J. Lygeros, K. H. Johansson, J. Zhang, S. Simić: Dynamical systems revisited: Hybrid systems with Zeno executions, in preparation

MH.   A.N. Michel and B. Hu, Towards a stability theory of general hybrid dynamical systems, *Automatica* 35 (1999), 371-384

PdM.   J. Palis, Jr. and W. de Melo, *Geometric Theory of Dynamical Systems*, Springer-Verlag, New York, 1982

SJSL.   S.N. Simić, K.H. Johansson, S. Sastry and J. Lygeros, Towards a geometric theory of hybrid systems, Technical Report UCB/ERL M00/3, University of California at Berkeley, December 1999

Controlled Invariance of Discrete Time Systems*

René Vidal, Shawn Schaffert, John Lygeros, and Shankar Sastry

Department of Electrical Engineering and Computer Sciences
University of California at Berkeley
Berkeley, CA 94720-1774
Phone: (510) 643 2382, Fax: (510) 642 1341
{rvidal,sms,lygeros,sastry}@eecs.berkeley.edu

Abstract. An algorithm for computing the maximal controlled invariant set and the least restrictive controller for discrete time systems is proposed. We show how the algorithm can be encoded using quantifier elimination, which leads to a semi-decidability result for definable systems. For discrete time linear systems with all sets specified by linear inequalities, a more efficient implementation is proposed using linear programming and Fourier elimination. If in addition the system is in controllable canonical form, the input is scalar and unbounded, the disturbance is scalar and bounded and the initial set is a rectangle, then the problem is decidable.

1   Introduction

The design of controllers is one of the most active research topics in the area of hybrid systems. Problems that have been addressed include hierarchical control [5, 19], distributed control [18], and optimal control using dynamic programming techniques [3, 4, 20, 23] or extensions of the maximum principle [11]. A substantial research effort has also been directed towards solving control problems with reachability specifications, that is designing controllers that guarantee that the state of the system will remain in a "good" part of the state space. Such control problems turn out to be very important in applications, and are closely related to the computation of the *reachable states* of a hybrid system and to the concept of *controlled invariance*. The proposed solutions extend game theory methods for purely discrete [21, 25] and purely continuous [2, 15] systems to certain classes of hybrid systems: timed automata [13, 17], rectangular hybrid automata [28] and more general hybrid automata [16, 26].

All of these techniques are concerned with hybrid systems whose continuous state evolves in continuous time, according to differential equations or differential inclusions. Unlike conventional continuous dynamical systems, little attention has been devoted to systems where the continuous state evolves in discrete time, according to difference equations. Besides being interesting in its own right, this class of hybrid systems can be used to approximate hybrid systems with

---

* Research supported by ONR under grant N00014-97-1-0946, by DARPA under contract F33615-98-C-3614, and by ARO under grant MURI DAAH04-96-1-0341.

N. Lynch and B. Krogh (Eds.): HSCC 2000, LNCS 1790, pp. 437–451, 2000.

differential equations. Indeed, most of the techniques that have been proposed for reachability computations for general continuous dynamics involve some form of discretization of the continuous space [8, 12, 26], followed by a reachability computation on the resulting discrete time system.

In Sect. 2, we formulate the problem of controller synthesis for discrete time systems under reachability specifications, introduce the concepts of maximal controlled invariant set and least restrictive controller, propose an algorithm for computing them, and show how the algorithm can be implemented using quantifier elimination. This immediately leads to a semi-decidability result for discrete time systems whose continuous dynamics can be encoded in a decidable theory of the reals. In Sect. 3, we implement the proposed algorithm for discrete time linear systems with all the sets defined by linear inequalities. The implementation is based on a more efficient method for performing quantifier elimination in the theory of linear constraints using linear programming and Fourier elimination. We also show that the problem is decidable when the single-input single-disturbance discrete time linear system is in controllable canonical form, the input is unbounded, and the safe set is a rectangle. Finally, in Sect. 4, we illustrate the proposed method with some examples. For the proofs we refer the reader to [27].

2   Discrete Time Systems and Safety Specifications

2.1   Basic Definitions

Let Y be a countable collection of variables and let $\mathbf{Y}$ denote its set of valuations, that is the set of all possible assignments of these variables. We refer to variables whose set of valuations is countable as *discrete* and to variables whose set of valuations is a subset of a Euclidean space $\mathbb{R}^n$ as *continuous*. For a set $\mathbf{Y}$ we use $\mathbf{Y}^c$ to denote the complement of $\mathbf{Y}$, $2^{\mathbf{Y}}$ to denote the set of all subsets of $\mathbf{Y}$, $\mathbf{Y}^*$ to denote the set of all finite sequences of elements of $\mathbf{Y}$, and $\mathbf{Y}^\omega$ to denote the set of all infinite sequences. Since the dynamical systems we will consider will be time invariant we will use $y = \{y[i]\}_{i=0}^N$ to denote sequences. We use $\wedge$ to denote conjunction, $\vee$ to denote disjunction, $\neg$ to denote negation, $\forall$ to denote the universal quantifier, and $\exists$ to denote the existential quantifier.

Definition 1 (Discrete Time System (DTS)). *A discrete time system is a collection $H = (X, V, \text{Init}, f)$ consisting of a finite collection of state variables, X, a finite collection of input variables, V, a set of initial states, $\text{Init} \subseteq \mathbf{X}$, and a reset relation, $f : \mathbf{X} \times \mathbf{V} \to 2^{\mathbf{X}}$.*

Definition 2 (Execution of DTS). *A sequence $\chi = (x, v) \in (\mathbf{X} \times \mathbf{V})^* \cup (\mathbf{X} \times \mathbf{V})^\omega$ is said to be an execution of the discrete time system H if $x[0] \in \text{Init}$, and for all $k \geq 0$, $x[k + 1] \in f(x[k], v[k])$.*

To ensure that every finite execution can be extended to an infinite execution we assume that $f(x,v) \neq \emptyset$ for all $(x,v) \in \mathbf{X} \times \mathbf{V}$. We call such a DTS *non-blocking*.[1]

We denote the set of all executions of H starting at $x_0 \in \mathbf{X}$ as $\mathcal{E}_H(x_0)$, and the set of all executions of H by $\mathcal{E}_H$. Clearly, $\mathcal{E}_H = \bigcup_{x_0 \in Init} \mathcal{E}_H(x_0)$.

Our goal here is to design controllers for DTS. We assume that the input variables are partitioned into two classes, $V = U \cup D$, where U are *control variables*, and D are *disturbance variables*. In this context a controller can be defined as a feedback map.

Definition 3 (Controller). *A controller, C, is a map $C : \mathbf{X}^* \to 2^{\mathbf{U}}$. A controller is called* non-blocking *if $C(x) \neq \emptyset$ for all $x \in \mathbf{X}^*$. A controller is called* memoryless *if for all $x, x' \in \mathbf{X}^*$ ending at the same state we have $C(x) = C(x')$.*

The interpretation is that, given the evolution of the plant state up to now, the controller determines the set of allowable controls for the next transition. With this interpretation in mind, we define the set of *closed loop causal executions* as

$$\mathcal{E}_{H_C} = \{(x,u,d) \in \mathcal{E}_H \mid \forall k \geq 0, u[k] \in C(x \downarrow_k)\},$$

where $x \downarrow_k$ denotes the subsequence of x consisting of its first k elements. Notice that a memoryless controller can be characterized by a map $g : \mathbf{X} \to 2^{\mathbf{U}}$, and its set of closed loop causal executions is simply

$$\mathcal{E}_{H_g} = \{(x,u,d) \in \mathcal{E}_H \mid \forall k \geq 0, u[k] \in g(x[k])\}.$$

Our goal is to use controllers to steer the executions of the plant, so that they satisfy certain desirable properties. In this paper we will restrict our attention to a class of properties known as *safety properties*: Given a set $F \subseteq \mathbf{X}$, we would like to find a non-blocking controller that ensures that the state stays in F for ever. We will say that a controller C *solves the problem* $(H, \Box F)$, if and only if C is non-blocking and for all $(x,u,d) \in \mathcal{E}_{H_C}$, $x[k] \in F$ for all $k \geq 0$. If such a controller exists we say that the problem $(H, \Box F)$ *can be solved*.

Even though safety properties are not the only properties of interest[2], they turn out to be very useful in applications. Many important problems, such as absence of collisions in transportation systems, mutual exclusion in distributed algorithms, etc., can be naturally encoded as safety properties. Fortunately, it can be shown that for this class of properties one can, without loss of generality, restrict attention to memoryless controllers.

Proposition 1. *The problem $(H, \Box F)$ can be solved if and only if it can be solved by a memoryless controller.*

Motivated by Proposition 1, we restrict our attention to memoryless controllers from now on.

---

[1] The condition is only sufficient. Although it can be refined to be necessary as well, we will not pursue this direction since the emphasis of this paper is controller synthesis.

[2] Other important properties are liveness properties (ensuring that the state eventually reaches a certain set, visits a set infinitely often, etc.), stability, optimality, etc.

2.2   Controlled Invariant Sets and Least Restrictive Controllers

The concept of controlled invariance turns out to be fundamental for the design of controllers for safety specifications [16]. Roughly speaking, a set of states, W, is called controlled invariant if there exists a controller that ensures that all executions starting somewhere in W remain in W for ever. More formally:

Definition 4 (Controlled invariant set). *A set $W \subseteq \mathbf{X}$ is called a controlled invariant set of H if there exists a non-blocking controller that solves the problem $(H', \Box W)$, where $H' = (X, V, W, f)$ (the same as H, but with $Init' = W$).*

We say that the controller that solves the problem $(H', \Box W)$ *renders the set W invariant*. Also, given a set $F \subseteq \mathbf{X}$, a set $W \subseteq F$ is called a *maximal controlled invariant subset of F*, if it is controlled invariant and it is not a proper subset of any other controlled invariant subset of F. The following lemma establishes the uniqueness of the maximal controlled invariant set.

Lemma 1. *The problem $(H, \Box F)$ can be solved if and only if there exists a unique maximal controlled invariant set, $\hat{W}$, with $Init \subseteq \hat{W} \subseteq F$.*

A useful and intuitive characterization of the concept of controlled invariance can be given in terms of the operator $Pre : 2^{\mathbf{X}} \to 2^{\mathbf{X}}$ defined by

$$Pre(W) = \{x \in W \mid \exists u \in \mathbf{U} \; \forall d \in \mathbf{D}, \; f(x, u, d) \cap W^c = \emptyset\}.$$

The following properties of the operator Pre are easy to establish and will be useful in the subsequent discussion.

Proposition 2. *The operator Pre has the following properties:*

1. *Pre is contracting, that is for all $W \subseteq \mathbf{X}$, $Pre(W) \subseteq W$;*
2. *Pre is monotone, that is for all $W, W' \subseteq \mathbf{X}$ with $W \subseteq W'$, $Pre(W) \subseteq Pre(W')$; and,*
3. *A set $W \subseteq \mathbf{X}$ is controlled invariant if and only if it is a fixed point of Pre, that is if and only if $Pre(W) = W$.*

Many memoryless controllers may be able to solve a particular problem. Controllers that impose less restrictions on the inputs they allow are in a sense better than controllers that impose more restrictions. For example, controllers that impose fewer restrictions allow more freedom if additional safety specifications are imposed, or if one is asked to optimize the performance of the (safe) closed loop system with respect to other objectives. To quantify this intuitive notion we introduce a partial order on the space of memoryless controllers. We write $g_1 \preceq g_2$ if for all $x \in \mathbf{X}$, $g_1(x) \subseteq g_2(x)$.

Definition 5 (Least restrictive controller). *A memoryless controller $g : \mathbf{X} \to 2^{\mathbf{U}}$ that solves the problem (H, F) is called least restrictive if it is maximal among the controllers that solve $(H, \Box F)$ in the partial order defined by $\preceq$.*

Lemma 2. *A controller that renders a set W invariant exists if and only if a unique least restrictive controller that renders W invariant exists.*

Notice that the least restrictive controller that renders a set W invariant must, by definition, allow $\hat{g}(x) = \mathbf{U}$ for all $x \notin W$. Summarizing Lemmas 1 and 2 we have the following:

Theorem 1. *The problem $(H, \Box F)$ can be solved if and only if there exists:*

1. *a unique maximal controlled invariant set $\hat{W}$ with $\text{Init} \subseteq \hat{W} \subseteq F$, and*
2. *a unique least restrictive controller, $\hat{g}$, that renders $\hat{W}$ invariant.*

Motivated by Theorem 1 we state the controlled invariance problem more formally.

Problem 1 (Controlled Invariance Problem (CIP)) *Given a DTS and a set $F \subseteq \mathbf{X}$ compute the maximal controlled invariant subset of F, $\hat{W}$, the least restrictive controller, $\hat{g}$, that renders $\hat{W}$ invariant, and test whether $\text{Init} \subseteq \hat{W}$.*

2.3 Computation of $\hat{W}$ and $\hat{g}$

We first present a conceptual algorithm for solving the CIP for general DTS. Even though there is no straightforward way of implementing this algorithm in the general case, in subsequent sections we show how this can be done for special classes of DTS.

Algorithm 1 (Controlled Invariance Algorithm)

> **initialization:** $W^0 = F$, $W^{-1} = \mathbf{X}$, $l = 0$
> **while** $W^{l-1} \cap (W^l)^c \neq \emptyset$ **do**
> $\qquad W^{l+1} = \text{Pre}(W^l)$
> $\qquad l = l + 1$
> **end while**
> **set** $\hat{W} = \bigcap_{l \geq 0} W^l$
> **set** $\hat{g}(x) = \begin{cases} \left\{ u \in \mathbf{U} \mid \forall d \in \mathbf{D}, \ f(x, u, d) \cap (\hat{W})^c = \emptyset \right\} & x \in \hat{W} \\ \mathbf{U} & x \notin \hat{W} \end{cases}$

Theorem 2. *$\hat{W}$ is the maximal controlled invariant subset of F and $\hat{g}$ is the least restrictive controller that renders $\hat{W}$ invariant.*

To implement the controlled invariance algorithm one needs to be able to (1) encode sets of states, perform intersection and complementation, and test for emptiness, (2) compute the Pre of a set, and (3) guarantee that a fixed point is reached after a finite number of iterations. For classes of DTS for which 1 and 2 are satisfied we say that the CIP is *semi-decidable*; if all three conditions are satisfied we say that the CIP is *decidable*. As an example, consider *finite state machines* (FSM), that is the class of DTS for which $\mathbf{X}$, $\mathbf{U}$ and $\mathbf{D}$ are finite. In this case, one can encode sets of states, perform intersection, complementation, test for emptiness and compute Pre by enumeration (or other more efficient representations). Moreover, by the monotonicity of W^l and the fact that $\mathbf{X}$ is finite, the algorithm is guaranteed to terminate in a finite number of steps. Therefore, the CIP is decidable for finite state machines.

In subsequent sections we show how the computation can be performed for DTS with state and input taking values on a Euclidean space and transition relations given by certain classes of functions of the state and input.

2.4   CIP for Definable Discrete Time Systems

In this section we consider the case where all the sets involved in the CIP can be expressed by means of a logic formula that belongs to the language of a certain logic theory. For example, we denote by $\text{Lin}(\mathbb{R})$ the theory of linear constraints and by $\text{OF}(\mathbb{R})$ the theory of polynomial constraints.

For some theories, it is possible to determine the sentences that belong to the theory. The Tarski-Seidenberg decision procedure provides a way of doing this for $\text{OF}(\mathbb{R})$. It can be shown that $\text{OF}(\mathbb{R})$ is decidable [22, 24], in other words, there exists a computational procedure that after a finite number of steps determines whether an $\mathcal{R}$-sentence belongs to $\text{OF}(\mathbb{R})$ or not. The decision procedure is based on quantifier elimination, an algorithm that converts a formula $\phi(x_1, \dots, x_n)$ to an equivalent quantifier free formula. Notice that this provides a method for testing emptiness. A set $Y = \{(x_1, \dots, x_n) \mid \phi(x_1, \dots, x_n)\}$ is empty if and only if the sentence $\exists x_1 \dots \exists x_n \mid \phi(x_1, \dots, x_n)$ is equivalent to false.

To relate this to the problem at hand, we restrict our attention to CIP which are "definable" in an appropriate theory.

Definition 6 (Definable CIP). *A CIP, $(H, \Box F)$, is definable in a theory if* $\mathbf{X} = \mathbb{R}^n$, $\mathbf{U} \subseteq \mathbb{R}^{n_u}$, $\mathbf{D} \subseteq \mathbb{R}^{n_d}$ *and the sets* $\mathbf{U}$, $\mathbf{D}$, Init, $f(x, u, d)$ *for all* $x \in \mathbf{X}$, $u \in \mathbf{U}$ *and* $d \in \mathbf{D}$, *and* F *are definable in the theory.*

If $(H, \Box F)$ and W^l are definable in $\text{OF}(\mathbb{R})$, then

$$\psi^l(x) \equiv \exists u \, \forall d \, \forall x' \mid [x \in W^l] \wedge [u \in \mathbf{U}] \wedge [(d \notin \mathbf{D}) \vee (x' \notin f(x, u, d)) \vee (x' \in W^l)] \ (1)$$

is a first order formula in the corresponding language. Therefore, each step of the controlled invariance algorithm involves eliminating the quantifiers in (1) to obtain a quantifier free formula defining W^{l+1}. The fact that $\text{OF}(\mathbb{R})$ is decidable immediately leads to the following:

Theorem 3. *The class of CIP definable in $\text{OF}(\mathbb{R})$ is semi-decidable.*

Moreover, if $(H, \Box F)$ is definable in $\text{OF}(\mathbb{R})$ and W is a controlled invariant set also definable in $\text{OF}(\mathbb{R})$, then the set $\{(x, u) \mid \forall d \in \mathbf{D} \ \forall x' \in f(x, u, d), \ x' \in W\}$ describing the least restrictive controller that renders W invariant is also definable in $\text{OF}(\mathbb{R})$. Furthermore, quantifier elimination can be performed in this formula, to obtain an explicit expression for the least restrictive controller. Finally, the question $W \cap \text{Init}^c = \emptyset$ can be decided. Therefore, if the algorithm happens to terminate in a finite number of steps, the CIP can be completely solved.

Although different methods have been proposed for performing quantifier elimination in $OF(\mathbb{R})$ [1, 22, 24], and the process can be automated using symbolic tools [9], the quantifier elimination procedure is in general hard, both in theory and in practice, since the solvability may be doubly exponential [14]. For the theory $Lin(\mathbb{R})$, a somewhat more efficient implementation can be derived using techniques from linear algebra and linear programming. The next section shows how quantifier elimination in the theory $Lin(\mathbb{R})$ can be performed more efficiently for the formula (1) used in the controlled invariance algorithm.

3   CIP for Discrete Time Linear Systems

A *linear CIP* (LCIP) consists of

- a Linear DTS (LDTS), i.e. a DTS with $\mathbf{X} = \mathbb{R}^n$, $\mathbf{U} = \{u \in \mathbb{R}^{n_u} \mid Eu \leq \eta\} \subseteq \mathbb{R}^{n_u}$, $\mathbf{D} = \{d \in \mathbb{R}^{n_d} \mid Gd \leq \gamma\} \subseteq \mathbb{R}^{n_d}$, Init $= \{x \in \mathbf{X} \mid Jx \leq \theta\}$ and a reset relation given by $f(x, u, d) = \{Ax + Bu + Cd\}$, where $A \in \mathbb{Q}^{n \times n}$, $B \in \mathbb{Q}^{n \times n_u}$, $C \in \mathbb{Q}^{n \times n_d}$, $E \in \mathbb{Q}^{m_u \times n_u}$, $G \in \mathbb{Q}^{m_d \times n_d}$, $\eta \in \mathbb{Q}^{m_u}$, $\gamma \in \mathbb{Q}^{m_d}$, $J \in \mathbb{Q}^{n \times m_i}$ and $\theta \in \mathbb{Q}^{m_i}$ with m_u, m_d and m_i being the number of constraints on the control, disturbance and initial conditions, respectively; and,
- a set $F = \{x \in \mathbb{R}^n \mid Mx \leq \beta\}$ where $M \in \mathbb{Q}^{m \times n}$, $\beta \in \mathbb{Q}^m$ and m is the number of constraints on the state.

Notice that LDTS are non-blocking and deterministic, in the sense that for every state x and every input (u, d) there exists a unique next state. Since the sets F, $\mathbf{U}$ and $\mathbf{D}$ are all convex polygons, and the dynamics f are given by a linear map, the LCIP is definable in the theory $Lin(\mathbb{R})$, and therefore, according to the discussion in Sect. 2.4, it is semi-decidable. We assume that the sets F and $\mathbf{U}$ can be either bounded or unbounded, but $\mathbf{D}$ is bounded[3].

For the LCIP it turns out that, after the l-th iteration, the set W^l can be described by m^l linear constraints as $\{x \in \mathbb{R}^n \mid M^l x \leq \beta^l\}$, that is, W^l remains a convex polygon. Obviously, $m^0 = m$, $M^0 = M$ and $\beta^0 = \beta$. Letting $\hat{A}^l = M^l A$, $\hat{B}^l = M^l B$ and $\hat{C}^l = M^l C$, (1) becomes

$$\psi^l(x) \equiv [M^l x \leq \beta^l] \wedge [\exists u \mid (Eu \leq \eta) \wedge (\forall d \mid (Gd > \gamma) \vee (\hat{A}^l x + \hat{B}^l u + \hat{C}^l d \leq \beta^l))].$$

Thus, in each step of the algorithm, we need to be able to eliminate variables u and d from the inner formulae, intersect the new constraints with the old ones and check if the new set is empty. Notice that not all of the new constraints generated by quantifier elimination may be necessary to define the set W^{l+1}. Also, some of the old constraints may become redundant after adding the new ones. Hence we need to check the redundancy of the constraints when doing the intersection.

---

[3] The theoretical discussion can be extended to unbounded $\mathbf{D}$ sets, but the computational implementation is somewhat more involved.

3.1 Quantifier Elimination

We first perform quantifier elimination on d over the formula

$$\phi^l(x, u) \equiv \forall d \mid (Gd > \gamma) \vee (\hat{A}^l x + \hat{B}^l u + \hat{C}^l d \leq \beta^l).$$

Let $\hat{a}_i^T$, $\hat{b}_i^T$ and $\hat{c}_i^T$ be the i-th row of $\hat{A}^l$, $\hat{B}^l$ and $\hat{C}^l$, respectively. Then, parsing ϕ^l leads to

$$\phi^l(x, u) \equiv \forall d \mid \bigwedge_{i=1}^{m^l} (Gd > \gamma) \vee (\hat{c}_i^T d \leq \beta_i^l - \hat{a}_i^T x - \hat{b}_i^T u).$$

Consider $\delta : \mathbb{R}^{m^l \times n_d} \to \mathbb{R}^{m^l}$ defined by $\delta_i(\hat{C}^l) = \max\limits_{d:Gd \leq \gamma} (\hat{c}_i^T d)$ for $i = 1, \dots, m^l$.

Proposition 3. $\phi^l(x, u)$ *is equivalent to* $\varphi^l(x, u) \equiv \hat{A}^l x + \hat{B}^l u \leq \beta^l - \delta(\hat{C}^l)$.

Therefore, the elimination of the $\forall$ quantifier can be done by solving a finite collection of linear programming problems. Since we have assumed that $\mathbf{D}$ is bounded, such an optimization problem is guaranteed to have a solution, and hence $\delta(\cdot)$ is well defined. Since $\delta(\cdot)$ is applied to each row of $\hat{C}^l$, in the sequel we will use $\delta_i(\hat{C}^l)$ and $\delta(\hat{c}_i^T)$ interchangeably. Notice that, strictly speaking, $\delta(\cdot)$ is not part $\mathrm{Lin}(\mathbb{R})$, but we use it as a shorthand for the constant obtained by solving the linear programs.

Next, we perform quantifier elimination on u over the formula

$$\phi^l(x) \equiv \exists u \mid (Eu \leq \eta) \wedge (\hat{A}^l x + \hat{B}^l u \leq \beta^l - \delta(\hat{C}^l)). \tag{2}$$

We will discuss two methods to eliminate u. The first is known as *Fourier Elimination* [10], and the second, attributed to Cernikov [6], is an application of Farkas Lemma on duality [7].

For the first method, assume we want to eliminate u_1 first. Let e_i be the i-th unit vector in $\mathbb{R}^{m^l + m_u}$,

$$H^l = \begin{pmatrix} \hat{B}^l \\ E \end{pmatrix} \quad \text{and} \quad \xi^l(x) = \begin{pmatrix} \beta^l - \delta(\hat{C}^l) - \hat{A}^l x \\ \eta \end{pmatrix}.$$

Thus $\phi^l(x)$ is equivalent to $\exists u \mid H^l u \leq \xi^l(x)$. Also define $P^l = \{p \mid H_{p1}^l > 0\}$, $Q^l = \{q \mid H_{q1}^l < 0\}$ and $R^l = \{r \mid H_{r1}^l = 0\}$, where H_{ij}^l refers to the i, j element of the matrix H^l. Then $\phi^l(x)$ is equivalent to

$$\exists u \mid \bigwedge_{p \in P^l} \bigwedge_{q \in Q^l} \left[\frac{1}{H_{q1}^l} \left(\xi_q^l(x) - \sum_{j=2}^{m} H_{qj}^l u_j \right) \leq u_1 \leq \frac{1}{H_{p1}^l} \left(\xi_p^l(x) - \sum_{j=2}^{m} H_{pj}^l u_j \right) \right]$$

$$\wedge \bigwedge_{r \in R^l} \left[0 \leq \left(\xi_r^l(x) - \sum_{j=2}^{m} H_{rj}^l u_j \right) \right].$$

Hence, after the elimination of u_1 we obtain

$$\exists u \mid \bigwedge_{p \in P^l} \bigwedge_{q \in Q^l \cup R^l} (H_{p1}^l \ -H_{q1}^l) \begin{pmatrix} \hat{e}_q^T \\ \hat{e}_p^T \end{pmatrix} \begin{pmatrix} \hat{A}^l x \\ 0 \end{pmatrix} \leq (H_{p1}^l \ -H_{q1}^l) \begin{pmatrix} \hat{e}_q^T \\ \hat{e}_p^T \end{pmatrix} \begin{pmatrix} \beta^l - \delta(\hat{C}^l) \\ \eta \end{pmatrix}$$

$$- (H_{p1}^l \ -H_{q1}^l) \begin{pmatrix} \sum_{j=2}^m H_{qj}^l u_j \\ \sum_{j=2}^m H_{pj}^l u_j \end{pmatrix}. \quad (3)$$

Therefore, the elimination of the $\exists$ quantifier is performed by taking nonnegative linear combinations of all pairs of constraints so as to cancel the quantified variable. Note that if all the coefficients of the quantified variable are positive (negative), then ϕ^l is true, and we need not to eliminate the remaining variables. Otherwise, after u_1 has been eliminated, we apply the same procedure to the constraints in (3), so as to eliminate $u_2, \ldots, u_{n_u}$. Since the procedure is based on nonnegative row operations, it is clear that

$$\phi^l(x) \equiv \Lambda^l \begin{pmatrix} \hat{A}^l x \\ 0 \end{pmatrix} \leq \Lambda^l \begin{pmatrix} \beta^l - \delta(\hat{C}^l) \\ \eta \end{pmatrix} \equiv (\tilde{M}^l x \leq \tilde{\beta}^l) \wedge (0 \leq \Lambda_2^l \eta), \quad (4)$$

where $\Lambda^l = [\Lambda_1^l \ \Lambda_2^l] \in \mathbb{Q}^{\tilde{m}^l \times (m^l + m_u)}$ is a matrix with nonnegative entries such that $\Lambda^l H^l = 0$, $\tilde{m}^l$ is the number of new constraints obtained through quantifier elimination, $\tilde{M}^l = \Lambda_1^l \hat{A}^l \in \mathbb{Q}^{\tilde{m}^l \times n}$ and $\tilde{\beta}^l = \Lambda_1^l (\beta^l - \delta(\hat{C}^l)) \in \mathbb{Q}^{\tilde{m}^l}$. Notice that if the condition $\Lambda_2^l \eta \geq 0$ is violated, then $\hat{W} = \emptyset$. Otherwise, we just need to add the new constraints $\tilde{M}^l x \leq \tilde{\beta}^l$ to the original set W^l.

Although *Fourier Elimination* is attractive because of its simplicity, it is quite inefficient. In general, it generates many new constraints in the intermediate steps, and in the worst case the method is exponential. This difficulty can be partially remedied since many of the inequalities are likely to be redundant [7].

An alternative method [6] computes the rows of Λ^l directly as the extreme points of the set $\{\lambda^l \in \mathbb{R}^{m+m_u} \mid \lambda^{l^T} H^l = 0 \wedge \lambda^l \geq 0 \wedge \sum \lambda_i^l = 1\}$, where the last constraint is added to ensure that the set is a polytope. Although the extreme points method is better than Fourier elimination, because it eliminates the costly intermediate steps, the computation of the extreme points is still costly and also generates a lot of redundant constraints. A more efficient method [14] uses a generalized linear programming formulation and an on-line convex hull construction to obtain an incremental inner approximation of the set defined by ϕ^l. The method considerably reduces the number of constraints defining the resulting set.

3.2   Intersection, Emptiness and Redundancy

Provided that $\Lambda_2^l \eta \geq 0$, the quantifier elimination procedure presented above computes the set of states $\tilde{W}^l \equiv \{x \mid \tilde{M}^l x \leq \tilde{\beta}^l\}$ that can be forced by u to transition into W^l. To obtain W^{l+1}, such a set must be intersected with W^l. Since both sets are convex, the intersection can be carried out by simply appending $\tilde{M}^l$ and $\tilde{\beta}^l$ to M^l and β^l, respectively. However, this method of performing the intersection is likely to lead to a description of the set which is larger than necessary since many of the constraints may be redundant. Algorithm 2 is aimed at checking the emptiness of the intersection and then eliminate redundant constraints. In the algorithm, $[]$ denotes an empty matrix, $\mathbf{1} = (1 \ldots 1)^T \in \mathbb{Q}^{\tilde{m}^l + m^l}$, and $m_i'^T$ and β_i' are the i-th rows of $M_0' = \begin{pmatrix} \tilde{M}^l \\ M^l \end{pmatrix}$ and $\beta_0' = \begin{pmatrix} \tilde{\beta}^l \\ \beta^l \end{pmatrix}$, respectively. Initially, $M' = M_0'$ and $\beta' = \beta_0'$.

The idea behind the algorithm is that $W^l \cap \tilde{W}^l \neq \emptyset$ if and only if $\exists x | M'x \leq \beta'$, which is equivalent to saying that $\min\{t \mid M'x \leq \beta' + \mathbf{1}t\} \leq 0$. Afterwards, if the problem $\max\{m_i'^T x \mid M'x \leq \beta'\}$ is feasible, and the constraint $m_i'^T x \leq \beta_i'$ is not redundant, then the optimal value of the problem is β_i'. Moreover, if the non-redundant constraint $m_i'^T x \leq \beta_i'$ is removed from the optimization problem, then the new optimal value m^* satisfies $m^* > \beta_i'$.

Algorithm 2 (Emptiness and Redundancy Algorithm)

> **initialization** $M' = M_0'$, $\beta' = \beta_0'$, $M^{l+1} = []$, $\beta^{l+1} = []$.
> $m^* = \min\{t \mid M'x \leq \beta' + \mathbf{1}t\}$
> **if** $m^* > 0$ **or** $\Lambda_2^l \eta \not\geq 0$ **then**
> > $\hat{W} = \emptyset$, **terminate** controlled invariance algorithm
> **else**
> > **for** $i = 1$ **to** $\tilde{m}^l + m^l$ **do**
> > > remove $m_i'^T$ from M' and β_i' from β'
> > > $m^* = \max\{m_i'^T x \mid M'x \leq \beta'\}$
> > > **if** $m^* > \beta_i'$ **then**
> > > > add $m_i'^T$ to M^{l+1} and M',
> > > > add β_i' to β^{l+1} and β'
> > > **end if**
> > **end for**
> **end if**
> **if** $M^{l+1} = M^l$ and $\beta^{l+1} = \beta^l$ **then**
> > $\hat{W} = W^l$, **terminate** controlled invariance algorithm
> **end if**

The controlled invariance algorithm terminates if the redundancy algorithm concludes that either $\Lambda_2^l \eta \not\geq 0$ or $W^l \cap \tilde{W}^l = \emptyset$ (in which case $\hat{W} = \emptyset$), or if all the new constraints are redundant (in which case $W^l = W^{l+1} = \hat{W}$)[4]. Otherwise, upon termination of the redundancy algorithm, the process is repeated for W^{l+1}. An obvious optimization of the code involves terminating both algorithms if after all new constraints in $\tilde{M}^l x \leq \tilde{\beta}^l$ have been tested, M^{l+1} and β^{l+1} are still empty. Notice that for all l the set W^l is a convex polygon as claimed. Summarizing:

---

[4] Note that any redundant constraint in the original description of F will be eliminated the first time the redundancy algorithm is invoked by the controlled invariance algorithm.

Theorem 4. *The LCIP is semi-decidable.*

In the next section we study situations where the algorithm is guaranteed to terminate in a finite number of steps. In Sect. 4, we will provide and example which actually converges after an infinite number of iterations.

3.3 Decidable Special Cases

We first summarize some of the observations made so far about situations where the algorithm terminates in a finite number of steps.

Proposition 4. *For an LCIP with* $\mathbf{U} = \mathbb{R}^{n_u}$*, if either one of the columns of* MB *is componentwise positive (negative), or if* $\mathrm{rank}(MB) = \min\{m, n\}$*, the algorithm terminates in a finite number of steps.*

Next, we limit our attention to the case $F = [\alpha_1, \beta_1] \times \ldots \times [\alpha_n, \beta_n] \subset \mathbb{R}^n$ with $\alpha_i \leq \beta_i$ and $[\alpha_i, \beta_i] \subset \mathbb{R}, i = 1 \ldots n$, $u \in \mathbb{R}$, and $d \in [d_1, d_2] \subset \mathbb{R}$. To remind ourselves of the fact that u and d are scalar, we use b and c instead of B and C. We also assume that (A, b) is in controllable canonical form, that is

$$x[k+1] = \begin{pmatrix} 0 & 1 & 0 & 0 & \cdots & 0 \\ 0 & 0 & 1 & 0 & \cdots & 0 \\ \vdots & & & \ddots & & \vdots \\ 0 & & & & & 1 \\ a_{n1} & a_{n2} & \cdots & & & a_{nn} \end{pmatrix} x[k] + \begin{pmatrix} 0 \\ 0 \\ \vdots \\ 0 \\ 1 \end{pmatrix} u[k] + \begin{pmatrix} c_1 \\ c_2 \\ \vdots \\ c_{n-1} \\ c_n \end{pmatrix} d[k]. \quad (5)$$

In this case $\psi^1(x)$ is equivalent to

$$\exists u \mid \bigwedge_{j=1}^{n} (\alpha_j \leq x_j \leq \beta_j) \wedge \bigwedge_{j=2}^{n} (\alpha_{j-1} - \delta(-c_{j-1}) \leq x_j \leq \beta_{j-1} - \delta(c_{j-1})) \wedge$$

$$\left(\alpha_n - \sum_{j=1}^{n} a_{nj} x_j - \delta(-c_n) \leq u \leq \beta_n - \sum_{j=1}^{n} a_{nj} x_j - \delta(c_n) \right). \quad (6)$$

From the last expression, it is clear that given $x_1 \in [\alpha_1, \beta_1]$, x_j exists if and only if $\alpha_j^1 = \max(\alpha_j, \alpha_{j-1} - \delta(-c_{j-1})) \leq \min(\beta_j, \beta_{j-1} - \delta(c_{j-1})) = \beta_j^1$, $j = 2 \ldots n$, and u exists if and only if $\alpha_n - \delta(-c_n) \leq \beta_n - \delta(c_n)$. It is straightforward to see that in the l-th iteration $(0 \leq l \leq n)$ W^l is defined by:

$$W^l = [\alpha_1^0, \beta_1^0] \times \ldots \times [\alpha_{l+1}^l, \beta_{l+1}^l] \times [\alpha_{l+2}^l, \beta_{l+2}^l] \times \ldots \times [\alpha_n^l, \beta_n^l],$$

where $\alpha_j^l = \max(\alpha_j^{l-1}, \alpha_{j-1}^{l-1} - \delta(c_{j-1}))$, and $\beta_j^l = \min(\beta_j^{l-1}, \beta_{j-1}^{l-1} - \delta(c_{j-1}))$, for $2 \leq l+1 \leq j \leq n$, with $\alpha_j^0 = \alpha_j$ and $\beta_j^0 = \beta_j$, for $1 \leq j \leq n$.

This means that after n iterations, the maximal controlled invariant set remains unchanged, and the least restrictive controller is given by the last constraint in (6), but with α_n and β_n replaced by α_n^{n-1} and β_n^{n-1}, respectively. This result can be summarized as follows:

Lemma 3. *Given system (5) with $F = [\alpha_1, \beta_1] \times \ldots \times [\alpha_n, \beta_n] \subset \mathbb{R}^n$, $\mathbf{U} = \mathbb{R}$ and $\mathbf{D} = [d_1, d_2] \subset \mathbb{R}$, the solution to the CIP, obtained after at most n iterations of the algorithm, is given by:*

$$\hat{W} = \begin{cases} \left\{ x \mid \bigwedge_{j=1}^{n} \alpha_j^{j-1} \le x_j \le \beta_j^{j-1} \right\} & if \ \bigwedge_{j=2}^{n} \left(\alpha_j^{j-1} \le \beta_j^{j-1} \right) \wedge \left(|c_n| \le \frac{\beta_n^{n-1} - \alpha_n^{n-1}}{d_2 - d_1} \right) \\ \emptyset & otherwise \end{cases}$$

$$\hat{g}(x) = \begin{cases} \left\{ u \mid \alpha_n^{n-1} - \delta(-c_n) \le u + \sum_{j=1}^{n} a_{nj} x_j \le \beta_n^{n-1} - \delta(c_n) \right\} & if \ x \in \hat{W} \\ \mathbf{U} & otherwise \end{cases}$$

Theorem 5. *For systems of the form (5) with $F = [\alpha_1, \beta_1] \times \ldots \times [\alpha_n, \beta_n] \subset \mathbb{R}^n$, $\mathbf{U} = \mathbb{R}$ and $\mathbf{D} = [d_1, d_2] \subset \mathbb{R}$, the LCIP is decidable.*

The conditions of Theorem 5 for decidability are somewhat demanding. If, for example, u is bounded, that is, $\mathbf{U} = [u_1, u_2] \subset \mathbb{R}$, then the new constraints added to x during each iteration may change the bounds on x to a non-rectangular polyhedron. In this case, the CIP is no longer decidable, and the system falls into the more general class of systems described at the beginning of the section. We conjecture that the LCIP is decidable in a much more general setting, using a completely different algorithm that exploits the stabilizability of the pairs (A, B) and (A, C) and the observability of the pair (A, M).

4   Experimental Results

The algorithm proposed in Sect. 3 was implemented in MATLAB. Here, we present two examples that were solved using this implementation. The first example is also worked out analytically to complete the semi-decidability result.

Example 1. The LDTS is defined by $\mathbf{U} = \mathbb{R}$, $\mathbf{D} = [-1, 1]$,

$$A = \begin{pmatrix} 0 & 1 \\ 1 & 1 \end{pmatrix}, B = \begin{pmatrix} 0 \\ 1 \end{pmatrix}, C = \begin{pmatrix} 1 \\ 1 \end{pmatrix}, M = \begin{pmatrix} 1 & 1 \\ -1 & -3 \\ 1 & -1 \\ -3 & 1 \end{pmatrix}, \text{ and } \beta = \begin{pmatrix} 100 \\ -50 \\ 100 \\ -50 \end{pmatrix}.$$

It is straightforward to see that the only new constraint added in the l-th iteration is $[0 \ m_l]x \le \beta_l$, where $m_l = -10 \cdot 3^{l-1}$, and $\beta_l = -210 - 265(3^{l-1} - 1)$. Therefore after an infinite number of iterations, $\hat{W}$ and $\hat{g}(x)$ converge to

$$\hat{W} = \left\{ x \mid \begin{pmatrix} M \\ 0 & -2 \end{pmatrix} x \le \begin{pmatrix} \beta \\ -53 \end{pmatrix} \right\}$$

$$\hat{g}(x) = \begin{cases} \{ u \in \mathbf{U} \mid u \ge \max(18 - x_1 - \frac{4x_2}{3}, -100 - x_1, -\frac{55}{2} - x_1 - x_2) \\ \quad u \le \min(98 - x_1 - 2x_2, -52 - x_1 + 2x_2) \} & if \ x \in \hat{W} \\ \mathbf{U} & else \end{cases}$$

Example 2. The LDTS is defined by

$$A = \begin{pmatrix} -1 & -8 & -1 \\ 1 & -4 & -1 \\ -5 & -3 & -1 \end{pmatrix}, B = \begin{pmatrix} 2 & 1 \\ 4 & 1 \\ 1 & -1 \end{pmatrix}, C = \begin{pmatrix} 3 & 2 & 1 \\ 2 & 1 & 7 \\ 1 & 2 & 1 \end{pmatrix}, E = \begin{pmatrix} 1 & 1 \\ 1 & -1 \\ 0 & 1 \\ 0 & -1 \end{pmatrix}, \eta = \begin{pmatrix} 1000 \\ 1000 \\ 1000 \\ 1000 \end{pmatrix},$$

$$M = \begin{pmatrix} 3 & 1 & 0 \\ -1 & 3 & 0 \\ 1 & -1 & 0 \\ -1 & -1 & 0 \\ -1 & 0 & 1 \\ 0 & 0 & -1 \end{pmatrix}, \beta = \begin{pmatrix} 100 \\ 100 \\ 100 \\ 100 \\ 100 \\ 100 \end{pmatrix}, G = \begin{pmatrix} 1 & 1 & 0 \\ -1 & 0 & 0 \\ 0 & 1 & 0 \\ 0 & -1 & 0 \\ 0 & 0 & 1 \\ 0 & 0 & -1 \end{pmatrix} \text{ and } \gamma = \begin{pmatrix} 1 \\ 1 \\ 1 \\ 1 \\ 1 \\ 1 \end{pmatrix}.$$

Using MATLAB, this example converges in two iterations. Information about the intermediate calculations of each iteration is shown in Table 1.

Table 1. Results of Example 2

Iteration	1	2
Number of LP problems for quantifier elimination on d	6	10
Number of constraints on (x, u) before elimination of u	10	14
Number of new constraints on x after elimination of u	281	614
Number of new non-redundant constraints on x	4	0
Total number of constraints on x after iteration	10	10

5   Conclusions and Future Work

We showed that the problem of computing the maximal controlled invariant set and the least restrictive controller for discrete time systems is well posed and proposed a general algorithm for carrying out the computation. We then specialized the algorithm to discrete time linear systems with convex polygonal constraints, and showed how it can be implemented using linear programming and Fourier elimination. The decidability of the problem was also analyzed, and some simple, but interesting cases were found to be decidable.

We are currently working on sufficient conditions under which the problem is decidable. So far, it seems that the decidability property is not only dependent on the system itself, but also on the initial set, as shown by Example 1. Another topic of further research, is the application of these algorithm to discrete time hybrid systems, where some states and inputs take values in finite sets, while others in subsets of a Euclidean space. It is easy to show how this class of systems is a special case of the more general class of DTS. Therefore, all the conclusions of Sect. 2 directly extend to them. Unfortunately the implementation of the controlled invariance algorithm is more complicated, even in the case where the continuous state evolves according to a linear difference equation.

References

[1] D.S. Arnon, G.E. Collins, and S. McCallum. Cylindrical algebraic decomposition I: the basic algorithm. *SIAM Journal on Computing*, 13(4):865–877, 1984.

[2] T. Başar and G. J. Olsder. *Dynamic Non-cooperative Game Theory.* Academic Press, 2nd edition, 1995.

[3] A. Bensoussan and J.L. Menaldi. Hybrid control and dynamic programming. *Dynamics of Continuous, Discrete and Impulsive Systems*, (3):395–442, 1997.

[4] M.S. Branicky, V.S. Borkar, and S.K. Mitter. A unified framework for hybrid control: Model and optimal control theory. *IEEE Transactions on Automatic Control*, 43(1):31–45, 1998.

[5] P.E. Caines and Y.J. Wei. Hierarchical hybrid control systems: A lattice theoretic formulation. *IEEE Transactions on Automatic Control*, 43(4):501–508, April 1998.

[6] R.N. Cernikov. The solution to linear programming problems by elimination of unknowns. *Soviet Mathematics Doklady*, 2:1099–1103, 1961.

[7] V. Chandru. Variable elimination in linear constraints. *The Computer Journal*, 36(5):463–472, 1993.

[8] T. Dang and O. Maler. Reachability analysis via face lifting. In *Hybrid Systems: Computation and Control*, vol. 1386 of *LNCS*, pp. 96–109. Springer Verlag, 1998.

[9] A. Dolzmann and T. Sturm. REDLOG: Computer algebra meets computer logic. *ACM SIGSAM Bulletin*, 31(2):2–9, 1997.

[10] L.B.J. Fourier. Analyse des travaux de l'Academie Royale des Sciences, pendant l'annee 1824, Partie matematique. Histoire de l'Academie Royale des Sciences de l'Institut de France 7, 1827.

[11] G. Grammel. Maximum principle for a hybrid system via singular perturbations. *SIAM Journal of Control and Optimization*, 37(4):1162–1175, 1999.

[12] M.R. Greenstreet and I. Mitchell. Integrating projections. In *Hybrid Systems: Computation and Control*, vol. 1386 of *LNCS*, pp. 159–174. Springer Verlag, 1998.

[13] M. Heymann, F. Lin, and G. Meyer. Control synthesis for a class of hybrid systems subject to configuration-based safety constraints. In *Hybrid and Real Time Systems*, vol. 1201 of *LNCS*, pp. 376–391. Springer Verlag, 1997.

[14] C. Lassez and J.-L. Lassez. Quantifier elimination for conjunctions of linear constraints via a convex hull algorithm. In *Symbolic and Numeric Computation for Artificial Intelligence*, pages 103–122. Academic Press, 1992.

[15] J. Lewin. *Differential Games.* Springer-Verlag, 1994.

[16] J. Lygeros, C. Tomlin, and S. Sastry. Controllers for reachability specifications for hybrid systems. *Automatica*, pages 349–370, March 1999.

[17] O. Maler, A. Pnueli, and J. Sifakis. On the synthesis of discrete controllers for timed systems. In *Theoretical Aspects of Computer Science*, vol. 900 of *LNCS*, pp. 229–242. Springer Verlag, 1995.

[18] A. Nerode and W. Kohn. Multiple agent hybrid control architecture. In *Hybrid Systems*, vol. 736 of *LNCS*, pp. 297–316. Springer Verlag, New York, 1993.

[19] G. Pappas, G. Lafferriere, and S. Sastry. Hierarchically consistent control systems. In *IEEE Conference on Decision and Control*, pages 4336–4341, December 1998.

[20] B. Piccoli. Necessary conditions for hybrid optimization. In *IEEE Conference on Decision and Control*, pages 410–415, December 7-10 1999.

[21] P. J. G. Ramadge and W. M. Wonham. The control of discrete event systems. *Proceedings of the IEEE*, Vol.77(1):81–98, 1989.

[22] A. Seidenberg. A new decision method for elementary algebra. *Annals of Mathematics*, 60:387–374, 1954.

[23] H.J. Sussmann. A maximum principle for hybrid optimal control problems. In *IEEE Conference on Decision and Control*, pages 425–430, December 7-10 1999.

[24] A. Tarski. *A decision method for elementary algebra and geometry.* University of California Press, 1951.

[25] W. Thomas. On the synthesis of strategies in infinite games. In Ernst W. Mayr and Claude Puech, editors, *Proceedings of STACS 95, vol. 900 of LNCS*, pp. 1–13. Springer Verlag, Munich, 1995.

[26] C. Tomlin, J. Lygeros, and S. Sastry. Computing controllers for nonlinear hybrid systems. In *Hybrid Systems: Computation and Control*, vol. 1569 of *LNCS*, pp. 238–255. Springer Verlag, 1999.

[27] R. Vidal, S. Schaffert, J. Lygeros, and S. Sastry. Controlled invariance of discrete time systems. Technical Report UCB/ERL M99/65, Electronics Research Laboratory, University of California, Berkeley, 1999.

[28] H. Wong-Toi. The synthesis of controllers for linear hybrid automata. In *IEEE Conference on Decision and Control*, pages 4607–4613, December 10-12 1997.

Dynamical Systems Revisited: Hybrid Systems with Zeno Executions⋆

Jun Zhang, Karl Henrik Johansson⋆⋆, John Lygeros, and Shankar Sastry

Department of Electrical Engineering and Computer Sciences
University of California, Berkeley, CA 94720-1770
{zhangjun,johans,lygeros,sastry}@eecs.berkeley.edu
http://www.eecs.berkeley.edu/~{zhangjun,johans,lygeros,sastry}

Abstract. Results from classical dynamical systems are generalized to hybrid dynamical systems. The concept of ω limit set is introduced for hybrid systems and is used to prove new results on invariant sets and stability, where Zeno and non-Zeno hybrid systems can be treated within the same framework. As an example, LaSalle's Invariance Principle is extended to hybrid systems. Zeno hybrid systems are discussed in detail. The ω limit set of a Zeno execution is characterized for classes of hybrid systems.

1 Introduction

Systems with interacting continuous-time and discrete-time dynamics are used as models in a large variety of applications. The rich structure of such hybrid systems allow them to accurately predict the behavior of quite complex systems. However, the continuous–discrete nature of the system calls for new system theoretical tools for modeling, analysis, and design. Intensive recent activity have provided a few such tools, for instance, Lyapunov stability results [1,14]. However, as will be shown in this paper, in many cases the results come with assumptions that are not only hard to check but also unnecessary. There are several fundamental properties of hybrid systems that have not been sufficiently studied in the literature. These include questions on existence and uniqueness of executions, which have only recently been addressed [12,7]. Another question is when a hybrid system exhibits an infinite number of discrete transitions during a finite time interval, which is referred to as Zeno. The significance of these questions has been pointed out by many researchers, e.g., He and Lemmon [3]

⋆ This work was supported by ARO under the MURI grant DAAH04-96-1-0341, the Swedish Foundation for International Cooperation in Research and Higher Education, Telefonaktiebolaget LM Ericsson's Foundation, ONR under grant N00014-97-1-0946, and DARPA under contract F33615-98-C-3614.
⋆⋆ Corresponding author.

N. Lynch and B. Krogh (Eds.): HSCC 2000, LNCS 1790, pp. 451–464, 2000.

write "An important issue [...] concerns necessary and sufficient conditions for a switched system to be live, deadlock free, or nonZeno."

The main contribution of the paper is to carefully generalize concepts from classical dynamical systems like ω limit sets and invariant sets, in a way so that Zeno executions are treated within the same framework as regular non-Zeno executions. It is then straightforward to extend existing results, for instance, Lyapunov stability theorems for hybrid systems [1,14]. We illustrate this by proving LaSalle's Invariance Principle for hybrid systems.

Zeno is an interesting mathematical property of some hybrid systems, which does not occur in smooth dynamical systems. Real physical systems are not Zeno. Models of physical systems may, however, be Zeno due to a too high level of abstraction. In the latter part of the paper, we characterize Zeno executions and their Zeno states, where the Zeno states are defined as the ω limit points of a Zeno execution. We are able to completely characterize the set of Zeno states for a few classes of hybrid systems. It is shown that the features of the reset maps are important. For example, if the resets are identity maps or the resets are contractions, the continuous part of the Zeno state is a singleton.

The outline of the paper is as follows. In Section 2 notation and some basic definitions of hybrid automata and executions are introduced. Some recent results on existence and uniqueness of executions for classes of hybrid automata are also given. Section 3 introduces invariants sets and ω limit sets for hybrid automata and gives a generalization of LaSalle's Invariance Principle. Finally, results on Zeno hybrid automata are given in Section 4, where for instance the ω limit set for Zeno executions are discussed and some necessary and sufficient conditions for Zenoness are given.

2   Hybrid Automata and Executions

2.1   Notation

For a finite collection V of variables, let $\mathbf{V}$ denote the set of valuations of these variables. We use lower case letters to denote both a variable and its valuation. We refer to variables whose set of valuations is finite or countable as *discrete* and to variables whose set of valuations is a subset of a Euclidean space as *continuous*. For a set of continuous variables X with $\mathbf{X} = \mathbb{R}^n$ for $n \geq 0$, we assume that $\mathbf{X}$ is given the Euclidean metric topology, and use $\|\cdot\|$ to denote the Euclidean norm. For a set of discrete variables Q, we assume that $\mathbf{Q}$ is given the discrete topology (every subset is an open set), generated by the metric $d_D(q, q') = 0$ if $q = q'$ and $d_D(q, q') = 1$ if $q \neq q'$. We denote the valuations of the union $Q \cup X$ by $\mathbf{Q} \times \mathbf{X}$, which is given the product topology, generated by the metric $d((q, x), (q', x')) = d_D(q, q') + \|x - x'\|$. Using the metric d, we define the distance between two sets $U_1, U_2 \subseteq \mathbf{Q} \times \mathbf{X}$ by $d(U_1, U_2) = \inf_{(q_i, x_i) \in U_i} d((q_1, x_1), (q_2, x_2))$. We assume that a subset U of a topological space is given the induced topology, and we use $\overline{U}$ to denote its closure, U^o its interior, ∂U its boundary, U^c its complement, $|U|$ its cardinality, and $P(U)$ the set of all subsets of U.

2.2   Basic Definitions

The following definitions are based on [8,4,7].

Definition 1 (Hybrid Automaton). *A hybrid automaton H is a collection $H = (Q, X, \text{Init}, f, \text{Dom}, \text{Reset})$, where*

- *Q is a finite collection of discrete variables;*
- *X is a finite collection of continuous variables with $\mathbf{X} = \mathbb{R}^n$;*
- *Init $\subseteq \mathbf{Q} \times \mathbf{X}$ is a set of initial states;*
- *$f : \mathbf{Q} \times \mathbf{X} \rightarrow T\mathbf{X}$ is a vector field;*
- *Dom $\subseteq \mathbf{Q} \times \mathbf{X}$ is the domain of H;[1]*
- *Reset $: \mathbf{Q} \times \mathbf{X} \rightarrow P(\mathbf{Q} \times \mathbf{X})$ is a reset relation.*

We refer to $(q, x) \in \mathbf{Q} \times \mathbf{X}$ as the *state* of H. Unless otherwise stated, we introduce the following assumption, to prevent some obvious pathological cases.

Assumption 1 *$|\mathbf{Q}| < \infty$ and f is Lipschitz continuous in its second argument.*

Note that, under the discrete topology on $\mathbf{Q}$, f is trivially continuous in its first argument. A hybrid automaton can be represented by a directed graph $(\mathbf{Q}, E)$, with vertices $\mathbf{Q}$ and edges

$$E = \{(q, q') \in \mathbf{Q} \times \mathbf{Q} : \exists x, x' \in \mathbf{X}, (q', x') \in \text{Reset}(q, x)\}.$$

With each vertex $q \in \mathbf{Q}$, we associate a set of continuous initial states

$$\text{Init}(q) = \{x \in \mathbf{X} : (q, x) \in \text{Init}\},$$

a vector field $f(q, \cdot)$, and a set

$$I(q) = \{x \in \mathbf{X} : (q, x) \in \text{Dom}\}.$$

With each edge $e = (q, q') \in E$, we associate a guard

$$G(e) = \{x \in \mathbf{X} : \exists x' \in \mathbf{X}, (q', x') \in \text{Reset}(q, x)\},$$

and a reset map

$$R(e, x) = \{x' \in \mathbf{X} : (q', x') \in \text{Reset}(q, x)\}.$$

Since there is a unique graphical representation for each hybrid automaton, we will use the corresponding graphs as formal definitions for hybrid automata in most examples.

---

[1] The set Dom is often called the invariant set in the hybrid system literature in computer science. We reserve this term for later in the paper, where we will discuss sets invariant in the usual dynamical systems sense.

Definition 2 (Hybrid Time Trajectory). *A hybrid time trajectory τ is a finite or infinite sequence of intervals $\tau = \{I_i\}_{i=0}^{N}$, such that*

- *$I_i = [\tau_i, \tau_i']$ for $i < N$, and, if $N < \infty$, $I_N = [\tau_N, \tau_N']$ or $I_N = [\tau_N, \tau_N')$; and*
- *$\tau_i \le \tau_i' = \tau_{i+1}$ for $i \ge 0$.*

A hybrid time trajectory is a sequence of intervals of the real line, whose end points overlap. The interpretation is that the end points of the intervals are the times at which discrete transitions take place. Note that $\tau_i = \tau_i'$ is allowed, therefore multiple discrete transitions may take place at the same "time". Since the dynamical systems we will be concerned with are time invariant we will sometimes, without loss of generality, assume $\tau_0 = 0$. Hybrid time trajectories can extend to infinity if τ is an infinite sequence or if it is a finite sequence ending with an interval of the form $[\tau_N, \infty)$. We denote by $\mathcal{T}$ the set of all hybrid time trajectories and use $t \in \tau$ as shorthand notation for that there exists i such that $t \in I_i \in \tau$. For a topological space K we use $k : \tau \to K$ as a short hand notation for a map assigning a value from K to each $t \in \tau$; note that k is not a function on the real line, as it assigns multiple values to the same $t \in \mathbb{R}$: $t = \tau_i' = \tau_{i+1}$ for all $i \ge 0$. Each $\tau \in \mathcal{T}$ is fully ordered by the relation $\prec$ defined by $t_1 \prec t_2$ for $t_1 \in [\tau_i, \tau_i']$ and $t_2 \in [\tau_j, \tau_j']$ if and only if $i < j$, or $i = j$ and $t_1 < t_2$.

Definition 3 (Execution). *An execution χ of a hybrid automaton H is a collection $\chi = (\tau, q, x)$ with $\tau \in \mathcal{T}$, $q : \tau \to \mathbf{Q}$, and $x : \tau \to \mathbf{X}$, satisfying*

- *$(q(\tau_0), x(\tau_0)) \in \text{Init}$ (initial condition);*
- *for all i with $\tau_i < \tau_i'$, $q(\cdot)$ is constant and $x(\cdot)$ is a solution[2] to the differential equation $dx/dt = f(q, x)$ over $[\tau_i, \tau_i']$, and for all $t \in [\tau_i, \tau_i')$, $(q(t), x(t)) \in \text{Dom}$ (continuous evolution); and*
- *for all i, $(q(\tau_{i+1}), x(\tau_{i+1})) \in \text{Reset}(q(\tau_i'), x(\tau_i'))$ (discrete evolution).*

We say a hybrid automaton *accepts* an execution χ or not. For an execution $\chi = (\tau, q, x)$, we use $(q_0, x_0) = (q(\tau_0), x(\tau_0))$ to denote the initial state of χ. The *execution time* $\tau_\infty(\chi)$ is defined as $\tau_\infty(\chi) = \sum_{i=0}^{N}(\tau_i' - \tau_i)$, where $N + 1$ is the number of intervals in the hybrid time trajectory. The argument χ will sometimes be left out. An execution is *finite* if τ is a finite sequence ending with a compact interval, it is called *infinite* if τ is either an infinite sequence or if $\tau_\infty(\chi) = \infty$, and it is called *Zeno* if it is infinite but $\tau_\infty(\chi) < \infty$. The execution time of a Zeno execution is also called the Zeno time. We use $\mathcal{E}_H(q_0, x_0)$ to denote the set of all executions of H with initial condition $(q_0, x_0) \in \text{Init}$, $\mathcal{E}_H^\infty(q_0, x_0)$ to denote the set of all infinite executions of H with initial condition $(q_0, x_0) \in \text{Init}$. We define $\mathcal{E}_H = \bigcup_{(q_0, x_0) \in \text{Init}} \mathcal{E}_H(q_0, x_0)$ and $\mathcal{E}_H^\infty = \bigcup_{(q_0, x_0) \in \text{Init}} \mathcal{E}_H^\infty(q_0, x_0)$. To simplify the notation, we will drop the subscript H whenever the automaton is clear from the context.

---

[2] "Solution" is interpreted in the sense of Caratheodory.

2.3   Classes of Automata

The notation previously introduced gives a convenient way to express existence and uniqueness of executions.

Definition 4 (Non-Blocking Automaton). *A hybrid automaton H is non-blocking if $\mathcal{E}_H^\infty(q_0, x_0)$ is non-empty for all $(q_0, x_0) \in \mathrm{Init}$.*

Definition 5 (Deterministic Automaton). *A hybrid automaton H is deterministic if $\mathcal{E}_H^\infty(q_0, x_0)$ contains at most one element for all $(q_0, x_0) \in \mathrm{Init}$.*

Note that if a hybrid automaton is both non-blocking and deterministic, then it accepts a unique infinite execution for each initial condition. In [7] conditions were established that determine whether an automaton is non-blocking and deterministic. The conditions require one to argue about the set of states reachable by a hybrid automaton, and the set of states from which continuous evolution is impossible. A state $(q, x) \in \mathbf{Q} \times \mathbf{X}$ is called *reachable by H*, if there exists a finite execution $\chi = (\tau, q, x)$ with $\tau = \{[\tau_i, \tau_i']\}_{i=0}^N$ and $(q(\tau_N'), x(\tau_N')) = (q, x)$. We use Reach_H to denote the set of states reachable by a hybrid automaton, and $\mathrm{Reach}_H(q)$ the projection of Reach_H to discrete state q. We will drop the subscript H whenever the automaton is clear from the context. The set Reach is in general difficult to compute. Fortunately, the conditions of the subsequent results will not require us to do so: any outer approximation of the reachable set will be sufficient. In [2,7] methods for computing such outer approximations using simple induction arguments are outlined.

The set of states from which continuous evolution is impossible is given by

$$\mathrm{Out}_H = \{(q^0, x^0) \in \mathbf{Q} \times \mathbf{X} : \forall \epsilon > 0, \ \exists t \in [0, \epsilon), \ (q^0, x(t)) \notin \mathrm{Dom}\},$$

where $x(\cdot)$ is the solution to $dx/dt = f(q^0, x)$ with $x(0) = x^0$. Note that if Dom is an open set, then Out is simply Dom^c. If Dom is closed, then Out may also contain parts of the boundary of Dom. In [7] methods for computing Out were proposed, under appropriate smoothness assumptions on f and the boundary of Dom. As before, we will use $\mathrm{Out}_H(q)$ to denote the projection of Out to discrete state q, and drop the subscript H whenever the automaton is clear from the context. With these two pieces of notation one can show the following two results [7].

Proposition 1. *A (deterministic) hybrid automaton is non-blocking if (and only if) for all $(q, x) \in \mathrm{Out} \cap \mathrm{Reach}$, $\mathrm{Reset}(q, x) \neq \emptyset$.*

Proposition 2. *A hybrid automaton is deterministic if and only if for all $(q, x) \in \mathrm{Reach}$, $|\mathrm{Reset}(q, x)| \leq 1$ and, if $\mathrm{Reset}(q, x) \neq \emptyset$, $(q, x) \in \mathrm{Out}$.*

We characterize the hybrid automata such that the state remains in the closure of the invariant along all executions.

Definition 6 (Domain Preserving). *A hybrid automaton is domain preserving if $\mathrm{Reach} \subseteq \overline{\mathrm{Dom}}$.*

The following result is now straightforward.

Proposition 3. *A hybrid automaton is domain preserving if and only if* Init $\subseteq$ Dom *and for all* $(q, x) \in$ Dom $\cap$ Reach, Reset$(q, x) \subseteq$ Dom.

Note that the use of Reach is again not limiting. Note also that the conditions of the lemma do not depend on the vector field f. This is because, by the definition of an execution, the state can never end up outside the closure of the domain along continuous evolution.

Definition 7 (Transverse Domain). *A hybrid automaton H is said to have transverse domain if there exists a function* $\sigma : \mathbf{Q} \times \mathbf{X} \to \mathbb{R}$ *continuously differentiable in its second argument, such that*

$$\text{Dom} = \{(q, x) \in \mathbf{Q} \times \mathbf{X} : \sigma(q, x) \geq 0\}$$

and for all (q, x) *with* $\sigma(q, x) = 0$, $L_f \sigma(q, x) \neq 0$.

Here $L_f \sigma : \mathbf{Q} \times \mathbf{X} \to \mathbb{R}$ denotes the Lie derivative of σ along f defined as

$$L_f \sigma(q, x) = \frac{\partial \sigma}{\partial x}(q, x) \cdot f(q, x)$$

In other words, an automaton has transverse domain if the set Dom is closed, its boundary is differentiable, and the vector field f is pointing either inside or outside of Dom along the boundary.[3] If H has transverse domain the set Out$_H$ admits a fairly simple characterization.

Proposition 4. *If H has transverse domain, then*

$$\text{Out}_H = \{(q, x) \in \mathbf{Q} \times \mathbf{X} : \sigma(q, x) < 0\}$$
$$\cup \{(q, x) \in \mathbf{Q} \times \mathbf{X} : \sigma(q, x) = 0 \text{ and } L_f \sigma(q, x) < 0\}.$$

3   Invariant Sets and Stability

We first recall some standard concepts from dynamical system theory, and discuss how they generalize to hybrid automata.

Definition 8 (Invariant Set). *A set $M \subseteq$ Init is called invariant if for all* $(q_0, x_0) \in M$, $(\tau, q, x) \in \mathcal{E}_H(q_0, x_0)$, *and* $t \in \tau$, *it holds that* $(q(t), x(t)) \in M$.

The class of invariant sets is closed under arbitrary unions and intersections. Invariant sets are such that all executions starting in the set remain in the set for ever. We are interested in studying the stability of invariant sets, i.e., determine whether all trajectories that start close to an invariant set remain close to it.

---

[3] Under appropriate smoothness assumptions on σ and f the definition of transverse domain can be relaxed somewhat by allowing $L_f \sigma(q, x) = 0$ on the boundary of Dom and taking higher-order Lie derivatives, until one that is non-zero is found. Even though many of the results presented here extend to this relaxed definition, the proofs are slightly more technical. We will therefore limit ourselves to the notion of transverse domain given in Definition 7.

Definition 9 (Stable Invariant Set). *An invariant set $M \subseteq$ Init is called stable if for all $\epsilon > 0$ there exists $\delta > 0$ such that for all $(q_0, x_0) \in$ Init, with $d((q_0, x_0), M) < \delta$, all $(\tau, q, x) \in \mathcal{E}_H(q_0, x_0)$, and all $t \in \tau$, $d((q(t), x(t)), M) < \epsilon$.*

An invariant set is called (locally) asymptotically stable if it is stable and in addition there exists $\Delta > 0$ such that for all $(q_0, x_0) \in$ Init, with $d((q_0, x_0), M) < \Delta$, and all $(\tau, q, x) \in \mathcal{E}_H^\infty(q_0, x_0)$, $\lim_{t \to \tau_\infty} d((q(t), x(t)), M) = 0$.

Note that since τ is fully ordered the above limit is well defined. The asymptotic behavior of an infinite execution is captured in terms of its ω limit set.

Definition 10 (ω limit set). *The ω limit point $(\hat{q}, \hat{x}) \in \mathbf{Q} \times \mathbf{X}$ of an execution $\chi = (\tau, q, x) \in \mathcal{E}_H^\infty$ is a point for which there exists a sequence $\{\theta_n\}_{n=0}^\infty$, $\theta_n \in \tau$, such that as $n \to \infty$, $\theta_n \to \tau_\infty$ and $(q(\theta_n), x(\theta_n)) \to (\hat{q}, \hat{x})$. The ω limit set $S_\chi \subseteq \mathbf{Q} \times \mathbf{X}$ is the set of all ω limit points of an execution χ.*

The following lemma establishes a relation between ω limit sets and invariant sets. For convenience the assumptions on the reset relation and the domain are given in the graphical notation introduced in Section 2.2.

Lemma 1. *Consider a deterministic hybrid automaton H with transverse domain. Assume it is domain preserving and that $f(q, \cdot)$ is C^1 for all $q \in \mathbf{Q}$. Furthermore, assume that for all $e = (q, q') \in E$, $R(e, \cdot)$ is continuous, and $G(e) \cap I(q)$ is an open subset of $\partial I(q)$. Then, for any execution $\chi = (\tau, q, x) \in \mathcal{E}_H^\infty$, if $x(\cdot)$ is bounded, then S_χ is (i) nonempty, (ii) compact, and (iii) invariant. Further, (iv) for all $\epsilon > 0$ there exists $T \in \tau$ such that $d((q(t), x(t)), S_\chi) < \epsilon$, $t \in \tau$, for all $t \geq T$.*

Proof. See [15]. The proofs of (i), (ii), and (iv) are similar to the corresponding result for continuous dynamical systems [10,13].

The conditions of the lemma are sufficient. They can also be shown to be tight: one can construct hybrid automata that violate any one of the conditions of the lemma that accept infinite executions whose ω limit set is not invariant. The conditions of the lemma are also sufficient to establish continuity of executions with respect to initial conditions, see [15].

LaSalle's Invariance Principle is a useful tool when studying the stability of conventional, continuous dynamical systems. Lemma 1 allows us to extend this tool to hybrid systems.

Theorem 1 (LaSalle's Invariance Principle). *Consider a hybrid automaton H that satisfies the conditions of Lemma 1. Assume there exists a compact invariant set $\Omega \subseteq \mathbf{Q} \times \mathbf{X}$ and let $\Omega_1 = \Omega \cap \text{Out}^c$ and $\Omega_2 = \Omega \cap \text{Out}$. Furthermore, assume there exists a continuous function $V : \Omega \to \mathbb{R}$, such that*

- *for all $(q, x) \in \Omega_1$, V is continuously differentiable with respect to x and $L_f V(q, x) \leq 0$; and*
- *for all $(q, x) \in \Omega_2$, $V(\text{Reset}(q, x)) \leq V(q, x)$.*

Define

$$S_1 = \{(q, x) \in \Omega_1 : \ L_f V(q, x) = 0\}$$
$$S_2 = \{(q, x) \in \Omega_2 : \ V(\text{Reset}(q, x)) = V(q, x)\},$$

and let M be the largest invariant subset of $S_1 \cup S_2$. Then, for all $(q_0, x_0) \in \Omega$ every execution $(\tau, q, x) \in \mathcal{E}_H^\infty(q_0, x_0)$ approaches M as $t \to \tau_\infty$.

Proof. Consider an arbitrary state $(q_0, x_0) \in \Omega$ and let $\chi = (\tau, q, x) \in \mathcal{E}_H^\infty(q_0, x_0)$. Since Ω is invariant, $(q(t), x(t)) \in \Omega$ for all $t \in \tau$. Since Ω is compact and V is continuous, $V(q(t), x(t))$ is bounded from below. Moreover, $V(q(t), x(t))$ is a non-increasing function of $t \in \tau$ (recall that τ is fully ordered), so therefore the limit $c = \lim_{t \to \tau_\infty(\chi)} V(q(t), x(t))$ exists.

Since Ω is bounded, x is bounded, and therefore the ω limit set S_χ is nonempty. Moreover, since Ω is closed, $S_\chi \subset \Omega$. By definition, for any $(\hat{q}, \hat{x}) \in S_\chi$, there exists a sequence $\{\theta_n\}_{n=0}^\infty$, $\theta_n \in \tau$, such that $\theta_n \to \tau_\infty$ and $(q(\theta_n), x(\theta_n)) \to (\hat{q}, \hat{x})$ as $n \to \infty$. Then,

$$V(\hat{q}, \hat{x}) = V(\lim_{n \to \infty}(q(\theta_n), x(\theta_n)) = \lim_{n \to \infty} V(q(\theta_n), x(\theta_n)) = c,$$

by continuity of V. Since S_χ is invariant (Lemma 1), it follows that $L_f V(\hat{q}, \hat{x}) = 0$ if $(\hat{q}, \hat{x}) \notin$ Out, and $V(\text{Reset}(\hat{q}, \hat{x})) = V(\hat{q}, \hat{x})$ if $(\hat{q}, \hat{x}) \in$ Out. Therefore, $S_\chi \subset S_1 \cup S_2$, which implies that $S_\chi \subset M$ since S_χ is invariant. Moreover, by (iv) in Lemma 1, the execution χ approaches S_χ, and hence M, as $t \to \tau_\infty$.

4   Zeno Hybrid Automata

Zeno hybrid automata accept executions with infinitely many discrete transitions within a finite time interval. Such systems are hard to analysis and simulate in a way that gives constructive information about the behavior of the real system. It is therefore important to be able to determine if a model is Zeno and in applicable cases remove Zenoness. These problems have been discussed in [4,5]. In this section, some further characterization of Zeno executions are made. Recall that an infinite execution χ is Zeno if $\tau_\infty(\chi) = \sum_{i=0}^\infty (\tau_i' - \tau_i)$ is bounded.

Definition 11 (Zeno Hybrid Automaton). *A hybrid automaton H is Zeno if there exists $(q_0, x_0) \in$ Init such that all executions in $\mathcal{E}_H^\infty(q_0, x_0)$ are Zeno.*[4]

Example 1. The hybrid automaton in Figure 1 is Zeno. This is easily checked by explicitly deriving the time intervals $\tau_i' - \tau_i$, which in this case gives a converging geometric series. Figure 2 shows an execution accepted by the automaton.

We make the following two straightforward observations.

---

[4] An alternative definition is to say that a hybrid automaton is Zeno if there is *at least one* Zeno execution in $\mathcal{E}_H^\infty(q_0, x_0)$. In that case, a non-deterministic Zeno hybrid automaton may accept both Zeno and non-Zeno executions, which may be an undesirable feature for instance in Reach set calculations. For deterministic hybrid automata the two definitions coincide.

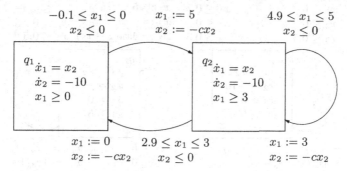

Fig. 1. An example of a Zeno hybrid automaton.

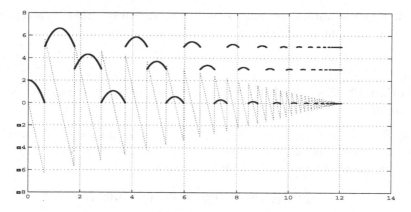

Fig. 2. An example of an execution for the hybrid automaton in Example 1. The continuous part of the state is shown: x_1 (solid) and x_2 (dotted).

Proposition 5. *A hybrid automaton is Zeno only if the graph* $(\mathbf{Q}, E)$ *has a cycle.*

Proposition 6. *If there exists a finite collection of states* $\{(q_i, x_i)\}_{i=1}^N$ *such that*

- $(q_1, x_1) = (q_N, x_N)$;
- $(q_i, x_i) \in \text{Reach}_H$ *for some* $i = 1, \ldots, N$; *and*
- $(q_{i+1}, x_{i+1}) = \text{Reset}(q_i, x_i)$ *for all* $i = 1, \ldots, N - 1$;

then there exists a Zeno execution.

Zenoness is critically dependent on the reset relation. For example, if in Example 1 the reset maps $x_2 := -cx_2$ are replaced by $x_2 := x_2/(dx_2 - 1)$, where $d = 1/\sqrt{20x_1(\tau_0)}$, then the time intervals $\tau_i' - \tau_i$ decrease as $\{1/i\}_{i=0}^\infty$. This is a diverging series, so the new hybrid automaton is not Zeno.

If the continuous part of the Zeno execution is bounded, then it has an ω limit point. We introduce the term Zeno state for such a point.

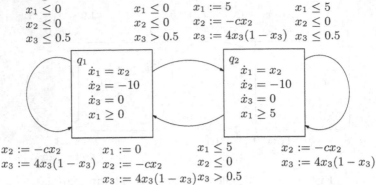

Fig. 3. A hybrid automaton that accepts Zeno executions that do not periodically jump between the discrete states.

Definition 12 (Zeno State). *The ω limit point of a Zeno execution is called the Zeno state.*

We use $Z_\infty \subset \mathbf{Q} \times \mathbf{X}$ to denote the set of Zeno states, so that Z_∞ is the ω limit set of the Zeno execution. We write $\mathbf{Q}_\infty$ for the discrete part of Z_∞ and E_∞ for the corresponding edges. In Example 1, we have

$$Z_\infty = \{(q_1, (0,0)), (q_2, (3,0)), (q_2, (5,0))\},$$

$\mathbf{Q}_\infty = \{q_1, q_2\}$, and $E_\infty = E$.

It is easy to construct an example with a Zeno executions that do not have a Zeno state. The idea is to let the continuous part of the execution become unbounded as $t \to \tau_\infty(\chi)$. It is also straightforward to derive examples where the set of Zeno states have any number of elements, as well as an infinite but countable or uncountable number of elements. An interesting question is if for a Zeno execution $\chi = (q, x, \tau)$, the discrete part q must become periodic for $t \in \tau$ sufficiently close to $\tau_\infty(\chi)$, as in Example 1. The answer is no as illustrated by the following example.

Example 2. Consider the Zeno hybrid automaton in Figure 3 (cf. Example 1). This system does not accept Zeno executions that periodically jump between the two discrete states. A simulation is presented in Figure 4, where x_1 and x_2 are shown. The third continuous state is initialized at $x_3(\tau_0) = 0.9$. The reason for the quasi-periodic behavior is that the reset map of x_3 is the logistic map and iteration of this map will give any value in $(0, 1)$, e.g., [10].

A reset relation Reset is *non-expanding*, if there exists $\delta \in [0, 1]$ such that $(q', x') \in \text{Reset}(q, x)$ implies $\|x'\| \leq \delta\|x\|$. It is *contracting*, if there exists $\delta \in [0, 1)$ such that $(q', x') \in \text{Reset}(q, x)$ and $(q', y') \in \text{Reset}(q, y)$ imply $\|x' - y'\| \leq \delta\|y - x\|$. Note that the reset relation has to be a function in the second case.

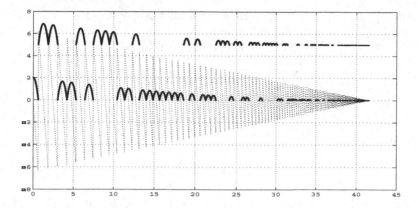

Fig. 4. An example of an execution for the hybrid automaton in Example 2. The continuous states x_1 (solid) and x_2 (dotted) are shown. Note how they illustrate the quasi-periodicity.

For smooth dynamical systems, a Lipschitz assumption on the vector field excludes finite escape time. This is not a sufficient condition for hybrid systems. However, if the reset relation is non-expanding (in addition to the Lipschitz assumption on $f(q, \cdot)$), then the continuous state is bounded along executions.

Lemma 2. *Consider a hybrid automaton with non-expanding reset relation. Then, there exists $c > 0$ such that for all executions $\chi = (\tau, q, x) \in \mathcal{E}_H$ and $t \in \tau$,*

$$\|x(t)\| \leq (\|x(\tau_0)\| + 1) e^{c(t - \tau_0)} - 1.$$

Proof. The proof, see [15], is similar to the corresponding result for continuous systems [10, Proposition 5.3].

When $x(\cdot)$ is bounded, the Bolzano–Weierstrass Property implies that there exists at least one Zeno state for each Zeno execution. If the continuous part of the reset relation is the identity map, then the continuous part of the Zeno state is a singleton, as proved next.

Theorem 2. *Consider a hybrid automaton such that $(q', x') \in \mathrm{Reset}(q, x)$ implies $x' = x$. Then, for every Zeno execution $\chi = (\tau, q, x)$, it holds that $Z_\infty = \mathbf{Q}_\infty \times \{\hat{x}\}$ for some $\mathbf{Q}_\infty \subseteq \mathbf{Q}$ and $\hat{x} \in \mathbf{X}$.*

Proof. For all sequences $\{\theta_i\}_{i=0}^\infty$, $\theta_i \in \tau$, such that $\theta_i \to \tau_\infty$, suppose $\theta_i \in [\tau_{n_i}, \tau'_{n_i}]$, where $n_i \to \infty$ as $i \to \infty$. We have

$$x(\theta_i) = x(\tau_{n_i}) + \int_{\tau_{n_i}}^{\theta_i} f\big(q(\tau_{n_i}), x(\tau)\big) \, d\tau$$

$$= x(\tau_{n_i}) + (\theta_i - \tau_{n_i}) f\big(q(\tau_{n_i}), (x_1(\xi_{n_i}^1), \ldots, x_n(\xi_{n_i}^n))^T\big),$$

for some $\xi_{n_i}^1, \ldots, \xi_{n_i}^n \in [\tau_{n_i}, \tau_{n_i}']$. Hence, for all $k > \ell \geq 0$,

$$
\begin{aligned}
x(\theta_k) = x(\theta_\ell) &+ (\tau_{n_\ell}' - \theta_\ell) f\big(q(\tau_{n_\ell}), (x_1(\xi_{n_\ell}^1), \ldots, x_n(\xi_{n_\ell}^n))^T\big) \\
&+ \sum_{i=n_\ell+1}^{n_k-1} (\tau_i' - \tau_i) f\big(q(\tau_i), (x_1(\xi_i^1), \ldots, x_n(\xi_i^n))^T\big) \\
&+ (\theta_k - \tau_{n_k}) f\big(q(\tau_{n_k}), (x_1(\xi_{n_k}^1), \ldots, x_n(\xi_{n_k}^n))^T\big),
\end{aligned}
$$

which gives that

$$
\|x(\theta_k) - x(\theta_\ell)\| \leq K \sum_{i=n_\ell}^{n_k} (\tau_i' - \tau_i),
$$

where $K > 0$ is a constant such that $\|f(q, x)\| \leq K$ for all $(q, x) \in \mathbf{Q} \times \mathbf{X}$. Such constant exists due to Lemma 2. By the fact that $\sum_{i=0}^{\infty}(\tau_i' - \tau_i) < \infty$, we know that $\{x(\theta_i)\}_{i=0}^{\infty}$ is a Cauchy sequence. The space $\mathbf{X} = \mathbb{R}^n$ is complete, so the sequence has a limit $\hat{x} = \lim_{i \to \infty} x(\theta_i)$. This limit is independent of the choice of sequence $\{\theta_i\}_{i=0}^{\infty}$, as follows from the following argument. Consider two sequences $\{\alpha_i\}_{i=0}^{\infty}$ and $\{\beta_i\}_{i=0}^{\infty}$, $\alpha_i, \beta_i \in \tau$, such that $\alpha_i \to \tau_\infty$ and $\beta_i \to \tau_\infty$. Suppose $\alpha_i \in [\tau_{m_i}, \tau_{m_i}']$ and $\beta_i \in [\tau_{n_i}, \tau_{n_i}']$, where $m_i \to \infty$ and $n_i \to \infty$ as $i \to \infty$, and $m_i \geq n_i$. Then,

$$
\begin{aligned}
x(\alpha_i) = x(\beta_i) &+ (\tau_{n_i}' - \beta_i) f\big(q(\tau_{n_i}), (x_1(\xi_{n_i}^1), \ldots, x_n(\xi_{n_i}^n))^T\big) \\
&+ \sum_{j=n_i+1}^{m_i-1} (\tau_j' - \tau_j) f\big(q(\tau_j), (x_1(\xi_j^1), \ldots, x_n(\xi_j^n))^T\big) \\
&+ (\alpha_i - \tau_{m_i}) f\big(q(\tau_{m_i}), (x_1(\xi_{m_i}^1), \ldots, x_n(\xi_{m_i}^n))^T\big).
\end{aligned}
$$

This gives that $\|x(\alpha_i) - x(\beta_i)\| \leq K \sum_{j=n_i}^{m_i}(\tau_j' - \tau_j)$. Hence, $\|x(\alpha_i) - x(\beta_i)\| \to 0$ as $i \to \infty$, which shows that both sequences have the same limit. This completes the proof.

Note that Theorem 2 gives the structure of the Zeno state for the large class of hybrid systems called switched systems [9], since these systems can be modeled as hybrid automata with identity reset relation.

If the reset relation is contracting and $(q', x') \in \text{Reset}(q, 0)$ implies that x' is the origin, then the continuous part of the Zeno state is also the origin.

Theorem 3. *Consider a Zeno hybrid automaton with contracting reset relation and such that $(q', x') \in \text{Reset}(q, 0)$ implies $x' = 0$. Then, for every Zeno execution $\chi = (\tau, q, x)$, it holds that $Z_\infty = \mathbf{Q}_\infty \times \{0\}$ for some $\mathbf{Q}_\infty \subseteq \mathbf{Q}$.*

Proof. For all sequences $\{\theta_i\}_{i=0}^{\infty}$, $\theta_i \in \tau$, such that $\theta_i \to \tau_\infty$, suppose $\theta_i \in [\tau_{n_i}, \tau_{n_i}']$, where $n_i \to \infty$ as $i \to \infty$. We have

$$
\begin{aligned}
\|x(\theta_i)\| &\leq \|x(\tau_{n_i})\| + \left\| \int_{\tau_{n_i}}^{\theta_i} f\big(q(\tau_{n_i}), x(\tau)\big) \, d\tau \right\| \\
&\leq \|x(\tau_{n_i})\| + K(\tau_{n_i}' - \tau_{n_i}),
\end{aligned}
$$

where $K > 0$ is the same constant as in the proof of Theorem 2. Using the fact that $\|x(\tau_{n_i})\| \leq \delta\|x(\tau'_{n_i-1})\|$, it follows that

$$\|x(\theta_i)\| \leq \delta\|x(\tau'_{n_i-1})\| + K(\tau'_{n_i} - \tau_{n_i})$$

$$= \delta\left\|x(\tau_{n_i-1}) + \int_{\tau_{n_i-1}}^{\tau'_{n_i-1}} f\big(q(\tau_{n_i-1}), x(\tau)\big)\, d\tau\right\| + K(\tau'_{n_i} - \tau_{n_i})$$

$$\leq \delta\|x(\tau_{n_i-1})\| + K\delta(\tau'_{n_i-1} - \tau_{n_i-1}) + K(\tau'_{n_i} - \tau_{n_i}).$$

By induction,

$$\|x(\theta_i)\| \leq \delta^{n_i}\|x(\tau_0)\| + K\sum_{m=0}^{n_i} \delta^{n_i-m}(\tau'_m - \tau_m).$$

Since

$$\sum_{n_i=0}^{\infty} K\sum_{m=0}^{n_i} \delta^{n_i-m}(\tau'_m - \tau_m) = K\sum_{m=0}^{\infty}(\tau'_m - \tau_m)\sum_{n_i=0}^{\infty}\delta^{n_i} = \frac{K\tau_\infty}{1-\delta} < \infty,$$

it holds that $K\sum_{m=0}^{n_i} \delta^{n_i-m}(\tau'_m - \tau_m) \to 0$ as $n_i \to \infty$. This yields that $\|x(\theta_i)\| \to 0$ as $i \to \infty$, which, hence, completes the proof.

A generalization of Theorem 3 holds if we change the assumption to that $(q', x') \in \mathrm{Reset}(q, x^*)$ implies $x' = x^*$ for some $x^* \in \mathrm{Dom}$, see [15].

For a large class of Zeno hybrid automata, the continuous part of the Zeno state is located on the intersection of the boundaries of $\mathrm{Dom}(q, \cdot)$ for $q \in \mathbf{Q}_\infty$. Next this result is stated for hybrid automata with non-expanding reset relation. Recall that $I(q) = \{x \in \mathbf{X} : (q, x) \in \mathrm{Dom}\}$.

Proposition 7. *Consider a hybrid automaton H with non-expanding reset relation. Assume it accepts a Zeno execution $\chi = (\tau, q, x) \in \mathcal{E}_H^\infty$ with set of Zeno states $Z_\infty = \{(q_i, x_i)\}_{i=1}^N$, $N \geq 1$. If, for all $i \in \{1,\ldots,N\}$ and $x \in I(q_i)^\circ$, $\mathrm{Reset}(q_i, x) = \emptyset$, then $x_i \in \partial I(q_i)$ for all $i \in \{1,\ldots,N\}$. Furthermore, if there exists $\hat{x} \in \mathrm{Dom}$ such that for all $i \in \{1,\ldots,N\}$, $x_i = \hat{x}$, then $\hat{x} \in \bigcap_{i=1}^N \partial I(q_i)$.*

Proof. See [15].

It follows from Proposition 7 that if the boundaries of $I(\cdot)$ are not intersecting, then there exist no Zeno executions with non-empty Zeno state and $N > 1$. Proposition 7 is thus a refinement of the condition given in Proposition 5, which states that a hybrid automaton is non-Zeno if the graph $(\mathbf{Q}, E)$ has no cycle.

5 Conclusions

Motivated by numerous assumptions like "In this paper, we assume that the switched system is live and nonZeno" [3] and suggestions like "Additional work is needed in determining the role that Zeno-type control might play in hybrid system supervision" [6], we have extended some classical results to hybrid systems, using tools that capture both non-Zeno and Zeno executions. We have also tried to illustrate some of the nature of Zeno by characterizing Zeno executions and Zeno states for a few quite broad classes of hybrid systems. Zeno hybrid automata are characterized from a geometric point of view in [11].

Acknowledgments

The authors would like to thank Magnus Egerstedt and Slobodan Simić for helpful discussions.

References

1. M. Branicky. Multiple Lyapunov functions and other analysis tools for switched and hybrid systems. *IEEE Transactions on Automatic Control*, 43(4):475–482, April 1998.
2. M. Branicky, E. Dolginova, and N. Lynch. A toolbox for proving and maintaining hybrid specifications. In P. Antsaklis, W. Kohn, A. Nerode, and S. Sastry, editors, *Hybrid Systems IV*, number 1273 in LNCS, pages 18–30. Springer Verlag, 1997.
3. K. X. He and M. D. Lemmon. Lyapunov stability of continuous-valued systems under the supervision of discrete-event transition systems. In *Hybrid Systems: Computation and Control*, volume 1386 of *Lecture Notes in Computer Science*. Springer-Verlag, Berlin, 1998.
4. K. H. Johansson, M. Egerstedt, J. Lygeros, and S. Sastry. On the regularization of Zeno hybrid automata. *Systems & Control Letters*, 38:141–150, 1999.
5. K. H. Johansson, J. Lygeros, S. Sastry, and M. Egerstedt. Simulation of Zeno hybrid automata. In *IEEE Conference on Decision and Control*, Phoenix, AZ, 1999.
6. M. D. Lemmon, K. X. He, and I Markovsky. Supervisory hybrid systems. *IEEE Control Systems Magazine*, 19(4):42–55, 1999.
7. J. Lygeros, K. H. Johansson, S. Sastry, and M. Egerstedt. On the existence of executions of hybrid automata. In *IEEE Conference on Decision and Control*, Phoenix, AZ, 1999.
8. J. Lygeros, C. Tomlin, and S. Sastry. Controllers for reachability specifications for hybrid systems. *Automatica*, 35(3), March 1999.
9. A. S. Morse. Control using logic-based switching. In Alberto Isidori, editor, *Trends in Control. A European Perspective*, pages 69–113. Springer, 1995.
10. S. Sastry. *Nonlinear Systems: Analysis, Stability, and Control*. Springer-Verlag, New York, 1999.
11. S Simić, K H Johansson, S Sastry, and J Lygeros. Towards a geometric theory of hybrid systems. In *Hybrid Systems: Computation and Control*, Pittsburgh, PA, 2000.
12. A. J. van der Schaft and J. M. Schumacher. Complementarity modeling of hybrid systems. *IEEE Transactions on Automatic Control*, 43(4):483–490, April 1998.
13. S. Wiggins. *Introduction to Applied Nonlinear Dynamical Systems and Chaos*. Springer-Verlag, New York, 1990.
14. H. Ye, A. Michel, and L. Hou. Stability theory for hybrid dynamical systems. *IEEE Transactions on Automatic Control*, 43(4):461–474, April 1998.
15. J. Zhang. Dynamical systems revisited: Hybrid systems with Zeno executions. Master's thesis, Dept of EECS, University of California, Berkeley, 1999.

Author Index